MATHEMATICS of MONEY

WITH ALGEBRA

Cheryl Clayton
Richardson High School
Richardson, Texas

SOUTH-WESTERN PUBLISHING CO.

Managing Editor: Eve Lewis
Developmental Editor: Enid M. Nagel
Production Editor: Shannon O'Connor
Coordinating Editor: Patricia Matthews Boies
Associate Photo Editor/Stylist: Fred M. Middendorf
Marketing Manager: Carol Ann Dana
Design: Pronk & Associates
Production Services: PC&F, Inc.

ISBN: 0–538–63471–5
 3 4 5 6 7 8 KI 01 00 99 98 97 96 95
Printed in the United States of America

I(T)P
International Thomson Publishing

South-Western Publishing Co. is an ITP Company. The ITP trademark is
used under license.

ABOUT THE AUTHOR
Cheryl Clayton, M.Ed.

Since 1979, the author has taught at Richardson High School, in Richardson, Texas where she teaches the popular mathematics of money course on which this book is based. Ms. Clayton's career has included teaching fundamentals of mathematics, algebra, geometry, trigonometry, analytic geometry and consumer mathematics at the junior high and high school levels. She received a Bachelor's degree from Texas Christian University and a Master's degree in Education from North Texas State University. Ms. Clayton is married, has two sons, and lives in Dallas.

REVIEWERS

Korita M. Azopardi
Chairperson
Mathematics Department
Calallen High School
Corpus Christi, Texas

George Benedetti
Business Education Instructor
West Valley High School
Yakima, Washington

Deborah D. Carruthers
Teacher, Computer Science
and Mathematics
Ross S. Sterling High School
Baytown, Texas

Amy Davidson
Mathematics Teacher
Westfield High School
Houston, Texas

Arne Engebretsen
Chair, Mathematics and
Computer Science
Greendale High School
Greendale, Wisconsin

Bruce E. Engelhard
Mathematics Teacher
Newport Alternative School
Newport, Rhode Island

Mary Hansen
Mathematics Teacher
South View Senior High School
Hope Mills, North Carolina

Brenda Beck Lynch
Mathematics Teacher
Willis Independent
School District
Willis, Texas

Shirley D. Oliver
Mathematics Teacher
Tomball High School
Tomball, Texas

Randy L. Pippen
Mathematics Department Chair
Lisle High School
Lisle, Illinois

Charles W. Tillerson
Chair, Department of
Mathematics
Highland Park High School
Dallas, Texas

Thurman Watson
Mathematics Teacher
Robert E. Lee High School
Baytown, Texas

CONTENTS

LETTER TO THE STUDENT

Many of you are entering the job market even before completing high school. Some of you are finding the realities of acquiring, spending, and saving money to be a challenging and sometimes overwhelming task. *Mathematics of Money With Algebra* will use and extend the skills you learned in Algebra I to help you manage your own finances.

This program encourages your active involvement by constructing and applying mathematical ideas through business and consumer themes. Since decision making is a major thrust, you are encouraged to use calculators and computers as tools for learning and doing mathematics.

The readings in *Mathematics of Money With Algebra* give you sketches of life that might seem familiar. You, like the characters in the book, are approaching adulthood and are faced with many financial decisions. This program offers assistance in financial planning so that the choices you make to establish an independent life-style will be informed choices.

Mathematics of Money With Algebra contains information you can use now and in the future. For example, when buying a car, you are encouraged to consider the cost of maintaining a car as well as the purchase price. By including up-to-date data and common situations, this text will help you gain confidence in your mathematical abilities as you solve problems which affect you and your family every day.

In order to use *Mathematics of Money With Algebra* completely and effectively, you must be able to use the basic skills of algebra automatically. This text is designed to help you succeed because your Algebra I skills will be completely reviewed in the special features *Algebra Refresher* and *Algebra Review.*

Graphing calculators and computer spreadsheet applications show you how graphs can communicate concepts about data more effectively. You will have opportunities to work with other students and learn more about businesses in your community as you complete the projects at the end of each chapter.

A final word of encouragement: Your knowledge of a variety of techniques to approach and solve problems will be greatly enhanced in this course if you read carefully and enjoy the challenge of mathematical problem solving. When you value mathematics, doors open to a variety of related disciplines and careers.

Simplify each expression. When you simplify a *numerical expression*, simplify within parentheses first. Then do all multiplications and divisions from left to right followed by all additions and subtractions from left to right.

Example $6.3 + 0.4(2.7) - (6.3 - 2.7 + 1)$
$= 6.3 + 0.4(2.7) - 4.6$ Simplify within parentheses first.
$= 6.3 + 1.08 - 4.6$ Multiply before adding and subtracting.
$= 7.38 - 4.60$ Add and subtract from left to right.
$= 2.78$

1. $5.1 + (11.4 - 6)$

2. $9.7 - (12.5 - 10)$

3. $3 \cdot 4.2 - 0.1 \cdot 8$

4. $0.06 \cdot 850$

5. $0.06 \cdot 850 + 0.45 \cdot 730$

6. $0.765 \cdot 5760 + 0.145 \cdot 13{,}500$

Each of the following *algebraic expressions* represents the product of 7 and a number *h*.

$7h \qquad 7 \cdot h \qquad 7(h) \qquad (7)(h)$

Evaluate an expression by replacing the variable by a number. Then simplify. Evaluate each expression for *h* = 6.7 or *x* = 8.6.

Example $6.2 \div 2 + 8h$
$= 6.2 \div 2 + 8(6.7)$ Substitute 6.7 for h.
$= 3.1 + 53.6$ Divide and multiply first, then add.
$= 56.7$

7. $x - 3.6$

8. $2 \cdot h + 4$

9. $2 + 4(h)$

10. $2 + 3(10 - x)$

11. $350 \cdot h \cdot 15$

12. $(1 + x \div 100)(1.5)$

Solve each *equation* for the variable. To solve an equation, ask yourself what operation was performed on the variable. Then ask how you can undo the operation. Remember to always keep the equation in balance.

Example $\begin{aligned} 7h - 316 &= 594 \\ 7h - 316 + 316 &= 594 + 316 \\ 7h + 0 &= 910 \\ 7h &= 910 \\ \frac{7h}{7} &= \frac{910}{7} \\ 1h &= 130 \\ h &= 130 \end{aligned}$

316 is subtracted from 7h.
Add 316 to each side.
$-316 + 316 = 0$
h is multiplied by 7.

Divide each side by 7.

$7 \div 7 = 1$
$1h = h$

13. $13x = 143$

14. $0.055P = 660$

15. $1116 = t - 261$

16. $0.7x + 100 = 37$

17. $12x - 59 = 841$

18. $5.2e + 498 = 2500$

PERSONAL EARNING
POWER

IN THIS CHAPTER YOU WILL MEET ALEX and his friends. These young people in their late teens are fictitious, but the decisions that they face are very real. Every year, real young people make choices similar to those that you will see Alex and his friends making.

You will see how Alex carefully considers the balance between the time and effort required in taking a part-time job and the financial rewards of working. He sees how the information that he reads in newspaper job ads translates into a paycheck for regular and overtime hours.

Daphne, a year younger than Alex and his classmates, watches her older brother Darrin make career decisions and begins to think about making her own. She learns about sources of help in exploring the world of work. She also learns how a person's income is enhanced by company benefits other than pay.

Alex's friend Betty will help us compare the consistent income of a salaried job with the variable income of a job paid by commission. A salesperson on a commission may make recommendations to the customer that differ from those a salaried salesperson would make. Betty tells us why.

Another friend, Luis, will see how and why deductions are made from his paycheck. His first paycheck from the grocery store where he works comes as a surprise to Luis. He had expected to receive his full hourly rate, but he received considerably less.

Alex, Daphne, Betty, and Luis look forward to the time when they will no longer have all their decisions made for them. They are glad to begin taking charge of their own lives. At the same time the importance of some of those decisions is becoming clear to them. Accepting the responsibility for their choices is not easy.

SALARIES: MAKING MONEY TAKES TIME

Alex and his classmates are entering their last year of high school. Many of his friends have already had summer jobs. Some have started saving money for college or for other schooling.

Alex's parents have been asking him what he plans to do after graduation. His girlfriend, Alice, is planning to go to State University next fall. He hates the thought of being away from Alice for a long time. They plan to get married after she gets her law degree, but that could take a while. If Alex goes to State to be with Alice, he may decide to study accounting, but he is not really handy with numbers. What Alex really enjoys is tinkering around with the van that his parents bought him when he learned to drive.

Alex could use some extra cash, since the van needs a few parts and he would like to take Alice to the Senior Prom. Alex's older sister Karen rode to the prom last year in a chauffeured limousine—Alex can just imagine what that must have cost!

The $483 that Alex saved from mowing lawns this summer is almost gone. Alex has been thinking about getting a job after school, partly to earn extra cash but also to get a taste of what the world of work will be like. If he can find a way to choose a career that he would really like, he may decide that he needs more training. Then getting a job might bring in the extra money to pay for it.

OBJECTIVES: *In this lesson, we will help Alex to:*

- *Compute the earnings from a part-time job.*
- *Compare an hourly wage with a weekly or monthly salary.*

EARNING INCOME

Alex already knows a few facts about money. He knows that his family's standard of living is determined by the amount of income his parents earn. **Income** is the money received from investments or from a person's activities at work or in business. His family, like most American families, depends on its labor for its primary income. Approximately nine out of ten American workers earn income in the form of wages or salary. A **wage** is an hourly or daily rate of pay; a **salary** is a weekly, monthly, or yearly rate of pay. Most clerical workers and people who do physical labor are wage earners. Professional and technical workers are usually salary earners. About 10% of all workers are in business for themselves and earn **self-employment income** instead of wages or salaries. These include such people as physicians, shopkeepers, writers, photographers, and farmers.

You may have wondered why a few baseball players have actually been paid over one million dollars for a season while the President of the United States earns $200,000 a year or why an actor who becomes a star earns far more than a professional accountant or teacher. Income from labor reflects social values, the monetary worth of the labor, and the demand for the labor in relation to the number of people who can do the job.

Alex began examining the classified ads in the employment section of the Sunday newspaper. He did not see any ads offering million-dollar salaries to baseball players. He did find, however, that part-time jobs available to students pay a specific amount for each hour worked. That amount is an **hourly rate.** The hours the employee is required to work each week are the employee's **regular hours.**

Some weeks an employee will be asked to work more than the required amount of hours. These extra hours are called **overtime hours** or simply overtime. The hourly rate for these overtime hours is generally greater than the regular hourly rate. Salaried employees receive the same salary each pay period whether they work the minimum number of hours required or put in extra time during weekends and evenings.

Ask Yourself

1. If an employee is paid by the hour and works more than the required number of hours in a week, the employee is generally paid more for the extra hours. What are the extra hours called?

2. If an employee is paid a monthly salary, will the employee be paid more for working on the weekend?

3. In which kinds of jobs do workers earn hourly or daily wages?

ALGEBRA *REVIEW*

Evaluate the expression $rh + t$ for the given values of h, r, and t.

1. $h = 5$
   $r = 4.4$
   $t = 7$

2. $h = 8.5$
   $r = 5.60$
   $t = 11$

3. $h = 19$
   $r = 6$
   $t = 23.50$

4. $h = 12$
   $r = 5.80$
   $t = 14$

Evaluate the expression $20r + (h - 20) \cdot 1.5r + t$ for the given values of h, r, and t.

5. $h = 21$
   $r = 6.20$
   $t = 34.50$

6. $h = 27$
   $r = 4.00$
   $t = 16.00$

7. $h = 20$
   $r = 5.15$
   $t = 17.50$

8. $h = 30$
   $r = 5.50$
   $t = 20$

SHARPEN YOUR SKILLS

—————— **SKILL 1** ——————

Alex's friend Ed delivers pizzas. Ed receives tips from customers if the pizza is still hot and if the order is properly filled.

Ed's earnings can be expressed by the following equation.

Earnings = hourly rate • number of hours + tips
$E = rh + t$ where r = hourly rate
h = number of hours
t = tips

EXAMPLE 1   During the first week of February, Ed worked $16\frac{1}{2}$ hours and received $28.50 in tips. His hourly rate is $3.40.

QUESTION   How much did Ed earn that week?

SOLUTION

$E = rh + t$ Substitute values in the equation.
$E = 3.40 • 16.5 + 28.50$ $r = 3.40,\ h = 16.5,\ t = 28.50$
$E = 84.60$ Use your calculator.

Ed earned $84.60 that week.

—————— **SKILL 2** ——————

EXAMPLE 2   If Ed works more than 20 hours in one week, he receives $1\frac{1}{2}$ times his regular wage rate of $3.40 for each of the extra hours. During the second week of February there was an ice storm. Many people ordered pizza to be delivered. That week, Ed worked 26 hours and received $43.75 in tips.

QUESTION   How much money did Ed make that week?

SOLUTION
Ed's earnings that week is the sum of his regular earnings, his overtime earnings, and his tips.

Overtime hours = total hours − regular hours = $h - 20$
Total earnings = regular earnings + overtime earnings + tips

$$\text{Total earnings} = \frac{\text{number of}}{\text{regular hours}} • \frac{\text{hourly}}{\text{rate}} + \frac{\text{overtime}}{\text{hours}} • \frac{\text{overtime}}{\text{rate}} + \text{tips}$$

$E = 20r + (h - 20) • 1.5r + t$ The overtime rate is $1\frac{1}{2}$ or
                                      1.5 times the regular rate.

$E = 20 • 3.4 + (26 - 20) • 1.5 • 3.4 + 43.75$
$E = 142.35$ Use your calculator.

Ed earned $142.35 that week.

EXAMPLE 3 Alex wondered how salaries stated by the week, month, and year compare with each other and with hourly wages. He devised a plan to find out by using his computer **spreadsheet program.**

QUESTION How much are the hourly, weekly, monthly, and yearly salaries of each of the jobs shown in these advertisements?

1.

RECEPTIONIST must have good phone skills. $6.50/h to start. Hours: 8:30am–5pm. Equal Oppty Employment. Call Monday–Thursday only.

2.

COMMUNITY WORK
Start a new career with the state's oldest consumer group. Great future! No experience nec. Paid training, holidays, vacations. Up to $350 per wk.

3.

ADMINISTRATIVE ASSISTANT
Social service agency seeks Admin. Asst. for data entry. Must have exp. with dBase III systems. Typing 40 wpm. Sal. $15 K + benefits.

SOLUTION

Alex knows that he could quickly compute the answers using his calculator. However, he is trying to learn how to use his computer spreadsheet program, which can do many calculations quickly if he sets up the spreadsheet properly.

Open the spreadsheet program to a new blank spreadsheet filled with blank rows and columns. The intersection of a *row* and a *column* is called a **cell.** A cell is labeled by the column and row that forms it. Begin by entering the headings Hourly, Weekly, Monthly, and Yearly in cells A1, B1, C1, and D1.

Column

	A	B	C	D
1	Hourly	Weekly	Monthly	Yearly	Row
2				
3				
4			Cell C5	
5				
6				

Alex enters the salaries that he knows and makes the computer calculate the others by using formulas.

Enter $6.50, the hourly salary for the receptionist, in cell A2.

For the formula for the weekly salary, use the hourly salary, which is located in cell A2. Multiply it by 40, since there are 40 hours in a workweek.

Enter +A2*40 in cell B2.

For the formula for the yearly salary, use the weekly salary, which is located in cell B2. Multiply it by 52, since there are 52 weeks in a year.

Enter +B2*52 in cell D2.

Since there are 12 months in a year, divide the yearly salary in cell D2 by 12.

Enter +D2/12 in cell C2.

40 hours = 1 workweek	Remember that * is a symbol
52 weeks = 1 year	for multiplication and / is a
12 months = 1 year	symbol for division.

Alex would like his answers rounded to two decimal places, since they represent money. Highlight the portion of the spreadsheet to be formatted and choose the option for fixed format of two decimal places. Although the spreadsheet will then display the numbers rounded to two decimal places, the computer will retain the numbers to many decimal places in its memory and use these values in computations.

	A	B	C	D
1	Hourly	Weekly	Monthly	Yearly
2	6.50	260.00	1126.67	13520
3		+A2*40	+D2/12	+B2*52
4				

Enter $350, the weekly salary for community work, in cell B3.

To determine the hourly salary, use the formula +B3/40 in cell A3.

To determine the yearly salary, use the formula +B3*52 in cell D3.

For the monthly salary, use the formula +D3/12 in cell C3.

Enter $15,000, the yearly salary for the administrative assistant, in cell D4.

To determine the monthly salary, use the formula +D4/12 in cell C4.

To determine the weekly salary, use the formula +D4/52 in cell B4.

For the hourly salary, use the formula +B4/40 in cell A4.

	A	B	C	D
1	Hourly	Weekly	Monthly	Yearly
2	6.50	260.00	1126.67	13520
3	8.75	350.00	1516.67	18200
4	7.21	288.46	1250.00	15000

The community work is the highest-paying job, and the receptionist is the lowest-paying job.

TRY YOUR SKILLS

Write an equation of the form $E = rh + t$ where r represents the regular hourly rate, h represents the number of hours worked, and t represents any tip or bonus. Then find the weekly earnings for each of the following "employees."

1. I am the tape recorder in the French language lab. I make $11.50 per hour, and I worked $29\frac{1}{2}$ hours this week. I received a bonus of $3.50 when I taught the students how to say "I love you" in French.

2. I am the button that turns on the light when you open the refrigerator door. My door was open $22\frac{1}{2}$ hours this week. My salary is $10.00 per hour, and I did not receive a bonus because I was on when my owner knocked over a bottle of juice onto his toe.

3.–4. Assume that the employees in Exercises 1 and 2 received overtime pay of $1\frac{1}{2}$ times their regular rate for all hours they worked over 20. Write an equation to represent their weekly earnings of the form

$$E = 20r + (h - 20) \bullet 1.5r + t$$

where r represents the regular hourly rate, h represents the number of hours worked, and t represents any tip or bonus. Determine the total earnings for each employee.

Create a spreadsheet to calculate the hourly, weekly, monthly, and yearly earnings for each of the jobs described below.

> 40 hours = 1 workweek
> 52 weeks = 1 year
> 12 months = 1 year

5.

HEALTH CARE Better Life Health Care is looking for a respiratory therapist. Earn as much as $17.85/h.

6.

Mold Maker
Jr. MOLD MAKER We pay as high as $8.50/h after 1 year.

7.

Data Processing Computer Operator
Must know Word-Perfect and want to learn other programs. $700/wk.

EXERCISE YOUR SKILLS

KEY TERMS

cell
hourly rate
income
overtime hours
regular hours
salary
self-employment
    income
spreadsheet
    program
wage

1. What is the major difference between the way earnings are computed for salaried employees and for hourly employees?

2. Why would an employer be willing to pay extra money for overtime hours?

3. Why would salaried employees be willing to work extra hours if their salary remained the same?

4. Why do many professional baseball players earn more money each year than the President of the United States?

5. If a student takes a part-time job, what do you think is the maximum number of hours the student should be allowed to work per week? Be sure to take into consideration the time need to keep up with school work.

Write an equation of the form $E = rh + t$, and find the weekly earnings for each of the following "employees."

6. I am the pay telephone in the hall by the principal's office. I work all the time but got paid $12.00 per hour for 29 hours this week. I got a $25.00 bonus from a guy whose girlfriend told him through me that she had decided not to break up with him.

7. I am the gate that goes up and down at the tollbooth. I made $11.50 per hour for $29\frac{1}{2}$ hours of work this week and received a $0.50 bonus from a taxi driver who got past me before I could close after the car in front of him.

8. I dispense tokens for video games at Showbiz. I eat $5 bills and spit out 20 tokens at a time. I worked 27 hours last week at $13.00 per hour and received an extra $1.00 when the tokens got stuck and a little boy kicked me.

9. I am the coffee machine in the teachers' lounge at the high school. I work many extra hours each week keeping these folks awake, but I was paid for only $26\frac{1}{2}$ hours at $11.00 per hour. I got no bonus because I ran out of coffee during the last staff meeting.

10. I am the golf cart that Karl pulls around on the golf course. I carry all his clubs and his lucky red rabbit's foot that his brother sold to him when he lost his. I worked 31 hours this week and made $14.00 per hour and received a $20.00 bonus when Karl made his first par on the number 2 hole at Keyton Park.

11. I am the button that you push when you want to ride your bicycle on the bike path across the freeway. I make the "Walk" sign come on. I make $12.50 per hour, and I worked $31\frac{1}{2}$ hours last week. I did not receive a bonus because I am not fast enough for the speed-racers.

12.–17. Assume that the "employees" of Exercises 6–11 received overtime pay of $1\frac{1}{2}$ times their regular rate for all hours they worked over 20. Write an equation to represent their weekly earnings of the form

$$E = 20r + (h - 20) \bullet 1.5r + t$$

where r represents the regular hourly rate, h represents the number of hours worked, and t represents any tip or bonus. Determine the total earnings for each employee.

Create a spreadsheet to calculate the hourly, weekly, monthly, and yearly earnings for each of the jobs described below.

40 hours = 1 workweek
52 weeks = 1 year
12 months = 1 year

18.
_____

Health Care
NURSES
RNs UP TO $35/h.
_____

19.
_____

Maintenance
MECHANIC Center Manuf. is in need of a person with 3–5 years exp. in machine repair/ punch press. $10.70/h.
_____

20.
_____

Sales!
Interesting inside sales position. Start at $5.75/h.
_____

21.
_____

Hairdresser
English and Spanish speaking. Guaranteed pay $300/wk. to start. Apply in person only.
_____

22.
_____

General Retail Clerk
Are you getting no-where with that same old job? Earn $600 per week. No exp. nec.
_____

23.
_____

DRIVERS
Drivers for local mes-senger. Earn $450 per week. Must have own car and good driving record.
_____

24.
_____

Pharmacist
Unlimited opportunity with growing drug store chain. Starting salary $33,000.
_____

25.
_____

Executive Asst.
Investment comp. seeks experienced and motivated individual to run office. Submit resume. From $40K.
_____

26.
_____

CHIEF CHEMIST
Will pay up to $55,000 for the right person with at least 15 years experience.
_____

Daphne is one year younger than Alex. Her older brother Darrin has had trouble finding a career. Daphne and Darrin have had long talks about "life after high school," and Daphne hopes to learn from Darrin's experience.

Ever since he and his family visited Cape Canaveral when he was 11, Darrin's ambition was to become an astronaut. When he knew that his eyesight was not good enough for astronaut training, he decided to try to participate in the space program in some other way.

He went to State University to major in aeronautical engineering. About mid-October, he began to wish that he had not skimmed through the science and mathematics courses he had taken in high school. Actually, he had taken them only because his friends had.

After only a year at State he came home and enrolled in a community college. Then he dropped out of college, took a job, and moved into an apartment. He had planned to save his money and return to State, but he discovered that living in an apartment was too expensive for him. He moved back home.

Daphne is glad to have Darrin at home but is troubled by the upsetting effect that all the changes have had on the family. She would like to have a clearer picture of what is waiting "out there" before she wastes a lot of time and money. She wants to train for a career that interests her and for which she has the required skills. She also wants to have some reason to believe that she will be able to find a job at the end of her training.

OBJECTIVES: In this lesson, we will help Daphne discover how to:

- *Identify career fields suited to talents and interests.*
- *Investigate jobs and their educational requirements.*
- *Predict what the job market will be like in several years.*
- *Understand what benefits an employer may offer in addition to a paycheck.*

CHOOSING A CAREER

Daphne wonders why Darrin is having such a hard time choosing a career. She knows that it is probably one of the most important decisions in a person's life. What she may not know is that surveys indicate that fewer than 50% of people who work enjoy their jobs. This figure would surely be much higher if young people entering the workplace took more time to assess their abilities and interests. Fortunately, Daphne does realize that her choice of career is important, not only because it directly affects how she will be spending most of her waking hours, but also because it will have a direct impact on the overall quality of her personal life. Unlike many people, Daphne knows that she can make choices. She intends to focus on fields that not only will offer her an attractive income, but also will provide her with a sense of satisfaction and self-esteem.

The types of occupations in the United States can be counted in the thousands. Some of these require long periods of education or training. In fact, most jobs require post–high school education or training. Perhaps some workers, like Darrin, missed an opportunity to benefit from the education they were being offered. Of the 40 occupations with the largest projected job growth in the next decade, only one in four will require a college degree or specialized training, according to employment projections published by the U.S. Bureau of Labor Statistics. This agency groups occupations in 13 clusters of related jobs, as follows:

- Industrial production
- Service
- Sales
- Transportation
- Technical and repair jobs
- Social science
- Art, design, and communication
- Office
- Education
- Construction
- Scientific and technical
- Health
- Social service

ASSESS YOURSELF

With the large number of occupational choices in front of her, Daphne wonders, "Where do I begin?" If you are asking yourself that same question, start with what you know about your own interests and abilities. Do you like frequent contact with other people, or do you prefer to spend a lot of time alone? Are you a good follower, or do you prefer directing others in a work effort? Identify your personal skills and interests.

- Do you communicate effectively with others?
- Are you comfortable with computational tasks?
- Do you enjoy conducting a research project or investigation?
- Do you prefer working with your hands?
- Have others complimented you on your creative talent?
- Do you relate unusually well with other people?
- Have you shown talent in leading others in social or athletic activities?

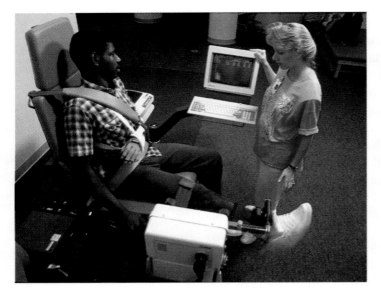

You also need to match your individual talents, interests, and goals with those demanded by various fields of work. A good place to begin your research is with the *Occupational Outlook Handbook* published by the U.S. Bureau of Labor Statistics. Many other sources are also available in school and public libraries. You may also find answers to your questions by interviewing people in fields that interest you.

EDUCATION AND INCOME

The amount and kind of education you receive greatly affects your choice of occupation and thus your current and lifetime income. People who have received postsecondary education can expect to earn more than $1,400,000 during their lifetimes. This is nearly two and a half times the $600,000 that is likely to be earned by workers with fewer than eight years of schooling and more than one and a half times as much as high school graduates. To maximize income, it pays for you to acquire additional education and training.

As the demand for goods and services changes, workers often have to change jobs. It is estimated that college-educated workers change jobs from four to eight times in their lifetimes. Workers with only a high school education change jobs even more frequently.

TECHNOLOGY

Job opportunities and occupational trends are being changed by technology. **Technology** enables goods and services to be produced more efficiently through the use of machines and labor-saving methods. As a percent of the total work force, the number of blue-collar workers (carpenters, plumbers, and factory workers, for example) is decreasing while the demand for white-collar workers (office workers, salespeople, and technicians, for example) is increasing. Technology requires workers to have better training to fill new types of jobs.

FRINGE BENEFITS

Wages and salary are just one kind of financial reward for work. Many companies also offer other benefits such as paid vacations, life and health insurance, free uniforms, retirement plans, stock options, and discounts on company products. These are examples of **fringe benefits.** A fringe benefit is like a hidden paycheck within a paycheck, amounting to between 10% and 25% of total employee compensation.

What do you get for this 10% or 25%? First of all, there are benefits that the company is compelled to offer by law, such as taxes to fund your state's unemployment compensation system, which provides temporary income for workers who are temporarily out of work through no fault of their own. A company must also pay for worker's compensation insurance (to cover work-related injuries or illnesses) and for part of your Social Security/Medicare tax. This last tax is called the **FICA tax**, since it is imposed as a consequence of the *Federal Insurance Contributions Act.*

Examples of nonmandatory employee benefits include paid sick leave, paid holidays, and paid vacation time. However, many companies go well beyond these commonly offered benefits. For example, some companies pay part or all of the premiums for your health insurance or life insurance. A company might also supplement funds that you put into a company retirement plan by matching your contributions with funds of its own.

Many companies encourage employees to enroll in study programs such as college courses, seminars, workshops, or training programs. Often, a company will partly or entirely subsidize education or training programs because it considers such programs to be a benefit for the company as well as the employee.

As Daphne continues her search for employment and career opportunities, she will take into account not only the fact that fringe benefits are a part of her total compensation package, but also the fact that they are a significant part of her potential employer's costs.

Ask Yourself

1. According to the U.S. Bureau of Labor Statistics, 40 occupations have the largest projected job growth in the next decade. How many of these will require a college degree or specialized technical training?

2. You may begin considering the choice of a career by identifying your own interests and abilities. Name eight skills in which you are strong.

3. One step in deciding on a career is to match your interests and talents with those required by various fields of work. What publication by the U.S. Bureau of Labor Statistics can you consult for this information?

4. How has technology affected the percent of blue-collar workers in the total work force?

5. What fringe benefits would you consider to be the most important in choosing a company to work for?

ALGEBRA *REVIEW*

Solve each proportion.
To solve a *proportion,* use the *Rule of Proportions:*

If $\dfrac{a}{b} = \dfrac{c}{d}$, then $ad = bc$.

Example $\dfrac{x}{14} = \dfrac{65}{7}$

$x \cdot 7 = 14(65)$ $x \cdot 7$ is "ad," and $14(65)$ is "bc."

$7x = 910$ $x \cdot 7$ and $7x$ are equivalent expressions.

$x = 130$ Divide both sides by 7.

1. $\dfrac{x}{6} = \dfrac{4}{5}$ 2. $\dfrac{x}{6.2} = \dfrac{3}{31}$

3. $\dfrac{4}{5} = \dfrac{x}{15}$ 4. $\dfrac{2.1}{9} = \dfrac{x}{6}$

5. $\dfrac{3.6}{9} = \dfrac{80}{x}$ 6. $\dfrac{132}{x} = \dfrac{6}{7}$

7. $\dfrac{52}{5200} = \dfrac{x}{17{,}500}$

8. $\dfrac{26}{x} = \dfrac{260}{34{,}000}$

SKILL 1

Dinotech Engineering needs to hire a new engineer this year to help design some of the computer components used in the U.S. space program. The cost of the following extra taxes and fringe benefits will figure into their plans.

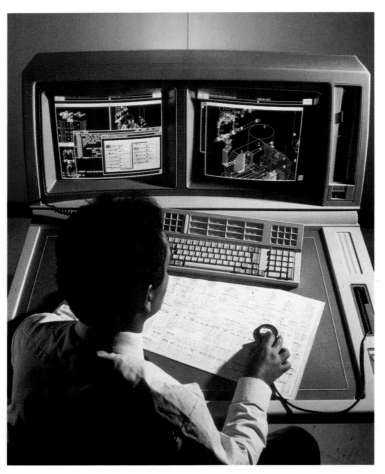

a. 7.65% of gross pay for Social Security and Medicare (FICA)

b. 5% of gross pay to match retirement contributions

c. $80 per month to help pay for health insurance

d. $10 per month for a life insurance policy

Dinotech operates 5 days a week, 52 weeks a year, or 260 days. Of the 260 workdays each year, the company grants its employees 10 vacation days, 6 holidays, and up to 10 sick-leave days.

EXAMPLE 1 Dinotech plans to offer the new engineer $45,000 per year as a starting salary.

QUESTIONS **1.** How much must the company budget to cover the engineer's fringe benefits?

2. How much of the engineer's $45,000 salary is paid for the nonworking days?

SOLUTIONS

1. **a.** $0.0765 \cdot 45{,}000 = 3442.50$ 7.65% of $45,000 for FICA

b. $0.05 \cdot 45{,}000 = 2250$ 5% of $45,000 for the retirement plan

c. $80 \cdot 12 = 960$ $80 per month for health insurance

d. $10 \cdot 12 = 120$ $10 per month for life insurance

Total additional costs: $6772.50

2. The number of nonworking days is

    10 vacation days + 10 sick-leave days + 6 holidays, or 26 days

Write a proportion and solve it.

$$\frac{\text{Nonworking days}}{\text{total days}} = \frac{\text{dollars earned on nonworking days}}{\text{amount earned for entire year}}$$

$$\frac{26}{260} = \frac{x}{45{,}000}$$

$$\frac{1}{10} = \frac{x}{45{,}000} \qquad \text{Simplify the left side.}$$

$$1 \cdot 45{,}000 = 10x \qquad \text{Use the rule of proportions.}$$

$$\frac{45{,}000}{10} = \frac{10x}{10} \qquad \text{Divide each side by 10.}$$

$$4500 = x$$

Dinotech pays $4500 for nonworking days.

SKILL 2

EXAMPLE 2 Daphne's older friend Chloe has completed her engineering training and is considering accepting a position at Aerotech. Chloe has also been offered a position at Electron Research and has described the two offers to Daphne.

	Aerotech	**Electron Research**
Salary	$48,000	$48,000
Medical insurance	$50/month	$60/month
Retirement	3% of gross pay	$2\frac{1}{2}$% of gross pay
Education	$1000/year	$1200/year
Travel Allowance	$120/month	$100/month

QUESTION Which of the two companies is offering Chloe more money including fringe benefits?

SOLUTION
Aerotech is offering:

Salary	$48,000	
Medical insurance: $50/month	600	50 • 12
Retirement: 3% of 48,000	1,440	0.03 • 48,000
Educational expenses	1,000	
Travel allowance	1,440	120 • 12
Total offer	$52,480	

Electron Research is offering:

Salary	$48,000	
Medical Insurance: $60/month	720	60 • 12
Retirement: $2\frac{1}{2}\%$ of 48,000	1,200	0.025 • 48,000
Educational expenses	1,200	
Travel allowance	1,200	100 • 12
Total offer	$52,320	

Aerotech's offer is slightly higher than Electron Research's.

TRY YOUR SKILLS

Fast Facts is a company that copies documents and creates brochures, charts, booklets, and other types of materials for other businesses. The company's compensation committee is submitting four possible fringe benefits options to the company's board of directors. For each employee and fringe benefits option,

    a. find the amount that Fast Facts must pay for extra taxes, retirement, health and life insurance, and other fringe benefits.

    b. find how much salary is paid for the given number of nonworking days in a total of 260 days.

	EMPLOYEE	SALARY	FRINGE BENEFITS OPTION
1.	Copier/collater	$12,000	7.65% of gross for FICA taxes
			$65/month for life and health insurance
			26 nonworking days
2.	Clerk/receptionist	$14,500	7.65% of gross for FICA taxes
			$65/month for life and health insurance
			2% of gross for retirement
			24 nonworking days
3.	Graphics designer	$33,000	7.65% of gross for FICA taxes
			$65/month for life and health insurance
			5% of gross for retirement
			$1000/year for training
			30 nonworking days
4.	Accountant	$48,000	7.65% of gross for FICA taxes
			5% of gross for retirement
			$100/month for life and health insurance
			15 vacation days and 6 holidays

For each of the following groups of employment opportunities,

    a. calculate the total amount of money being offered, including fringe benefits.

    b. determine which of the three positions offers the most money.

5.

Company	A	B	C
Salary	$37,500	$38,000	$35,800
Retirement benefits	3.5% of gross pay	3% of gross pay	5% of gross pay
Medical insurance	$55/month	$45/month	$50/month
Educational expense	$500/year	$800/year	$1000/year
Travel allowance	$2400/year	$100/month	$150/month

6.

Company	A	B	C
Salary	$42,500	$41,800	$44,000
Stock options	5% of gross pay	6.5% of gross pay	4.5% of gross pay
Medical and life insurance	$100/month	$115/month	$100/month
Car lease	$250/month	$280/month	$240/month
Travel allowance	$350/month	$375/month	$300/month

7. Mr. Grant receives a salary of $34,000 and takes vacation days and holidays for 28 of the company's 242 operating days. How much of his salary is paid for the days during which he does not work?

EXERCISE YOUR SKILLS

1. What are some examples of jobs that are available even though people who are looking for employment cannot find jobs?

2. Why do you think that workers with a high school education change jobs more frequently than college-educated workers?

3. As an aid in answering some of your questions about a particular kind of job, it may be helpful to interview someone who is already working in that field. What are six questions that you might ask in such an interview?

KEY TERMS
FICA tax
fringe benefits
technology

For each company fringe-benefits policy described below,

    a. find the amount that the company must pay for extra taxes, retirement, health and life insurance, and other fringe benefits.

    b. find how much salary is paid for the given number of nonworking days in a total of 240 operating days.

	EMPLOYEE	SALARY	FRINGE-BENEFITS POLICY
4.	Computer operator	$40,000	7.65% of gross for FICA taxes
			5% of gross for retirement
			$100/month for life and health insurance
			$2000/year for training
			24 nonworking days
5.	Delivery person	$12,000	7.65% of gross for FICA taxes
			$50/month for health insurance
			8 holidays and 4 sick days
6.	Office manager	$56,000	7.65% of gross for FICA taxes
			$150/month for life and health insurance
			$200/month per diem expense allowance
			$250/month car lease
			10% of gross in stock options
			18 nonworking days
7.	Lab technician	$28,000	7.65% of gross for FICA taxes
			$200/month for health insurance
			26 nonworking days
8.	Theatre manager	$35,000	7.65% of gross for FICA taxes
			$800/year for training
			15 vacation days and holidays
9.	Executive secretary	$42,000	7.65% of gross for FICA taxes
			$225/month for life and health insurance
			8% of gross in stock options
			15 vacation days, 6 holidays
10.	Copy Editor	$25,000	7.65% of gross for FICA taxes
			$175/month for health insurance
			10 holidays and 10 vacation days
11.	Pharmacist	$52,000	7.65% of gross for FICA taxes
			$275/month for health insurance
			6% of gross for retirement
			22 nonworking days

For each of the three employment opportunities **A, B,** and **C,**

   a. calculate the total amount of money being offered, including fringe benefits.

   b. determine which of the three positions offers the most money.

12.

Company	A	B	C
Salary	$22,000	$22,000	$22,000
Retirement benefits	2% of gross pay	3% of gross pay	2.5% of gross pay
Medical insurance	$25/month	$20/month	$40/month
Expense allowance	$200/month	$250/month	$300/month

13.

Company	A	B	C
Salary	$27,000	$30,000	$23,500
Retirement benefits	4% of gross pay	2% of gross pay	5% of gross pay
Medical insurance	$100/month	none	$150/month
Expense allowance	none	$200/month	$350/month

14.

Company	A	B	C
Salary	$47,000	$39,000	$35,500
Retirement benefits	none	3.5% of gross pay	1% of gross pay
Medical insurance	$50/month	$125/month	$200/month
Expense allowance	$300/month	$300/month	$500/month

MIXED REVIEW

Write an equation of the form $E = rh + t$ where r represents the regular hourly rate, h represents the number of hours worked, and t represents any bonus or tip. Then find the weekly earnings for each of the following "employees."

   1. I am the gremlin who arranges for you to be stopped by every traffic light between your home and your job on any day that you are running a little late. I get paid only when I actually make you late for work. This week I worked 28 hours and got paid $14.50 per hour. I get no bonuses because no one likes me.

   2. I am the auto-focus camera that you took with you on your once-in-a-lifetime vacation to the South Seas. I worked for 32 hours at $10.50 an hour helping you take 125 well-focused photos out of 128. I also received a $35 bonus for one photo that earned you a prize in a local photography contest.

3.–4. Assume that the "employees" of Exercises 1–2 received overtime pay of $1\frac{1}{2}$ times their regular rate for all hours they worked over 20. Write an equation to represent their weekly earnings of the form

$$E = 20r + (h - 20) \bullet 1.5r + t$$

Where r represents the regular hourly rate, h represents the number of hours worked, and t represents any tip or bonus. Determine the total earnings for each employee.

Create a spreadsheet to calculate the hourly, weekly, monthly, and yearly earnings for each job described below.

40 hours = 1 workweek
52 weeks = 1 year
12 months = 1 year

5.

Do you know how to sell? Start with us at $6.25/h and GROW!

6.

Laboratory Assistant
We are looking for a qualified worker trained in the daily operating procedures of a biotechnical laboratory. Two years experience desired but not necessary. Starting salary $420/ week.

7.

COMPUTER PROGRAMMER
Experience required. Flexible hours can be arranged. Starting salary $3,800/month.

DO MORE, MAKE MORE: COMMISSIONS AND PAYMENTS BY ITEM

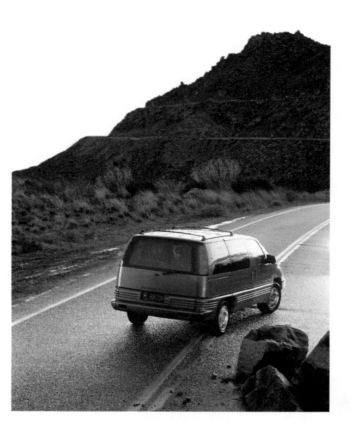

Betty goes to high school with Alex. She knows that Alex is trying to decide whether getting a part-time job is worth the effort and time involved. She and Alex were dating before Alex met Alice. In fact, Betty was with Alex when he was shopping around for his van.

Alex and his parents seemed to know a lot about car salespeople when they were picking out the van. At one of the lots they visited, a saleswoman kept trying to push Alex's father and mother into choosing a gorgeous, current-year model. Betty noticed that the saleswoman kept saying that it was "loaded." It had not only a very powerful engine but all the extras: "power this" and "power that," a really plush interior, and all kinds of built-in storage space. Of course, it had a "powerful" price tag also.

Betty guessed that it cost more than Alex's parents wanted to spend. After all, Alex just needed transportation to and from school.

As they left the car lot and drove toward another lot, where Alex would eventually buy his van, Betty asked why the saleswoman had seemed so eager to sell the most expensive van they had. Alex's father explained, "Oh, she works on commission, Betty. You see, the more money we spend, the more she makes! Perhaps she has not stopped to realize that if she does not sell us what we want, she will not make anything from us!"

Betty was not sure she understood how a person's earnings could be affected by the price of a van. She decided to find out.

Determine how much a salesperson earns when his or her pay is based on a commission.

Determine how much a person earns when his or her pay is based on a piece rate, or item rate.

Understand that the earnings of a person who works on commission or piece rate can vary from month to month.

EARNING COMMISSIONS

People whose salaries depend on how much they sell in a given period of time are said to be paid on **commission.** These salespeople receive a percent of each amount that they bring in. For example, Jason, who works at a used-car lot, earns a 15% commission on every car he sells. Today, Jason sold a car for $6000. His earnings are 15% of $6000.

$$6000 \cdot 0.15 = 900$$

Jason earned $900. The more cars Jason sells, the more money he makes. If he sells no cars on a particular day, he earns no money that day.

Some salespeople earn a regular salary plus a commission, as illustrated in the ad below. If the headline is to be believed, the salesperson could earn $50,000 in a year, but he or she would probably have to make a lot of sales to earn that much.

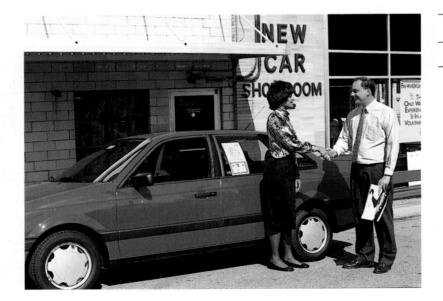

Auto

$50,000+ POTENTIAL

We are seeking 2 aggressive self-motivated people to add to our highly successful team. Previous auto-sales experience helpful. We are especially interested in those with HEAVY RETAIL SALES EXP. We have an intensive in-house training program to launch you onto a successful sales career. We offer:

- Weekly salary & Comm.
- New demo
- Top medical ins. program
- Heavy advertising
- Excellent service dept. to keep your customers happy

Suppose that the weekly salary is $300. For 52 weeks the salary will be $15,600 ($300 • 52 weeks). If the salesperson earns 10% commission on all sales and sells $344,000 worth of cars during the year, the total earnings will be

$15,600	$300 • 52 = regular salary
+ 34,400	$344,000 • 0.10 = commission
$50,000	Total earnings for the year

If these are new cars, then between 25 and 40 of them are probably being sold a year. If the commission is less than 10%, the salesperson will have to sell more cars to make the same amount of money.

Some companies also encourage their salespeople to sell more by offering them a **graduated commission.** If they sell up to a specified amount, their commission rate will be a certain percent. If they sell more than that amount, the percent will increase.

PIECE RATE

Instead of receiving a salary for working a certain amount of time, some people are paid a **piece rate,** or **item rate** for the number of items that they produce or sell. If you work fast or work on your own schedule, being paid a piece rate may be advantageous to you. Like commission, a piece rate is sometimes combined with a **base salary,** which you would receive regardless of how many items you produced or sold.

PUSH MONEY

Consumers are easily influenced when they lack adequate information about products that they buy. Unfortunately, in many stores, salespeople are not informed enough about a product to provide adequate information to customers. They are trained just to take a customer's order.

Another problem is the salesperson who is paid **push money,** a cash incentive provided by a *manufacturer* for selling the manufacturer's products. Push money is not the same as a commission, which is money paid by the *store* to salespeople to sell *all* the products in the store. If you notice a salesperson aggressively pushing a certain product, that salesperson may be receiving push money. Let the buyer beware!

ALGEBRA *REVIEW*

Solve for x or a.

1. $14x - 8 = 20$

2. $3.5 + 2.5a = 18$

3. $(3.5 + 2.5)a = 18$

4. $26x = 78$

5. $\dfrac{1}{26}x = 78$

6. $300 + 0.10x = 50,000$

Simplify.
Example
$$2m + 3.5m - 2.7p + 8$$
$$= (2 + 3.5)m - 2.7p + 8$$
$$= 5.5m - 2.7p + 8$$

7. $9x + 3.5x$

8. $0.05x - 0.03x + 7$

9. $(2.1 + 3.9)b + 4b$

10. $1y + 14y$

11. $0y + 14y + 1 • 5$

12. $y + 14y + 0 • 5$

CLASSIFIED ADS

As Betty was seeking information about salespeople whose earnings depend on commissions, she discovered the following classified ads in the employment section of her Sunday newspaper. The jobs seemed very attractive. But Betty thought that there was more to the matter than she could see from the ads.

Sales

ONE-CALL CLOSERS

If your are willing to learn and want to earn

$650–$1200
(weekly commission)

- CASH PAID DAILY
- NO INVESTMENT
- DAYTIME SALES
- COMPLETE TRAINING
- AUTO REQUIRED
- MANAGEMENT OPPTY.

CALL MONDAY ONLY

SALES/GOLF

Earn more than you require. $40-$120,000 per year commissions. Self-starting closer.

TIRED OF FALSE PROMISES?

(I was)

- Join one of the fastest growing industries in the country.
- Enjoy success as hundreds of our representatives do.
- Earn over $100,000 per year.
- FRESH FREE LEADS
- NO SATURDAYS
- 10% OF BACK-END PAID/MO.

ADVENTURE INN BIG MONEY

Direct sales experience required. Seminar sales. Sell vacation club. Prospects furnished. Work evenings and weekends. 5-day work week required. Straight commission. Earn up to $1600 a week. Closed Monday.

CAREER SALES OPPORTUNITY

Would You Believe $144/day?

General Research is expanding its headquarters to the suburbs. We are hiring 2-3 people of legal age for marketing and management training program. Minimum $1500/month guaranteed commissions and bonuses to start. All fringe benefits including hospitalization, medical and life insurance. Must have dependable auto. Call today.

Betty remembered talking with a friend who sells cars. She found out that if a rebate or discount is given, that fact affects how much commission is earned. Suppose that a car is priced at $6000 and a $420 discount is offered to make the customer want to buy the car. The salesperson receives a commission on only $5580 (that is, on 6000 − 420). Because of details such as this, Betty realizes that finding a job requires a lot of research. She cannot just read the ads.

Ask Yourself

1. How can people earn the high pay offered in ads such as those shown?

2. How does a graduated commission work?

3. What is push money?

4. How is push money different from a commission?

SHARPEN YOUR SKILLS

SKILL 1

You can use a formula to find the commission C when you are given the rate of commission r, the price p, and the number of items n sold. Remember to round commission to the nearest cent.

Commission = price • commission rate • number sold

Commission Formula

$C = prn$ where p = price of one item
r = commission rate
n = number of items sold

EXAMPLE 1 Betty's friend Rosalie is a salesperson at a department store. She receives a 7% commission for selling small appliances and housewares. Among the many items that she sells is a fancy ceramic mug. Rosalie would like to know how many mugs she needs to sell to earn at least $50.

COMPUTER MUG of glazed
earthenware with terminal and
keyboard as handle. Dishwasher
safe. For hot or cold beverages,
pencils, and more. Holds 14 oz.
#45678 mug $12.95

QUESTION How can Rosalie use her graphing calculator to find out how many mugs she needs to sell to earn $50?

SOLUTION
Rosalie uses the commission formula.

$C = prn$ $p = 12.95, r = 7\% = 0.07$
$C = 12.95 \cdot 0.07 \cdot n$
$C = 0.9065n$

To use her graphing calculator to graph this equation, Rosalie has to change the variable C to y and the variable n to x:

$y = 0.9065x$

Rosalie enters the equation into her graphing calculator. She sets the range as shown and graphs the equation.

Xmin: 0   Ymin: −10
Xmax: 60   Ymax: 60
Xscl: 10   Yscl: 10

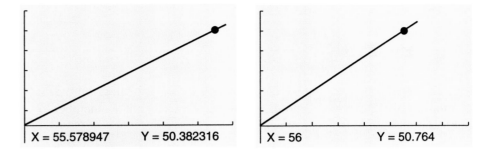

| X = 55.578947 | Y = 50.382316 | | X = 56 | Y = 50.764 |

The graph is a line as shown at the left above. Rosalie uses the TRACE function to move the cursor along the line until the y value is close to $50. She concludes that she can earn over $50 in commissions by selling 56 mugs.

She would like x values to be integers, so she uses the INTEGER function of her graphing calculator or sets the range to a "friendly window" and redraws the graph as shown at the right above.

To confirm the calculator results, Rosalie substitutes $n = 56$ into the commission equation.

$C = 0.9065n$
$C = 0.9065 \cdot 56 \qquad n = 56$
$C = 50.764$

The equation method confirms the results of the graphing calculator. Rosalie's commission is $50.76 for selling 56 mugs.

───────── **SKILL 2** ─────────

You can use a formula to find the earnings E when you are given the monthly salary s, the commission rate r, and the total sales t.

Earnings with Commission Formula

$E = s + rt$ where s = monthly salary
r = commission rate
t = total sales for the month

EXAMPLE 2 William earns a salary of $1000 per month selling cars plus a commission of 6% on each car that he sells. In his first month on the job, he sold 3 cars for a total of $29,500 in car sales.

QUESTION How much did William earn in all?

SOLUTION

$E = s + rt$ Use the earnings formula.
$E = 1000 + 0.06(29{,}500)$ $s = 1000, r = 6\% = 0.06, t = 29{,}500$
$E = 2770$

You can enter the calculation in your graphing calculator as written, since the calculator will multiply before adding.

William earned $2770 in salary and commissions during the month.

EXAMPLE 3 Yvette is a real-estate salesperson. Her earnings are based on a *graduated commission*. She receives 4% of any sale up to $100,000. For any amount over $100,000 she receives a 6% commission. Last Sunday she sold the house described below.

> **MINT CONDITION!!** $109,000
> You'll appreciate the love and care that went into this stunning 3-bdrm, 2-bath home. Vaulted ceilings, custom drapes, mst. bath w/whirlpool/double vanity.

QUESTION How much did she make selling the home?

SOLUTION

Commission = price • rate

Find the commission C_1 (read "C sub one") on $100,000. Then find the commission C_2 (read "C sub two") on the amount over $100,000.

$C_1 = 100,000 • 0.04 = 4000$

$C_2 = 9000 • 0.06 = 540$ $109,000 - 100,000 = 9000$

Total commission = $C_1 + C_2$
Total commission = $4000 + 540 = 4540$

Yvette's commission is $4540.

───── **SKILL 3** ─────

To solve problems involving piece rate, use the following formula.

> **Piece Rate Formula**
> $E = rn$ where r = the piece or item rate
> n = the number of items

EXAMPLE 4 Carole decorates wedding cakes at Sloane's Bakery. Mrs. Sloane pays her $16.00 for each cake that she decorates.

QUESTION How much money did Carole earn in a week in which she decorated 5 wedding cakes?

SOLUTION

Substitute in the formula for piece work.

$$E = rn$$
$$E = 16.00 \cdot 5$$
$$E = 80$$

Carole earned $80 for 5 wedding cakes.

EXAMPLE 5 Mrs. Sloane needs Carole to help her with customers when she is not busy decorating cakes. She has offered Carole a choice of one of two payment plans.

> Plan 1: Salary of $120 for a 20-hour week and $12 for each cake
> Plan 2: Salary of $105 for a 20-hour week and $14.50 for each cake

QUESTION Which plan should Carole choose?

SOLUTION

Carole is not sure which plan is better for her. She thinks that she can decorate at least 7 cakes a week. Carole writes two equations, one for each plan.

> Plan 1: $E = 120 + 12n$ Earnings = salary + piece-rate pay
> Plan 2: $E = 105 + 14.50n$

Carole wants to know the least value of n that will make the same money with either plan. She solves the *system of equations* by setting the two earnings equal to each other.

$$120 + 12n = 105 + 14.5n$$

When solving an equation, you may subtract the same quantity from each side of the equation. Carole decides to subtract $12n$ to obtain a simpler equation.

$120 + 12n = 105 + 14.5n$	
$120 + 12n - 12n = 105 + 14.5n - 12n$	Subtract $12n$ from both sides.
$120 = 105 + 2.5n$	Combine similar terms: $12n - 12n = 0$; $14.5n - 12n = 2.5n$
$120 - 105 = 105 + 2.5n - 105$	Subtract 105 from both sides.
$15 = 2.5n$	Simplify.
$\dfrac{15}{2.5} = \dfrac{2.5n}{2.5}$	Divide both sides by 2.5.
$6 = n$	Simplify.

Carole will earn the same money under both plans if she decorates 6 cakes.

Plan 1: $E = 120 + 12n = 120 + 12(6) = 192$ $n = 6$
Plan 2: $E = 105 + 14.50n = 105 + 14.50(6) = 192$ $n = 6$

She also calculates her earnings under both plans if she decorates 7 cakes.

Plan 1: $E = 120 + 12(7) = 204$ $n = 7$
Plan 2: $E = 105 + 14.5(7) = 206.50$ $n = 7$

Plan 2 is better for her when $n = 7$. Carole concludes that she earns more money under Plan 2 if $n > 6$ (n is greater than 6) and more money under Plan 1 if $n < 6$ (n is less than 6).

TRY YOUR SKILLS

The commission rate, the price of each item, and the number of items sold are given for each salesperson.

   a. Calculate the commission earned by selling the first number of items sold (for example, 10 items in Exercise 1).

   b. Write an equation to show the commission y earned by selling x number of items.

   c. Graph the equation on your graphing calculator. Then use the TRACE function to find the commissions for all four amounts. Remember to use an integer function or a "friendly window."

	Person	Price	Rate	Number Sold
1.	Carole	$16.75	9%	10; 18; 24; 32
2.	José	21.90	8.5%	8; 16; 22; 28
3.	Gabe	21.95	7.5%	5; 12; 15; 18

Tracey and several of her friends have jobs working for a monthly salary plus commission. Find the month's earnings for each person. Monthly salary, commission rate, and the amount of sales are shown. Use a computer spreadsheet.

	Name	Monthly Salary	Commission Rate	Total Sales	Commission Earned	Total Earnings
4.	Tracey	$ 350	12%	$6125		
5.	Isabel	475	11.5%	6250		
6.	Damien	1000	8.5%	7100		

7. Tom assembles radios and tape recorders in a factory. He receives $9 for each radio and $12 for each tape recorder. How much did he earn in a week in which he assembled 26 radios and 38 tape recorders?

8. Alice has her own barber shop. She charges $9 for a regular haircut and $16 for a personalized haircut. Last week, 26 of her customers had a regular haircut, and 8 customers had the personalized haircut. What were her earnings for the week?

9. Patrick sells annual memberships in a health club. His boss has offered him a choice between Plan 1, no salary but a straight $30 for each membership sold, and Plan 2, a weekly base salary of $300 plus $15 for each membership he sells. How many memberships would Patrick have to sell to make the same amount of money under either plan?

EXERCISE YOUR SKILLS

1. If you are earning a commission, why is it to your advantage to sell a relatively expensive item?

2. A person working on commission cannot predict exactly what next month's income is going to be. What is a good way to handle money so that there is enough to pay the bills?

3. Imagine that you work on commission in a large department store and that during the month before Christmas and Hanukkah, your department has a high volume of sales. As a result, your salary is three times what it was the previous month. Would you assume that next month's salary will be equally high and so be tempted to spend most of your earnings? Why or why not?

4. Reread each ad that Betty read in this lesson. Decide what is positive and what is negative about each position being advertised. Make a list of questions that you would ask if you went for an interview for each job. Which jobs appeal to you the most? Which jobs appeal the least?

KEY TERMS

base salary
commission
graduated
    commission
item rate
piece rate
push money

The commission rate, the price of each item, and the number of items sold are given for each salesperson.

    a. Calculate the commission earned by selling the first number of items sold (for example, 18 items in exercise 5).

    b. Write an equation to show the commission y earned by selling x number of items.

    c. Graph the equation on your graphing calculator. Then use the TRACE function to find the commissions for all four amounts. Remember to use an integer function or a "friendly window."

	Person	Price	Rate	Number Sold
5.	Ellen	$ 3.75	10%	18; 28; 38; 48
6.	Ivy	22.80	8.2%	6; 10; 14; 18
7.	Peter	7.99	9.5%	15; 25; 35; 45

8. Jenny earns a base monthly salary of $2150 plus a commission of $7\frac{1}{2}\%$ on her sales. In July her sales were $7350. How much did she earn in July?

9. Samantha sells used cars. Her earnings are based on a graduated commission. She receives 3% of her total sales up to $20,000 and $4\frac{1}{4}\%$ of all sales over $20,000. Find Samantha's total earnings for a week in which she sold cars priced at $6,399, $2,495, $7,450, and $10,188.

Eduardo is a real-estate salesperson who specializes in selling houses. His earnings are based on a graduated commission. He receives 2% on each sale up to $100,000 and 3% of the amount above $100,000. How much would Eduardo earn for selling each of these houses?

10.
    BEAUTY & SPACE
    $199,900 5 bedrooms, 3 1/2 baths, formal living and dining. Study. Master bedroom. Lots of storage. Jacuzzi.

11.
    LOCATION PLUS $149,900
    4 bedrooms, 2 1/2 baths, formal living and dining. Family room, game room.

The salespeople below are offered a choice of two monthly payment plans.

    Plan 1:   Base salary + commission
    Plan 2:   A higher rate of commission and no base salary

Let t represent the amount of sales and E represent the earnings. Write an equation for each payment plan. By solving the equations together, find the amount of sales needed to make the same earnings with either plan.

12. Eric
    Plan 1:   9% of sales + $900
    Plan 2:   12% of sales

13. Wanda
    Plan 1:   7% of sales + $2400
    Plan 2:   10% of sales

Boris repairs TV and VCR remote-control units at $7.50 an hour in an electronic equipment repair store. He also gets $4.00 for each unit that he repairs. He works a 40-hour week.

14. How many units does he have to repair to earn at least $500 per week?

15. It takes Boris an average of $\frac{1}{2}$ hour to repair a unit. What is the most money that he can earn in one week?

16. How many units does Boris have to repair to earn $700 in one week? Explain your answer.

MIXED REVIEW

Determine the total weekly earnings for each employee described in each classified ad below. Use an equation of the form $E = rh + t$, where r represents the regular hourly rate, h represents the number of hours worked, and t represents the tips or bonus, if any.

1.

Health Care
LAB TECHNICIAN $10/h
45 hours/week

2.

Manicurist
English and Spanish speaking. Guaranteed pay $6.50/h to start. Tips can be up to $90 for a 35-hour week!

For each company fringe benefits policy described below, find

a. the amount that the company must pay for extra taxes, retirement, health and life insurance, and other fringe benefits.

b. how much salary is paid for the number of nonworking days in a total of 260 operating days.

	EMPLOYEE	SALARY	FRINGE BENEFITS POLICY
3.	Bookkeeper	$40,000	7.65% of gross for FICA taxes
			3% of gross for retirement
			$100/month for life and health insurance
			26 nonworking days
4.	Cashier	$18,500	7.65% of gross for FICA taxes
			$50/month for health insurance
			8 holidays, 4 sick days, 10 vacation days
5.	Bank manager	$55,000	7.65% of gross for FICA taxes
			$120/month for life and health insurance
			$200/month expense allowance
			18 nonworking days

1-4 DEDUCTIONS: WHO GETS WHAT?

Luis has known Alex since they were in the seventh grade together. He and Alex have been comparing notes about the increasing costs of dating, keeping a car running, and getting any kind of schooling after high school. Luis realized three years ago that the only way to come up with the money he needed was to find a job. He started out at the supermarket, first stocking the shelves, then bagging the groceries. Now he runs the cash register and sometimes gets to work behind the deli counter.

What a shock it was to Luis the first time he received a weekly paycheck. He knew that he had worked 20 hours and that he was supposed to get $4.75 an hour. He figured that the $95.00 was just enough to buy one new tire for his car.

When he saw that the check was for only $86.73, he immediately wanted to know what had happened to the rest of the money.

Luis knew that all the deductions from the check were somehow related to the mass of papers that got spread around the family dining room every April. His mother would perform numerous calculations with her calculator and transcribe some of them onto sheets of paper about taxes. It was all very mysterious, so Luis didn't ask questions. After April 15, all the papers vanished for another year.

Luis has heard his parents mention Medicare frequently in the last year. His grandfather has been hospitalized for a chronic illness and has had numerous medical tests and treatments. Some of these costs are covered by Medicare.

Understand why income taxes are deducted from a paycheck and what they are used for.

Learn what FICA taxes are for.

Realize why a new employee fills out a W-4 form.

Calculate take-home pay by subtracting deductions from gross pay.

TAXES: A TOOL OF GOVERNMENT

When Luis began to investigate the matter, he learned that the taxes collected by the federal government every year are used for many purposes, the largest of which are the nation's defense, the funding of the Social Security system and Medicare, and interest on the national debt. Federal taxes also support a variety of other activities including national parks, airport construction, agriculture, nutritional programs, school lunch programs, and libraries.

Luis's taxes—and yours, too, of course—are used to support not only federal activities, but also activities of state and local governments. In our society,

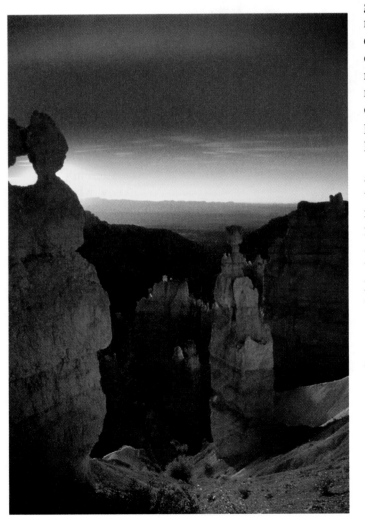

government performs several broad functions, many of which significantly affect the general economy. These include the upkeep and funding of national parks and public highways, the redistribution of income to care for the needy, regulation of certain industries such as communications, and the establishment and preservation of the legal framework for the protection of the rights of citizens.

Luis knows that parks, highways, schools, and the police department are supported by taxes. He does not know how the government redistributes income so that some citizens pay for government benefits to other citizens. Taxes are the government's major means of making these transfer payments to support federal programs such as Social Security, programs for disabled veterans, disaster relief, and various welfare services.

Luis knows that, as a citizen, he is entitled to suggest changes in the way his taxes are spent. The legislators representing him are the people who can actually make the changes in the tax laws. In doing so, they must take into account the effect that new tax legislation will have on workers, consumers, business and industry, and the economy in general. This may sound easy, but it is not. Legislators realize that no one likes tax increases, and so they avoid raising taxes if they possibly can.

WHY SOCIAL SECURITY?

So government operations are supported by income taxes. Are Social Security taxes just another form of income tax?

From the moment we are born until we die, people of all ages are affected if **income security** is absent. Children are affected if their parents lose their ability to earn an income. Young adults are concerned about how they will survive if they are disabled by illness or injury. Middle-aged people are concerned about how their children would be supported and educated if they were to die or become disabled. Older retired people often worry that their retirement income will not cover their living expenses.

In 1935 the U.S. government established **Social Security** to help protect citizens from economic insecurity. Simply put, the Social Security system is a combination of disability insurance and a pension plan that is compulsory for most workers. It provides a base on which individuals may build protection for themselves and their dependents. The cost of the system is paid for by both workers and employers.

Most employed people have **FICA** (Federal Insurance Contributions Act) taxes deducted from their paychecks. Employers also pay into the program by matching the amounts paid by the employees. A portion of the FICA taxes is paid into the Social Security trust fund, which entitles the employee eventually to receive Social Security benefits. A portion is used to pay for **Medicare.**

A few of the benefits paid for by FICA taxes include the following:

SOCIAL SECURITY
- **Disability insurance**   Income to individuals who become disabled and thus cannot work. Disability-income benefits are paid after a five-month waiting period during which it is determined whether the physical or mental disability is severe enough to prevent the person from working for at least a year.
- **Survivors insurance**   Money for monthly living expenses to survivors of deceased workers who were eligible for Social Security benefits. These benefits go to children and in some cases to a widow or widower.
- **Retirement benefit**   Monthly Social Security payments made to a retired worker and to the worker's spouse.

MEDICARE
- **Health insurance**   A two-part health insurance program for people over age 65. Medicare benefits consist of hospital insurance and medical (doctor) insurance.

FORMS

When Luis began his job at the supermarket, he was asked to fill out a **Form W-4,** *Employee's Withholding Allowance Certificate.* The information that he put on this form determined the amount that his employer deducts from Luis's paycheck for income taxes. The amount depends on Luis's income level, his marital status, and the number of withholding allowances that he claims.

Form **W-4** Department of the Treasury Internal Revenue Service	**Employee's Withholding Allowance Certificate** ▶ For Privacy Act and Paperwork Reduction Act Notice, see reverse.	OMB No. 1545-0010 19--

1 Type or print your first name and middle initial
    Luis Estevez   Last name   **2** Your social security number   **123-45-6789**

Home address (number and street or rural route)
    456 Cedar Street

3 [X] Single [ ] Married [ ] Married, but withhold at higher Single rate.
Note: *If married, but legally separated, or spouse is a nonresident alien, check the Single box.*

City or town, state, and ZIP code
    Inland, In. 12345

4 If your last name differs from that on your social security card, check here and call 1-800-772-1213 for more information ▶ [ ]

5 Total number of allowances you are claiming (from line G above or from the worksheets on page 2 if they apply) .   **5** *1*
6 Additional amount, if any, you want withheld from each paycheck . . . .   **6** $
7 I claim exemption from withholding for 1993 and I certify that I meet **ALL** of the following conditions for exemption:
   ● Last year I had a right to a refund of **ALL** Federal income tax withheld because I had **NO** tax liability; **AND**
   ● This year I expect a refund of **ALL** Federal income tax withheld because I expect to have **NO** tax liability; **AND**
   ● This year if my income exceeds $600 and includes nonwage income, another person cannot claim me as a dependent.
   If you meet all of the above conditions, enter "EXEMPT" here . . . . ▶ **7**

Under penalties of perjury, I certify that I am entitled to the number of withholding allowances claimed on this certificate or entitled to claim exempt status.

Employee's signature ▶   Date ▶   , 19

8 Employer's name and address (Employer: Complete 8 and 10 only if sending to the IRS)   **9** Office code (optional)   **10** Employer identification number

Cat. No. 10220Q

His employer does not have to withhold any tax from Luis's income if he did not have any income-tax liability (that is, did not owe any taxes) last year and if he does not expect to have any tax liability this year. Students often qualify for this no-withholding status because they have only part-time or summer jobs.

READING YOUR PAYCHECK

When he got his first job, Luis really looked forward to his first paycheck. He knew that his income from this job would be the first step toward financial

---

Evans Foods   Pay Statement
1000 Center Street
Inland, IN 12345

Check No.	Name		Social Security Number
3705	Luis Estevez		123-45-6789

Pay Date	Pay Period	Type
07/10/--	07/03/-- - 07/09/--	Salary

	Gross Pay	Federal Inc. Tax	FICA	State Inc. Tax	Other	Net Pay
	$95.00	$1.00	$7.27			$86.73

independence. That check is shown above. A pay statement summarizing his deductions is shown at the bottom of page 38.

Young workers are often surprised by the difference between total pay, or **gross pay,** and net pay, or take-home pay. **Take-home pay** is the amount remaining after federal and state taxes and other deductions have been taken out of the check. A **deduction** is an amount of money subtracted from a person's gross pay for such items as FICA taxes, federal and state income taxes, insurance premiums, and union dues. After deductions your take-home pay is likely to be about 20% to 30% less than your gross pay.

Ask Yourself

1. What are five broad economic functions of our government?

2. What are four examples of the redistribution of income that the government accomplishes through taxation and spending?

3. Who pays for the federal Social Security system?

4. How does your employer determine how much of your income to withhold for taxes?

5. When you first get a job, which federal tax form are you asked to complete and submit to your employer?

6. How is take-home pay computed?

ALGEBRA *REVIEW*

Percent means "per hundred."

Example $7\% = 0.07 = \frac{7}{100}$

Express each as a decimal.
   1. 5%   2. 20%   3. 7.5%

   4. 0.5%   5. 7.65% 6. 125%

Express each as a percent.
   7. 0.30   8. 0.02

   9. 1.5   10. 0.004

Find each amount.
   11. 5% of 120

   12. 7% of $4.56

   13. 120% of $808

   14. 7.5% of $9.10

   15. 7.65% of $4920

SHARPEN YOUR SKILLS

SKILL 1

To find the amount to withhold for federal income tax, use a table like the one shown for a single person with weekly wages.

SINGLE Persons—WEEKLY Payroll Period

		And the number of withholding allowances claimed is—					
If the wages are—		(For Wages Paid in 19--)					
At least	**But less than**	**0**	**1**	**2**	**3**	**4**	**5**
		The amount of income tax to be withheld is—					
$0	$50	$0	$0	$0	$0	$0	$0
50	55	1	0	0	0	0	0
55	60	1	0	0	0	0	0
60	65	2	0	0	0	0	0
65	70	3	0	0	0	0	0
70	75	4	0	0	0	0	0
75	80	4	0	0	0	0	0
80	85	5	0	0	0	0	0
85	90	6	0	0	0	0	0
90	95	7	0	0	0	0	0
95	100	7	1	0	0	0	0
100	105	8	1	0	0	0	0
105	110	9	2	0	0	0	0

To find FICA taxes (Social Security and Medicare), use the following.

> If annual income is less than $57,600, then the deduction for FICA taxes is 7.65% of gross pay.

EXAMPLE 1 Luis's salary for this week is $95. Luis entered one withholding allowance on his W-4 form.

QUESTIONS 1. How much will be withheld for income tax?

2. How much will be deducted for Social Security and Medicare?

3. How much is Luis's take-home pay?

SOLUTIONS

1. Find the amount to withhold on the table shown above. Notice that Luis's wages, $95.00, are on the eleventh line, "at least 95, but less than 100." Read across to the withholding allowance column headed 1. The number in that column is the dollar amount to be withheld. The amount withheld for income tax from Luis's check will be $1.

2. Luis's salary is less than $57,600, so the deduction for FICA taxes is 7.65% of the gross pay.

$$0.0765 \bullet 95 = 7.2675 \qquad 7.65\% = 0.0765$$

The deduction for FICA taxes is $7.27.

3. Luis can find his take-home pay by writing and solving an equation for his take-home pay T, where g = gross pay, w = amount withheld for income taxes, and f = deduction for FICA taxes.

Take-home pay = gross pay − all deductions
$$T = g - (w + f)$$
$$T = 95 - (1 + 7.27)$$
$$T = 95 - 8.27$$
$$T = 86.73$$

Luis's take-home pay is $86.73.

SKILL 2

EXAMPLE 2 Luis wanted to see how he could use a spreadsheet to represent his take-home pay.

SOLUTION

He set up a spreadsheet in the following way.

	A	B	C	D	E
1	Name	Luis Estevez			
2					
3	Gross	Withholding	Income Tax	FICA	Take-Home
4	Pay	Allowances	Withholding	Withholding	Pay
5					
6					

He used row 1 for the Name line and rows 3 and 4 for the five column headings. He completed row 6 by typing the following into cells A6–E6. He formatted columns D and E for two decimal places.

6	95	1	1	0.0765*A6	+A6−(C6+D6)

On the computer screen, row 6 actually appears as follows.

6	95	1	1	7.27	86.73

By using formulas in columns D and E, Luis was able to find the take-home pay for any value of the gross pay that he wishes by filling in columns A and C with the correct numbers and letting the computer calculate the values in cells D6 and E6.

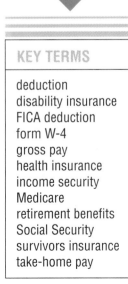

TRY YOUR SKILLS

Use the income tax withholding table in the Reference Section to find the take-home pay for the given wages and withholding allowances. Assume that each person is married. For the FICA deduction, find 7.65% of gross pay. Use the formula for take-home pay

$$T = g - (w + f)$$

where g is the gross pay, w is the amount withheld for taxes, and f is the deduction for FICA.

	Monthly Salary	Withholding Allowances		Monthly Salary	Withholding Allowances
1.	$2400	0	**2.**	$ 700.00	3
3.	3400	5	**4.**	4380.00	1
5.	1280	5	**6.**	4600.00	1
7.	2000	3	**8.**	2208.33	5
9.	2550	4	**10.**	1666.67	0

11.–20. Create a computer spreadsheet to calculate the take-home pay for the people in Exercises 1–10. Use formulas for the columns for the FICA deduction and take-home pay.

EXERCISE YOUR SKILLS

1. Why is it difficult to make changes in tax policies?
2. Why did the federal government start the Social Security system?
3. The Social Security taxes that you are paying now are being transferred to current beneficiaries. Who will pay for the benefits that you will receive in the future?
4. Why do you suppose that the Social Security system, unlike private insurance, is not voluntary for the majority of workers?
5. The provisions for retirement benefits under the Social Security system eliminate or alleviate certain risks for its recipients. What are some of these risks?

Use the income tax withholding table in the Reference Section to find the take-home pay for the given wages and withholding allowances. Assume that each person is married. For the FICA deduction, take 7.65% of gross pay. Use the formula for take-home pay

$$T = g - (w + f)$$

where g is the gross pay, w is the amount withheld for taxes, and f is the deduction for FICA (Social Security and Medicare).

	Monthly Salary	Withholding Allowances			Monthly Salary	Withholding Allowances
6.	$2640	1		**7.**	$2500.00	1
8.	4720	2		**9.**	1500.00	5
10.	4000	5		**11.**	1500.00	4
12.	3000	3		**13.**	4166.67	2
14.	4700	1		**15.**	3500.00	3

16.–25. Create a computer spreadsheet to calculate the take-home pay for the people in Exercises 6–15. Use formulas for the columns for Social Security/Medicare and take-home pay.

MIXED REVIEW

1. Marsha's job at the library pays her $12.50/h plus $1\frac{1}{2}$ times her regular wage rate for each hour over 35 hours. How much did she earn in a week during which she worked 42 hours?

2. A law firm is offering a new employee the following fringe benefits: $1100 in annual health insurance premiums, 6% of gross pay to match the employee's retirement contributions, and a monthly travel allowance of $100. The annual salary is $53,000, and the FICA taxes are 7.65% of gross pay. Find the total annual cost that the law firm has to budget for the new employee.

3. Alvin receives a 4% commission on his total automobile sales in a given month plus an additional 1.5% on all sales over $60,000 during the same month. In April, Alvin sold $87,000 worth of automobiles. What were his total earnings for April?

4. Inez receives a weekly salary of $280 at the real-estate office where she works and also receives a 3.6% commission for each house that she sells. In a recent year she worked for 45 weeks and was credited with sales of $600,000. What was her annual income?

5. Ricardo owns a tour-bus company. He offers his tour guides a choice between receiving $1.50 for each tourist who rides on the guide's tour bus and receiving a salary of $150 for a five-day week plus $1.00 for each tourist. Each guide conducts two tours a day, and each bus holds 40 tourists. On the average, what is the least percent of the bus that should be full for the straight-commission option (no salary) to be the more attractive choice?

6. Katrina receives a salary of $46,000 and takes vacation days and holidays for 24 of the company's 260 operating days. How much of her salary is paid for the days during which she does not work?

7. Diane assembles electronic beepers in a factory. She receives $15 for each deluxe model and $11 for each standard model. How much did she earn in a week when she assembled 17 deluxe models and 13 standard models?

1. If your company requires a lot of overtime, would you prefer to work for an hourly rate or for a monthly salary. Why?

2. Why would you not trust the advice of a salesperson whom you knew was getting push money?

3. Why is your take-home pay less than your gross earnings?

4. Why would you expect a person with little education, job training, or both to have more difficulty getting a job than a person with a college degree and work experience?

Find the total weekly pay based on an overtime rate of $1\frac{1}{2}$ times the regular hourly rate for all working time over 20 hours.

	Number of Hours	Wages/Hour	Bonuses
5.	42.0	$ 6.50	$10.00
6.	35.5	18.75	20.00
7.	40.0	21.50	41.00

Find the total number of hours you would work during the week if you worked on the following time schedule.

8. Saturday 3–8 P.M.   Sunday 3–8 P.M.   Monday 4–11 P.M.
   Tuesday off   Wednesday 4–8 P.M.   Thursday 6–10 P.M.

9. Assume that you work on the schedule in Exercise 8. Find your total weekly salary if you make $5.25 per hour and an overtime rate of $1\frac{1}{2}$ times your regular hourly rate for any time over 20 hours.

10. Find the total annual cost to an employer for a new employee hired at $30,000 per year. The annual insurance premiums are $900 for health insurance and $120 for life insurance. The FICA taxes are 7.65% of gross pay.

11. Find the total annual cost to an employer to hire a lab technician at $44,000 per year. Health insurance is $260 per month. Retirement is 3% of gross pay and the FICA taxes are 7.65% of gross pay.

Find the amount of total sales and total commissions made on the basis of the given information.

	Number of Items	Unit Price	Commission Rate	Total Sales	Commission
12.	25	$ 108.92	15%		
13.	3	3405.00	8.2%		
14.	42	23.50	6.3%		
15.	150	19.50	7%		

Solve each problem.

16. Your salary is $300 per week plus 15% commission. What is your week's pay if your total sales for the week are $4175?

17. Katy assembles craft kits. She receives $8 for each regular kit and $12 for each advanced kit. How much did she earn in a week in which she assembled 35 regular kits and 22 advanced kits?

18. Shawn sells annual memberships in an exercise club. His boss has offered him a choice between Plan 1, no salary but a straight $45 for each membership sold, and Plan 2, a weekly base salary of $250 plus $20 for each membership sold. How many memberships would Shawn have to sell to make the same amount of money under either plan?

19. Your graduated commission pays you 3% on all sales up to $7000 and 5% on all sales over $7000. What is your pay if your total sales are $10,800?

20. Your piece rate at a local toy factory is $2.75 for each assembled toy, and your weekly salary is $170. What are your earnings in a week during which you assembled 110 toys?

Use the income tax withholding tables in the Reference Section to find the income tax and FICA withholding for each single person making the monthly salaries given below. What will be the take-home pay for each person?

	Monthly Salary	Withholding Allowances	Income Tax Withholding	FICA Withholding	Take-Home Pay
21.	$1500.00	0			
22.	2728.00	2			
23.	1731.20	5			

For each company fringe benefits policy described below, find

    a. the amount that the company must pay for extra taxes, retirement, health and life insurance, and other fringe benefits.

    b. how much salary is paid for the indicated number of nonworking days in a total of 260 operating days.

	Employee	Salary	Fringe Benefits Option
24.	Auto mechanic	$32,000	7.65% of gross for FICA taxes
			$150/month for health insurance
			3% of gross for retirement
			28 nonworking days
25.	Bank teller	$23,500	7.65% of gross for FICA taxes
			$120/month for health insurance
			22 nonworking days

CHAPTER 1 TEST

1. If you work 60 hours each week, would you make more money working for $6 an hour or for $1200 a month?

2. How does your take-home pay differ from your gross earnings?

3. Which of the two following employment positions provides more money in salary and fringe benefits?

	Atwood Electronics	Technology, Inc.
Salary	$38,500	$36,000
Medical insurance	$85/month	$115/month
Retirement benefits	3% of gross pay	3.5% of gross pay
Expense account	$450/month	$130/month

4. If your pay is $500 per week plus a 12% commission, what is your week's pay during a week in which your total sales are $2723?

5. Your graduated commission pays 3% for sales up to $8500 and 5% on sales over $8500. Find your pay for total sales of $12,000.

Find the total weekly pay based on an overtime rate of $1\frac{1}{2}$ times the regular hourly rate for all working time over 20 hours.

6. You work $43\frac{1}{2}$ hours at $4.75 per hour and receive $19.50 in tips.

7. You work 37 hours at $4.60 per hour and receive $28.00 in tips.

Assume that 40 hours = 1 workweek, 52 weeks = 1 year, and 12 months = 1 year. Find the hourly, weekly, monthly and yearly salaries for the following pay rates.

8. $6392 per month   9. $1276.80 per week   10. $31.70 per hour

Find the total sales and total earnings based on the information shown.

11. You sell 17 sewing kits at $35 each. Your piece rate is $4 per kit.

12. You sell 78 VCRs at $383.60 each. Your commission rate is 10%.

13. You sell 5 virtual-reality consoles at $2050 each. Your commission rate is 10%.

Use the tax tables in the Reference Section to complete the following chart for each single person making the monthly salaries given.

	Monthly Salary	Withholding Allowances	Income Tax Withholding	FICA Withholding	Take-Home Pay
14.	$3456	3			
15.	2880	4			
16.	4000	1			

46   CHAPTER 1   TEST

CUMULATIVE REVIEW

Determine the total weekly earnings for each employee described in each classified ad below. Use an equation of the form $E = rh + t$, where r represents the regular hourly rate, h represents the number of hours worked, and t represents the tips or bonus, if any.

1.
Dental Care
LAB TECHNICIAN $13/h
40 hours/week

2.
Hair Stylist
English and Spanish speaking. $15.50/h to start. Tips can be up to $150 for a 35-hour week!

3. Find the weekly earnings of a server who earned $4.80 an hour for 25 hours at a restaurant and who received $130 in tips during that time.

4. Keith earns $7.50 an hour and $1\frac{1}{2}$ times that amount for time over 20 hours in 1 week. What is his pay for 23 hours of work?

5. Does the size of a fringe benefit necessarily increase when your salary is raised? Explain your answer.

6. An auto parts store is offering a new manager the following fringe benefits: $180/month in health insurance premiums and 5% of gross pay to match the employee's retirement contributions. The annual salary is $47,500 and the FICA taxes are 7.65% of gross pay. Find the annual cost that the auto parts store has to budget for the manager.

7. Carmen sells real estate. She receives 5% of the first $100,000 of the selling price of a house and 6% of the amount over $100,000. How much commission does Carmen receive on a house that sells for $225,000?

8. Sylvia has a part-time business videotaping weddings. Her rate is $75 for the ceremony and $150 if both the ceremony and the reception are covered. Last year she was able to cover 25 weddings, 10 of which included the reception. Her expenses for the year were $200. What were her earnings for the year?

9. James has a monthly base salary of $900 and a commission rate of 9% on his sales. Last month he sold 25 stereo systems at $1450 each and 18 VCRs at $425 each. What were his earnings last month?

Use the income tax withholding tables in the Reference Section to find the income tax and FICA withholding for each single person making the monthly salaries given below. What will be the take-home pay for each person?

	Monthly Salary	Withholding Allowances	Income Tax Withholding	FICA Withholding	Take-Home Pay
10.	$4440	2			
11.	2990	1			
12.	1250	0			
13.	2200	3			

PROJECT 1–1: College Information

Obtain information from colleges you might wish to attend.

1. Call or write three colleges for a copy of their college catalog.

2. Make a chart comparing costs of tuition; room and board; transportation to school; personal expenses; and so on. **Note:** Transportation to school may or may not be important depending upon how many times during each year you return home.

3. List the field(s) of study you are considering.

4. List the degrees or majors offered by each school.

5. Bring in three college catalogs. Turn in your cost comparison chart. Turn in a list of majors and degrees offered.

6. Write about which college you think you would like to attend and explain why.

Extension

Interview people who have attended college. Write a one-page report on their experiences.

PROJECT 1–2: Career Interview

1. Make arrangements for an interview with someone who has a job in a career field or occupational area that interests you.

2. Prepare for the interview by making a list of questions that you would like answered. The following list of job topics should help you prepare questions for the interview.

Place of work and daily tasks	Hours worked
Working environment	Education required
Appearance needed for job	Experience needed
Special talents needed	Strength needed
Hazards at work	Time pressures on job
Entry-level earnings	Top-level earnings
Type of work—easy/hard	Future of job potential
Fringe benefits of job	Security of job
Work as part of a team	Work individually

3. Complete the interview. Write a report that summarizes what you have learned. Include specific questions and answers from your interview.

Solve each system of equations for x and y using substitution.

Example

$$x + y = 10 \qquad (1)$$
$$2x - y = 8 \qquad (2)$$

$y = 10 - x$	Solve (1) for y.
$2x - (10 - x) = 8$	Substitute $10 - x$ for y in (2).
$2x - 10 + x = 8$	Simplify.
$3x = 18$	Collect terms.
$x = 6$	
$6 + y = 10$	Substitute 6 for x in (1).
$y = 4$	

The solution is $x = 6$ and $y = 4$. Check the solution in both original equations.

1. $x - 2y = 11$
$2x + y = 7$

2. $2x + 3y = -1$
$-5x - y = 9$

3. $x + y = 4$
$x - y = 1$

4. $y = -2.5$
$2x + 3y = 2.5$

A system of linear equations can also be solved using the addition and subtraction method. First multiply or divide one or both equations by numbers so that one pair of like terms has opposite coefficients. Then add or subtract the equations to eliminate one variable.

Solve each system of equations using the addition and subtraction method.

Example

$$x + y = 1 \qquad (1)$$
$$2x + 5y = -4 \qquad (2)$$

$-2x - 2y = -2$	Multiply equation (1) by -2.
$2x + 5y = -4$	Equation (2)
$3y = -6$	Add equations.
$y = -2$	Divide by 3.
$x + (-2) = 1$	Substitute -2 for y in equation (1).
$x = 3$	Add 2 to both sides.

The solution is $(3, -2)$. Check the solution in both original equations.

5. $x + y = 7$
$2x - 3y = -1$

6. $2x + y = 3$
$x - 2y = 9$

7. $2x + 3y = 7$
$3x - 2y = 4$

8. $x - y = 1$
$-3x + 2y = 0$

9. $x + 5y = 3$
$2x - y = 6$

10. $x - y = 0$
$5x + 4y = -9$

Check Out the Account

IN THIS CHAPTER THE young people in our story will learn about checking accounts. Jeff will examine the differences among services offered with checking accounts at different banks and will compare charges. He will calculate new checking account balances based on the cost of checks, service charges, extra fees, and interest earned.

Latoya will show us how to write a check, endorse it correctly, and keep a proper check register. She imagines herself the treasurer of the Eldorado Open Golf Tournament, writing checks for thousands of dollars to the winners. If she pursues her interest in business and finance, she may some day be responsible for large amounts of money.

Larry will discover the necessity of keeping a close watch on how much money he has in his checking account and will reconcile his check register with a bank statement. He is fascinated by automated teller machines, or ATMs, which give him money from his checking account at convenient locations all over town–just for entering the correct codes. But he is still a little uncertain about exactly what happens at the other end–in his checking account at the bank. He needs to learn more.

Many young people open their own checking accounts when they begin working at regular jobs and handling money that is truly their own. A checking account can help you use money safely and efficiently. But it needs to be managed carefully so that you always know how much money you have!

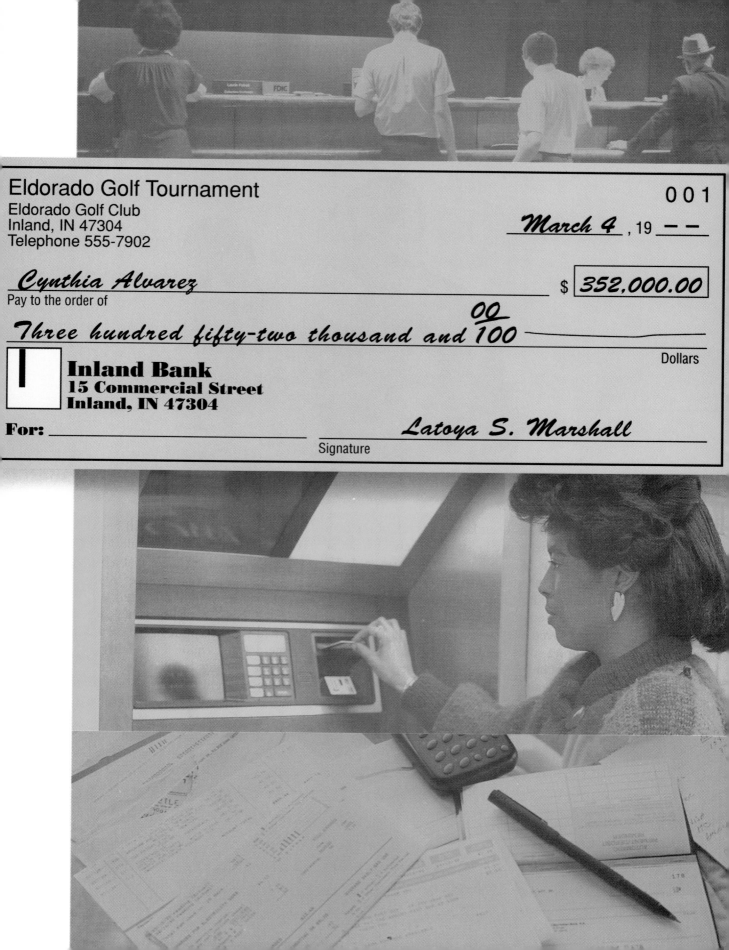

Eldorado Golf Tournament
Eldorado Golf Club
Inland, IN 47304
Telephone 555-7902

001

March 4 , 19 – –

Pay to the order of   *Cynthia Alvarez* $ 352,000.00

Three hundred fifty-two thousand and 00/100 ——————————

Dollars

Inland Bank
15 Commercial Street
Inland, IN 47304

For: _____   *Latoya S. Marshall*

Signature

Jeff has been keeping track of his own money ever since he started receiving a dollar a week in allowance when he was seven. Jeff's older brother, Jeremy, never seems to know where all his money goes, but Jeff keeps very careful records of what he has earned, spent, and saved. Jeff earned $305 last summer tutoring some of his classmates as they took algebra in summer school. He spent only $15 of his summer earnings and saved the rest.

Jeremy has asked to borrow money from Jeff several times. The last time, he borrowed $25 for a gift he wanted to buy for his girlfriend, Ellen. Jeremy did not pay the money back when he said he would. Jeff has decided to open a checking account so that his money

will be a little safer. He will be doing some more tutoring now that semester exams are coming up, and he knows that a bank account will help him keep track of his income in preparation for filing a tax return next spring.

Jeff also has bills to pay now, too, since he applied for a gasoline credit card in his own name. Jeff does quite a lot of driving when he does the tutoring. He also pays the automobile insurance premium.

Jeff knows that different banks offer many of the same services but that service charges, interest rates, and minimum balance requirements can vary from bank to bank. Before he chooses which bank to go to, he will investigate the differences.

OBJECTIVES: *In this lesson, we will help Jeff to:*

Learn what different types of checking accounts are available.
Calculate the interest and bank charges for a checking account.
Compare the costs of banking services offered by various banks.
Use a spreadsheet to compare costs of various checking account options.

CHECKING ACCOUNTS

The check is the most widely used means of transferring money. Nearly 90% of all monetary transactions, such as buying goods and paying rent, are made with checks. **Checks** are orders written by a depositor directing a bank to pay out money. Checks are safe and convenient. They provide proof of payment and records for tax and budget purposes.

Most young people open their first checking account shortly after they get a job or go to college. When a person opens a checking account, the bank and the customer establish a contractual agreement that allows the customer to deposit money in the bank and to write checks on the account. The bank agrees to maintain the account, provide records, and honor checks when they are presented for collection. When a customer writes a check, it is a demand on the customer's deposits so checks are called **demand deposits.**

Following is an explanation of the different types of checking accounts that are available.

Cost-Per-Check Account Cost-per-check accounts are considered thrift accounts or minimum use accounts. Generally, one has to maintain a minimum balance and pay a fee for each check. The fee may range from $0.02 to $0.25 per check. Some banks also charge a service charge or maintenance fee that varies from $0.50 to $2.50 or more per month even if there is no activity in the account.

Minimum-Balance Account Minimum-balance accounts require the customer to maintain a certain balance. The minimum balance may be a low balance or an average balance. With a **low-balance account** the customer is charged a service fee even if the account falls below the minimum only one day in a month. An **average-balance account** can drop to zero as long as the customer deposits enough money during the month to bring the account average for the month up to the minimum required. Some banks advertise minimum-balance accounts as "free checking accounts." They may be low-cost accounts, but they are not free.

Free Checking Account Some banks provide totally free checking; that is, there are no minimum balance requirements or service charges. Some banks provide free checking in the belief that checking account customers will also use the bank's other services such as savings accounts, consumer loans, and safe-deposit boxes.

NOW (Negotiable Order of Withdrawal) Account Another type of checking service is an interest-bearing account known as a **NOW account.** One drawback to NOW accounts is the minimum balance requirement. If the account balance falls below the required balance, the bank charges a monthly service fee or a check-handling fee.

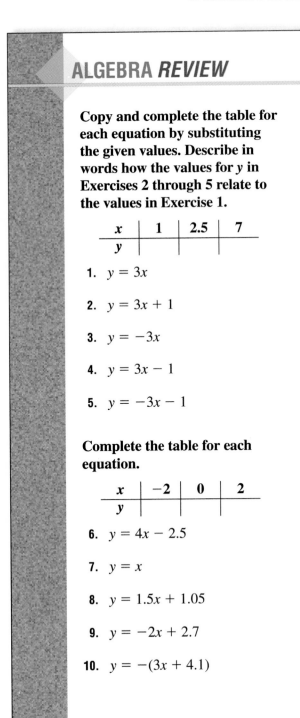

ALGEBRA *REVIEW*

Copy and complete the table for each equation by substituting the given values. Describe in words how the values for *y* in Exercises 2 through 5 relate to the values in Exercise 1.

x	1	2.5	7
y			

1. $y = 3x$

2. $y = 3x + 1$

3. $y = -3x$

4. $y = 3x - 1$

5. $y = -3x - 1$

Complete the table for each equation.

x	-2	0	2
y			

6. $y = 4x - 2.5$

7. $y = x$

8. $y = 1.5x + 1.05$

9. $y = -2x + 2.7$

10. $y = -(3x + 4.1)$

Ask Yourself

1. Why is a check referred to as a demand deposit?

2. Are minimum-balance accounts free?

3. Does a NOW account require a minimum balance?

4. Which is the least expensive account?

SHARPEN YOUR SKILLS

SKILL 1

In Chapter 1 you worked with percentages to find commissions. In this lesson you will work with percents less than 1%.

1% means 1 part out of 100.
0.5% means $\frac{1}{2}$ of 1%.
As a decimal, 0.5% = 0.005.

Interest is an amount of money paid for the use of money. Interest is based on percent, which means parts out of one hundred.

To find 0.5% of $246, convert 0.5% to a decimal and multiply.

$0.005 \cdot 246 = 1.23$ 0.5% = 0.005
0.5% of $246 is $1.23

At the first bank he visits, Jeff reads a brochure about a Presto account. The brochure states that a checking account earns interest of 0.45% per month and that there is a charge of 5 cents per check.

Jeff remembers that interest is an amount paid based on percentage. Jeff does a quick mental calculation and sees that if he left $100 in the bank for a month, it would earn $0.0045 \cdot 100 = \$0.45$ in interest. Not much, he thinks, but it is something.

Each month the bank pays any interest due and charges any fees due. The interest that is paid is determined by multiplying the monthly average balance by the monthly interest rate. The charge for checks is determined by multiplying the number of checks by the charge for each check. The amount earned or charged can be expressed by the following formula.

Amount earned or charged = interest − charge for checks

$A = br - cn$ where A = amount earned or charged
                              b = monthly average balance
                              r = monthly interest rate
                              n = number of checks
                              c = charge for each check

EXAMPLE 1 Presto checking accounts pay 0.45% interest per month and charge $0.05 per check.

QUESTION What is the amount earned or charged if 17 checks are written during one month and the average balance is $620?

SOLUTION

Remember to add the interest paid and subtract the charges for checks.

$A = br - cn$ Use the formula.
$A = 620(0.0045) - 0.05(17)$ $b = 620$, $r = 0.0045$, $c = 0.05$, $n = 17$
$A = 1.94$ Use your calculator.

The amount earned is $1.94.

SKILL 2

Jeff visits another bank to compare costs. This bank charges a service charge in addition to the charge for checks. So we need to adjust the formula.

> Amount earned or charged = interest − charge for checks
> − service charge
>
> $A = br - cn - s$ where A = amount earned or charged
> b = monthly average balance
> r = monthly interest rate
> n = number of checks
> c = charge for each check
> s = service charge

EXAMPLE 2 Super checking accounts earn interest of 0.4% per month and have the following service charges.

For balances of $500 or more, there is no service charge and no charge for checks.

For balances under $500, there is a charge of $0.025 per check.

For balances of $200 to $499.99, there is a service charge of $4.00 per month.

For balances under $200, the service charge is $6.00 per month.

QUESTION What amount will be earned or charged for the following average balances?

a. $650 with 20 checks

b. $400 with 18 checks

c. $150 with 16 checks

SOLUTION

Jeff used the formula above.

$A = br - cn - s$

a. $A = 650(0.004) - 0(20) - 0 = 2.6$
The account earned $2.60.

b. $A = 400(0.004) - 0.025(18) - 4 = -2.85$
The account was charged $2.85.

c. $A = 150(0.004) - 0.025(16) - 6 = -5.8$
The account was charged $5.80.

SKILL 3

EXAMPLE 3 Jeff learned that there are a number of differences in the interest rates and charges that banks use for checking accounts. He decided to make a spreadsheet to compare various interest rates and charges.

QUESTION What column headings and formulas should be used in a spreadsheet to calculate the interest and charges on Super checking accounts?

SOLUTION

For Super checking accounts there are three different ways of charging fees as shown in the spreadsheet.

	A	B	C	D	E	F	G	H
1		Interest	Interest	Number	Cost per	Cost of	Service	Amount Earned
2	Balance	Rate	Earned	of Checks	Check	Checks	Charge	or Charged
3	500 or more						+C4−F4−G4	
4	650	0.004	2.6	20	0	0	0	2.60
5	200 to 499.99		+A4*B4					
6	400	0.004	1.6	18	0.025	0.45	4	(2.85)
7	under 200					+D6*E6		
8	150	0.004	0.6	16	0.025	0.4	6	(5.80)
9								
10								
11								

The cost of checks and the service charge are different for all three cases. The formulas for the case of a $500 balance are shown below.

Interest earned: $+A4*B4$

Interest rate times the average balance.

Cost of checks: $+D4*E4$

Number of checks times the cost per check.

Amount earned or charged: $+C4-F4-G4$

Interest earned minus the cost of checks minus the service charge.

You can use the COPY function of your spreadsheet to create the formulas for rows 6 and 8 of the spreadsheet. For example, if you copy cell C4 (interest earned) to cell C6, then the program will create the formula $+A6*B6$ for that cell. You will see the value 1.6 in the cell.

You may wish to format the column for "Amount Earned or Charged" for currency as shown. Usually a negative amount will be shown in parentheses as in the spreadsheet on page 57.

SKILL 4

EXAMPLE 4 Jeremy's father owns a small business and has to decide between two checking accounts. Bank A does not pay interest and does not charge for checks but has a $4 per month service charge. Bank B pays 0.4% interest per month and charges $0.075 per check. Jeremy's father expects to maintain a balance of $1000 but is not sure how many checks he will write.

QUESTION How does the monthly number of written checks affect the interest and charges for the two accounts?

SOLUTION
Write equations like the ones used before for the interest and charges. Let y be the total of the interest and charges. Let x be the number of checks. The interest for Bank B is $0.004(1000) = 4$.

For Bank A: $y = -4$
For Bank B: $y = 4 - 0.075x$

Graph the equations on a graphing calculator using these range values.

Xmin: -5 Ymin: -6
Xmax: 125   Ymax: 6
Xscl: 10   Yscl: 1

The trace function shows that the graphs intersect at approximately $x = 107$, $y = -4$.

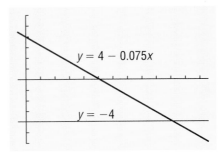

To confirm the calculator results, solve the system of equations by substituting the value for y in the second equation.

$$-4 = 4 - 0.075x$$
$$-8 = -0.075x$$
$$x = 106.67 \qquad \text{To the nearest hundredth}$$

Both methods show that the interest-bearing account in Bank B will be less expensive if fewer than 106 checks are written per month, and more expensive if 107 or more checks are written per month.

TRY YOUR SKILLS

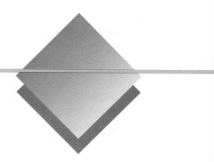

Three banks offer the following arrangements for checking accounts.

National: Interest 0.5% per month, a charge of $0.09 per check for balances under $500.

Central: Interest 0.54% per month, a charge of $0.02 per check, $4 service charge for balances under $200.

Western: Interest 0.35% per month, a charge of $0.06 per check for balances under $1000.

Find the amount earned or charged for each of the following average balances for one month.

1. $400 at National, with 22 checks written

2. $500 at Western, with 9 checks written

3. $175 at Central, with 31 checks written

4. $760 at National, with 40 checks written

5. $340 at Central, with 28 checks written

6. $1280 at Western, with 60 checks written

7. Copy and complete the table comparing the interest earned, charges, and changes in balance at each of the three banks for a balance of $400 with 20 checks written.

Balance	Interest Earned	Cost of Checks	Service Charge	Amount Earned or Charged
National				
Central				
Western				

8. Complete a table like that in Exercise 7 for an average balance of $1220 with 30 checks written.

EXERCISE YOUR SKILLS

1. Write a general algebraic equation that can be used to find the amount earned or charged in a checking account in a bank that pays interest, charges a monthly fee, and charges for each check written.

2. What are two things that a bank agrees to do as its part of the agreement when a person opens a checking account?

3. Explain how a spreadsheet will be useful to Jeff in organizing data and explaining checking accounts to his friends.

4. Why might a person choose a bank even though it does not have high interest rates or low costs per check?

5. What might lead a person to pay special attention to the cost per check that a bank charges?

6. What is the difference between a low-balance account and an average-balance account?

7. Set up a spreadsheet with headings and formulas as explained in Skill 3 and shown.

	A	B	C	D	E	F	G
1		Interest	Interest	Number of	Cost per	Cost of	Amount
2	Balance	Rate	Earned	Checks	Check	Checks	Earned
3			[+A3*B3]			[+D3*E3]	[+C3−F3]

8. To use the spreadsheet correctly, in which cells must you enter numbers?

9. Which cells do you leave blank, letting the spreadsheet fill them in?

10. Write an algebraic expression to calculate the amount that will appear in cell C3.

11. Write an algebraic expression to calculate the amount that will appear in cell F3.

12. Write an algebraic equation to calculate the amount that will appear in cell G3.

13. Show how the equation that you wrote is related to the formulas used in the spreadsheet.

For Exercises 14–16, use a spreadsheet with headings like the ones shown in the tables to calculate the interest, costs, and new balance for the banks as described. To find the new balance, it is necessary to add monthly interest and subtract monthly charges.

14. At First Bank:

No interest is paid on the balance.

There are no service charges.

The cost per check is $0.0225. If the beginning balance is below $1000, First Bank charges an additional $0.23 for each check written.

	Beginning Balance	Number of Checks	Interest Earned	Service Charges	Cost of Checks	Extra Charges	New Balance
a.	$650	18					
b.	625	26					
c.	480	20					
d.	190	31					

15. At Second Bank:

Interest is paid on the beginning balance at the rate of 0.5% per month.

The service charge is $3.00 per month.

The cost per check is $0.0225. There are no additional charges whether or not a minimum balance is maintained.

	Beginning Balance	Number of Checks	Interest Earned	Service Charges	Cost of Checks	Extra Charges	New Balance
a.	$650	18					
b.	625	26					
c.	480	20					
d.	190	31					

16. At Third Bank:

Interest is paid at the rate of 0.52% per month on the beginning balance.

There are no service charges.

The cost per check is $0.0225. If the beginning balance is below $500, Third Bank charges an additional $0.25 for every check written.

	Beginning Balance	Number of Checks	Interest Earned	Service Charges	Cost of Checks	Extra Charges	New Balance
a.	$650	18					
b.	625	26					
c.	480	20					
d.	190	31					

17. Write a formula for the amount earned or charged in a checking account for which interest is paid but for which there is no service charge or charge per check.

18. Sarah's checking account has an average balance of $120 for one month, and it earns 42 cents in interest. Find the monthly interest rate at that bank.

19. Jason's bank has a monthly service fee of $3. He finds that an average balance of $180 causes a decrease in his account of $2.28. Find the monthly interest rate at the bank.

20. Marcia's bank has no service charge for a checking account but pays interest and charges $0.04 for each check written. With a balance of $160 and an interest rate of 0.3%, Marcia finds that her account shows a bank charge of $0.04 for the month. How many checks did she write?

21. Bank S pays interest of 0.55% and charges $0.07 per check. Bank T pays interest of 0.25% and charges $0.025 per check. Susan expects to maintain a balance of $800 but is not sure how many checks she will write. Write and graph equations as in Example 4 to determine which account is less expensive. Explain your results.

22. Henry expects to maintain a balance of $500 in his checking account. Write and graph equations to determine which account is less expensive. Bank K pays interest of 0.35% and charges $0.08 per check. Bank M pays no interest and charges $0.025 per check. Explain your results.

MIXED REVIEW

1. At $7.25 an hour, how much does Jessica earn for 38 hours of work?

2. Richard earns $6.80 an hour but gets $1\frac{1}{2}$ times his regular pay for every hour that he works over 40 hours. What is his pay for 46 hours of work?

3. Jorge receives a salary of $39,000 and takes vacation days and holidays for 24 of the company's 260 operating days. How much of his salary is paid for days during which he does not work?

4. What is the amount of a 7.5% commission on sales of $1245?

5. Karen is paid a commission of $40 on a sale of $200. What is her rate of commission?

In a spreadsheet, cell B2 contains an hourly rate of pay, cell B3 contains the number of hours, cell B5 contains FICA deductions at 7.65%, and cell B6 contains all other deductions. Use this information to find each of the following.

6. a spreadsheet formula for the gross wages

7. a spreadsheet formula for the FICA withholding

8. a formula for the total deductions

9. a formula for the take-home pay

2-2 USING CHECKS: WRITE IT DOWN

Latoya receives a weekly check from the Hometown Diner, the restaurant where she works after school. The weekly paychecks are printed on a computer, but they are all signed by the manager. Latoya imagines how good it would feel to sign thousands of dollars' worth of checks every month. Latoya's parents write checks to pay the bills that come to their house every month. They do not seem to enjoy the task all that much, but maybe that is because there are usually more bills than there is money to pay them with.

Some of Latoya's earnings are used to help pay those bills. Latoya would like to help write the checks, too, and keep track of how much money her parents have in their checking account. What Latoya really wants to do is to write checks for a large corporation for thousands of dollars of someone else's money! But for the moment she will settle for learning how to write a proper check and maintain an accurate check register. She plans to open her own checking account when she feels comfortable with these procedures.

OBJECTIVES: *In this lesson, we will help Latoya to:*

- *Write checks and maintain a check register.*
- *Endorse checks properly.*
- *Make a bank deposit.*
- *Use a spreadsheet to maintain a check register.*

WRITING AND ENDORSING CHECKS

Most people make money transactions using checks. When checks are properly written, as shown on page 65, they provide proof of payment and are an excellent record of transactions. However, it is important to prevent someone from altering your checks. You must carefully fill in the **payee** (the person to whom the check is written), sign it as the **drawer** (the person from whose account the funds are to be withdrawn), and keep a careful record in your check register. A **check register** is a separate form on which a checking account holder keeps a record of deposits and checks written. The small, portable check register is called a checkbook.

Endorsements   It is also important to sign and endorse checks properly. Banks and other businesses that cash or accept checks for payment are careful that the signature on the front of the check or the endorsement on the back is that of the person presenting the check for payment. That is why businesses often ask for identification before accepting a check from a person they do not know.

An **endorsement** is a signature and a message to the bank telling it to cash a check, deposit it, or transfer your right to the check to someone else. It is written on the back of the check, at the left-hand end. Endorsements are made on the back of the check exactly as your name appears on the face of the check. Even if your name is misspelled, you should sign it as it is written on the face and then add the correct spelling. There are three common types of endorsements: blank, restrictive, and full.

A **blank endorsement** is your signature only. A check with a blank endorsement is like cash. Anyone who has the check can present it for payment at a bank. For this reason you should not blank endorse a check unless you are at the bank.

A **restrictive endorsement** is your signature and a message that limits the use of the check. A restrictive endorsement usually reads "For deposit only." This type of endorsement allows you to send the check by mail for deposit without fear of loss. If a check endorsed "for deposit only" is lost or stolen, it cannot be cashed.

A **full endorsement** is your signature and a message that directs the transfer of the check to someone else whom you designate. A full endorsement is written: "Pay to the order of . . . " followed by the name of the recipient and your signature. This endorsement transfers the right of payment to the new payee.

Latoya follows the procedures listed below when writing a check and keeping a record in her check register. The numbers of the steps correspond to the circled numbers in the illustration. It is important to use permanent ink to write checks, never pencil or erasable ink.

These are the steps to be followed when writing a check.

1. Before writing the check, record in the check register: the check number, the date, the payee, the purpose of the check, and the amount of the check.

2. On the check itself, write the date. This is important for your own records and in case the check is ever needed as proof of payment.

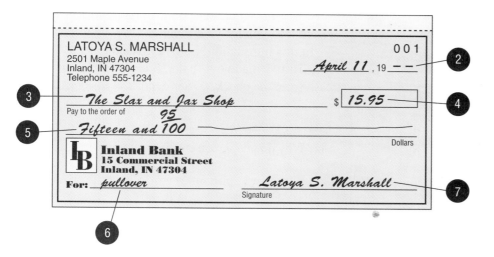

CHECK NO.	DATE	CHECKS DRAWN OR DEPOSITS MADE	BALANCE FORWARD ➡	✓		
D	4/10	TO Deposited paycheck	DEDUCT CHECK – ADD DEPOSIT +		258	49
		FOR for 3/16 – 3/31	BALANCE ➡		258	49
001	4/11	TO The Slax and Jax Shop	DEDUCT CHECK – ADD DEPOSIT +		15	95
		FOR pullover	BALANCE ➡		242	54
002	4/13	TO Cash	DEDUCT CHECK – ADD DEPOSIT +		40	00
		FOR current expenses	BALANCE ➡		202	54
D	4/16	TO Deposit	DEDUCT CHECK – ADD DEPOSIT +		20	50
		FOR	BALANCE ➡		223	04
003	4/22	TO Spangler Gifts, Inc.	DEDUCT CHECK – ADD DEPOSIT +			88
		FOR postage due on 4/8 order	BALANCE ➡		222	16
AT	4/29	TO Cash from automatic teller	DEDUCT CHECK – ADD DEPOSIT +		30	00
		FOR game tickets and dinner	BALANCE ➡		192	16

PLEASE BE SURE TO DEDUCT ANY PER CHECK CHARGES OR MAINTENANCE CHARGES THAT AFFECT YOUR ACCOUNT

3. Designate the payee; that is, write the name of the person or organization to whom you are paying the money.

4. Write the amount of the check, in numbers, next to the dollar sign.

5. Write the amount of the check, in words, on the middle line and draw a line to fill in the space to the word "Dollars."

DEPOSIT TICKET

CASH	CURRENCY		
	COIN		
LIST CHECKS SINGLY		160	42

70-7293/2719 810

TOTAL FROM OTHER SIDE			
TOTAL		160	42
LESS CASH RECEIVED		20	00
NET DEPOSIT		140	42

DATE ___ *April 3* ___ 19 – –
DEPOSITS MAY NOT BE AVAILABLE FOR IMMEDIATE WITHDRAWAL

Latoya S. Marshall

USE OTHER SIDE FOR ADDITIONAL LISTING

BE SURE EACH ITEM IS PROPERLY ENDORSED

Inland Bank
15 Commercial Street
Inland, IN 47304

CHECKS AND OTHER ITEMS ARE RECEIVED FOR DEPOSIT SUBJECT TO THE PROVISIONS OF THE UNIFORM COMMERCIAL CODE OR ANY APPLICABLE COLLECTION AGREEMENT

ALGEBRA *REVIEW*

Write an equation to express the relationship shown. Ask yourself how the y values can be derived from the x values.

1.

x	1	3	5
y	2	6	10

2.

x	−1	0	1
y	1	0	−1

3.

x	2	4	6
y	5	9	13

4.

x	1	2	3
y	2	5	8

5.

x	−2	0	2
y	5	1	−3

Write the equation of a line through each set of points.

6. (2, 3), (3, 4), (4, 5)

7. (2, 4), (3, 6), (4, 8)

8. (1, −3), (2, −6), (3, −9)

9. (1, 2.5), (2, 5.5), (3, 8.5)

6. Write the purpose for which each check is written on the line at the bottom of the check.

7. Sign your name as the drawer on the signature line.

DEPOSIT TICKETS

When you deposit money in the bank, you must fill out a deposit ticket as shown above. Notice above how a deposit ticket is completed. Generally, if you withdraw cash, you must sign the deposit ticket as shown. The bank will stamp the deposit slip and return it to you for your records.

In this transaction, Latoya gives the bank a check for $160.42 made out to her. She also asks for $20 in cash from the check, so her deposit is $140.42.

Ask Yourself

1. Why do people use checks for money transactions?

2. Why is it important to write checks properly?

3. Why does the bank require an endorsement when cashing a check?

SHARPEN YOUR SKILLS

SKILL 1

EXAMPLE 1 You are about to mail out checks to pay the following bills: Central Telephone Co., $43.78; Harvey's Computer Store, $102.05; Downtown Music, $18.75. The checks you are using begin with number 101. The date is October 6. Your initial balance is $201.68.

QUESTION How would you complete the check register, entering each transaction and finding the balance after each transaction?

SOLUTION

Following is the check register with the entries. Payments are subtracted to find each new balance; deposits are added.

CHECK NUMBER	DATE	CHECKS/DEPOSITS		AMOUNT	BALANCE
101	10/6	TO: *Central Telephone*			201.68
		FOR: *Phone bill*		43.78	−43.78
102	10/6	TO: *Harvey's Computer*			157.90
		FOR: *Software*		102.05	−102.05
103	10/6	TO: *Downtown Music*			55.85
		FOR: *CD*		18.75	−18.75
		TO:			37.10
		FOR:			

SKILL 2

EXAMPLE 2 Latoya read in the paper about the Eldorado Golf Tournament played in her city on March 4. She imagined what it would be like to write checks to the prize winners. The paper listed the top five as follows:

Cynthia Alvarez	$352,000
Mark Louis	$240,000
Peggy Race	$150,000
Pablo Chosa	$ 75,000
Tina Marin	$ 50,000

QUESTION How would Latoya make out the largest check?

SOLUTION

The following is a check correctly made out to the winner.

Eldorado Golf Tournament	0 0 1
Eldorado Golf Club Inland, IN 47304 Telephone 555-7902	*March 4* , 19 _ _

Cynthia Alvarez $ *352,000.00*
Pay to the order of

               00
Three hundred fifty-two thousand and 100   Dollars

L B | **Inland Bank**
15 Commercial Street
Inland, IN 47304

For: _____   *Latoya S. Marshall*
                                Signature

SKILL 3

EXAMPLE 3   Latoya has learned that she should endorse a check on the back exactly as her name appears on the face of the check. Even if her name is misspelled, she should sign it as it is written on the face and then write her full name correctly. Her grandfather recently gave her a check using "Toya," his pet name for her. Latoya is mailing the check to the bank for deposit.

QUESTION   How should Latoya endorse the check?

SOLUTION

Latoya should write her name as it appears on the face of the check and then write her full name. She should add the restriction, as shown, so that no one will be able to cash the check.

For deposit only
Toya Marshall
(Latoya Marshall)

	4 0 7
	July 23 , 19 _ _

rshall $ *25.00*

         00
and 100   Dollars

                 Howard Marshall
                 Signature

Some people use a spreadsheet as a check register. A spreadsheet is useful for this purpose because it is set up almost like a checkbook and because of the computations that can be done by using a spreadsheet.

EXAMPLE 4 Latoya has the following transactions written down and will enter them as she starts her spreadsheet. All of the transactions were done on May 3. Balance: $178.85; Check 301, Home Hardware, $37.12; Check 302, Campus Bookstore, $45.60; Check from Mark Li, $40.00 for deposit.

QUESTION How do you set up a spreadsheet as a check register?

SOLUTION

To set up a check register spreadsheet, you must set the columns needed and write the formulas for automatic computation. A model is shown below.

	A	B	C	D	E
1	Number	Date	Checks and Deposits	Amount	Balance
2					178.85
3	301	5/3	Home Hardware	−37.12	141.73
4	302	5/3	Campus Bookstore	−45.60	96.13
5		5/3	gift from Mark Li, deposit	40.00	136.13
6					

In the "Amount" column, D, it is important to enter deposits as positive amounts and checks as negative amounts. Then, when the formulas are entered for addition, the spreadsheet will compute the new balance correctly for each transaction. Each new balance is found by adding the amount (negative for a check; positive for a deposit) to the previous balance.

When you set up your check register spreadsheet, you may insert a formula in the "Balance" column, as shown below. On your computer screen, cell E3 will show a true balance, not the formula.

	A	B	C	D	E
1	Number	Date	Checks and Deposits	Amount	Balance
2					178.85
3					+E2+D3
4					

You insert the information for columns A to D yourself. The formula will calculate the new balance, which will appear in cell E3. When you proceed to row 4, use the COPY command to copy the formula from cell E3 to cell E4. A new formula, +E3+D4, will be automatically created for that cell.

TRY YOUR SKILLS

Enter the following transactions in a check register, finding the balance after each transaction. Each exercise will be a separate line on your simulated check register. Use a check register form supplied by your teacher, or draw up your own check register form.

1. Starting balance $400
2. Check 101, Feb 2, to Josh Harmon, $175.60 for electrical work
3. Check 102, Feb 3, to Sunrise Shop, $27.90 for a jacket
4. Check 103, Feb 5, to Bright Spot, $46.23 for a lamp
5. Feb 6, deposit, $45.60
6. Check 104, Feb 6, to Garden Center, $19.25 for plants
7. Write the check for the Garden Center (Exercise 6).

EXERCISE YOUR SKILLS

1. Why should you write the information in the check register before you make out a check?
2. What is meant by "writing your name the same way" each time you sign a document?
3. If you have done all your computations correctly, what does the checkbook balance tell you?
4-7. For the golf tournament of Example 2, write checks to the second through fifth place finishers listed and for the amounts listed.
8. Write an endorsement as you would on the back of a check that you intend to mail to the bank for deposit.
9-39. Create a check register spreadsheet, and enter each of the following transactions, finding the balance after each transaction. Record an initial balance of $660.

CHECK NUMBER	DATE	CHECKS/DEPOSITS			AMOUNT
9. 301	1/15	To: Panasonic Bike Shop	For:	Bicycle	$ 181.50
10. 302	1/18	To: Brian's Sports	For:	Helmet	25.00
11. 303	1/20	To: Friendly Feet	For:	Shoes	25.00
12. 304	1/21	To: 50K Ride-A-Thon	For:	Entry Fee	5.00
13. 305	1/23	To: Gargantua's Dept. Store	For:	Socks	14.26
14. 306	1/23	To: Pandora's Hi Fashion	For:	Shorts	15.04
15. 307	1/25	To: Amalgam Dept. Store	For:	Bike Rack	115.00
16.	1/25	Deposit—Paycheck			311.19
17. 308	1/25	To: Sweatshirts Outrageous	For:	Jacket	71.81
18. 309	1/25	To: Bike World	For:	Tires	111.26
19. 310	1/25	To: Bruce's Sports	For:	Racing Pedals	40.17
20.	1/25	Deposit—Savings			200.00
21. 311	1/25	To: Camper's Delight	For:	Back Pack	98.50
22. 312	1/25	To: Light The Night	For:	Lantern	24.00
23. 313	1/25	To: Army Surplus	For:	Mess Kit	25.00
24. 314	1/25	To: Cash			50.00
25. 315	1/25	To: Sounds Galore	For:	Tapes	63.00
26.	1/25	Deposit—Loan			120.00
27. 316	1/25	To: Rent-A-Tent	For:	Tent	125.00
28. 317	1/26	To: Cash			25.00
29. 318	1/28	To: The Grocer	For:	Groceries	55.30
30. 319	1/30	To: Energy Source	For:	Batteries	9.50
31.	2/1	Deposit—Tutoring			100.00
32. 320	2/1	To: County Vehicles	For:	Registration	5.00
33. 321	2/1	To: Biking News	For:	Magazine	18.28
34. 322	2/3	To: Ozurka	For:	Water	4.50
35. 323	2/4	To: Familiar Pharmacy	For:	First Aid	7.55
36. 324	2/5	To: Emergency	For:	Clinic	15.00
37. 325	2/5	To: Familiar Pharmacy	For:	Medicine	9.11
38. 326	2/6	To: Sticks and Stones	For:	Crutches	95.00
39. 327	2/6	To: Forever Photo	For:	Album	11.71

40. You have received a check from your part-time job. Your boss used your full name, including your middle initial. You are at the bank. Show how you would endorse the check.

41. You have received a check from your part-time job. Your boss used your full name, including your middle initial. This time you are going to mail the check to the bank. Show how you would endorse the check.

42. Your aunt mailed you a check. She used your nickname, "Rusty," instead of your real name. You are at the bank. Show how you would endorse the check.

43. Your father gave you a check. He used your first initial and last name only. You want to give the check to your friend, Judy Bruns, to repay a loan she made you. Show how you would endorse the check.

44. Your friend Tracy has a checking account but says that she doesn't bother with the check register. She says that the bank keeps records of her deposits and expenditures. What do you say to convince Tracy to keep an accurate record of transactions in her check register?

MIXED REVIEW

1. What is the rate for time-and-a-half based on an hourly rate of $9.40?

2. Write an equation for total pay at x hours for regular time and y hours of overtime at $9.40 an hour and time-and-a-half for overtime.

3. If Sharon works 45 hours at the rates given in Exercise 2, what will be her total pay if 40 hours are regular hours?

4. Lou gets paid $16 an hour plus time-and-a-half for any hours over 40. What does he earn if he works for 49 hours?

5. Jack receives a base pay of $240 a week and 9% commission on a sale of $200. What are his earnings for the week?

6. Leslie receives 4.5% commission on her real estate sales. This month she sold two houses, one for $113,000 and one for $79,500. What was her commission this month?

7. Jeffrey sells real estate. This month he sold a house for $225,000. The commission was 5% but he only received half of the commission because he had to split the commission with another agent. What was his commission?

8. Alfredo's weekly take-home pay is $103 and his deductions are $27. What percent of his gross pay is his take-home pay?

Northern Bank offers a checking account that pays interest of 0.3% per month for all accounts and charges $0.08 per check for balances of less than $500. Find the amount earned or charged for each of the following average balances for one month.

9. $350 with 18 checks written

10. $550 with 20 checks written

2-3 RECONCILIATION: DO THE BANK AND I AGREE?

we got what's hot!

Larry still watches in fascination as the $10 and $20 bills come tumbling down out of the automated teller machine at the convenience store down the street. The first time Larry used the machine, he wasn't sure he would know what buttons to push, but the instructions on the computer screen were very simple; after one visit, Larry knew what to do. In fact, one weekend Larry spent Saturday afternoon driving around town locating other machines that would accept his card, and he came home with $240 in his pocket in tens and twenties. Driving around so much had made Larry a little hungry; he spent some money for pizza, some for a beef sandwich, some more for popcorn, and finally some for frozen yogurt.

Next to the yogurt store was one of Larry's favorite haunts—the music store. Larry spotted a new CD he wanted and bought two copies—the extra one for his girlfriend, Lorrie Anne. Larry considered making some notes on

the amounts of money he had spent, but he forgot about it while he was thinking how much Lorrie Anne would like the present.

As Larry was gathering up his keys and change and getting ready to drive to Lorrie Anne's house, one of the slips of paper from the ATM, which Larry had stuck in his pocket, fell to the floor and caught Larry's attention. Printed near the end of the receipt was "Available Balance: $14.78." "Wait a minute, now," Larry thought. "I wrote two checks yesterday that haven't reached the bank yet. They totaled more than $150. How can I only have $14.78 left? Why, I deposited $300 two days ago! What became of the $300?"

Larry had a sinking feeling that he knew what had become of the money. What would happen at the bank tomorrow if those checks he wrote went through his account? Larry wondered whether there was any way to stuff some of that cash back into the machine.

LESSON 2–3   RECONCILIATION: DO THE BANK AND I AGREE?   73

Assess the advantages and disadvantages of using electronic fund transfers.
Reconcile a bank statement with the corresponding check register balance.

AUTOMATED TELLER MACHINES (ATMS)

No longer do customers have to rush to the bank before it closes. **Automated teller machines (ATMs)** allow customers easy access to their accounts during banking or nonbanking hours. By using a special ATM card and punching in a personal identification number, a customer can deposit or withdraw money and even obtain a loan at the site of an ATM. Automated teller machines are placed in convenient locations such as airports, shopping malls, and street corners. They may also be located on the premises of the bank.

But what if your ATM card is lost, stolen, or used without permission? If you lose or misplace your ATM card, you must notify the bank immediately. Usually, if you notify the bank within two business days, the most you will have to pay is $50.

Follow these rules of bookkeeping when using electronic fund services:

1. Always keep the record of your transaction. Check the date, amount, location, and type of transaction. You will later use this information to verify your monthly statement.

2. If a mistake is made at the time of the transaction, call your bank for direct customer service. If you cannot get service at the time, contact the bank as soon as possible.

3. Enter your debit transactions (that is, money taken out) in your check register just as you would a check transaction. This practice allows you to maintain an accurate running total of how much money you have.

The Checkless Society A few years ago, people thought that the checkless society, in which all transactions are made electronically, would soon be upon us. But this was not to be. Many bank customers are still reluctant to use ATMs and computers. They fear breakdowns of the electronic equipment or computer errors that could tie up their funds for days or weeks. Nonetheless, electronic banking is widely used and will continue to grow.

KEEPING A RECORD OF YOUR MONEY

Customers receive monthly **statements** from their banks that reflect all checking account transactions: deposits, checks cleared, service charges, and the ending balance of the account at the close of the statement period. Several days elapse from the time the statement is prepared to the time the customer receives the statement. During this time, additional checks may be written and deposits made and recorded in the register. Also, if the owner of the checking account has written a check but the recipient has not yet cashed it, the bank will not have subtracted the amount of that check from the account. Such checks are called **outstanding checks.** Each of these reasons will cause the customer's check register and statement to differ. The process of finding the correct balance is called **reconciliation.** (The procedure to follow for reconciling your bank statement and check register is described in detail in Sharpen Your Skills.)

After the reconciliation procedure is completed, the adjusted balances should agree. They represent the correct amount of money that remains in the checking account.

Each month you should reconcile the bank's statement with the balance in your register. Then make a brief notation in the register to record the fact that your account is in balance.

If the check register and statement do not agree, you should first check your own arithmetic and then—even though the statement was probably created on a computer and is therefore likely to be accurate—check the bank's arithmetic as well. You will have received your checks back from the bank with your statement. These are called **canceled checks.** Compare these checks with your register to see whether you missed any outstanding checks. Compare the amount you recorded for each check with the register amount and the statement amount for that check. When you find an error, recalculate the balance affected, and compare it with the other balance.

If there is still no agreement, report the matter to the bank. You should do this within ten working days of receiving the statement. Bank bookkeeping departments work with customers to determine whether there has been an error by the bank and will help customers find the reason for the differences. Some banks charge a fee for assistance with reconciliation.

Ask Yourself

1. What should you do if you lose your ATM card?

2. What is a monthly statement?

3. What are outstanding checks?

ALGEBRA *REVIEW*

Solve each equation for x.

1. $x - 2 = 0$

2. $x + 5 = 2$

3. $2x - 8 = 0$

4. $2x - 1 = 5$

5. $3x = 2x + 1$

6. $5x - 1 = 2x + 8$

7. $x - 1 = 8 - 2x$

8. $1.2x = 7 + 1.4x$

9. $2(x - 1) = 3(x - 2)$

10. $2.5x - 1.75 = 3.25 - 1.25x$

SKILL 1

EXAMPLE 1 The bank statement below shows the monthly transactions for Larry Lender. Notice the following entries on the statement.

1. Account balance at the beginning of the period, $186.43.

2. Service charge, $1.90. It is important to record this in your checkbook.

3. The record of checks written, received, and processed by the bank.

4. Deposits made by Larry Lender during the month.

5. Balance at the close of the statement period.

INLAND BANK
BANK STATEMENT

TO:
LARR Y LENDER
4115 N. LINCOLN
INLAND, IN 47304

ACCOUNT NUMBER:
75-197-66

DATE	YOUR BALANCE WAS	WE SUBTRACTED NO.	CHECKS	SERVICE CHARGE	WE ADDED NO.	DEPOSITS	MAKING YOUR PRESENT BALANCE
8/31/-	186.43	14	586.65	1.90	2	706.09	303.97

DATE	CHECKS			DEPOSITS	BALANCE
					186.43
8/2				286.75	473.18
8/6	125.00				348.18
8/9	23.46	40.00			284.72
8/10	9.45	15.00			260.27
8/12	15.74	139.00			105.53
8/15	34.10				71.43
8/16				419.34	490.77
8/17	21.19				469.58
8/19	8.00				461.58
8/22	14.86	10.00	45.00		391.72
8/27	85.85				305.87
8/31	1.90SC				303.97

Please notify bank immediately of any change of address. The account will be considered correct if errors are not reported immediately.

Symbol code:
SC Service Charge
OD Overdrawn Account
SP Stop Payment
EC Error Correction

The balance in Larry's check register is $405.87. There are no outstanding checks, but he has made one deposit for $100 that does not appear on the statement.

QUESTION   What might be the reasons why Larry's balance is not the same as the balance on the bank statement?

SOLUTION

The first reason is that the deposit is not on the bank statement. Add the deposit to the bank balance:

$303.97 + 100.00 = 403.97$

Compare the adjusted bank balance with Larry's balance:

$405.87 - 403.97 = 1.90$

Look on the bank statement for an amount of 1.90. There is a service charge of 1.90. Larry discovers that he has not subtracted this charge from his balance. When he does, the adjusted balances are the same.

SKILL 2

In the process of reconciling the balance on the bank statement with the balance in your checkbook, it generally happens that you must adjust both balances. This adjustment or correction does not mean that one or the other of the balances is wrong; it simply means that with new information the balances must be brought up to date.

How is the process of adjusting the bank statement balance similar to the process of adjusting the checkbook balance and how is it different?

The procedures for making the adjustments are similar. In both cases, amounts that were not available earlier must be added to or subtracted from the initial balance to find the adjusted balance. Adjustments of the bank statement and checkbook follow the equation

Balance + deposits not recorded − checks and fees not recorded
= new balance

Differences are found in the numbers used. The beginning balance and the amounts of the unrecorded deposits, checks, and fees might all be different. But if all of the information is correct and the computations are done correctly, then the adjusted balances will be the same.

With each bank statement the bank also sends the customer a reconciliation form. This form is set up to help you compare the bank's records with your records and reconcile the account. That is, you adjust your balance and the bank's balance to arrive at the correct current balance.

EXAMPLE 2   In another month Larry received a bank statement showing a balance of $254.78. Larry's checkbook balance is $211.23. Larry compares the statement with his checkbook and finds the following, not shown on the statement: deposit, $60; checks, No. 226, $23.65, and No. 231, $82.30. On the statement, but not in his checkbook, is a bank charge of $2.40.

QUESTION How does Larry use the reconciliation form to reconcile his checkbook with the bank and to find new adjusted balances?

SOLUTION

The transactions in question are listed on the reconciliation form below. Notice where each transaction appears and how it is used.

1. The balance on the bank statement and in the checkbook.
2. Deposits not on the statement.
3. Checks not on the statement.
4. Service charges not in the checkbook.
5. Adjusted balances.

YOU CAN EASILY
BALANCE YOUR CHECKBOOK
BY FOLLOWING THIS PROCEDURE

FILL IN BELOW AMOUNTS FROM YOUR CHECKBOOK AND BANK STATEMENT

BALANCE SHOWN ON BANK STATEMENT **(1)** $ **254.78**	BALANCE SHOWN IN YOUR CHECKBOOK **(1)** $ **211.23**
ADD DEPOSITS NOT ON STATEMENT **(2)** $ **60.00**	
	ADD ANY DEPOSITS AND OTHER CREDITS NOT ALREADY ENTERED IN CHECKBOOK $
TOTAL $ **314.78**	
	TOTAL

SUBTRACT CHECKS ISSUED BUT NOT ON STATEMENT

#	$	
226	**23.65**	**(3)**
231	**82.30**	

SUBTRACT SERVICE CHARGES AND OTHER BANK CHARGES NOT IN CHECKBOOK

$ **2.40** **(4)**

TOTAL $ **105.95**	TOTAL $ **2.40**
BALANCE **(5)** $ **208.83**	BALANCE **(5)** $ **208.83**

THESE TOTALS REPRESENT THE CORRECT AMOUNT OF MONEY YOU HAVE IN THE BANK AND SHOULD AGREE. DIFFERENCES, IF ANY, SHOULD BE REPORTED TO THE BANK WITHIN TEN DAYS AFTER THE RECEIPT OF YOUR STATEMENT

TRY YOUR SKILLS

1. What amounts must be added to the bank statement balance or to the check register balance in preparing the adjusted balance?

2. What amounts must be subtracted from the bank statement or from the check register balance in preparing the adjusted balance?

3. Your bank statement shows a closing balance of $75.65. There are no outstanding checks or deposits. Your checkbook shows a balance of $77.95. What might account for the different balances?

Your bank statement shows a closing balance of $102.60. The following are not on the statement: deposit, $50.00; check, $23.88; ATM withdrawal, $40.00. There is a service charge of $1.50 on the statement but not in your checkbook.

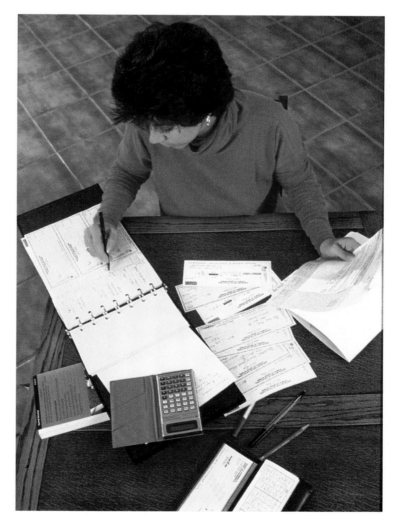

4. What amount(s) must be added to the bank statement balance?

5. What amount(s) must be subtracted from the bank statement balance?

6. What should be the balance in your checkbook before you do a reconciliation?

7. What should be the balance after you do a reconciliation?

8. Roger has a balance of $218.52 in his check register. The balance on his bank statement is $487.35. Not reported on his bank statement are a deposit for $335.56 and checks for $572.00, $39.83, and $77.11. The service charge was $7.43. Roger noticed checks for $52.12 and $25.00 were not in his check register. He had not recorded the service charge in his check register. What should the balance be in his check register after he reconciles his account?

EXERCISE YOUR SKILLS

1. What are two advantages of ATMs?
2. What is a disadvantage of ATMs?
3. Why is the current balance on the bank's statement often different from the balance in your check register?
4. What can you do to balance your account if the initial use of the reconciliation form does not yield matching balances?

Larry received the following monthly statement from his bank.

INLAND BANK

BANK STATEMENT

TO:
LARR Y LENDER
4115 N. LINCOLN
INLAND, IN 47304

ACCOUNT NUMBER:
75-197-66

DATE	YOUR BALANCE WAS	WE SUBTRACTED NO.	CHECKS	SERVICE CHARGE	WE ADDED NO.	DEPOSITS	MAKING YOUR PRESENT BALANCE
9/30/– –	303.97	11	561.45	2.30	2	400.00	140.22

DATE	CHECKS			DEPOSITS	BALANCE
					303.97
9/1	100.25				203.72
9/3	75.00				128.72
9/9				200.00	328.72
9/11	214.00				114.72
9/16	12.95	15.45			86.32
9/19				200.00	286.32
9/21	11.15	20.00	34.50		220.67
9/24	5.40	7.75			207.52
9/28	65.00				142.52
9/30	2.30SC				140.22

Please notify bank immediately of any change of address. The account will be considered correct if errors are not reported immediately.

Symbol code:
SC   Service Charge
OD   Overdrawn Account
SP   Stop Payment
EC   Error Correction

5. How many check amounts are reported on the statement?
6. What is the initial balance for the account?
7. What is the closing balance for the account?
8. What is the total dollar amount of the deposits reported on the statement?
9. What service charge is reported on the statement?
10. Not reported on Larry's bank statement are two checks: No. 238, $25.00, and No. 241, $38.05. Also not reported is a deposit for $50.00. The service charge listed on the statement is not recorded in Larry's check register. The balance in Larry's check register is $129.47. Use this information to reconcile Larry's account using a reconciliation form provided by your teacher, or copy the one shown earlier in the lesson.
11. Cheryl has a balance of $478.83 in her check register. The balance on her bank statement is $345.75. Not reported on her bank statement are a deposit for $250 and three checks for $85.00, $54.32, and $129.75. She forgot to record two ATM withdrawals of $50 and $100 in her check register. She also needs to record the service charge of $2.15 in her check register. What should the balance be in her check register after she reconciles her account?
12. Following are a portion of Rachel Rosen's check register and the bank statement that she received. Reconcile her bank statement with her check register. Obtain a reconciliation form from your teacher, or make a copy of the one shown earlier in the lesson.

Check Register

CHECK NUMBER	DATE		CHECKS/DEPOSITS	AMOUNT	BALANCE $1840 63
201	3/7	TO:	FOREVER PHOTOS	183 70	
		FOR:	CAMERA		1656 93
202	3/8	TO:	LIGHTNING DEVELOPING	63 50	
		FOR:	PICTURES		1593 43
203	3/8	TO:	WE-HAVE-IT DEPT. STORE	23 80	
		FOR:	ALBUMS		1569 63
204	3/9	TO:	FLASH BACK	12 95	
		FOR:	CAMERA CASE		1556 68
	3/10	TO:	DEPOSIT	363 90	
		FOR:	PAYCHECK		1920 58
205	3/12	TO:	M-MART	12 50	
		FOR:	FILM		1908 08
ATM	3/15	TO:	ATM	50 00	
		FOR:	CASH		1858 08
206	3/17	TO:	CAMERA CASE	85 60	
		FOR:	LENSES		1772 48
	3/17	TO:	DEPOSIT	250 00	
		FOR:			2022 48
ATM	3/17	TO:	ATM	50 00	
		FOR:	CASH		1972 48
207	3/19	TO:	LIGHTNING DEVELOPING	26 85	
		FOR:	CHEMICALS		1945 63
208	3/21	TO:	ELECTRIC COMPANY	95 75	
		FOR:	ELECTRIC BILL		1849 88

Checking Account Statement

ACCT. 190-12566
DATE 3/31/- -
PAGE 1

National City Bank

RACHEL ROSEN
717 NOR TH WILSON PLACE
INLAND, IN 47304

BALANCE FORWARD	NO. OF CHECKS	TOTAL CHECK AMOUNT	NO. OF DEP.	TOTAL DEPOSIT AMOUNT	SERVICE CHARGE	BALANCE THIS STATEMENT
1840.63	7	577.80	2	613.90	6.50	1870.23

CHECKS AND OTHER DEBITS		DEPOSITS AND OTHER CREDITS	DATE	BALANCE
201	183.70		3/10	1656.93
202	63.50		3/10	1593.43
203	23.80		3/11	1569.63
204	12.95	363.90	3/11	1920.58
205	12.50		3/15	1908.08
206	85.60		3/20	1822.48
208	95.75	250.00	3/22	1976.73
	50.00ATM		3/23	1926.73
	50.00ATM		3/24	1876.73
	6.50SC		3/25	1870.23

PLEASE EXAMINE AT ONCE.
IF NO ERRORS ARE REPORTED
WITHIN 10 DAYS, THE ACCOUNT WILL
BE CONSIDERED CORRECT.
PLEASE ADVISE US
IN WRITING OF ANY CHANGE
IN YOUR ADDRESS

KEY TO SYMBOLS

AD	AUTOMATIC DEPOSIT	DM	DEBIT MEMO
AP	AUTOMATIC PAYMENT	EC	ERROR CORRECTED
AR	AUTOMATIC REVERSAL	IE	INTEREST EARNED
CB	CHARGE BACK	OD	OVERDRAWN
CC	CERTIFIED CHECK	RC	RETURN CHECK CHG
CM	CREDIT MEMO	RT	RETURN ITEM
CO	CHARGE OFF	SC	SERVICE CHARGE

13. Jose has a balance of $899.10 in his check register. The balance on his bank statement is $569.93. Not reported on his bank statement are two checks for $120.60 and $53.77 and a deposit for $175. He had forgotten to record in his check register a transfer payment to his credit card for $125 and two ATM withdrawals of $100 each. He had also not recorded the service charge of $3.54. What should the balance be in his check register after he reconciles his account?

14. Both the bank and the customer who has an account at the bank want records to be accurate. But their reasons are not exactly the same. Think of yourself as a banker, and write a paragraph on why you want all accounts to be accurate and up to date. Then think of yourself as a person with a bank account, and write a paragraph about why you try to keep your banking records accurate and up to date.

1. At $8.50 per hour, how much does Nancy earn for 26.5 hours of work?

2. If you earn a commission of 3% on sales up to $80,000 and 5% on sales over $80,000, what is your commission on sales of $96,000?

3. Norma Torres is paid a base salary of $375 a month and a commission of 9%. Find her monthly earnings if her total sales are $8700.

4. Jeremy works from 3:00 P.M. until 6:30 P.M., five days a week. How many hours does he work a week?

5. Sarah's checking account pays interest of 0.35% a month and charges 6 cents per check. What amount will be earned or charged if her average balance is $235 and she writes 17 checks?

6. Find the balance for a checking account after the following transactions are made: initial balance, $260.50; check written for $76.45; check written for $19.87; deposit, $35.00.

7. For most employees, the FICA tax is 7.65%, which includes a 6.2% tax for Social Security. The remainder of the tax is for Medicare. What is the Medicare tax on Basilio's gross weekly income of $300?

8. For most employees, the FICA tax is 7.65%, which includes a 6.2% tax for Social Security. The remainder of the tax is for Medicare. What is the Medicare tax on Tricia's gross monthly income of $2900?

9. Find the balance for a checking account after the following transactions are made: initial balance $549.13; check written for $137.58; check written for $89.60; check written for $111.25; deposit, $100.00; ATM withdrawal, $60.00.

10. Enid has a balance of $1235.54 in her check register. The balance on her bank statement is $689.66. Not reported on her bank statement are a deposit for $575, a transfer from her savings account for $250, and two checks for $195.00 and $437.85. She forgot to record two ATM withdrawals of $150 and $200 in her check register. She also needs to record the service charge of $3.73 in her check register. What should the balance be in her check register after she reconciles her account?

11. Pete has a balance of $99.10 in his check register. The balance on his bank statement is $432.51. Not reported on his bank statement are two checks for $325.50 and $22.25, an ATM withdrawal for $200.00 and a deposit for $155. He had forgotten to record in his check register a transfer payment to his credit card for $60. He had also not recorded the interest earned of $0.66. What should the balance be in his check register after he reconciles his account?

12. Ernestine sells real estate. This month she sold a house for $375,000. The commission was 5% but she only received one-third of the commission because she had to split the commission with two other agents. What was her commission?

1. Find the interest for one month on $120 at 0.35% interest per month.

2. A bank pays interest of 0.45% per month and charges a finance fee of $4 on accounts with balances of less than $500. Find the total payment or charge for an account with a balance of $467.

3. If you had $487.51 in the bank and paid someone $63.94 by check, what would be your new bank balance?

4. Give three reasons why the balance on a bank statement might be different from the balance in your check register.

5. The Ridgewood Savings Bank pays 0.65% monthly interest and charges $0.10 per check. The Second National Bank pays 0.35% monthly interest and charges $0.04 per check. Write equations for the monthly charge and payment at each bank for a balance of $1000 but an unknown number of checks. Graph the equations using a graphing calculator. Use the graph to find the number of checks for which each bank is more economical.

6. The cost of Presto+ checks is $0.0225 per check. There is an additional charge of $0.15 for each check over 20. No service charge is charged for beginning balances of $500 and above; $5.00 is charged for beginning balances below $500. Interest of 0.45% per month is paid on the beginning balance. Copy and complete the chart.

	Beginning Balance	Number of Checks	Cost of Checks	Extra Charges	Service Charge	Interest Earned	New Balance
a.	$ 500	20					
b.	1000	10					
c.	600	16					
d.	400	24					
e.	300	30					
f.	100	32					

7. Find the new balance at the end of each transaction. The beginning balance is $0.00.

Date	Check No.	Check/Deposit	Amount	Balance
2–17		Deposit	$455.78	
2–18	405	Ravine Variety	85.95	
2–18	406	Power & Light	38.53	
2–18	407	Douglas Water	7.45	
2–19	408	Folly Gasoline	14.76	
2–20	409	Folly Service Station	53.90	
2–20		Deposit	104.88	
2–21	410	Bay Oil Company	16.28	

8. Reconcile the bank statement and check register summarized below.

BANK STATEMENT			
Date	**Transaction**	**Amount**	**New Balance**
6–11	Balance forward	$400.00	$400.00
6–12	ATM cash withdrawal	50.00	350.00
6–12	Check #435	19.37	330.63
6–14	Check #437	35.00	295.63
6–17	Check #436	40.55	255.08
6–19	Deposit	50.00	305.08
6–19	ATM cash withdrawal	25.00	280.08
6–21	Check #439	11.16	268.92
6–22	Check #438	1.00	267.92
6–25	Service charge	5.00	262.92
6–26	Check #440	85.25	177.67
6–26	ATM cash withdrawal	80.00	97.67
6–27	ATM transfer payment	50.00	47.67

CHECK REGISTER			
Date	**Transaction**	**Amount**	**New Balance**
6–05	Balance forward	$400.00	$400.00
6–11	Check #435	19.37	380.63
6–12	ATM cash	50.00	330.63
6–12	Check #436	40.55	290.08
6–13	Check #437	35.00	255.08
6–14	Check #438	1.00	254.08
6–15	Check #439	11.16	242.92
6–19	ATM cash	25.00	217.92
6–19	Deposit	50.00	267.92
6–29	Deposit	40.00	307.92

9. Set up a spreadsheet to calculate charges and new balances for the checking account described in Exercise 6. Find the interest earned and new balance for the following: $155, 11 checks; $278, 17 checks; $505, 21 checks; $789, 27 checks.

10. Write a restrictive endorsement as you would to deposit a check in the bank.

1. The cost of Presto+ checks is $0.0225 per check. There is an additional charge of $0.15 for each check over 20. No service charge is charged for beginning balances of $500 and above; $5.00 is charged for beginning balances below $500. Interest of 0.45% per month is paid on the beginning balance. Copy the chart, and complete it using the information given.

	Beginning Balance	Number of Checks	Cost of Checks	Extra Charges	Service Charge	Interest Earned	New Balance
a.	$ 400	20					
b.	150	20					
c.	1000	32					

2. Find the new balance at the end of each transaction.

Date	Check No.	Check/Deposit	Amount	Balance
4–15		Deposit	$1214.85	
4–15	415	VISA	101.97	
4–18	416	Craft Supply Co.	34.85	
4–19	417	Village Variety	83.66	
4–20		Deposit	320.00	

3. Reconcile the bank statement and check register shown below.

BANK STATEMENT			
Date	Transaction	Amount	New Balance
4–2	Balance forward	$200.00	$200.00
4–4	Check #426	35.00	165.00
4–6	Check #425	13.19	151.81
4–15	Deposit	230.00	381.81
4–20	Service charges	8.30	373.51
4–22	ATM cash withdrawal	150.00	223.51

CHECK REGISTER			
Date	Transaction	Amount	New Balance
4–1	Balance forward	$200.00	$200.00
4–1	Check #425	13.19	186.81
4–2	Check #426	35.00	151.81
4–22	Deposit	500.00	651.81
4–22	Check #427	413.00	238.81
4–25	Check #428	187.18	51.63

CUMULATIVE REVIEW

1. You worked 10 hours at $5.25 per hour. You received $11.50 in tips. Find your total earnings.

2. You worked 24 hours at $4.90 per hour. You received $1\frac{1}{2}$ times your hourly rate for any hours over 20. Find your total wages.

3. You worked the schedule shown below. You earned the same as in Exercise 2. Find your total wages for the week.

Monday	Tuesday	Wednesday	Thursday	Friday	Saturday
4–7 P.M.	4–7 P.M.	5–10 P.M.	4–7 P.M.	4–7 P.M.	9 A.M.–2 P.M.

4. You earn a base salary of $400 per month plus commission of 8% of sales. If your sales for one month are $4000, how much will you earn that month?

5. As a realtor, you sold a house for $210,000. You earn 4% commission on the first $100,000 and 5% commission on any amount over $100,000. How much commission did you earn on the house?

6. You earned $125.75. Your income tax withholding is $5, and your FICA withholding is 7.65% of your gross pay. Find your take-home pay.

7. You earned $181.25. Your income tax withholding is $13, and your FICA withholding is 7.65% of your gross pay. Find your take-home pay.

8. You earn $7.10 per hour. You work 40 hours per week for 50 weeks. You get two weeks of paid vacation plus $2500 worth of fringe benefits. Find your yearly earnings, including benefits.

9. You wrote 25 checks one month. The bank charges $0.025 per check for the first 20 checks and $0.10 for each check over 20. How much are the charges for that month?

10. You earn $46,000 per year plus 16% of your salary in fringe benefits. Find your yearly earnings, including benefits.

11. Your uncle wrote you a check and used your full legal name. Write your name as you would to endorse the check.

12. What is a blank endorsement? When is the only time you should use it?

13. You wrote a check for $24.65. You had $95.16 in your checking account. Show how you would enter this check in your check register.

14. Use a graphing calculator to compare the following kinds of accounts. For each one, assume a balance of $1000 but an unknown number of checks. Bank A pays interest of 0.4% per month and charges $0.085 per check. Bank B pays interest of 0.2% per month and charges $0.055 per check. Use the trace function to find the approximate point of intersection for the two graphs. Under what conditions is each bank better?

PROJECT 2–1: **Banking Institutions**

Gather information about a commercial bank and a savings and loan institution in your community. Compare your information with that of your classmates. Below are some possible questions.

1. Are checking accounts available?
2. What are the service charges?
3. Are savings accounts available?
4. What method is used to determine interest paid?
5. Are special checks provided?
6. Are ATM cards available?
7. Does the institution have deposit insurance?
8. For what amount are the accounts insured?
9. Can electronic fund transfers be made?
10. What types of loans are available?
11. Are safe-deposit boxes available?
12. Is estate planning guidance available?
13. Is tax assistance or counseling provided?

PROJECT 2–2: **Automated Teller Machines**

Automated teller machines allow customers 24-hour access to banking services. Because anyone can walk or drive up to a machine and use it, banking institutions take security precautions to protect against theft. To use your card, you must insert it into the machine and also enter your personal identification number (PIN) into the machine. The bank sometimes allows you to choose your own four- or five-digit PIN. If anyone should have your card and know your PIN, he or she could take money from your account.

1. Visit local banks, and find how you obtain a PIN. Some people use their birthdate as their PIN. Is this a good idea? Why or why not?
2. Visit local banks and find what security precautions they take with regard to ATM cards.

The *commutative* and *associative* properties of addition and multiplication can help you simplify a numerical or algebraic expression.

	Addition	Multiplication
Commutative property	$a + b = b + a$	$a \cdot b = b \cdot a$
Associative property	$(a + b) + c = a + (b + c)$	$(a \cdot b) \cdot c = a \cdot (b \cdot c)$

The associative property allows you to express $3 + (0.06 + 8)$ and $(5.4 \cdot 2) \cdot 10$ without parentheses.

$$3 + (0.06 + 8) = 3 + 0.06 + 8 = 3.06 + 8 = 11.06$$
$$(5.4 \cdot 2) \cdot 10 = 5.4 \cdot 2 \cdot 10 = 10.8 \cdot 10 = 108$$

Simplify.

EXAMPLE $(11.23 + x) + 1.77$

$= (x + 11.23) + 1.77$ Commutative property for addition
$= x + (11.23 + 1.77)$ Associative property for addition
$= x + 13$ Simplify.

1. $8a + (12 + 3a)$ **2.** $(200)(50y)(5)(2x)$

3. $2x + 3y - x + 4y$ **4.** $0.5(8x)(2a)$

Two other properties that can help you simplify expressions are:

Distributive property	$a(b + c) = ab + ac$
Definition of subtraction	$a - b = a + (-b)$

Remember to subtract a number is to add its opposite.

EXAMPLE $11x - 6(x - 1)$

$= 11x + (-6)[x + (-1)]$ Definition of subtraction
$= 11x + [(-6)x + (-6)(-1)]$ Distributive property
$= [11x + (-6x)] + 6)$ Associative property of addition
$= 5x + 6$ Add like terms.

5. $4.03x + 5.97 - 0.03x$ **6.** $28 - (8.5 - 10c)$

7. $40a - 20b - 9a$ **8.** $(4.2 - 3p) - 2p$

9. $-1(x^2 - 1.06) + x$ **10.** $y^3 - (7y^3 + 7)$

A DOLLAR *Saved.*

MARIA IS IN THE SAME HIGH SCHOOL as Alex, Jeff, and their friends. She and her friends face some of the same tough choices as their other classmates. Most of them have earned some money of their own and are enjoying the sense of independence it has brought them. They are eager to be out of school and completely on their own. They also realize the importance of choosing a career that they will find satisfying. For many this choice means attending college for further training, thus delaying total independence for a while. Many of these students' parents have already been saving money toward this extra education and have been encouraging their sons and daughters to save as well.

3–1 **Savings: Save Now — Buy Later**

3–2 **Compound Interest: Money That Grows**

3–3 **The Federal Reserve: The Bank's Bank**

In this chapter, Maria and her friend Nelson will help us examine how people go about saving money for purchases that require more cash than can be obtained easily from a regular salary. Maria dreams of buying her own car and saves money toward that goal each month. As a result, she learns about various savings institutions and how they pay interest on her money. Nelson looks into interest in greater detail, discovering how often interest may be paid on a savings account and what is meant by an annual interest rate.

Their friend Olivia will help explain to us the impact of the Federal Reserve System on both savers and consumers. She discovers that transferring money over great distances takes only minutes. She also learns that the Federal Reserve System plays a key role in managing the country's money supply.

Maria does not think of herself as extravagant but ever since she began earning money on her own, she has noticed several items that she would like to buy. She was even able to buy a small car just three months ago after she persuaded her father that she would be able to pay for the monthly installment charges, insurance, gasoline, and regular maintenance. He helped her pick out the car, which had only 14,000 miles on it and had been owned and maintained by a nationwide car-rental chain. So far she has not had to buy so much as a new windshield wiper blade.

It was not easy for Maria to convince Dad that she could keep up the car payments. For six months she put half of her weekly paycheck from a fast-food restaurant into her savings account at the bank. Dad had told her to keep the money for her college expenses next year, but what he really wanted to find out was whether Maria would be able to save up money for the car. After six long months, Dad was convinced that she could.

Maria will share with us some of what she has learned about saving money. She has decided that saving isn't so hard after all and that maybe she will save even more so that she can buy some of the things she did without during the six months. Perhaps she can even take her friend Melody out to lunch to celebrate Melody's birthday.

In this lesson, we will help Maria to:

- *Recognize why people save money.*
- *Identify the places where people commonly deposit savings.*
- *Explain factors such as interest rates and liquidity that influence the return that your money can earn at financial institutions.*
- *Explain the differences between regular savings accounts, money market accounts, and certificates of deposit.*

WHY PEOPLE SAVE

People have many reasons for saving—to pay for a vacation, stereo or sports equipment, or a major appliance, for example. Many young people also save to help pay for college.

Maria's father had two reasons for saving. First, soon after Maria began working, he realized that, apart from Maria's own wishes, the family actually did need another car, since arranging transportation for everyone too often required skillful coordination. Second, he realized that by helping Maria follow through with her plan to pay for the car, he could not only ease the family's transportation problem but, perhaps more important, also help his daughter to develop the self-discipline she needed to save regularly.

As most people do, Maria's family knows that unforeseen events such as an accident, illness, or the loss of a job can cause a sudden loss of income. For this reason, financial advisors recommend a savings reserve of at least three to six months' salary.

It takes self-discipline and consistency of purpose to maintain a habit of saving each month, so developing these admirable qualities is a strong reason for acquiring that habit. In addition to its positive effect on your character, the saving habit also is directly rewarding. Whenever you let a financial institution such as a bank use your money, it will pay you interest on that money.

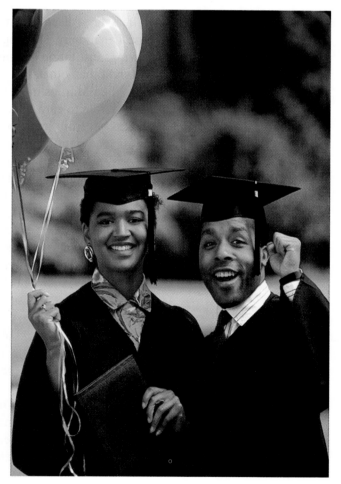

WHERE PEOPLE SAVE

Once Maria had made a firm decision to open a savings account, she had to decide what kind of institution she would choose. She knew that she had several options such as commercial banks, savings banks, savings and loan associations, and credit unions. Saving in these institutions is usually safe and offers a relatively fixed rate of return on money.

Commercial banks offer a wide variety of services that persuade many families to select this kind of bank as the center of their financial affairs.

These services include offering checking and savings accounts, making loans, issuing credit cards, and renting safe-deposit boxes. Commercial banks usually also sell traveler's checks and money orders. They also offer financial counseling and trust and investment services.

Savings banks operate in much the same way as commercial banks; their services include savings accounts, check cashing, safe-deposit boxes, and savings bank life insurance. They also specialize in real-estate loans. Savings banks are common in the northeastern part of the United States.

Savings and loan associations, sometimes also called *thrift institutions,* generally lend money for home purchases and home construction.

Credit unions are not-for-profit financial institutions that pay interest on members' savings and use these savings to make loans to their members. Credit unions have membership requirements. The members are people who belong to a professional organization, church, company, or other group that has a common social, economic, political, or religious basis. Maria's father, along with 7500 other federal employees, is a member of their county's Federal Employees Credit Union.

HOW TO SELECT A SAVINGS ACCOUNT

After Maria discovered the many types of savings institutions that are eager to use her money, she began to wonder why she should choose one kind of savings institution over another. One reason is that some institutions pay more **interest** than others.

When Maria's money is put to work, she should expect it to work as hard as possible to earn interest with a high degree of safety. *Interest* is the return savers get from letting someone else use their money. The *interest rate* is the most important factor in determining the interest a savings account will earn. By calling around to various institutions, Maria found out that interest rates vary from place to place. She also learned that the *liquidity* of a savings instrument influences the interest rate that she can get on her savings. **Liquidity** means the ease and speed with which savings can be converted to cash. The longer the savers are willing to tie up their money, the higher the interest rate they will receive.

Regular Savings Accounts   The most liquid or flexible choice for Maria is a **passbook** or **regular savings account,** an account that allows her to make deposits and withdrawals at any time. A regular savings account accepts deposits of any size and allows you to withdraw any amount that does not exceed your total balance. Maria's family already has its emergency-fund savings in a regular savings account.

Money Market Accounts   If you can do without the convenience of a regular savings account, you may be able to obtain a higher rate of interest than would otherwise be the case. One way you can do this is to open a **money market account**. In exchange for a higher rate of interest, you may have to accept some restrictions on your access to the account. For more details on money market accounts see Lesson 8–1.

Certificates of Deposit   Another way to obtain a higher rate of interest is to purchase a **certificate of deposit (CD)**. A certificate of deposit is less liquid than a regular savings account. It is purchased at a bank by filling out a special form that records the amount of savings a person chooses to place in a special account. The CD specifies a fixed amount of money that must be deposited and a period of time during which the saver promises not to withdraw money from the account. The interest rates for six-month certificates are tied to the weekly U.S. Treasury bill interest rate.

Since the bank has the use of your CD money for a guaranteed period of time, it can afford to pay a higher rate of interest than on a regular savings account. The federal government requires the bank to impose stringent penalties if you make an early withdrawal from a certificate of deposit account. For example, you can be penalized three months' interest on certificates of less than one year if you withdraw the money before the date that you originally agreed to.

Maria's family has placed her school tuition money in certificates of deposit because it is willing to accept the restrictions on its ability to withdraw the money in exchange for the relatively high interest that the certificates pay.

Ask Yourself

1. What emergencies do people usually save for?

2. How much money do financial advisors recommend that people save to cover emergencies such as sickness or the loss of a job?

3. What are three types of institutions in which people usually keep their savings?

4. What are four of the services offered by commercial banks?

5. What examples can you give of the kind of people who might belong to the same credit union?

6. What is interest?

ALGEBRA *REVIEW*

In each formula, substitute the given values. Then solve to find the value of the remaining variable.

Example

$$p = 2l + 2w; p = 80, l = 15$$
$$80 = 2(15) + 2w$$
$$80 = 30 + 2w$$
$$50 = 2w \qquad \text{Subtract 30 from each side.}$$
$$25 = w \qquad \text{Divide each side by 2.}$$

1. $A = lw; l = 2.6, w = 7$

2. $i = 250rt; r = 0.07, t = 3$

3. $s = l + w; l = 16, w = 4.6$

4. $B = p + i, B = 302.5, p = 250$

5. $i = p(0.045)t; p = 5000, t = 15$

6. $i = prt; p = 1500, r = 0.05, t = 10$

7. $B = p + i; p = 750, i = 30.63$

8. $B = p + prt; p = 100, r = 3\%, t = 4$

SHARPEN YOUR SKILLS

SKILL 1

The amount of money invested is called the **principal** p. **Simple interest** i can be determined using the following formula. The **interest rate** r is a percent of the principal.

Simple Interest Formula

$i = prt$ where i = interest

p = principal

r = interest rate

t = time

EXAMPLE 1   Maria can save $85 per week from her paycheck. After saving for a year, she decides to buy a one-year CD that pays 4% simple interest.

QUESTIONS   **1.** How much will Maria save in 52 weeks?

                2. How much interest will the CD earn in 1 year?

                3. How much will Maria's CD be worth after 1 year?

SOLUTIONS

1. Total savings = weekly savings • number of weeks

$$= 85 \cdot 52$$
$$= 4420$$

In 52 weeks, Maria will save $4420.

2. To find the interest after 1 year, Maria uses the formula for simple interest.

$$i = prt$$
$$i = 4420 \cdot 0.04 \cdot 1$$
$$i = 176.80$$

The CD earns $176.80 in interest.

3. The balance in the CD will be the principal plus the interest.

Balance = principal + interest

$$B = p + i$$
$$B = 4420 + 176.80$$
$$B = 4596.80$$

The total amount in the CD at the end of one year is $4596.80.

If Maria leaves the money in the CD, then in the second year the bank will pay interest on both the original principal and on the previous year's interest. In other words, instead of being simple interest, as in year 1, the interest for year 2 and all later years will be *compound interest*. Compound interest is discussed in Lesson 3–2.

EXAMPLE 2   Maria's friend Dwight wants to buy a golf bag. He finds one on sale for $59.84. The sale price will be in effect for 1 month.

> **Golf Bag . . . . . . SALE $59.84**
> Less than half original price! Genuine leather, sturdy construction. Just like the ones the pros use!

QUESTION   If Dwight can save $22 per week, will he have enough money to buy the bag before the sale is over?

SOLUTION

Weeks needed = cost of golf bag ÷ amount saved each week
$$= 59.84 \div 22$$
$$= 2.72$$

Dwight will have enough money in three weeks, so Dwight will have enough money to buy the golf bag before the sale is over.

_____ **SKILL 3** _____

EXAMPLE 3   Maria's sister Sarita would like to buy one or more of the items shown below. She can save $25 each week from her salary.

1. **Audio Box**   3-bank EQ.
   Local-distance switch.
   Time/frequency display switch.
   Sale $59.87   Reg. $75.00

2. **Quartz Watch**
   Yellow case with black band.
   Measures elapsed and lap time.
   Sale $89.99   Reg. $135.00

3. **26" 10-speed Bicycle**
   Men's or women's mountain bike.
   Front and rear brakes.
   Sale $109.99   Reg. $119.95

4. **35mm Camera** $139.99
   Completely automatic. No focusing. Auto flash. Uses 2 AA batteries (not included).

QUESTION   How many weeks will Sarita have to save to buy each item?

SOLUTION

To find the number of weeks of savings required, Sarita prepares a bar graph using a spreadsheet program. She wants the graph to show equally spaced $25 levels so that she can easily see the number of weeks of savings required for each of the four items. Sarita begins by creating the following spreadsheet:

	A	B
1	stereo	59.87
2	watch	89.99
3	bicycle	109.99
4	camera	139.99

Most spreadsheet programs allow you to graph by selecting an option such as GRAPH or CHART. Choose the type of graph, in this case, a **bar graph.** Then use the option that allows you to choose minimum and maximum values on each axis.

Sarita chooses 0 for the minimum and 150 for the maximum on the y-axis.

You can also select an option to show horizontal gridlines. You may be able to select the spacing between gridlines by using a command called SCALE, INCREMENT, or something similar.

Sarita selects 25 for the scale.

The spreadsheet program will automatically label the y-axis (the left vertical axis) with the labels 0, 25, 50, 75, 100, 125, and 150 with or without dollar signs, as you choose. The spreadsheet menu will also let you choose an option with a name such as INTERIOR LABELS that lets you position each of the four sales prices above its graph. You will probably have to type the text labels yourself. Fortunately, you can instruct most spreadsheet programs to place certain kinds of text inserts (headings, for example) in their proper position.

Sarita chooses the INTERIOR LABELS option. Then she types in the labels for the top, bottom, and right side of the graph. The graph is shown below.

From the graph, it will take 3 weeks to save enough for the stereo, 4 weeks for the watch, 5 weeks for the bicycle, and 6 weeks for the camera.

TRY YOUR SKILLS

A bank pays 3.5% simple interest on a one-year CD. The table below shows the amounts that Maria and her friends have saved for a year to buy a CD. Find out how much is saved, how much interest is earned by the CD in one year, and the total amount of money in each account after the bank has credited the interest to the accounts.

	Amount Saved per Week	Number of Weeks	Total Saved	Interest Earned	Total in Account
1.	$10	52			
2.	15	48			
3.	20	50			
4.	25	52			
5.	30	50			
6.	40	48			

One of Maria's friends, Donna, decided to save up for a $330 vacation. Find out how many weeks she needs to save. How long will she have to save for each of the indicated weekly savings?

	Amount Saved per Week	Number of Weeks Needed	Total Saved
7.	$20		
8.	32		
9.	18		
10.	30		
11.	35		
12.	40		

Draw a bar graph to show the number of weeks it will take to save enough money to buy the items listed in Exercises 13–16. Assume that you save $25 per week.

13. Car stereo system, $139.99

14. Portable AM/FM stereo, $219.99

15. Computer, $499

16. 10-speed bicycle, $339.99

EXERCISE YOUR SKILLS

1. Why would a less liquid savings account earn higher interest than a more liquid account?

2. Why do you think that the federal government imposes heavy penalties for liquidating a certificate of deposit before its maturity date?

3. Why should savers deposit emergency savings in a regular savings account rather than in a certificate of deposit?

A bank pays 4% simple interest on a one-year CD. The following table shows amounts saved for a year to buy a CD. Find out how much is saved, how much interest is earned by the CD in 1 year, and the total amount of money in each account after the bank has credited the interest to the accounts.

	Amount Saved per Week	Number of Weeks	Total Saved	Interest Earned	Total in Account
4.	$ 30	52			
5.	50	48			
6.	100	50			

Dwight is considering buying some items. Find out how much he can save and how long he will have to save to be able to afford each item below when he saves at the stated rate.

	Amount Saved per Week	Number of Weeks Needed	Total Saved
7.	$15		
8.	20		
9.	25		
10.	50		

Sale: $139.99 . . Reg. $159.95 Murray's "Pro-Master" 3–9 irons, pitching wedge, 1–3–5 woods. Lightweight steel shafts. Stainless-steel woods.

	Amount Saved per Week	Number of Weeks Needed	Total Saved
11.	$15		
12.	20		
13.	25		
14.	50		

Diamond/Sapphire Ring 14K gold. Elegant setting.
Regular $275.00
Special Price $199.99

	Amount Saved per Week	Number of Weeks Needed	Total Saved
15.	$15		
16.	20		
17.	25		
18.	50		

Sale: $149.99 . . . . . Reg. $200 Teak Portable CD Player with 3-beam laser pick-up. Music shuffle for random playback. Uses 4 AA batteries (not included).

KEY TERMS

bar graph
certificate of
deposit (CD)
commercial banks
credit unions
interest
interest rate
liquidity
money market
account
principal
passbook or regular
savings account
savings and loan
associations
savings banks
simple interest

Draw a bar graph to show the number of weeks it will take to save enough money to buy the items listed in Exercises 19–21. Assume that you save $25 per week.

19. Auto-focus camera, $189.99

20. VCR, $199.99

21. Keyboard, $89.95

22.–24. Draw a bar graph like the one for Exercises 19–21 assuming you save $40 per week.

MIXED REVIEW

1. A real-estate salesperson receives 4% of the first $100,000 of the selling price of a home and 6% of any amount over $100,000. How much commission does the salesperson receive on a home that sells for $250,000?

2. Jared has a part-time business videotaping weddings. His rates are $60 for the ceremony and $110 if both the ceremony and the reception are covered. Last year he was able to videotape 29 weddings, 4 of which included the reception. His expenses for the year were $900. What were his net earnings for the year?

Fill in the table below using the following information about Essex National Bank checking accounts.

Interest is paid on the beginning balance at the rate of 0.4% per month.
The service charge is $3.50 per month.
The cost per check is $0.03.
There are no additional charges regardless of the amount of the balance.

	Beginning Balance	Number of Checks	Interest Earned	Service Charges	Cost of Checks	New Balance
3.	$ 800	14				
4.	16,000	27				
5.	166	8				

Suppose that on October 7 your checking account balance is $426.67.

6. What will be your balance after depositing a $75.81 check from another state on October 8?

7. What will be your balance after you write a check of $350 against your account on October 9?

8. Why might it not be a good idea to write a check for $80 on October 10?

9. Keith earns $7.50 an hour and $1\frac{1}{2}$ times that amount for time over 20 hours in 1 week. What is his pay for 23 hours of work?

3-2 COMPOUND INTEREST: MONEY THAT GROWS

Maria's friend Nelson was impressed by the discipline that she imposed on herself to save money to cover the operating expenses of a car. He knew that there were times when Maria was tempted to spend some of her savings, but he also saw how fast those savings were growing. In fact, he discovered that her total account balance was larger than the money she had put in the bank. Maria explained that her savings were earning interest. Nelson saw this as the bank's way of rewarding her for keeping her money in the bank.

Nelson spends his extra time writing programs in the computer lab. He received recognition at a science competition for a simulation program that projects the changing growth patterns of the great forests in the Western Hemisphere and the impact those changes could have on human lives. He became interested in the topic when a lumber company wanted to cut some of the trees from a national forest where he and his family have gone camping.

As a result of his study, Nelson learned how long it takes to grow new trees. He began to wonder if growing new money from old money was anything like that.

OBJECTIVES: *In this lesson, we will help Nelson to:*

- *Compute the total interest for a savings account when the interest is compounded annually, semiannually, or quarterly.*
- *Compute interest in a savings account using the compound interest formula.*

COMPOUND INTEREST

Just as Maria discovered that interest rates vary in different institutions, Nelson found that the way in which interest is calculated can vary, too.

The interest rate is always expressed as an annual percent. This means that if Maria kept $1000 in a CD for one year with an interest rate of 5%, she would earn $50 in simple interest. If p represents the principal, r represents the annual rate of interest, and t represents the time in years, then the interest i is found from the simple interest formula.

$$i = prt$$
$$i = 1000 \cdot 0.05 \cdot 1$$
$$i = 50$$

Her simple interest is $50.

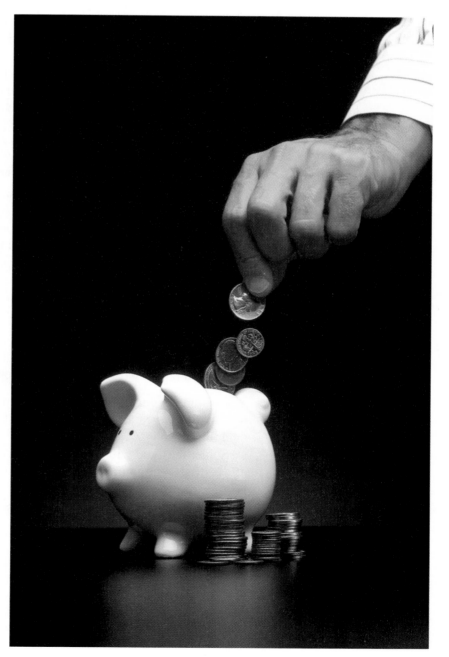

If Maria were to earn simple interest on her investment of $1000 for a second year, then she would earn $1000 \cdot 0.05 \cdot 1$, or $50 in the second year also. However, in practice, interest is not calculated that way. All savings institutions pay **compound interest,** that is, interest on the principal and on *the previously paid interest,* assuming that the interest is left in the account. In the second year, Maria will receive

$$1050 \cdot 0.05 \cdot 1, \qquad \text{or } \$52.50$$

in interest if the compounding is done only once a year.

However, most banks compound at least *semiannually.* This means that at least twice a year the interest earned is added to the previous balance, so the principal balance on which interest is paid grows a little faster than if the compounding were done only once a year. Some banks pay compound interest on whole dollar amounts only. But most large banks now compute interest on *exact amounts, compounded daily.* In this book, compound interest will be figured on exact amounts.

If Maria received 5% **compounded semiannually** on her original $1000, then her interest for the first year would be calculated as follows. To find half of the yearly interest, divide the yearly interest rate by 2. If the yearly rate is 5%, divide by 2 to obtain a semiannual rate of $2\frac{1}{2}\%$.

$$1000 \cdot 0.025 = 25.00 \qquad \text{First 6 months: interest on initial principal}$$
$$1025 \cdot 0.025 = \underline{25.63} \qquad \text{Second 6 months: interest on principal and interest}$$
$$50.63 \qquad \text{Total interest for 1 year}$$

Of course, the interest has been rounded to the nearest cent. Maria earned 63 cents more than she would have if the interest had not been compounded.

You can calculate the interest that is **compounded quarterly** in a similar manner. Since the interest is calculated and paid four times a year (every three months), the quarterly interest rate is found by dividing the yearly rate by 4. If the yearly rate is 5%, divide by 4 to obtain a quarterly rate of 1.25%.

Increasing the frequency of compounding does not greatly increase the amount of interest that you actually get. The table below shows the total interest that you would receive over ten years on $1000 at a rate of 5% under several compounding methods.

	Annually	Semiannually	Quarterly	Continuously
Interest received	$628.89	$638.62	$643.62	$648.72

Even if your original $1000 were compounded *continuously* (every split second), in ten years you would gain only about $20 more than by compounding annually. So it does not pay to put a lot of emphasis on the frequency of compounding when you choose a bank. However, *all* of the compounding methods give you much more money than you would receive if the interest were calculated just as simple interest, as shown below:

$$i = prt$$
$$i = 1000 \cdot 0.05 \cdot 10 \qquad \text{5\% interest paid on only the original \$1000}$$
$$i = 500 \qquad \text{10 years of simple interest}$$

You would get only $500 in simple interest at 5% on your original $1000 at the end of ten years. Compound interest would pay you at least $128 more than that.

Sometimes you will want to have an approximate idea of the long-range effect that compounding will have on your savings or investments. Fortunately, there is an easy calculation that you can perform to help you do this. It is explained at the top of the next page.

RULE OF 72

If you invest $1000 at an annually compounded rate of 3% and make no withdrawals, your balance will double to $2000 in about $23\frac{1}{2}$ years; the time for the $2000 to redouble to $4000 at a 3% rate is also about $23\frac{1}{2}$ years. The **Rule of 72** offers a fast and reasonably accurate way to determine how long it will take for a sum of money to double when it is compounded over several years. To find the time for an investment to double, divide 72 by the annual interest rate times 100. Since $10\% \cdot 100 = 10$, just ignore the percent symbol. For example, if your $5000 investment has an annual return of 10%, then in roughly 7 years ($72 \div 10 = 7.2$) your money will have doubled to $10,000.

Another use for the Rule of 72 is to determine how long it will take for a currency to lose half its purchasing power. If the annual inflation rate stays at 6% for 12 years ($72 \div 6 = 12$), then the currency's purchasing power will drop by 50%.

The Rule of 72 is useful in calculating how long it will take you to save for a special long-term purpose, such as college expenses or retirement.

Ask Yourself

1. What is simple interest?

2. What is compound interest?

3. Do you think you will earn more interest on a savings account if the compounding is done quarterly rather than semiannually? Will the difference in the interest be very large?

4. What advantages do you give up to earn interest in a savings account rather than in some other kind of investment?

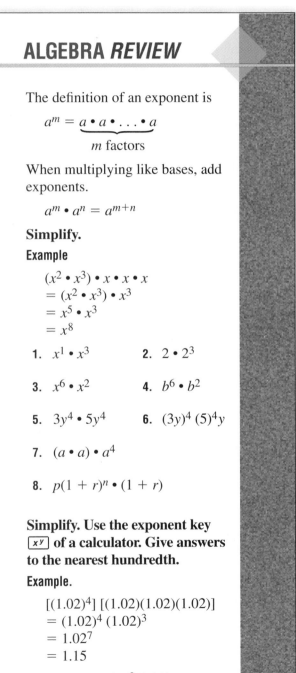

ALGEBRA *REVIEW*

The definition of an exponent is

$$a^m = \underbrace{a \cdot a \cdot \ldots \cdot a}_{m \text{ factors}}$$

When multiplying like bases, add exponents.

$$a^m \cdot a^n = a^{m+n}$$

Simplify.

Example

$$(x^2 \cdot x^3) \cdot x \cdot x \cdot x$$
$$= (x^2 \cdot x^3) \cdot x^3$$
$$= x^5 \cdot x^3$$
$$= x^8$$

1. $x^1 \cdot x^3$ 2. $2 \cdot 2^3$

3. $x^6 \cdot x^2$ 4. $b^6 \cdot b^2$

5. $3y^4 \cdot 5y^4$ 6. $(3y)^4 (5)^4 y$

7. $(a \cdot a) \cdot a^4$

8. $p(1 + r)^n \cdot (1 + r)$

Simplify. Use the exponent key $\boxed{x^y}$ of a calculator. Give answers to the nearest hundredth.

Example.

$$[(1.02)^4] [(1.02)(1.02)(1.02)]$$
$$= (1.02)^4 (1.02)^3$$
$$= 1.02^7$$
$$= 1.15$$

9. $15,000(1.01)^3(1.01)$

10. $(1.035)^2(1.035)(1.035)$

SHARPEN YOUR SKILLS

SKILL 1

To find semiannual interest, multiply the principal p by the semiannual rate $r \div 2$ since $i = prt$ and $t = \frac{1}{2}$.

EXAMPLE 1 Nelson's parents have a certificate of deposit in the amount of $10,000. It is held by a bank that pays 5% interest, compounded semiannually.

QUESTION How much will Nelson's parents have in this account after 2 years?

SOLUTION
Remember that the semiannual rate for 5% annual interest is $5\% \div 2$, or $2\frac{1}{2}\%$.

$i = $ principal • semiannual rate
$i = 10,000 • 0.025$
$i = 250$

This interest is added to the principal of $10,000. The new balance is $10,250, so the next time that interest is paid, it will be computed on a balance of $10,250. Each time, the interest is added to the balance before the next interest is calculated.

At the end of 2 years the family's balance sheet will look like this.

Time Period	Interest Earned	Principal Balance
Beginning	$ 0.00	$10,000.00
End of first half-year	250.00	10,250.00
End of first year	256.25	10,506.25
End of third half-year	262.66	10,768.91
End of second year	269.22	11,038.13

To obtain the interest earned in two years, use the formula

$p + i = B$

where p is the original principal, i is the interest and B is the balance at the end of two years.

$p + i = B$
$i = B - p$ Subtract p from both sides.
$i = 11,038.13 - 10,000$ $B = 11,038.13$ and $p = 10,000$
$i = 1038.13$

The interest earned over two years is $1038.13.

Nelson wanted to know whether he could predict how much interest the family would have obtained if the CD had been held for more than two years. He decided to see whether there was a pattern to the numbers in the table. First he wrote the bank balance at the end of six months in the following form.

$$p + i = 10{,}000 + 10{,}000 \cdot 0.025$$
$$= 10{,}000 \cdot 1 + 10{,}000 \cdot 0.025 \qquad x = x \cdot 1$$
$$= 10{,}000 \cdot (1 + 0.025) \qquad ab + ac = a(b + c)$$

The balance at the end of six months was $10{,}000 \cdot 1.025$. Nelson had found the following pattern:

New balance = old balance $\cdot (1 + 0.025)$
$$B_1 = p(1.025)$$

where p is the original principal and B_1 (read as "B sub one") is the total balance at the end of one six-month period.

Next Nelson noticed that at the end of two interest periods, the previous new balance became the new old balance. Thus at the end of two periods (1 year),

New balance = old balance $\cdot 1.025 \qquad 1 + 0.025 = 1.025$
$$B_2 = p(1.025) \cdot 1.025$$
$$B_2 = p(1.025)(1.025)$$
$$B_2 = p(1.025)^2 \qquad a \cdot a = a^2$$

where B_2 is the account balance at the end of two six-month periods.

Nelson continued this pattern and found that at the end of three six-month periods the account balance would be

$$B_3 = p(1.025)^2 \cdot (1.025)$$
$$= p(1.025)^2 \cdot (1.025)^1$$
$$= p(1.025)^3 \qquad a^2 \cdot a^1 = a^{2+1} = a^3$$

From this pattern, Nelson was able to discover the following **compound interest formula** for the balance B in the account at any future date.

Compound Interest Formula

$B = p(1 + r)^n$ where B = balance
p = original principal
r = interest rate for the time period
n = total number of time periods

Nelson wondered how he might use the new formula that he had discovered.

EXAMPLE 2 Nelson decided to try the formula to find how much the CD would be worth at the end of five years.

QUESTION How much will a $10,000 CD at 5% interest, compounded semiannually, be worth in five years?

SOLUTION

$B = p(1 + r)^n$ Compound interest formula

$B = 10,000 (1 + 0.025)^{10}$ $r = 5\% \div 2$, or 2.5%; $n = 10$ six-month periods

$B = 12,800.85$ Use your calculator.

The CD would be worth about $12,800 at the end of five years.

To enter the calculation in your graphing calculator use the exponent key $\boxed{x^y}$, sometimes labeled $\boxed{\wedge}$ or $\boxed{a^b}$. The Enter key is sometimes labeled $\boxed{\text{EXE}}$.

10000 $\boxed{(}$ 1 $\boxed{+}$ 0.025 $\boxed{)}$ $\boxed{x^y}$ 10 $\boxed{\text{ENTER}}$

SKILL 3

According to the Rule of 72, Nelson can find the speed at which his invested money will double by dividing 72 by 100 times the annual interest rate. Nelson would like to know whether he will be a millionaire by investing in CDs.

> ## Rule of 72
>
> $$\frac{72}{\text{Annual interest rate} \cdot 100} = \text{years to double}$$

EXAMPLE 3 Nelson invests $10,000 in a CD that pays 6% compounded quarterly.

QUESTION How long will it take his investment to double?

SOLUTION Using the Rule of 72,

$72 \div 6 = 12$ $6\% \cdot 100 = 6$

It will take 12 years to double the investment.

Nelson decided to check this result using the compound interest formula. If the interest is compounded quarterly, then the rate for each period is

$r = 0.06 \div 4 = 0.015$

If he invests the money for 12 years, then the number of periods is

$n = 12 \cdot 4 = 48$

Substitute in the compound interest formula.

$B = p(1 + r)^n$ Compound interest formula

$B = 10,000(1 + 0.015)^{48}$ Use a calculator.

$B = 20,435$ To the nearest dollar

Since 20,435 is approximately 2 • 10,000, the result is confirmed; $10,000 has approximately doubled to $20,435.

EXAMPLE 4 Next, Nelson decided to continue the doubling to see how many 12-year periods he needed to reach $1,000,000.

QUESTION How long will it take his investment to be worth more than 1 million dollars?

SOLUTION

He noticed that in 12 years, the value of the investment had grown to about 2.04 times its earlier value (20,435 ÷ 10,000 = 2.0435). This was not exactly a double, but since he did not need an exact result, he believed that he could get a very good estimate by using 2 instead of 2.04. He made a table to keep track of the doublings.

Doublings	Years Passed	Balance
1	12	$ 20,435
2	24	40,870
3	36	81,740
4	48	163,480
5	60	326,960
6	72	653,920
7	84	1,307,840

It will take 7 doublings or 84 years to reach one million dollars. If Nelson is to become a millionaire through investments, he should either invest more money to begin with or find an investment that will pay a higher rate of return.

─────── **SKILL 4** ───────────────────

EXAMPLE 5 Nelson has given up on the idea of becoming a millionaire quickly. Instead, he will invest $20,000 in an investment plan that he hopes will provide 10% interest, compounded quarterly.

QUESTION Nelson now wants to have $100,000 in his investment plan by the time he retires. How long will that take?

SOLUTION

Nelson substitutes all the values he can into the compound interest formula. The current value of his principal p is $20,000, the balance B that he hopes to have in the future is $100,000, and the quarterly rate of interest r that he hopes the investment will pay is 0.10 ÷ 4 = 0.025.

$$B = p(1 + r)^n$$
$$100,000 = 20,000(1 + 0.025)^n$$
$$5 = 1.025^n$$

Use the compound interest formula.
$B = 100,000; p = 20,000; r = 0.025$
Divide each side by 20,000.

Nelson is not sure how to solve this equation for n. He decides to use his graphing calculator. For each side of the above equation he writes a new equation, using the variable x for n.

$$y = 1.025^x$$
$$y = 5$$

He graphs them using these range values.

Xmin: 0
Xmax: 95
Xscl: 10
Ymin: 0
Ymax: 10
Yscl: 1

$x = 66$ $y = 5.1024072$

The horizontal line is the graph of $y = 5$ and the steep, rising curve (called an **exponential curve**) is the graph of $y = 1.025^x$. Nelson moves the trace cursor along the curve until he gets as close as he can to $y = 5$ without being less than 5. He finds that when $x = 66$, $y = 5.1024072$, so he substitutes 66 for n in the compound interest formula.

$$B = 20,000(1.025)^{66}$$
$$B = 102,048 \qquad \text{To the nearest dollar}$$

Nelson's $20,000 investment will grow to over $100,000 when $n = 66$. Since n represents the number of *quarters* (not years), Nelson needs to divide by 4 to find the number of years that it will take for him to retire.

$$\text{Number of years} = 66 \div 4 = 16\tfrac{1}{2}$$

If the annual interest rate remains at 10%, then Nelson can retire with $100,000 in $16\frac{1}{2}$ years.

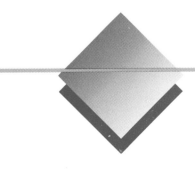

TRY YOUR SKILLS

Use a calculator to find the amount of interest and the new balance that will accumulate over two years on the following principal amounts at the given interest rate compounded as shown.

	Principal	Interest Rate	How Often Compounded	Interest Earned	New Balance
1.	$1000	6%	Annually	First period: _____	_____
				Second period: _____	_____
2.	$1000	6%	Semiannually	First period: _____	_____
				Second period: _____	_____
				Third period: _____	_____
				Fourth period: _____	_____

Suppose that in 1776, Benjamin Franklin and other founders of our nation each decided to invest a small amount of money for their descendants. Assume that they were able to find a bank willing to pay 4% interest, compounded semiannually for the indefinite future. Use the compound interest formula $B = p(1 + r)^n$ to find the amount that each investment will be worth after the indicated number of years have passed. Remember that for twice-a-year compounding the number of periods is twice the number of years and the interest rate for each 6-month period is $0.04 \div 2$, or 0.02.

3. Thomas Jefferson invested $100.00 for 30 years.

4. Paul Revere invested $50.00 for 90 years.

5. John Hancock invested $20.00 for 200 years.

6. Use the Rule of 72 and your calculator to find the year in which $1 that Benjamin Franklin invested in the year 1776 at 8% compounded quarterly would have become worth over $2000. Organize your results in a table like the one below.

Doublings	Years Passed	Balance
1		
2		
3		
4		
5		
6		

7. Assume that Nelson has $25,000 to invest and that he expects to earn 9% per year on his investment. The interest will be compounded quarterly. Use the Rule of 72 to approximate how long it will take for Nelson's money to double repeatedly until it has grown to more than 1 million dollars.

8. Use the compound interest formula and a graphing calculator as in Skill 4 to check your answer to Exercise 7. Which method do you think is easier? Which is more accurate? Tell which method you prefer, and give a reason for your answer.

EXERCISE YOUR SKILLS

KEY TERMS

compound interest
compounded
    quarterly
compounded
    semiannually
Rule of 72

Fill in the missing numbers. (Do not write in your book.)

1. Interest that is compounded annually is paid every ___?___ months.
2. Interest that is compounded semiannually is paid every ___?___ months.
3. Interest that is compounded quarterly is paid every ___?___ months.
4. 5% interest compounded annually means ___?___% paid every ___?___ months.
5. 6% interest compounded quarterly means ___?___% paid every ___?___ months.

Use a calculator to find the amount of interest and the new balance that will accumulate over two years on the following principal amount at the given interest rate compounded as shown. The first quarter is done for you as an example.

	Principal	Interest Rate	How Often Compounded	Interest Earned		New Balance
6.	$1000	6%	Quarterly	First period:	$15	_____
7.				Second period:	_____	_____
8.				Third period:	_____	_____
9.				Fourth period:	_____	_____
10.				Fifth period:	_____	_____
11.				Sixth period:	_____	_____
12.				Seventh period:	_____	_____
13.				Eighth period:	_____	_____

Suppose that in 1789, John Adams invested $7.50 in a bank at 4.5% compounded semiannually and stipulated in his will that the money was to be allowed to accumulate in the account after his death. Use the compound interest formula to find the value of the account in each case.

14. After 50 years   15. After 70 years   16. After 100 years

17. In Exercise 16, suppose that the interest rate is 6% instead of 4.5%. Find the value of the account after 100 years. Compare the two results. What conclusion do you draw from them?

Nelson has invested $5000 in a CD. The interest is compounded annually. Use the Rule of 72 to find how many years it will take for the investment to grow to $40,000 at the given rates of interest.

18. 2%   19. 6%   20. 8%   21. 10%

In the year that Nelson was born, Nelson's father invested $15,000 in an investment fund that has been paying 8% interest compounded quarterly. Use the graphing method outlined in Skill 4 to determine how long it will take for the $15,000 to grow enough to pay for Nelson's costs of attending college. Nelson estimates the costs of several colleges to be as given below.

22. $75,000   **23.** $45,000   **24.** $37,500   **25.** $52,500

26. Nelson plans to begin college at age 18. Which of the above amounts will be impossible to acquire by then?

27. At the age of 19, Robert placed $2000 into an investment at 6% compounded annually and added no money to the investment for the rest of his life. His friend Amos, who was the same age, waited until he was 30 to begin his investment program by also investing $2000 at 6% compounded annually. Later, at age 40, Amos invested another $2000 at 6% compounded annually, and still later, at age 50, he invested the final $2000 of his $6000 investment on the same interest terms. Use the compound interest formula to show that at age 60, when they both retire next year, Robert's investment will have grown to a larger amount than Amos's. How much will each investment be worth?

MIXED REVIEW

Find the commission earned and the total monthly earnings for each salesperson.

1. Paul, with a monthly base salary of $900 and a commission rate of 9% on sales of $6975

2. Gabe, with a monthly base salary of $2500 and a commission rate of 7% on sales of $7475

Suppose your bank statement shows a closing balance of $236.50 and the following are not on the statement: deposit: $80.00; check: $18.37; ATM withdrawal: $110.00. There is a service charge of $1.50 on the statement but not in your checkbook.

3. What amounts must be added to the bank statement balance?

4. What amounts must be subtracted from the bank statement balance?

5. What should be the balance in your checkbook before you do a reconciliation?

6. What should be the checkbook balance after you do a reconciliation?

7. Pauline thinks she will write about 120 checks a month in her new business. She expects to maintain an average balance of $1500. Which of the following banks offers her the better deal? Explain your answer.

Bank A: interest of 0.35% per month paid on the average balance and a charge of $0.06 for each check.

Bank B: no interest paid on the checking account and no charge for checks but a monthly service charge of $6.

Every year for as long as she can remember, Olivia has been receiving a birthday gift from her Aunt Millicent. Every November, a $20 check travels from California to Olivia's Indiana home. Last year Olivia's mother flew to California and stayed three weeks when Aunt Millicent had surgery.

Olivia knew that while in California, Mom needed more money than she had taken with her and that she was able to cash a check on an Indiana bank at Aunt Millicent's bank. Olivia thought that it was amazing that the California bank could somehow take the cash that Mom had in the bank in Indiana and give it to Mom in California in a matter of minutes. Dad said that it had something to do with the "Federal Reserve" and "computers" and

declared that "computers are taking over the world." Dad has a tendency to complain. Many times, he has told Olivia how much more you could buy with $20 when he was a kid.

Olivia herself has noticed that Aunt Millicent's birthday check doesn't buy as much as it used to. Olivia took that as a sign that she was growing up, since the larger-size clothes that she now needs to buy cost more than the clothes she wore when she was little. The way her father talks about inflation, taxes, government spending, and cost-of-living increases reminds her of something she heard about the Federal Reserve System in economics class. Olivia decided to find out more about how the Federal Reserve System works.

OBJECTIVES: In this lesson, we will help Olivia to:

- *Describe the organization of the Federal Reserve System.*
- *List four important functions of the Federal Reserve System.*
- *Describe how the Federal Reserve System controls monetary policy.*
- *Observe and calculate the multiplier effect on the nation's money supply.*

FUNCTIONS OF THE FEDERAL RESERVE SYSTEM

The **Federal Reserve System** consists of 12 Federal Reserve banks, 25 branch banks, a board of governors, and the Federal Open Market Committee, which directs the purchases and sales by the reserve banks of Federal government securities and other obligations in the open market. The following map of the Federal Reserve System shows the locations of the district banks and branches. Olivia has seen this map in her economics class. She noticed that she lives in District 7 and Aunt Millicent lives in District 12.

ORGANIZATION OF THE FEDERAL RESERVE SYSTEM

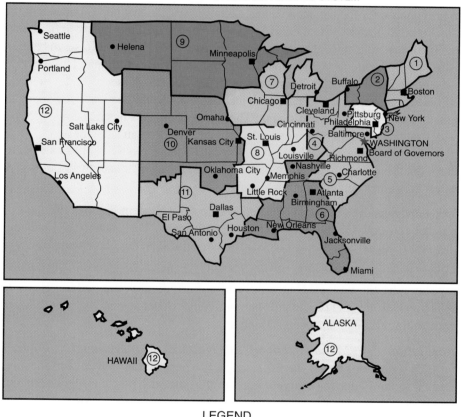

LEGEND

— BOUNDARIES OF FEDERAL RESERVE DISTRICTS

★ BOARD OF GOVERNORS OF THE FEDERAL RESERVE SYSTEM

● CITY WHERE A BRANCH OF FEDERAL RESERVE BANK IS LOCATED

— BOUNDARIES OF FEDERAL RESERVE BRANCH TERRITORIES

■ CITY WHERE FEDERAL RESERVE BANK IS LOCATED

③ FEDERAL RESERVE DISTRICT NUMBER. THIS NUMBER APPEARS ON THE CURRENCY ISSUED BY THE FEDERAL RESERVE BANK IN THE DISTRICT

The important functions of the Federal Reserve System are the following:

1. To act as a central bank.
2. To serve as a bank for the U.S. government.
3. To supervise financial institutions.
4. To regulate and manage the nation's money supply.

The Nation's Central Bank   The Federal Reserve System issues the country's main type of currency, the **Federal Reserve note.** Such notes are **legal tender;** that is, they are money that by law must be accepted for payments of debts and taxes. This currency is printed by the Bureau of Printing and Engraving.

In addition to issuing currency, the Federal Reserve banks act as a major clearinghouse for the collection and return of checks so that depositors' accounts are credited quickly. When Olivia deposits Aunt Millicent's check in her bank, the bank transmits this information electronically to the Federal Reserve bank in its district, District 7, which includes Olivia's town in Indiana. This Federal Reserve bank sends the information to the Federal Reserve bank in Aunt Millicent's district, District 12, which includes California. The District 12 bank in turn contacts Aunt Millicent's bank. The use of computers has made this process very fast, and the banks can now complete such transfers in a day or two.

In its role as a central bank the Federal Reserve System supplies cash to meet the temporary needs of its member banks, for instance during holidays when consumer demand for cash is high.

The Government's Bank   The Federal Reserve System handles the flow of income and expenditures for the federal government. The issuance of Medicare payments, for instance, may come from a Federal Reserve bank near you.

Supervision of Financial Institutions   All member banks are supervised and regulated by the Federal Reserve System. Periodically, the Federal Reserve System examines the records of member banks to determine whether they are conforming to Federal Reserve standards.

Regulation and Management of the Nation's Money Supply   Olivia's father's comments about the decrease in the purchasing power of money are related to the Federal Reserve's most important function: the regulation and management of the nation's **money supply**—the total amount of coins, paper currency, and demand deposits (checking-account money) in circulation in the economy. The Federal Reserve System tries to maintain the right amount of money in circulation. If there is too much money, **inflation** results; that is, the overall purchasing power of money declines. If the money supply is too low, economic activity will diminish and may even lead to a recession and an increase in unemployment.

If business activity is slowing down and unemployment does begin to rise, the Federal Reserve may try to expand the supply of money and credit. To do this, it will encourage financial institutions to borrow short-term money from the Federal Reserve banks at its relatively low interest rates. With such low rates, businesses will be encouraged to borrow money to expand their operations and wage earners will be encouraged to borrow money to build more homes and buy more goods. On the other hand, suppose that business activity expands too rapidly. This will be characterized by some positive things such as full employment but also by rising prices and too much borrowing by both businesses and consumers. Then the Federal Reserve banks will try to cool off the economy by reducing the money and credit supply. If the money supply is limited in this way, then interest rates will rise, and businesses and individuals will be discouraged from borrowing.

By maintaining a balance in the amount of money in circulation, the Federal Reserve tries to ensure that credit is plentiful enough to allow expansion of the economy but not so plentiful that rapid inflation occurs. A rough indication of the rate of inflation is provided by the **Consumer Price Index (CPI),** a statistic that is calculated monthly by the Bureau of Labor Statistics. This index reflects the price of a specific group of goods and services used by the average household. This "market basket" consists of about 400 goods and services in the areas of food, housing, transportation, clothing, entertainment, and medical and personal care. (You will learn more about the CPI in Chapter 9.)

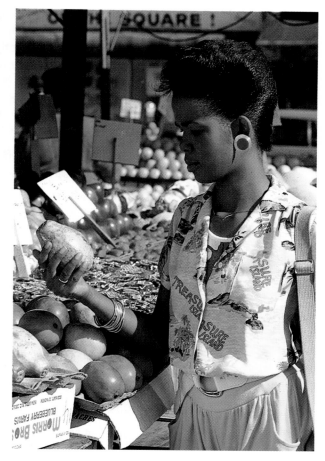

In summary, by controlling the money supply the Federal Reserve System affects the prices that consumers pay for goods and services. When interest rates are low and the money supply increases, people spend more freely. Suppliers can more easily raise their prices because demand for goods and services is greater than their supply. Thus higher inflation may occur. On the other hand, when interest rates are high and money is tight, people buy fewer goods and services. Therefore merchants have an oversupply of inventory and must lower their prices to attract buyers.

THE CREATION OF DEPOSITS

By law, depository institutions (banks) must deposit a specified percent of their customers' deposits with the Federal Reserve System. Such a deposit is known as a **required reserve.** It serves as a safeguard against a financial institution's placing too much of its customers' deposits into various investments. These investments might range from simple loans to the purchase of government securities.

The Federal Reserve can exert several minor controls on deposits in financial institutions; these controls are quite powerful tools. However, the real strength that the Federal Reserve exerts is due to its control over an institution's required reserves. This is due to the **multiplier effect** that these reserves create throughout the banking system.

For any given *demand deposit* (money in a checking account) a financial institution has the right to invest or extend loans from the **excess reserves.** As an example, assume that the reserve requirement is 20%; that leaves 80% in excess reserves. This is illustrated below by the simple balance sheet for a $1000 demand deposit. The financial institution actually creates an extra $800 of money or credit. This is done by lending the $800 rather than holding it in the vault as a reserve.

INITIAL DEPOSIT			
Assets		**Liabilities**	
Required reserves (20%)	$200	Demand deposit	$1000
Loans and investments (80%)	800		
Total	$1000	Total	$1000

If the $800 that is used in making loans is deposited by the borrower in the same or other institution, still more money or credit can be created. The second deposit creates an extra $640 of money or credit, as shown below.

SECOND DEPOSIT			
Assets		**Liabilities**	
Required reserves (20%)	$160	Demand deposit	$800
Loans and investments (80%)	640		
Total	$800		$800

We can also expect this $640 to be deposited by the new borrower. The multiplier effect continues as shown below.

THIRD DEPOSIT			
Assets		**Liabilities**	
Required reserves (20%)	$128	Demand deposit	$640
Loans and investments (80%)	512		
Total	$640	Total	$640

The cycle continues until there is not a penny left to deposit. By this time the original $1000 deposit will have enabled close to $4000 in new money to be created.

MANAGING THE MONEY SUPPLY

The Federal Reserve System uses the reserve requirements to manage the money supply. If the reserve requirement is increased, then the multiplier effect for creating new money or credit is reduced. This is known as a **tight-money policy.** If the reserve requirement is reduced, then the multiplier effect is increased. This is known as an **easy-money policy.**

Economic conditions can prevent the multiplier effect from reaching its full force. For example, if the economy is in a period of high inflation and interest rates are also quite high, businesses and consumers may not wish to borrow money. In this case the lack of demand for money means that not all the excess reserves available will be loaned out, and the multiplier effect will be reduced. If the economy is in a period of recession, businesses and consumers may be afraid to borrow even though interest rates are low. This, too, will tend to keep the multiplier effect from reaching its full potential, since financial institutions will not be able to make the loans that they would like. The multiplier effect can reach its full potential in creating new money and credit only when the demand for loans equals the supply of money available for those loans.

Another condition that can keep the multiplier effect from reaching its full potential is personal behavior regarding money and financial institutions. Some people do not trust depository institutions but keep their money at home or in safe-deposit boxes. If this money does not get into circulation (through spending or depositing), then the multiplier effect will be diminished.

Olivia and her brother Orson will certainly do their part to keep their money in circulation. They aren't very big savers; they spend whatever comes their way and wherever they can, for example, at the local music store. Perhaps the owner of the store will deposit its cash receipts in his checking account at the local bank, thus allowing the bank to multiply that money four or five times.

Ask Yourself

1. What states make up District 12 of the Federal Reserve System?

2. What are four important functions of the Federal Reserve System?

3. What is legal tender?

4. How does the Federal Reserve System try to expand the money supply?

5. What is a reserve requirement?

6. What is the difference between a tight-money policy and an easy-money policy?

7. What is the multiplier effect?

8. Describe the Federal Reserve System.

ALGEBRA REVIEW

For each series (sum of two or more terms), tell whether there is a common ratio between successive terms. If there is, give that ratio.

Examples

a. $1 + 2 + 4 + 8 + 16$
$2 \div 1 = 2$
$4 \div 2 = 2$
$8 \div 4 = 2$
$16 \div 8 = 2$
Yes; 2

b. $1 + 2 + 3 + 4$
$2 \div 1 = 2$
$3 \div 2 = 1.5$
No

1. $8 + 4 + 2 + 1 + \dfrac{1}{2}$

2. $x + 4x + 8x + 16x$

3. $0.04 + 0.08 + 0.12 + 0.16$

4. $0.01 + (0.01)(0.01) + (0.01)(0.01)(0.01)$

5. $0.04 + (0.04)^2 + (0.04)^3$

6. $0.3 + 2(0.3) + 3(0.3) + 4(0.3)$

7. $100(0.07)^2 + 100(0.07)^3 + 100(0.07)^4 + 100(0.07)^5$

SHARPEN YOUR SKILLS

SKILL 1

EXAMPLE 1 Olivia wants to find out the multiplier effect on an initial deposit of $1000 through five levels.

QUESTION What is the total amount of extra money or credit created through five levels of the multiplier effect?

SOLUTION

Olivia made the following table to show the first five levels and the total additional money available for loans and investments after five levels. The required reserve is 20% of the demand deposit.

$$0.20(1000.00) = 200.00$$

The amount of loans and investments is the demand deposit minus the required reserves.

$$1000.00 - 200.00 = 800.00$$

At each level beginning with Level 2, the demand deposit is the loan amount from the previous level.

Level	Demand Deposit	Required Reserve (20%)	Loans and Investments
1	1000.00	200.00	800.00
2	800.00	160.00	640.00
3	640.00	128.00	512.00
4	512.00	102.40	409.60
5	409.60	81.92	327.68
		Total after 5 levels	2689.28

This shows how the money supply will multiply if the initial $800 loan is all redeposited in the banking system and if the resulting $640 loan also is all redeposited, and so on through five levels of loans and redeposits. The result is the creation of $2689.28 that did not exist before the inital $1000 deposit.

SKILL 2

EXAMPLE 2 Olivia wants to use a spreadsheet program to find out the multiplier effect on an initial deposit of $1000 through ten levels.

QUESTION What is the total amount of extra money or credit created through ten levels of the multiplier effect?

SOLUTION

Olivia used formulas in the spreadsheet program to give the numerical values shown in the table in Example 1. The spreadsheet shows the actual values (not the formulas). Thus in cell D5 you will see "512.00," not "+B5−C5."

	A	B	C	D
1	Level	Demand	Required	Loans and
2		Deposit	Reserve (20%)	Investments
3	1	1000.00	@round(0.2*B3,2)	+B3−C3
4	2	+D3	@round(0.2*B4,2)	+B4−C4
5	3	+D4	@round(0.2*B5,2)	+B5−C5

In Column C, Olivia used a "rounding" function (often expressed as @ROUND) so that the spreadsheet would use the rounded value in the remaining calculations. If she did not do this, a calculated value would not actually be rounded; it would only be displayed as rounded and small errors would be created. For cell C3, Olivia typed @ROUND (0.2*B3,2) to obtain a value rounded to two decimal places.

Olivia used the spreadsheet's COPY command to extend the formulas created in Rows 3 and 4 to the other rows. The spreadsheet creates the formulas for the new rows but displays the values for those formulas.

To find the total amount of money created, Olivia used the spreadsheet's "sum" function (often expressed as @SUM) to add the entries of Column D from Row 3 to Row 12 and display that total in Column D, Row 13. For cell D3, Olivia typed @SUM(D3 . . D12).

	A	B	C	D
1	Level	Demand	Required	Loans and
2		Deposit	Reserve (20%)	Investments
3	1	1000.00	200.00	800.00
4	2	800.00	160.00	640.00
5	3	640.00	128.00	512.00
6	4	512.00	102.40	409.60
7	5	409.60	81.92	327.68
8	6	327.68	65.54	262.14
9	7	262.14	52.43	209.71
10	8	209.71	41.94	167.77
11	9	167.77	33.55	134.22
12	10	134.22	26.84	107.38
13			Total after 10 levels	3570.50 @SUM(D3..D12)

The total amount of extra money or credit created through ten levels of the multiplier effect is $3570.50.

If you extend the spreadsheet even further, to 20 levels, 30 levels, and so on, you will create more and more money as summarized below.

Level	Amount of New Money Created
5	2689.28
10	3570.50
20	3953.88
30	3995.05
40	3999.47

Even though the deposits and redeposits continue to ever higher levels without end, the total amount of new money created by the initial $1000 deposit seems to have a limiting value of $4000. In other words, it seems that the sum

$$800 + 640 + 512 + 409 + \ldots$$

with an infinite number of terms (indicated by the three dots at the end) has a finite sum, $4000. This is actually the case.

The sum can also be written as

$$800 + 800(0.8) + 800(0.8)^2 + 800(0.8)^3 + 800(0.8)^4 + \ldots$$

You can see that the ratio of each pair of successive terms is always the same, namely, 0.8. Any series that has a **common ratio** (such as 0.8) is called a **geometric series.** If the series has an infinite number of terms and if it has a sum (many series do not), then that sum is called the **sum of an infinite geometric series.** The series above *is* infinite and *does* have a sum. The formula for the sum of an infinite geometric series is

$$S = \frac{a}{1 - r}$$

where a is the first term 800 and r is the common ratio 0.8. Since $1 - 0.8 = 0.2$, the expression $1 - r$ is the reserve requirement 0.2.

EXAMPLE 3   Olivia wants to find out the maximum multiplier effect on an initial deposit of $1000.

QUESTION   What is the maximum amount of money a $1000 deposit can create and what is the multiplier?

SOLUTION
For a $1000 deposit, initial loan amount of $800, and 20% reserve requirement,

$$S = \frac{a}{1-r} \qquad a = 800;\ 1 - r = 0.2$$

$$S = \frac{800}{0.2} \qquad \text{Reserve requirement} = 0.2$$

$$S = 4000$$

The original $1000 deposit can create as much as $4000 in new money. To find the multiplier add the amount created to the original amount and divide by the original amount.

$$\frac{1000 + 4000}{1000} = 5$$

The multiplier is 5.

TRY YOUR SKILLS

Make a table to show the multiplier effect on the amounts below if the Federal Reserve requirement is 20%. Show the first ten levels and find the total extra money that is generated.

1. Initial deposit: $500

2. Initial deposit: $2500

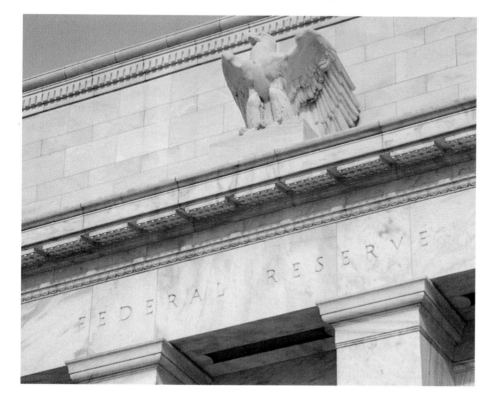

Make a table or use a spreadsheet to show the multiplier effect on the amounts below if the Federal Reserve requirement is 25%. Show the first ten levels and find the total extra money that is generated.

3. Initial deposit: $500

4. Initial deposit: $2500

5.–8. For the initial deposits of Exercises 1–4, find the maximum amount of new money created by the multiplier effect.

9.–12. What is the multiplier for Exercises 5–8?

13. Consider your answers to Exercises 1–12. How does changing the reserve requirement affect the extra money generated?

EXERCISE YOUR SKILLS

1. Does one bank actually send cash to another bank across the country to cash a check drawn on an account there? Why or why not?

2. What happens if there is too much money in circulation? too little money in circulation?

3. How does the multiplier effect work?

4. How can the Federal Reserve reduce or increase the multiplier effect?

Make a table or use a spreadsheet to show the multiplier effect on the amounts below if the Federal Reserve requirement is as given. Show the first ten levels and find the total extra money that is generated.

5. Reserve requirement: 15%; initial deposit: $500

6. Reserve requirement: 15%; initial deposit: $2500

7. Reserve requirement: 15%; initial deposit: $12,500

8. Reserve requirement: 18%; initial deposit: $12,500

9. Reserve requirement: 20%; initial deposit: $12,500

10. Reserve requirement: 25%; initial deposit: $12,500

11.–16. For the initial deposits of Exercises 5–10, find the maximum amount of new money created by the multiplier effect.

17.–22. What is the multiplier for Exercises 11–16?

23. Consider your answers to Exercises 1–22. How does changing the reserve requirement affect the extra money generated?

24. Find the maximum amount of extra money that can be generated by a savings account deposit of $4000 if the Federal Reserve System's reserve requirement is 18%.

25. Find the maximum amount of extra money that can be generated by a savings account deposit of $1000 if the Federal Reserve System's reserve requirement is 16%.

An employer has offered two of his salespeople the two options shown. What is the least amount of sales that each person would need for Option 2 to be the better choice?

1. Ellen: Option 1 is a base salary of $1224 and an 8% commission on all sales; Option 2 is no base salary and 12% commission on all sales.

2. Peter: Option 1 is a base salary of $724 and a 10% commission on all sales; Option 2 is no base salary and 12% commission on all sales.

For the given weekly savings, find out how much Dwain can save and how long he will have to save to be able to afford a business calculator that usually sells for $209.00 but that is on sale for $179.99. The sale ends nine weeks from now.

3. $50 per week
4. $25 per week

5. $20 per week
6. $15 per week

7. How much will $15,500 be worth at the end of three years if it earns 7% interest, compounded quarterly?

8. How much will $4900 be worth at the end of five years if it earns 4% interest, compounded semiannually?

9. Suppose that your checkbook shows a balance of $290.56. Your bank deducts a service charge of $2.50 each month but does not pay interest and does not charge for each written check. The only check that you wrote during the month was for $80.60 and was mailed the day before the bank's monthly statement arrived. What will the bank show as the closing balance?

Use the Rule of 72 to determine how long it will take your $3300 savings account to double in value if it is growing at a rate of

10. 4% per year
11. 6% per year

12. 10% per year
13. 12% per year

Susan's paycheck is for $998.43. There was an amount of $276 withheld for income taxes as well as 7.65% of the gross salary withheld for FICA taxes.

14. Let x represent the gross salary and write an equation for the gross salary.

15. What is Susan's gross salary?

16. How much money is withheld in all?

How much can be saved over the following time periods if $45 can be saved each week?

1. 4 weeks

2. 12 weeks

3. 26 weeks

4. 1 year (52 weeks)

How long will it take to save $165 if you can save the following amounts each week? What is the total amount saved in that time period?

	Amount Per Week	Number of Weeks	Total Amount Saved
5.	$15		
6.	25		
7.	50		

8. Ivan can save $30 a week to buy one or more of the following items.

A watch for $111.99
A camera for $199.99
A VCR for $260.00
A computer monitor for $309.00

Draw a graph or use a spreadsheet graphing program to show the number of weeks required to save up for each item.

A bank pays 3% simple interest on a one-year CD. Find out how much is saved, how much interest is earned by the CD in one year, and the total amount of money in each account after the bank has credited the interest to the accounts.

	Amount Saved per Week	Number of Weeks	Total Saved	Interest Earned	Total in Account
9.	$ 25.00	52			
10.	32.50	48			
11.	68.00	50			
12.	110.00	40			

Use a calculator to find the amount of interest and the new balance that will accumulate over two years on the following principal amount at the given annual interest rate compounded semiannually.

	Principal	Interest Rate	Interest Earned	New Balance
13.	$2000	5%	First period: _____	_____
14.			Second period: _____	_____
15.			Third period: _____	_____
16.			Fourth period: _____	_____

Use a calculator to find the amount of interest and the new balance that will accumulate over two years on the following principal amount at the given annual interest rate compounded quarterly.

	Principal	Interest Rate	Interest Earned		New Balance
17.	$2000	5%	First period:	_____	_____
18.			Second period:	_____	_____
19.			Third period:	_____	_____
20.			Fourth period:	_____	_____
21.			Fifth period:	_____	_____
22.			Sixth period:	_____	_____
23.			Seventh period:	_____	_____
24.			Eighth period:	_____	_____

25. Use the compound interest formula to find out how much $2000 will be worth at the end of two years, at an annual interest rate of 5% compounded semiannually and compounded quarterly. Do your results agree with Exercises 16 and 24?

Suppose that in 1750, when he was 44 years old, Benjamin Franklin invested 50 pounds (£50) in a bank at 3.5% compounded semiannually and stipulated in his will that the money was to be allowed to accumulate in the account after his death. Use the compound interest formula to find the value of the account in each case.

26. After 5 years   27. After 50 years   28. After 75 years

Peggy's grandmother opened a savings account for her that pays interest compounded semiannually. Use the Rule of 72 to find how many years it will take for the original principal of $500 to double at the given rates of interest.

29. 3%   30. 4%   31. 8%   32. 10%

Make a table to show the multiplier effect on the amounts below if the Federal Reserve requirement is as given. Show the first five levels, and find the total extra money that is generated.

33. Reserve requirement: 20%; initial deposit: $2000

34. Reserve requirement: 25%; initial deposit: $1000

35. Reserve requirement: 17%; initial deposit: $11,500

36. Reserve requirement: 15%; initial deposit: $25,000

37.–40. For the initial deposits of Exercises 33–36, find the maximum amount of new money created by the multiplier effect.

41.–44. What is the multiplier for Exercises 37–40?

45. Find the maximum amount of extra money that can be generated by a checking account deposit of $800 if the Federal Reserve System's reserve requirement is 18%.

How much can be saved over the following time periods if $75 can be saved each week?

1. 4 weeks
2. 12 weeks
3. 26 weeks
4. 1 year (52 weeks)

How long will it take to save $355 if you can save the following amounts each week? What is the total amount saved in that time period?

	Amount Per Week	Number of Weeks	Total Amount Saved
5.	$20		
6.	30		
7.	60		

A bank pays 4% simple interest on a one-year CD. Find out how much is saved, how much interest is earned by the CD in one year, and the total amount of money in each account after the bank has credited the interest to the accounts.

	Amount Saved per Week	Number of Weeks	Total Saved	Interest Earned	Total in Account
8.	$ 35.00	52			
9.	42.50	52			
10.	96.80	48			
11.	125.00	46			

Suppose that you have a savings account with $8500 in it. It pays 7% interest compounded as shown below. Find the value at the end of each year for the next four years.

12. Annually
13. Semiannually
14. Quarterly
15. Monthly

Use the Rule of 72 to find how many years it will take for $7000 to double at the given rates of interest.

16. 9%
17. 3%

A $3000 deposit is made in a checking account when the Federal Reserve System's reserve requirement is 16%.

18. Make a table to show the multiplier effect on the deposit. Show the first five levels, and find the total extra money that is generated.

19. Find the maximum amount of extra money that can be generated.

20. What is the multiplier?

For each company fringe benefits policy described below, find

    a. the amount that the company must pay for extra taxes, retirement, health and life insurance, and other indicated fringe benefits.

    b. how much salary is paid for the indicated number of nonworking days in a total of 240 operating days.

	EMPLOYEE	SALARY	FRINGE BENEFITS POLICY
1.	Office manager	$45,000	7.65% of gross for FICA taxes;
			4% of gross for retirement;
			$125/month for life and health insurance;
			24 nonworking days
2.	Restaurant cashier	$12,000	7.65% of gross for FICA taxes;
			$80/month for health insurance;
			10 holidays, 4 sick days, no vacation days

Basil's earnings as a real-estate salesperson are based on a graduated commission. He receives 4% on each sale up to $125,000 and 6% of the amount above $125,000. How much would he earn for selling each house below?

3.

PRICED TO SELL
$108,500 Imagine yourself in this 2000 square foot home.

4.

LOW-MAINTENANCE YARD
$139,950 Large living room, formal dining and breakfast area, 4 bedrooms, 2 baths.

5. Suppose you wrote 35 checks in July. Your bank charges $0.035 per check for the first 25 checks and $0.15 for each check over 25. How much were your July bank charges?

6. Find the maximum amount of extra money that can be generated by a checking account deposit of $2200 if the Federal Reserve System's reserve requirement is 18%.

7. How much will $9500 be worth at the end of five years if it earns 8% interest, compounded quarterly?

8. How much will $22,500 be worth at the end of five years if it earns 4% interest, compounded semiannually?

9. Suppose that you wrote a check for $243.50. Before writing the check, you had $982.61 in your checking account. Show how you would enter this check in your check register.

Peggy's grandmother opened a savings account for her that pays interest compounded semiannually. Use the Rule of 72 to find about how many years it will take for the original principal of $500 to double at the given rates of interest.

10. 3%   **11.** 4%   **12.** 8%   **13.** 10.5%

PROJECT 3–1: Savings Accounts

As you know, there are many different kinds of savings accounts. These accounts may vary greatly. Visit local banks and savings and loan associations to find information on various types of savings accounts.

1. Name of institution
2. Types of accounts available
3. Rate of interest for each type of account
4. Minimum balance requirements
5. How interest is compounded
6. When interest is paid

PROJECT 3–2: Federal Reserve System

As you learned in this chapter, the Federal Reserve System is a very important agency. It acts as a fiscal agent for the U.S. government. The Federal Reserve banks regulate our economy by regulating the money supply.

Gather information about the Federal Reserve System by doing research at your local library, by interviewing bankers, and/or by reading news magazines and newspapers. Answer the following questions.

1. What Federal Reserve District do you live in?
2. How does the Federal Reserve System actually process your checks?
3. What is the current condition of the money supply?
4. Is the Federal Reserve currently trying to expand or reduce the money and credit supply?
5. What are the current implications of the above activity?
6. What is the current required reserves for banks?

Remember that the *slope m* of a line is found by taking any two points on the line and dividing the change in y by the change in x.

$$m = \frac{y_2 - y_1}{y_2 - x_1}$$

For the graph and points shown the slope m is

$$m = \frac{5 - (-1)}{-2 - 3} = \frac{6}{-5} = -\frac{6}{5}$$

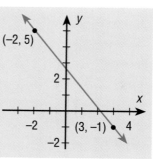

In an equation of the form $y = mx + b$, m is the slope and b is the y-intercept. Use the slope-intercept method to graph each equation.

Example $2y + 3x = 12$

$2y = -3x + 12$ Solve for y.

$y = -\frac{3}{2}x + 6$

When $x = 0$, $y = 6$. The y-intercept is 6.

The slope is the coefficient of x, $-\frac{3}{2}$.

1. $y = 2x - 3$

2. $3y = 2x - 6$

3. $2x - y = 10$

4. $x = 10 - 5y$

5. $3y + 2x = -12$

6. $y - x = 0$

To graph a *linear inequality,* first graph the related equation, then test points to see which side of the line satisfies the inequality. To graph several inequalities, shade the regions for each one and see where they overlap. Remember to include the line of the equation in the solution for ≤ or ≥.

Find the solution of each system.

Example $y \le 2x + 3$ (1)
             $y > -x + 1$ (2)

Equation (1) has slope 2 and y-intercept 3. Equation (2) has slope -1 and y-intercept 1. The solution is the points in the green region, including all points on or below $y = 2x + 3$ and all points above but not on $y = x + 1$.

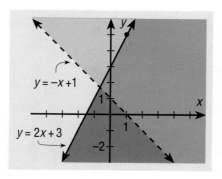

7. $y \le 3x - 4$
    $2 \ge -x$

8. $2x + 3y \le 12$
    $x - y \ge 5$

9. $2y < x - 4$
    $y > 2x - 2$

10. $x + y > 0$
     $x + y < 0$

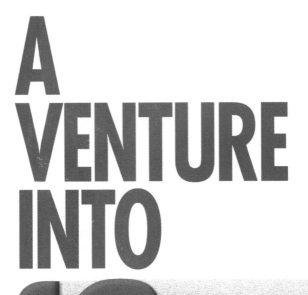

CHAPTER 4

A VENTURE INTO $ BUSINESS

EVELYN AND THREE OF HER FRIENDS have decided to operate a small business to earn money for college. They have read books with suggestions about everything from house– and pet–sitting to selling popcorn and soft drinks at ball games to singing telegrams and printing mailing labels.

They now realize that there is a lot to think about in starting a business. Will there be an interest in and demand for the product or service that they offer? Can they do the work efficiently enough to make money? What will be the costs of producing and marketing their service or product? How will they keep track of their sales revenue, costs, and profit? This chapter is about how the group of students answers these and other business questions.

In the process of starting their business, the students learn about payroll records and deductions for different kinds of taxes. They also discover that there are many costs and expenses that they had not previously considered. All that they learn will be useful to them in the future, whether they work for a small or large company, whether they work for a private business such as a store or a public agency such as the post office. Careful planning and care with expenses are always important.

Evelyn is a natural take-charge person and entrepreneur. She first thinks about starting a word-processing business. She has identified a market among fellow students who do not have access to computers or who do not type. She has already done some typing for people.

Greg will help with the typing. He is a design student who brings other talents to the business. He has already painted designs on sweatshirts and sold them. As it turns out, this leads to a more successful business than word processing.

Freda, a friend of Evelyn and Greg, is brought into the business because of her marketing ability. She can sell anything. Her record-keeping and financial skills are not as impressive.

They all need Hari, who can use a spreadsheet program to calculate the effect of different costs and sales on revenue and profit. He can find the break-even point and use linear programming to find how different factors affect the business. This will help the group decide how much to produce and what price to charge.

4-1 EMPLOYEES ARE PART OF THE COMPANY

Evelyn is a junior in high school. She and her friends have been studying business principles in their economics class. They have several ideas for services or products that they could sell without having to invest large sums of money.

One of these services is word processing. Evelyn's family bought a computer several years ago, and Evelyn has used it as a word processor to prepare reports and essays for school. She has also learned how to prepare a spreadsheet and has experimented with some functions that the spreadsheet program enables her to do. She realizes that the computer should help her in several ways as she considers starting a business. It could be used to provide a service, and it could also be used to organize her company's records.

Some of her classmates do not have access to a word processor. Or if they do have one, they do not know how to type. Perhaps they would be willing to pay to have research papers professionally typed and printed on a word processor. What is more, there is a college nearby that Evelyn now sees as being full of potential customers. There is no end to the term papers and research papers college students must write. Word processing is clearly a business worth considering.

If Evelyn and her friends decide to offer this service, they will have to charge a fair and reasonable price so that it will be worth the time and effort they will have to spend. Managing any business requires making many decisions. Evelyn will investigate some of these decisions.

Examine some of the functions of management in a business.

Prepare a payroll showing wages, salaries, deductions, and take-home pay for employees in a small company.

Use a spreadsheet to display payroll information.

LABOR

If Evelyn and her friends are going to invest their physical effort, mental effort, and skills in a new venture, they want to see a return on their investment. The efforts and skills that they invest are their **labor.** The word *labor* is also used for those workers who do not share in the ownership or the executive decision making of the company they work for. In a small company like Evelyn's, all the employees perform the functions of labor, management, and ownership. Even in very large firms, labor is sometimes represented on management boards. This is because workers have unique experience and insight into the production process.

MANAGEMENT

Management is one key to a successful business operation. Suppose Evelyn comes up with the best idea there is and acquires the most intelligent and industrious labor force. She cannot have a successful business without good organization and management. In all businesses, plans must be made, workers hired, materials obtained, and equipment purchased or built. Putting all these factors together is called **entrepreneurship.** This is the role Evelyn is playing now as she envisions a business and seeks to make it real.

MARKET ECONOMY

As manager, Evelyn must be aware that it is her responsibility to sell the best product at a fair price and still make the highest profit possible. In a **market economy,** profit is the incentive, or reason, for the **producers** to satisfy the wants of **consumers,** or users of the service or product. The owners of property are free to determine how they can most efficiently use available resources. Evelyn's competition—that is, other businesses that offer the same product or service—may be more efficient. She must find out what competition she has.

In a market economy, consumers influence what will be produced by the way they spend their income. Of course, businesses try to influence consumer demand through promotion and advertising, but the final decision is made by the consumers. Consumers decide whether or not to buy, in what quantities, and at what price. If consumers cannot obtain what they want from one producer, there is usually another producer that is happy to satisfy them.

If a product or service is of high quality, it will usually sell for a higher price than one of low quality. In other words, in a competitive market we pay for our goods and services according to how well they satisfy our needs and wants.

Evelyn and her friends are thinking about starting a word-processing business. Evelyn and her friend Greg would do the typing. Both are good typists, and Greg has learned how to use Evelyn's computer. Their friend Freda would be responsible for advertising. She would deliver flyers to neighborhood homes and students at school. She would also contact people at the community college and ask permission to put up posters in the student recreation center and some of the classroom buildings. Another friend, Hari, would be in charge of the finances. He has a special skill with figures and can show Evelyn how to set up their records.

They hope to provide a high-quality service at a competitive price and still be able to pay reasonable wages and manage their other resources as well.

Ask Yourself

1. What incentives do businesses have to satisfy the wants of consumers?

2. How do consumers influence what will be produced?

3. How do businesses attempt to influence consumers in their choices?

ALGEBRA *REVIEW*

Graph the line through each set of points. Give the equation of the line and its slope.

1. $(-2, -5), (0, -1), (2, 3)$

2. $(-1, 0), (2, 3), (5, 6)$

3. $(3, -2), (1, 0), (-2, 3)$

4.
x	3	7	9
y	-3	-7	-9

5.
x	-1	0	1
y	2	3	4

6.
x	-3	-1	1
y	7	3	-1

For Exercises 7–10, let x represent hours worked and y represent wages. Write an equation for wages received at the given rate.

7. $6.50 per hour

8. $5.20 per hour

9. Find the slope of the graph of the equation representing wages received at a rate of $7.25 per hour.

10. Find the hourly rate and equation for the wages received using the following values.

x	18	25	40
y	$89.10	$123.75	$198.00

SHARPEN YOUR SKILLS

EXAMPLE 1 Evelyn plans to keep accurate records of the wages earned by each person in the business. The first record will be a weekly listing of the gross pay for the employees. This list is shown in the spreadsheet developed by Evelyn and Hari. Each heading on the spreadsheet contains important information. Greg and Hari are supporting themselves, so that they each can claim one exemption and pay at lower tax rates than Evelyn and Freda, who are supported by their parents and do not claim themselves as exemptions. Taxes also depend on marital status.

	A	B	C	D	E	F
	Employee	Number of	Marital	Hourly	Hours	Gross
		Exemptions	Status	Rate	Worked	Pay
1						
2						
3	Evelyn	0	S	6.50	14	91.00
4	Greg	1	S	6.50	12	78.00
5	Freda	0	S	5.50	16	88.00
6	Hari	1	S	5.50	4	22.00
7	Total					279.00

PAYROLL REGISTER FOR WEEKLY EARNINGS

+D3∗E3

@SUM(F3..F6)

QUESTIONS **a.** What algebraic formula can be used to calculate gross pay?
b. What formula should be used to calculate gross pay for the spreadsheet?

SOLUTION
To calculate gross pay, multiply hourly rate by the number of hours worked.

a. The algebraic formula is

$$p = r \cdot h$$ where p = gross pay
r = hourly rate
h = number of hours worked

For Evelyn, $p = 6.5 \cdot 14 = 91$. Her gross pay is $91.00.

b. For the spreadsheet calculation, use the formula +D3∗E3, in cell F3. Then copy this formula in the F column for the gross pay of the other employees. To find the total gross pay in Row 7, use the following sum function.

@SUM(F3..F6)

---- **SKILL 2** ----

EXAMPLE 2 The next record that Evelyn and Hari prepare is for payroll deductions and take-home pay.

QUESTION How would you set up a spreadsheet for payroll deductions?

SOLUTION

Evelyn and Hari must consult tax tables as shown in Lesson 1–4 to determine the federal income tax withholding. Use the table for single persons with a weekly payroll period in the Reference Section. For example, Evelyn earned $91 and claims 0 exemptions. The amount of withholding for her is $7.00, found in the 90–95 row and the zero column.

FICA withholding is found by taking 7.65% of gross pay. For Evelyn, $91 \cdot 0.0765 = 6.96$. Her FICA withholding is $6.96. A formula is entered into the spreadsheet to calculate the FICA withholding.

PAYROLL REGISTER FOR WEEKLY DEDUCTIONS

	A	B	C	D	E	F
		Gross	Income Tax	FICA	Total	Take-Home
1	Employee	Pay	Withholding	Withholding	Deductions	Pay
3	Evelyn	91.00	7.00	6.96	13.96	77.04
4	Greg	78.00	0.00	5.97	5.97	72.03
5	Freda	88.00	6.00	6.73	12.73	75.27
6	Hari	22.00	0.00	1.68	1.68	20.32
7	Total	279.00	13.00	21.34	34.34	244.66

+B3*0.0765
+C3+D3
+B3−E3

For Evelyn, the formulas to be entered in the spreadsheet are as follows.

FICA taxes in cell D3:	+B3*0.0765
Total deductions in cell E3:	+C3+D3
Take-home pay in cell F3:	+B3−E3

For the other employees these formulas are copied into the appropriate cells. For row 7 the sum formula is used to add each column.

---- **SKILL 3** ----

There is a limit to the amount of your income that is subject to FICA tax. The current rule is as follows:

> Social Security: 6.2% of all income at or under $57,600
> Medicare: 1.45% of all income at or under $135,000

For most people, the percent of income collected is 6.2% + 1.45%, or 7.65%. From time to time, the above rule is modified by the government, usually by raising one or both of the income maximums.

EXAMPLE 3   Evelyn's father earns $68,000 a year and Hari's mother earns
$150,000 a year.

QUESTION   How much do Evelyn's father and Hari's mother pay in FICA taxes?

SOLUTION
The amount paid by Evelyn's father is

$$0.062 \cdot 57,600 = 3571.20 \qquad \text{6.2\% of \$57,600}$$
$$0.0145 \cdot 68,000 = \underline{986.00} \qquad \text{1.45\% of \$68,000}$$
$$\text{Total} \quad 4557.20$$

The amount paid by Hari's mother is

$$0.062 \cdot 57,600 = 3571.20 \qquad \text{6.2\% of \$57,600}$$
$$0.0145 \cdot 135,000 = \underline{1957.50} \qquad \text{1.45\% of \$135,000}$$
$$\text{Total} \quad 5528.70$$

Evelyn's father pays $4557.20 in FICA taxes and Hari's mother pays
$5528.70 in FICA taxes.

──────── **SKILL 4** ────────────────────────────

EXAMPLE 4   The third record that Evelyn and Hari must prepare is a monthly
payroll summary for each employee. This form contains the
payroll information for an individual, week by week.

QUESTION   How should Evelyn and Hari use a spreadsheet to prepare a
monthly payroll summary?

SOLUTION
The summary should show gross pay, deductions, and take-home pay, week
by week. A form is shown below.

	A	B	C	D	E
	MONTHLY PAYROLL SUMMARY				
	Employee: Evelyn				
1	Week	Gross	Income Tax	FICA	Take-Home
2	Ending	Pay	Withholding	Withholding	Pay
3	3/7	91.00	7.00	6.96	77.04
4	3/14	65.00	3.00	4.97	57.03
5	3/21	104.00	8.00	7.96	88.04
6	3/28	78.00	4.00	5.97	68.03
7	Totals	338.00	22.00	25.86	290.14

$+B3 - C3 - D3$

$+B3 * 0.0765$

In this spreadsheet, take-home pay is calculated by subtracting both
deductions from gross pay. Remember to use the sum function for the totals.

TRY YOUR SKILLS

One week, Freda works 25 hours. Freda is single and claims no exemptions.

1. Find her gross pay if she makes $5.50 per hour.
2. Find the amount of her federal income tax withholding.
3. Find the amount of her FICA withholding.
4. Find her take-home pay.

Jackie is hired to deliver work and run errands. One week, she works 15 hours at $4.80 an hour. She is single and does not claim herself as an exemption.

5. Find her gross pay.
6. Find the amount of her federal income tax withholding.
7. What will be her FICA withholding?
8. Find her take-home pay.

EXERCISE YOUR SKILLS

KEY TERMS

consumers
entrepreneurship
labor
management
market economy
producers

1. What are several qualities or factors that are needed to make a small business successful?
2. What are some of the responsibilities of a good manager?
3. Give two different meanings of the word *labor*.
4. What investment do workers make in a business?

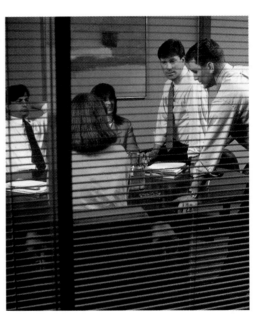

5. Construct a payroll register for weekly deductions to record the amount for federal income tax withholding based on the tax table in the Reference Section, for FICA withholding based on 7.65% of gross pay, and for take-home pay. Kendra worked 20 hours at a rate of $5.25 an hour, with no exemptions. Paulo worked 9 hours at $8.20 an hour, with one exemption. Michael worked 29 hours at $6.10 an hour, with no exemptions. Also record the totals for each category. All of the employees are single.

6. Use a computer spreadsheet to prepare a payroll register for weekly earnings like the one shown in Example 1. You will create four spreadsheets, one for each week in the month.

Employee	Number of Exemptions	Marital Status	Hourly Rate	Hours Worked Week Ending 4/4	4/11	4/18	4/25
Catlyn	0	S	$3.75	16	14	21	32
Sara	1	S	4.50	20	28	25	18
Joleen	0	S	8.00	26	16	16	24
Hernando	1	S	6.24	14	10	30	20

7. Use a computer spreadsheet with the headings shown below to prepare four weekly payroll registers for each of the employees in Exercise 6. Refer to the tables for Federal Income Tax Withholding in the Reference Section to find the federal income tax to be withheld. For FICA withholding, use 7.65% of gross pay.

PAYROLL REGISTER FOR WEEKLY DEDUCTIONS					
Employee	Gross Pay	Income Tax Withholding	FICA Withholding	Total Deductions	Take-Home Pay

8. Use a spreadsheet program to prepare a monthly payroll summary with the headings shown below for each employee from Exercises 6 and 7. Show the weekly totals for payroll information to prepare this summary.

MONTHLY PAYROLL SUMMARY				
Employee: _____				
Week	Gross Pay	Income Tax Withholding	FICA Withholding	Take-Home Pay

9. Construct a computer spreadsheet that will be a payroll register combining the information and calculations for earnings with those of deductions. Use the spreadsheet to compute gross pay, deductions, and take-home pay for the following employees. All of them are single.

Employee	Exemptions	Hourly Rate	Hours
Adikes	1	$7.85	10
Carney	1	5.30	18
Inez	0	6.78	27
Sun-Li	0	9.36	17

How would you need to adjust the FICA tax for a person who earns the following in a year.

10. More than $57,000, but less than $135,000.

11. More than $135,000.

Find the amount of FICA taxes that are to be paid by the following employees.

12. Ms. Dupont, a salesperson who earned $75,000 last year

13. Mr. DeCecco, a master carpenter who earned $7200 a month last year

14. Mr. Andersen, a tugboat captain who earned $143,000 last year

15. Mrs. Ramirez, a magazine publisher who earned $210,000 last year

16. What is the maximum amount of FICA taxes that a person could pay in one year?

17. A person who pays the maximum FICA tax in one year must make a salary greater than what amount?

MIXED REVIEW

Complete the missing portions of the check register shown.

	Number	Date	Checks/Deposits	Amount	Balance
					$335.60
1.	435	4/17	to Frank Brown	$110.00	
2.	——	4/20	from Sue Dunn	45.80	

Find the simple interest on each CD.

3. $1500 at 5% for one year

4. $2800 at 4.5% for one year

5. Liz is saving $15 a week. How long will it take her to save enough money for a blender that costs $149.50?

6. Write the formula for finding the value of an investment that is earning compound interest.

Find the interest on each investment.

7. $900 invested for 3 years at 6% compounded annually

8. $900 invested for 3 years at 6% compounded quarterly

9. Katy prepares flower arrangements for Rodger's Flower Store. She receives $8 for each budget arrangement, $12 for each regular arrangement, and $20 for each deluxe arrangement. How much did she earn in a week in which she assembled 40 budget arrangements, 20 regular arrangements, and 10 deluxe arrangements?

10. Find the maximum amount of extra money that can be generated by a checking account deposit of $4350 if the Federal Reserve System's reserve requirement is 20%.

After a few weeks, Evelyn and Greg discover that with only Evelyn's computer to use, the word-processing business cannot be run efficiently. Evelyn recalls how she typed a 15-page English paper for Greg last year. In return, Greg painted a beautiful eagle on a sweatshirt for Evelyn. Many of their friends admired the sweatshirt. That suggested another possible business idea. Greg is wondering what costs would be involved in painting designs on sweatshirts or T-shirts to sell to his friends. He also would need to know the prices at which they would have to sell sweatshirts to be competitive and still make a profit.

OBJECTIVES: *In this lesson, we will help Greg to:*

Examine the many costs of producing items for sale in a small business.
Establish prices that cover the costs of production.

ADD-ON COSTS

Greg has contacted a wholesale supplier who is willing to provide plain T-shirts and sweatshirts for a reasonable price. Transforming a plain T-shirt into a painted one is a matter of having the proper paints, a place to work, design ideas, and enough time to do the painting.

Greg is aware that it takes a long chain of events to turn a **commodity** into a consumer good. In this chain are costs—referred to as **add-on costs**—that contribute to the price of the end product. Add-on costs are those that add to the price of a raw commodity as it goes through the processing and marketing steps in the channel of distribution. Add-on costs cover such things as labor, advertising, energy, and transportation.

Labor   At every stage of the production process, labor costs are incurred. The business partners set wages for themselves that are then figured as costs of the business. These contribute to the add-on cost of any final product. Labor costs include not only wages but also employee benefits such as pension plans and medical, dental, and life insurance. Evelyn and Hari have already designed a spreadsheet to keep track of the basic labor costs they would incur. Greg will also have to take this into consideration when marketing T-shirts and sweatshirts.

Packaging   Packaging costs reflect both the material and the design of the package. Packaging plays an important role in marketing a product. Frequently, a product's package is responsible for a buyer's initial reaction. Therefore, packaging is often designed to project a particular image or attract a specific audience. Greg thinks the artistic design he will paint on his shirts, while not actually a package, will project an image that he can sell.

Advertising This category includes all kinds of materials and activities used to promote sales of a particular product. In addition to media promotion (television, magazine, and radio), promotional materials might include educational pamphlets, booklets, posters, and other publications. Advertising costs vary greatly from one product to another. Evelyn and Greg have discussed wearing some of the shirts as a kind of "walking advertisement" for their products. When Evelyn wears the shirt Greg made for her last year, friends often ask where they can get one like it.

Energy Energy expenses include electricity, fuel oil, natural gas, or any other energy source used in the manufacturing, distribution, and sales of a product. Energy is necessary to complete each step in the marketing and distribution chain. For example, storage may require energy for special heat and humidity controls. Evelyn and Greg have discussed with their parents the idea of paying a portion of the **utility** (energy) bills for the home they work in.

Transportation The cost of transportation is added to the cost of a product at practically every step in the marketing and distribution channel. Transportation costs include fuel, maintenance and upkeep, depreciation, and even labor costs. Transportation for this enterprise mainly involves Freda's driving to deliver flyers and posters and to talk with people, plus transporting their finished products to their customers.

Ask Yourself

1. What do labor costs include, besides wages?

2. What are several ways of advertising a product?

3. What do energy expenditures cover?

ALGEBRA *REVIEW*

Solve for x.

1. $3x = 30$

2. $2x + 50 = 280$

3. $2.5x - 3 = 12$

4. $x + 2.7 = 2x - 5.6$

5. $15.6x + 188 = 500$

6. $9x + 120 = 660$

7. $-3.8x - 12.2 = -6.5$

8. $8.1x + 11.7 = 7.2x - 9.9$

9. $55 - 37x = 75 + 23x$

10. $2(x - 4) = 13$

11. $25 = 5(3x - 7)$

12. $23x = 19 - 3(x + 2)$

Graph each equation. Determine the slope of each graph.

13. $y = 5x$

14. $2y + 4x = 8$

15. $y - 3x - 5 = 0$

16. $4x = 3y - 6$

SHARPEN YOUR SKILLS

SKILL 1

EXAMPLE 1 Evelyn, Greg, Freda, and Hari want to find the costs involved in producing and selling hand-painted T-shirts and sweatshirts. They have the following information.

> Materials
>> 12 T-shirts, $5.50 per shirt; 12 Sweatshirts, $7.50 per shirt; Paints, $1.25 per shirt
>
> Labor, $5.00 per hour
>> Evelyn and Greg, 15 hours each; Freda and Hari, 10 hours each
>
> Packaging
>> Small plastic bags, $2.00 per 100
>
> Advertising
>> Flyers, $5.00 per 100
>
> Energy
>> Part of utility bill at Evelyn's house, $2.00 per week
>
> Transportation
>> Freda's car, $25 per week

QUESTION What is the total cost to make 24 shirts (12 T-shirts and 12 sweatshirts) in one week?

SOLUTION

They use the information to determine the total cost.

Materials		
T-shirts	12 • 5.50	66.00
Sweatshirts	12 • 7.50	90.00
Paints	24 • 1.25	30.00
Labor at $5.00 per hour		
Evelyn and Greg	2 • 15 • 5.00	150.00
Freda and Hari	2 • 10 • 5.00	100.00
Packaging, 2.00 ÷ 100	= 0.02 per bag	
24 bags	0.02 • 24	0.48
Advertising, flyers		5.00
Energy, part of utilities		2.00
Transportation, Freda's car		25.00
	Total	468.48

The total cost to make 24 shirts in one week is $468.48.

One of a businessperson's most important tasks is establishing prices. Prices must be high enough to pay all the costs and earn money for the business but low enough to sell as many of the product as possible.

EXAMPLE 2 Evelyn, Greg, Freda, and Hari want to price the hand-painted T-shirts and sweatshirts in Example 1 so that they cover all of their costs.

QUESTION What should be the price of the T-shirts and sweatshirts to pay all costs?

SOLUTION

Evelyn, Greg, Freda, and Hari use the total cost and the number of items to be sold to determine the average cost of a shirt to the nearest cent.

$468.48 \div 24 = 19.52$ The total cost is $468.48 for 24 shirts.

To the nearest dollar, a price of $20 will pay all the costs.

Since the T-shirts cost less to produce than the sweatshirts, the students will put a lower selling price on the T-shirts. But the average of the two prices should be $20. Thus they might sell the T-shirts for $17 and the sweatshirts for $23. There is no absolute formula or rule for establishing prices. Businesses try different prices to find out what is successful.

TRY YOUR SKILLS

Three students are planning to make a video about their school. Find each of the following costs.

1. Materials: 25 video tapes at $2.40 each

2. Labor: 18 hours at $5.60 an hour

3. Advertising: 200 flyers at $0.035 each

4. Find the total costs for the items in Exercises 1–3.

5. To cover costs, what price to the nearest dollar should the students place on the tapes in Exercises 1–4?

6. Two students are planning to make kites for resale. Find the total of the following costs.

    | Materials: | 40 kite kits at $5.20 |
    | Labor: | 55 hours at $4.00 an hour |
    | Promotion: | 4 ads at $7.50 each |

7. To cover costs, what price to the nearest dollar should the students place on the kites in Exercise 6?

EXERCISE YOUR SKILLS

1. What add-on costs increase the price of a product?

2. How does packaging play a role in marketing a product or service?

3. When you purchase a nationally advertised item, are you paying for any advertising costs?

4. What are some of the things a business must consider when establishing the price of a product?

In Exercises 5–16, assume that the advertising, energy, and transportation costs are the following.

Advertising:	$ 5.00
Energy:	2.50
Transportation:	25.00
Total:	$32.50

5. Three students have decided to sell bumper stickers advertising school spirit slogans. Find the cost to produce 200 stickers each week if the costs include

Labor:	45 hours at $4.50 per hour
Materials:	$1.04 per sticker

Advertising, energy, and transportation costs

6. What selling price to the nearest dollar should be placed on the bumper stickers of Exercise 5 to cover costs?

7. Four students have decided to applique letters on towels to personalize them. They think they can prepare 100 towels per week. Find the cost to make 100 towels if the costs include

Labor:	60 hours at $4.00 per hour
Materials:	Towels at $2.00 each
	Letters at $0.20 per towel
Packaging:	Cardboard boxes at $0.10 each

Advertising, energy, and transportation costs

8. What selling price to the nearest dollar should be placed on the towels in Exercise 7 to cover costs?

9. Two students plan to sell pennants with the school logo on them. Find the cost to produce 200 pennants in a week if the costs include

Labor:	40 hours at $4.50 per hour
Materials:	Pennants at $1.00 each
	Paints at $0.20 per pennant

Advertising, energy, and transportation costs

10. To cover costs, what selling price to the nearest dollar should the students place on the pennants of Exercise 9?

11. Four students want to make hand-painted coffee mugs for use in cars. Find the costs to produce 50 mugs if the costs include

    Labor: 50 hours at $5.00 per hour
    Materials: Mugs at $3.00 each
                     Paints at $0.40 per mug
    Packaging: Boxes at $0.15 each
    Advertising, energy, and transportation costs

12. What selling price to the nearest dollar should be placed on the mugs of Exercise 11 to cover costs?

13. Two students plan to sell tennis balls at the tennis courts. They will purchase the tennis balls through a discount store and resell them. Find the cost of 100 cans of tennis balls if the costs include

    Labor: 20 hours at $4.00 per hour
    Materials: Balls at $2.15 per can
    Advertising, energy, and transportation costs

14. What selling price to the nearest dollar should be placed on the cans of tennis balls of Exercise 13 to cover costs?

15. Two students plan to order school book bags from a manufacturer and sell them in their school. They will paint the school's insignia on the bags. Find the cost to prepare and sell 50 book bags if the costs include

    Labor: 35 hours at $4.50 per hour
    Materials: Bags at $3.50 each
                     Paints at $0.50 per bag
    Advertising, energy, and transportation costs

16. What selling price to the nearest dollar should be put on the book bags in Exercise 15 to cover the costs and make $50 in profit?

The following are strategies from which the students might choose in different situations: raise prices, lower prices, increase the number of products for sale, find ways to decrease costs, enhance the quality of the product. Tell which of these strategies you would choose in each of the following situations, and give reasons for your choice. You may choose more than one strategy for a given situation.

17. Very few of the products sell in the first few days.

18. All of the products sell out on the first day.

19. People tell you that they like the product but it costs too much.

1. Alison earns a commission of 6.5% on each of her sales. What is her commission on a sale of $858?

2. Carol's take-home pay is $416.75. There was an amount of $45.00 withheld for income taxes as well as 7.65% of the gross salary withheld for FICA taxes (Social Security and Medicare). Let x represent the gross salary and write an equation for the gross salary. Then find how much money is withheld in all.

3. Tom has a balance of $413 in his checking account. What will be the new balance after monthly interest of 0.5% is credited to the account?

4. Sean has a bank balance of $235.67. What is his balance after he deposits $56.78 and writes a check in the amount of $15.49?

5. How much will $7200 be worth at the end of 3 years if it earns 5% interest, compounded semiannually?

Make a table to show the first two levels of new money that is created for the Federal Reserve requirement that is given.

6. Reserve requirement: 20%; Initial deposit: $2,400

7. Reserve requirement: 15%; Initial deposit: $1,800

8. Reserve requirement: 25%; Initial deposit: $1,800

9. Suppose that your checkbook shows a balance of $375.21. Your bank deducts a service charge of $5.50 each month but does not pay interest and does not charge for each written check. The only check that you wrote during the month was for $160.32 and was mailed the day before the bank's monthly statement arrived. What will the bank show as the closing balance?

10. Carmen's annual salary is $78,000. How much does Carmen pay for FICA taxes each year?

11. Michael's annual salary is $150,000. How much does Michael pay for FICA taxes each year?

12. Julie's annual salary is $55,000. How much does Julie pay for FICA taxes each year?

Use the Rule of 72 to find how many years it will take for $3000 to grow to $24,000 at the given rates of interest.

13. 4.5%   14. 6%   15. 8%

A $1600 deposit is made in a checking account when the Federal Reserve System's reserve requirement is 17%.

16. Find the maximum amount of extra money that can be generated.

17. What is the multiplier?

4-3 COST, REVENUE, AND PROFIT FUNCTIONS

Freda has always been a good salesperson. She has sold everything from sunglasses to video tapes, from book bags to personalized towels. She has a knack for understanding people and exactly what motivates them to buy something they might not know they want or need.

The desire to earn money keeps Freda selling, even though she never seems to be able to hold onto her profits. In fact, she doesn't always know whether she has made a profit or not. A lot of cash passes through her hands every month, but her savings account has not grown much.

When Freda joined Evelyn and Greg's business, she hoped that they could help her find out where all her money goes. In return, Freda could help Evelyn and Greg understand the effects of competition on sales. As a result they could all make a profit.

OBJECTIVES: *In this lesson, we learn how to:*

- *Explain how competition works.*
- *Find the profit or loss from the sales of a company's products.*
- *Use the cost function, revenue function, and profit function.*

COMPETITION

Competition is a rivalry among sellers for consumers' dollars or a rivalry among producers for the factors of production. Competition keeps producers from charging unreasonably high prices and making excessive profits. Certain conditions have to exist for competition to work as it should.

For Competition to work:

- There should be a large number of buyers and sellers.
- Buyers should be informed about the quality and price of products.
- Buyers should have choices among products.
- New products should have easy entry into the marketplace.

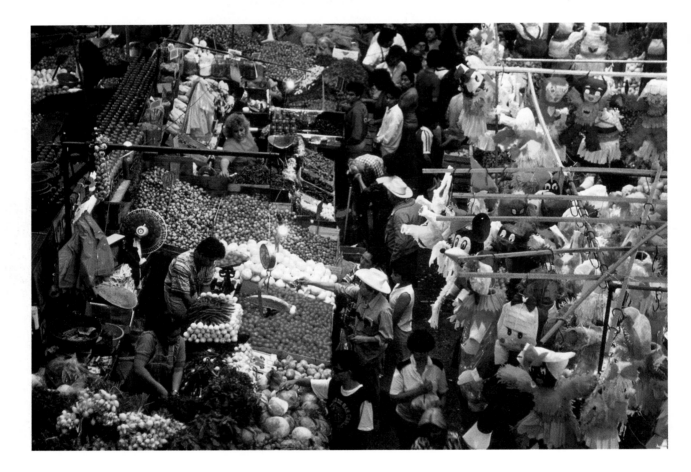

COST, REVENUE, AND PROFIT

Revenue is the money that a business receives from customers for its products or services. **Profit** is the difference between revenues and costs. If revenues are greater than costs, then a business makes a profit. If costs are greater than revenues, then the business suffers a loss.

Usually, if a company's profits are high, other companies will present similar products at lower prices and take away some of the business and the profits. Evelyn, Greg, and Freda hope to offer a quality product at a low price. To offer the lowest price possible, they analyze their costs. They separate costs into *fixed costs* and *variable costs*. **Fixed costs** are costs that remain the same each week. Based on the assumption that the employees work the same number of hours each week, these costs include labor, transportation, advertising, and energy. **Variable costs** vary according to how many items are produced; these include material and packaging. The cost per item is called the **unit cost.** Similarly the price per unit is called the **unit price.**

Ask Yourself

1. What conditions must be present for competition to work?

2. Without competition, what might happen to prices?

SHARPEN YOUR SKILLS

SKILL 1

The total cost incurred by a business includes both fixed and variable costs. Total cost can be expressed algebraically.

> ## Cost Function
> $c = un + f$ where c = total cost
> u = unit cost
> n = number of units
> f = fixed cost

In the above equation for the cost function, the expression un represents the variable cost. As an example of the cost function, if the students purchase and paint n shirts at a cost of $6.00 each, then $6.00 is the unit cost and an equation for the cost function is

$c = 6n + 0$ The fixed cost is assumed to be 0.
$c = 6n$

Revenue, the amount of money a company receives for its products or services, can also be expressed algebraically.

> ## Revenue Function
> $r = sn$ where r = revenue
> s = selling price per unit, or unit price
> n = number of units sold

For example, if n of the above shirts are sold at $22 each, then an equation for the revenue function is $r = 22n$.

The profit is the amount left when a company subtracts costs from revenue.

> ## Profit Function
> $p = r - c$ where p = profit
> r = revenue
> c = total cost

In the case of the painted shirts, an equation for the profit function is
$p = 22n - 6n$, or $p = 16n$.

EXAMPLE 1 Evelyn and her friends have decided to work a fixed number of hours and to start their business by doing only T-shirts. They have analyzed their costs and separated them into fixed and variable categories as shown.

FIXED COSTS: LABOR, ADVERTISING, ENERGY, TRANSPORTATION

Fixed Costs	
Labor, 50 hours at $5 per hour	$250.00
Advertising	5.00
Energy	2.00
Transportation	25.00
Total fixed cost	$282.00

VARIABLE COSTS: MATERIALS FOR PRODUCTION AND PACKAGING

Unit Cost per T-Shirt	
Materials: T-shirts	$5.50
Paint	1.25
Package	0.02
Total unit cost	$6.77

QUESTION
a. What is the cost function for the business?
b. With a selling price of $19, what is the revenue function?
c. What will be the profit in selling 30 T-shirts?

SOLUTION

a. The cost function includes the unit cost, the number purchased, and the fixed costs. For the purchase of n T-shirts the cost function is

$$c = un + f$$
$$c = 6.77n + 282 \qquad u = 6.77;\ f = 282$$

b. The revenue function is the product of the selling price and the number sold. For the sale of n T-shirts at $19 each,

$$r = sn$$
$$r = 19n \qquad s = 19$$

c. The profit function states that profits are found by subtracting the total cost from revenue,

$$p = r - c$$

For 30 T-shirts, the revenue is

$$r = 19n$$
$$= 30(19) = 570$$

The cost to make 30 shirts is

$$c = 6.77n + 282$$
$$= 6.77(30) + 282$$
$$= 203.10 + 282$$
$$= 485.10$$

The profit is the revenue minus the cost.

$$p = r - c$$
$$= 570 - 485.10$$
$$= 84.90$$

The weekly profit for selling 30 T-shirts is $84.90. This is not a lot, but the students are paying themselves hourly wages that are included in the fixed costs so they believe that they do not need a large profit.

———— **SKILL 2** ————

Profits in the T-shirt business depend on costs, selling price, and sales quantities. The spreadsheet can be a powerful tool for analyzing the effect of these factors on the business.

EXAMPLE 2 Hari wants to compare profits for different quantities of T-shirts.

QUESTION How can he use a spreadsheet to analyze the profits for various quantities?

SOLUTION
To compare profits for different sales quantities, Hari develops the following spreadsheet.

+A3*B3+C3

+B3*E3

+F3−D3

	A	B	C	D	E	F	G
1	Unit	Number	Fixed	Total	Unit		Profit
2	Cost	Produced	Cost	Cost	Price	Revenue	(Loss)
3	6.77	15	282.00	383.55	19.00	285.00	($98.55)
4	6.77	20	282.00	417.40	19.00	380.00	($37.40)
5	6.77	25	282.00	451.25	19.00	475.00	$23.75
6	6.77	30	282.00	485.10	19.00	570.00	$84.90
7	6.77	35	282.00	518.95	19.00	665.00	$146.05

Hari is pleased that the enterprise seems to be becoming more profitable as the number produced rises to 30.

Notice in column G that a loss is indicated by a number in parentheses. Spreadsheet programs allow you to represent a loss of $98.55 either as −98.55 or as ($98.55). Select the style you prefer using the menu choice for formatting.

We have seen how costs, sales, and profits are related. We have also learned that different tools are available to analyze business finances. These include algebraic equations and the computer spreadsheet that we used in the previous two skills. Graphing is another method that is useful for financial analysis.

EXAMPLE 3   Hari wants to graph the cost function.

QUESTION   How can you use the graphing feature of a graphing calculator to graph the cost function from Example 1?

SOLUTION

The equation for the cost function from Example 1 is $c = 6.77n + 282$. To enter this equation in a graphing calculator you need to use the variables x and y instead of n and c. When the equation is rewritten using x and y, the result is

$$y = 6.77x + 282$$

Choose the range values.

Xmin: -2	Ymin: -75
Xmax: 45.5	Ymax: 600
Xscl: 2	Yscl: 50

The graph of the cost function is shown.

TRY YOUR SKILLS

Use the costs in Example 1.
1. Determine the cost of producing 23 T-shirts.
2. Determine the profit or loss in selling 23 T-shirts at $19 each.

A mug is purchased for $3.75, painted for $0.37, and packaged for $0.08. The fixed costs are $75. The mugs sell for $8 each.
3. Find the unit cost.
4. Determine the cost function.
5. Find the cost of preparing 32 mugs.
6. What is the revenue function?
7. What is the profit function?
8. What is the profit if all 32 mugs are sold?

Students selling bumper stickers at $2.00 per sticker have the following costs.

Fixed costs: Labor, 45 hours at $4.50 per hour; Advertising, energy, and transportation, $32.50

Variable costs: $1.04 per sticker

A partial spreadsheet is shown.

	A	B	C	D	E	F	G
1	Unit	Number	Fixed	Total	Unit	Revenue	Profit
2	Cost	Produced	Cost	Cost	Price	Per Unit	(Loss)
3	1.04	200	235.00	443.00	2.00	400.00	
4	1.04	300	235.00		2.00		
5	1.04	400	235.00		2.00		

9. Determine the total cost formula for cell D4.
10. Determine the revenue formula for cell F4.
11. Determine the profit (loss) formula for cell G4.
12. Complete the spreadsheet.

EXERCISE YOUR SKILLS

1. What are some of the factors that limit the amount that a business can charge for its products or services?
2. Why do businesses have sales in which they reduce prices?
3. How are the cost and revenue functions related when a business does not make a profit?
4. Why might a business reduce prices so much that it does not make a profit?
5. Price fixing occurs when different companies agree on what they will charge for particular products. Why is there a law against price fixing?

KEY TERMS

competition
fixed costs
profit
revenue
unit cost
unit price
variable costs

Students selling personalized towels at $5.00 per towel have the following costs.

Fixed costs: Labor, 60 hours at $4.00 per hour
              Advertising, energy, and transportation, $32.50
Variable costs: Towels, $2.00 each
              Letters, $0.20 per towel
              Cardboard boxes at $0.10 each

6. Determine the cost function for n towels.
7. Find the cost of 100, 200, and 300 towels.
8. Determine the revenue function for n towels.
9. Find the revenue for selling 100, 200, and 300 towels.
10. Use the cost and revenue functions to determine a profit function.
11. Find the profit in selling 100, 200, and 300 towels.

Students selling pennants at $2.00 each have the following costs.

Fixed costs: Labor, 40 hours at $4.50 per hour
              Advertising, energy, and transportation, $32.50
Variable costs: Pennants, $1.00 each
              Paints, $0.20 per pennant

12. Determine the cost, revenue, and profit functions.
13. Set up a spreadsheet to show the cost, revenue, and profit for quantities of 200, 300, and 400.
14. Graph the cost function.

Students selling hand-painted mugs at $10.00 each have the following costs.

Fixed costs: Labor, 50 hours at $5.00 per hour
              Advertising, energy, and transportation, $32.50
Variable costs: Mugs, $3.00 each
              Paints, $0.40 per mug
              Packaging, $0.15 each

15. Determine the cost, revenue, and profit functions.
16. Set up a spreadsheet to show costs, sales, and profits for quantities of 25, 50, and 75.
17. Graph the cost function.

Students selling tennis balls at $4.00 per can have the following costs.

Fixed costs: Labor, 20 hours at $4.00 per hour
              Advertising, energy, and transportation, $32.50
Variable costs: Balls, $2.75 per can

18. Determine the cost, revenue, and profit functions.
19. Set up a spreadsheet to show costs, revenues, and profits for quantities of 100, 200, and 300 cans.
20. Graph the cost function.

Students selling book bags at $7.50 per bag have the following costs.

Fixed costs: Labor, 35 hours at $4.50 per hour
Advertising, energy, and transportation, $32.50
Variable costs: Bags, $4.00 each

21. Determine the cost, revenue, and profit functions.

22. Set up a spreadsheet to show costs, revenues, and profits for quantities of 25, 50, and 75.

23. Graph the cost function.

MIXED REVIEW

1. Tracy earns $130 for working 24 hours. Of that time, 20 hours are at her regular rate and the remainder is at $1\frac{1}{2}$ times the regular rate. What is her regular hourly rate?

2. Janice earned $29,670 last year. Her fringe benefits were 14.5% of her salary. What was the value of the fringe benefits?

3. Louis earns a salary of $145 a week and also a commission of 4.5% on his sales. What are his total weekly earnings in a week in which his sales are $1,324?

4. A house has been sold for $190,000. The real-estate agent who sold the house earned a commission of 3.5%. The broker for whom she worked earned 2.5%. How much did each receive?

Enter the following transactions in a check-register form. Make up your own form with the following headings: Check Number, Date, Checks/Deposits, Amount, and Balance. Under "Checks/Deposits" have two lines for each entry, the top one labeled "To:" and the bottom one labeled "For:". Find the new balance after each transaction. The starting balance is $620.

5. Check 701, February 17, to Farley Smith, $256.50 for VCR repair

6. Deposit on February 19 of $400

7. Check 702, February 19, to Abe's Supermarket, $94.77 for groceries

8. Earl's check register shows a balance of $104.56. He knows that he has a $20 check outstanding. His bank statement shows a balance of $125.96. Does his account balance? Explain your answer.

9. Tabitha sells annual memberships in an exercise club. Her boss has offered her a choice between Plan 1, no salary but a straight $50 for each membership sold, and Plan 2, a weekly base salary of $225 plus $25 for each membership sold. How many memberships would Tabitha have to sell to make the same amount of money under either plan?

10. Use the Rule of 72 to find how long it will take your $4550 savings account to double in value if it is growing at a rate of 6.6% per year.

4-4 BREAK-EVEN POINT

Evelyn and her friends, like all businesspeople, have no guarantees. When they began their business venture, they did not know whether they would have many, few, or even no customers. These students were determined but also a little anxious. They wanted to succeed. In business the primary indicator of success is making a profit. As we have seen, this will occur when revenues are greater than costs.

Usually, the more efficient a business is, the greater its profits will be. A business can increase its profits by offering consumers better products, lower prices, and better services than competitors. Thus consumers benefit from the efforts of businesses to make a profit.

Profits are important to a business, not because businesspeople are greedy, but because extra money is needed to investigate new business opportunities, to set aside reserves for times when the company does not make a profit, and to provide for emergencies.

Profit, then, is the incentive or reward for improved production and services. It was the desire for profit that led Evelyn and the others to decide on the T-shirt business rather than word processing, and it was the profit motive that helped them determine the appropriate quantities to buy and sell. Finally, it was their interest in profits that led these students to ask Hari to help them understand the finances of the businesses.

The first thing Hari taught the other students was that they should be aware of the break-even point that separates losses from profits.

Graph the cost and revenue functions.
Determine the break-even point algebraically.
Determine the break-even point using the graphing calculator.

BREAK-EVEN POINT

We have seen that establishing prices is a most important step for any business. Prices and the number of products sold determine a company's revenue. If a business's total cost is greater than its revenue, then the business incurs a *loss*. If the revenue is greater than the total cost, then the business makes a *profit*. The point at which cost and revenue are equal is called the **break-even point.** This point is given numerically by the number of products sold at a given price. It is also useful to illustrate the break-even point by graphing the line for cost and the line for revenue. The break-even point is where these two lines intersect.

The region between the lines and above the break-even point is the **profit region.** The region between the lines and below the break-even point is the **loss region.**

Ask Yourself

1. Why are profits important to a business?

2. What is the break-even point?

3. What factors determine the break-even point?

SHARPEN YOUR SKILLS

EXAMPLE 1 Hari wants to find the break-even point for the T-shirt business when the fixed cost is $282, the unit cost is $6.77, and the selling price is $19.

QUESTION How can a graph be used to show the break-even point for Hari's T-shirt business?

SOLUTION
From Lesson 4-3, Example 1, the equation for the cost function is

$$c = 6.77n + 282 \qquad n = \text{the number of T-shirts made}$$

and the equation for the revenue function is

$$r = 19n \qquad n = \text{the number of T-shirts sold}$$

Since c and r both refer to amounts of money, you can use the same axis to graph them. Since the amount of money depends on the number of T-shirts, use y for c and r, and x for n, the number of T-shirts. The equations are

$$y = 6.77x + 282$$
$$y = 19x$$

Graph these equations on a graphing calculator with the following range values.

Xmin: −2   Ymin: −75
Xmax: 45.5   Ymax: 600
Xscl: 2   Yscl: 50

The break-even point is the point where the graphs intersect. The region between the lines and above the break-even point is the *profit region*. In this region the revenue is greater than the cost. The region between the lines and below the break-even point is the *loss region*. In this region the cost is greater than the revenue. The break-even point is the point where the cost and the revenue are *equal*. Using the trace function the break-even point appears to be (23, 437.71).

EXAMPLE 2   Hari wonders if there is a way to determine the break-even point algebraically.

QUESTION   How can Hari solve the equations of Example 1 algebraically to find the break-even point?

SOLUTION

A system of two linear equations can be solved by the addition and subtraction method or by substitution. In this case, substitution appears to be the more direct method.

$$y = 6.77x + 282 \qquad \text{(1) Cost equation}$$
$$y = 19x \qquad \text{(2) Revenue equation}$$

$$19x = 6.77x + 282 \qquad \text{Substituting for } y \text{ in (1)}$$
$$12.23x = 282 \qquad \text{Subtract } 6.77x \text{ from both sides.}$$
$$x = 23.06 \qquad \text{To the nearest hundredth}$$

$$y = 6.77(23.06) + 282 \qquad \text{Substitute for } x \text{ in (1).}$$
$$y = 438.1 \qquad \text{To the nearest tenth.}$$

$$y = 19(23.06) \qquad \text{Substitute for } x \text{ in (2).}$$
$$y = 438.1 \qquad \text{To the nearest tenth.}$$

We need to interpret the meaning of the point (23.06, 438.1). Since only whole T-shirts can be made, x values need to be whole numbers. Since making 23 T-shirts would result in a loss, although small, the students need to make 24 T-shirts to break even. The break-even point is 24 T-shirts.

If $x = 23$
Cost: $6.77(23) + 282 = 437.71$
Revenue: $19(23) = 437$
Loss: $437 - 437.71 = -0.71$

If $x = 24$
Cost: $6.77(24) + 282 = 444.48$
Revenue: $19(24) = 456$
Profit: $456 - 444.48 = 11.52$

TRY YOUR SKILLS

Students selling bumper stickers at $2.00 each have the following costs.

Fixed costs: Labor, 50 hours at $4.20 per hour
                        Advertising, energy, and transportation, $32.50
Variable costs: $1.04 per sticker

1. What is the cost function?

2. What is the revenue function?

3. Using algebraic methods, solve the cost function and revenue function simultaneously to find the break-even point.

4. Graph the equations to show the regions of profit and loss.

EXERCISE YOUR SKILLS

1. The break-even point is the intersection of the graphs of what two functions?

2. Why is the break-even point important to a businessperson?

Students selling pennants at $2.00 each have the following costs.

Fixed costs: Labor, 40 hours at $4.50 per hour
                    Advertising, energy, and transportation, $32.50
Variable costs: Pennants, $1.00 each
                    Paints, $0.20 per pennant

3. What is the cost function?

4. What is the revenue function?

5. Find the break-even point algebraically.

Students selling tennis balls at $4.00 per can have the following costs.

Fixed costs: Labor, 20 hours at $4.00 per hour
                    Advertising, energy, and transportation, $32.50
Variable costs: Balls, $2.75 per can

6. What is the cost function?

7. What is the revenue function?

8. Graph the cost function and revenue function.

9. Find the break-even point using calculator or computer features.

Students selling book bags at $9.00 each have the following costs.

Fixed costs: Labor, 35 hours at $4.50 per hour
                    Advertising, energy, and transportation, $32.50
Variable costs: Bags, $4.00 each

10. What is the cost function?

11. What is the revenue function?

12. Find the break-even point using any method.

13. Graph the cost function and revenue function.

14. What is the profit function?

15. Find the point at which the business suffers a loss of $65.

16. Find the point at which the business gains a profit of $75.

Students selling personalized towels at $5.00 each have the following costs.

    Fixed costs: Labor, 60 hours at $4.00 per hour
                            Advertising, energy, and transportation, $32.50
    Variable costs: Towels, $2.00 each
                            Letters, $0.20 per towel
                            Cardboard boxes, $0.10 each

17. What is the cost function?

18. What is the revenue function?

19. Graph the cost and revenue functions.

20. Find the break-even point.

21. What is the profit function?

22. Find the point at which the business suffers a loss of $78.10.

23. Find the point at which the business gains a profit of $194.60.

MIXED REVIEW

1. Veronica earns a commission of $72 on a sale of $600. What is her percent of commission?

2. Find the value to which $500 grows when it is invested for 5 years at 6.5% interest compounded annually.

In Exercises 3–6, suppose that the Federal Reserve requires that 20% of an initial deposit of $5,000 be held in reserve but that the balance may be lent out by the bank.

3. Show the first two levels of extra money that is generated.

4. What is the total amount of new money that is potentially created by the multiplier effect?

5. What is the multiplier for the case in which the reserve requirement is 20%?

6. If the reserve requirement were raised to 25%, what would be the value of the multiplier?

Find the cost of producing the given amounts of the following items. In both cases, the cost of advertising, energy, and transportation is $47.50.

7. Five students are making personalized school calendars. They expect to prepare 25 of the calendars per week. What will be the cost of production if the time required is 20 hours at $4.75 per hour and the cost of materials is $1.20 per calendar?

8. Two students plan to produce coin dispensers for highway tolls. How much will it cost to produce 30 dispensers if the time for the labor is 25 hours at $5.50 per hour and the cost of materials is $0.83 per dispenser?

4-5 LINEAR PROGRAMMING

As the T-shirt company grew, the students came to appreciate Hari more and more. He knows, both from experience and from studying, how mathematics helps in planning and evaluating business situations.

Hari was especially helpful when his friends learned that there are limits to what they can and cannot do in business. For example, they may have only between 10 and 15 hours a week available to work at their business, and during those hours they can do only so much work.

These limits, or constraints, are found in every business. The students buying coffee mugs or school supplies for resale discovered that they could purchase only as many of these products as their money would allow. Sometimes they could not buy as many as they wanted because their suppliers had a limited number on hand to sell to them. They also discovered that there were limits to the demand for the products that they prepared. They might be able to sell 50 personalized lunch boxes, but they soon found that they could not expect to sell 500.

Smart businesspeople recognize the constraints under which their businesses must operate. They identify these constraints and make their plans within the boundaries of the constraints.

OBJECTIVES: *In this lesson, we will help Hari to:*

- *Write inequalities for some of the business's constraints.*
- *Graph the inequalities and find the points of intersection.*
- *Test the points to find the numbers that minimize costs or maximize profits.*

PROFIT

You have seen that, like all businesspeople, Evelyn and her friends want to make a profit. If they sell too few T-shirts, they will not cover their expenses. If they purchase too many, they will be left with unsold **inventory,** or stock on hand.

A more **efficient** business generally achieves greater profits at a lower cost. Thus consumers also benefit from efficiency.

CONSTRAINTS

Constraints are the conditions that limit business activities. Some examples of constraints are the amount of money available for investment, the time or materials available for production, and the demand for the product or service. There are many other constraints. Each business must discover its own limits and plan accordingly.

Linear programming is a mathematical method for planning within given constraints. The constraints are generally expressed as linear inequalities. A linear inequality is an inequality that has as its graph a half-plane bounded by a straight line.

Maximum and Minimum   We use linear programming models to maximize or minimize certain factors in a business situation. To **maximize** means to find the greatest value within the constraints; to **minimize** means to find the least value within the constraints. For example, a business wishes to maximize profits and minimize costs.

Ask Yourself

1. What are three of the constraints that a person starting a business must consider?

2. What does a business try to minimize?

3. What does a business try to maximize?

4. How does a business become more efficient?

ALGEBRA *REVIEW*

Graph each equation.

1. $y = 2x - 3$

2. $3x + 2y = 24$

Find the coordinates of the point of intersection.

3. $x + y = 5$
   $2x - y = 4$

4. $3y = 5x - 1$
   $y - x = 1$

Solve by graphing.

5. $x + 2y \geq 2$
   $2x - y \leq 3$

6. $x \leq 3y$
   $2y - 1 \leq x$

7. Solve the system by graphing. Find the coordinates of the corners of the region bounded by the inequalities.
   $x \leq 4$
   $y \leq 6$
   $x + 2y \geq 12$

SHARPEN YOUR SKILLS

SKILL 1

EXAMPLE 1 Evelyn and her family are planning a car trip that will take several days. They must consider the following constraints.

a. On the first day they are leaving at about noon and don't want to drive after dark. They have only 7 hours in which to drive.

b. They want to travel at least 300 miles that day.

c. They want to stay within the speed limit of 55 miles per hour.

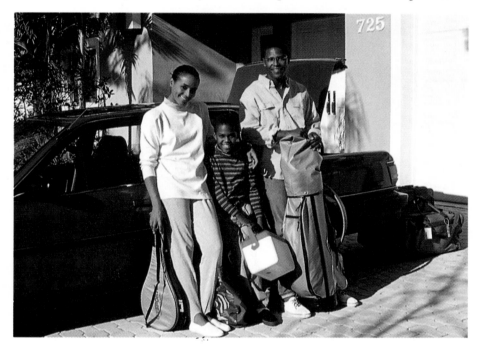

QUESTION How can you use linear programming to graph the constraints as inequalities and show different possible ways to make the trip?

SOLUTION

Let x represent hours and y represent miles. Write inequalities for each constraint.

a. The trip must be no more than 7 hours. $\qquad x \leq 7; x \geq 0$

b. The distance must be at least 300 miles. $\qquad y \geq 300$

c. Obey the speed limit of 55 miles per hour. $\qquad$ Distance = rate • time

$$y \leq 55x$$

Note that x must be greater than zero, since time is positive.

Graph the system of inequalities as shown. The triangular shaded region satisfies all three inequalities. You may not be able to graph $x \leq 7$ on your graphing calculator since it is not a function.

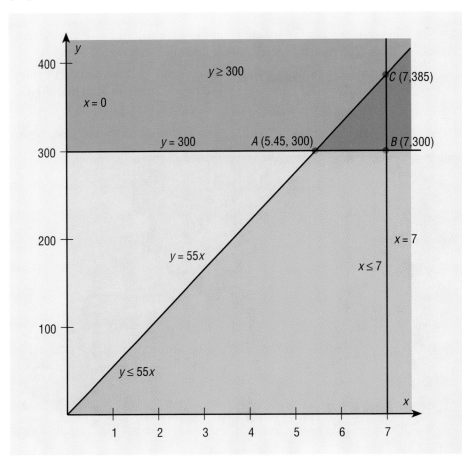

We can find the corners of the triangular region by solving each pair of equations that intersect at the given point.

Point A	Point B	Point C
$y = 300$	$y = 300$	$x = 7$
$y = 55x$	$x = 7$	$y = 55x$
$55x = 300$		$y = 55x$
$x = 5.45$		$y = 55(7)$
$y = 300$		$y = 385$
$A(5.45, 300)$	$B(7, 300)$	$C(7, 385)$

Although all the points in the shaded region satisfy the three inequalities, the vertices have special meaning.

$A(5.45, 300)$ represents the shortest driving time, 5 h 27 min. It would give the least amount of time in the car.

$B(7, 300)$ represents the lowest speed possible, 43 miles per hour. It would offer the most leisurely drive.

$C(7, 385)$ represents the greatest distance, 385 miles. You would use this to travel as far as possible before dark.

──── **SKILL 2** ────

EXAMPLE 2 Evelyn, Freda, Greg, and Hari have begun selling both T-shirts and sweatshirts. They purchase the T-shirts for $5.50 and the sweatshirts for $7.50. They have become successful but have also discovered constraints that affect their business.

    a. To satisfy the demand and not disappoint customers, they must produce a total of at least 52 shirts a week.

    b. Their supplier can supply them with no more than 45 blank sweatshirts per week.

    c. Because of the coming cool weather, they must be prepared to sell at least as many sweatshirts as T-shirts, and possibly more.

    d. Because of the time available, they cannot prepare more than a total of 70 shirts of both kinds per week.

QUESTION How can they use linear programming to find the lowest costs within the given constraints?

SOLUTION

To solve the problem using linear programming, you must first do two things:

    1. Choose a quantity that you want to maximize or minimize and write an equation for that quantity. This quantity is called the **objective function.**

    2. Identify the constraints.

In this case we want to minimize the total cost of the T-shirts and sweatshirts. This cost is expressed by the following equation.

$$c = 5.50x + 7.50y$$

    where c = the cost
    x = the number of T-shirts
    y = the number of sweatshirts

Expressing the constraints as inequalities, we have the following.

a. The students must sell a total of more than 52 shirts.

$$x + y \geq 52$$

In slope-intercept form, this is written as

$$y \geq -x + 52$$

b. They can obtain no more than 45 blank sweatshirts.

$$y \leq 45$$

c. They will sell at least as many sweatshirts as T-shirts.

$$y \geq x$$

d. They cannot prepare more than a total of 70 shirts.

$$y + x \leq 70 \qquad \text{or} \qquad y \leq -x + 70$$

We now graph the lines for each of these inequalities. This can be done by using a graphing calculator.

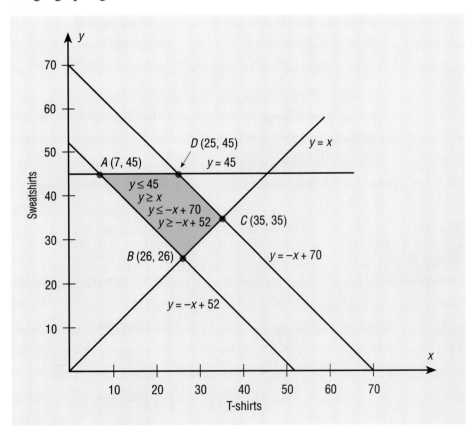

The shaded region satisfies all four inequalities.

Solve each pair of equations that intersect at the given point.

Point A	Point B	Point C	Point D
$y = 45$	$y = x$	$y = x$	$y = 45$
$x + y = 52$	$x + y = 52$	$x + y = 70$	$x + y = 70$
$x + 45 = 52$	$2x = 52$	$2x = 70$	$x + 45 = 70$
$x = 7, y = 45$	$x = 26, y = 26$	$x = 35, y = 35$	$x = 25, y = 45$
$A(7, 45)$	$B(26, 26)$	$C(35, 35)$	$D(25, 45)$

Substitute these coordinates in the objective function, $c = 5.50x + 7.50y$.

For point A, $5.50(7) + 7.50(45) = 376$
For point B, $5.50(26) + 7.50(26) = 338$
For point C, $5.50(35) + 7.50(35) = 455$
For point D, $5.50(25) + 7.50(45) = 475$

The lowest cost will be achieved by using the coordinates at point B; that is, by making and selling 26 T-shirts and 26 sweatshirts.

─────── **SKILL 3** ───────────────────────────────

The constraints listed and graphed in Example 2 will be the same whether we are considering cost or revenue.

EXAMPLE 3   Evelyn's group can raise some additional money to invest, so they want to maximize profit. They sell the T-shirts for $17 each and the sweatshirts for $23 each.

QUESTION   How can they find the sales quantities that will give the maximum profit within the given constraints?

SOLUTION
We use the same linear programming model that was used in Example 2. However, we use a different objective function for revenue. This is the revenue function, the number of items sold multiplied by unit price.

$r = 17x + 23y$ where r = the total revenue
17 = the selling price of the T-shirts
x = the number of T-shirts sold
23 = the selling price of the sweatshirts
y = the number of sweatshirts sold

For intersection points on the graph the revenues are

$A(7, 45)$: $17(7) + 23(45) = 1154$
$B(26, 26)$: $17(26) + 23(26) = 1040$
$C(35, 35)$: $17(35) + 23(35) = 1400$
$D(25, 45)$: $17(25) + 23(45) = 1460$

The maximum revenue is at point D. But this is not the maximum profit.

Profit equals revenue minus cost: $p = r - c$.

We now use the costs from Example 2 to find profit at each point.

	revenue	$-$	cost	$=$	profit
$A(7, 45)$	1154	$-$	376	$=$	778
$B(26, 26)$	1040	$-$	338	$=$	702
$C(35, 35)$	1400	$-$	455	$=$	945
$D(25, 45)$	1460	$-$	475	$=$	985

Within the given constraints the greatest profit occurs at point $D(25, 45)$, which represents 25 T-shirts and 45 sweatshirts. We can also find the maximum profit by writing an objective function for profit.

$$p = 17x + 23y - (5.50x + 7.50y)$$
$$p = 11.5x + 15.5y$$

TRY YOUR SKILLS

1. Find the coordinates of the point of intersection of the graphs $y = 8$ and $3x + y = 20$.

2. Draw the triangle bounded by the inequalities $y \leq 8$, $x \leq 12$, $2y + x \geq 20$.

3. Find the corners of the triangle drawn in Exercise 2.

4. Write inequalities for the following constraints.

   a. You want to work no more than 10 hours per week.

   b. You want to earn at least $45 per week.

   c. You won't be able to get a job earning more than $6.00 per hour.

5. Draw the triangle for the constraints listed in Exercise 4, and explain what each corner of the triangle means.

EXERCISE YOUR SKILLS

Students selling pennants have joined those selling bumper stickers. They can purchase pennants for $1.00 each and bumper stickers for $1.25 each. They want to earn as much money as possible but have discovered that they must work under the following constraints. Let x represent pennants, and y represent bumper stickers.

   a. They can obtain no more than 200 pennants.

   b. They can obtain no more than 400 bumper stickers.

   c. They must sell a combined total of at least 300.

1. Write inequalities to express the constraints. Let x represent the number of pennants and y represent the number of bumper stickers.

2. Graph the constraint inequalities, and shade the region that satisfies all the inequalities.

3. Find the coordinates of the corners of the region in Exercise 2.

4. Write the objective function to minimize costs.

5. Write the objective function to maximize revenue if the pennants and bumper stickers sell for $2 each.

6. Write the objective function to maximize profit.

7. Substitute the corner coordinates in the objective function to find the number of pennants and stickers that they should sell to maximize profit.

KEY TERMS

constraints
efficient
inventory
linear programming
maximize
minimize
objective function

A friend of Hari's goes to work for a computer company that manufactures computer chips. To meet demands, the company must manufacture at least 4000 chips of type A per day and 6000 chips of type B per day. The testing department cannot process more than a total of 15,000 per day. The company earns a profit of $0.50 for each chip of type A sold and $0.40 for each chip of type B sold. Let x stand for the number of A chips and y stand for the number of B chips.

8. Write the objective function for the profit.

9. Write three inequalities to express the constraints.

10. Graph the inequalities.

11. Find the coordinates of the corners of the region defined by the inequalities.

12. Test the coordinates of the corners in the objective function to find the pair that gives the greatest profit.

Students selling coffee mugs have joined with those selling desk organizers. Costs are $3.55 for each coffee mug and $6.40 for each desk organizer. To make a profit, they must sell a combined total of at least 50 coffee mugs and desk organizers. The total spent on the supplies for both items must be less than $350. Because of arrangements with suppliers, the number of mugs must be less than four times the number of organizers. The number of organizers must be less than 40. Let x represent the coffee mugs and let y represent the organizers.

13. Write the objective function to minimize costs.

14. Write inequalities to express the constraints.

15. Graph the constraint inequalities and shade the region that satisfies all the inequalities.

16. Find the corners of the region in Exercise 15.

17. What is the number of coffee mugs and desk organizers they should sell to minimize their cost?

Freda's father is a farmer with 210 acres on which to plant. From crop x he will earn $400 an acre; from crop y he will earn $350 an acre. To meet demands, he must plant at least 40 acres of crop x and 50 acres of crop y. Because of the conditions of the soil, he may not plant more than 80 acres of crop x. According to state regulations, the acres for crop y must be no more than twice the number of acres for crop x.

18. Write the objective function to maximize revenue.

19. Write inequalities to express the constraints.

20. Graph the constraint inequalities and shade the region that satisfies all the inequalities.

21. Find the corners of the region in Exercise 20.

22. How many acres of each crop should he plant to maximize his earnings?

Some students who are starting a business plan to sell earrings for $4 a pair and necklaces for $10 each. Their unit costs are $2.50 for the earrings and $9 for the necklaces. They must work within the following constraints. The store through which they are selling will take only up to a total of 45 pairs of earrings and necklaces combined. Their source of supplies is such that the number of necklaces that they can make will be less than twice the number of pairs of earrings. They must sell at least 10 necklaces to maintain their agreement with the store. Demand will not permit them to sell more than 30 pairs of earrings. Let x represent the earrings and y represent the necklaces.

23. Write the objective function to minimize cost.

24. Write the objective function to maximize revenue.

25. Write the objective function to maximize profit.

26. Write inequalities to express the constraints.

27. Graph the constraint inequalities and shade the region that satisfies all the inequalities.

28. Find the corners of the region in Exercise 27.

29. How many of each should they sell to minimize cost?

30. How many of each should they sell to maximize revenue?

31. How many of each should they sell in order to maximize profit?

MIXED REVIEW

Joseph repairs watches at $10 an hour and receives $2 for each watch that he repairs. He works 30 hours a week.

1. How many watches does he have to repair in order to earn at least $350 per week?

2. It takes Joseph an average of 20 minutes to repair a watch. What is the most money that he can reasonably expect to make in one week?

Suppose that $100 is invested at 5% annually. Find the value of the investment after the given number of years and at the given frequency of compounding. Use the calculator's exponent key $\boxed{x^y}$.

3. 50 years, compounded annually

4. 50 years, compounded quarterly

5. 100 years, compounded annually

A partially completed monthly payroll summary is shown below. Complete the table. Use a spreadsheet program, if available.

	Week Ending	Gross Pay	Income Tax Withholding	FICA Withholding	Take-Home Pay
6.	9/3	$ 87.00	$10		
7.	9/10	123.00	19		
8.	9/17	167.50	23		
9.	9/24	201.00	28		
10.	Totals				

11. What is the limit of your liability if you immediately report the loss or theft of your ATM card?

12. The fixed costs in Amanda's business are $325 per week. The variable costs are $3.50 per item. What are her total costs if she produces 150 items in one week?

Find the following earnings:

	Number of Hours	Hourly Rate	Weekly Earnings		Number of Hours	Hourly Rate	Weekly Earnings
1.	38	$ 3.35		2.	26	$ 7.50	
3.	41	22.70		4.	22	14.80	

Find the weekly take-home pay for each of these single people if each claims one exemption. Use the tables for federal income tax withholding in the Reference Section.

	Gross Pay	Income Tax Withholding	FICA Withholding	Take-Home Pay
5.	$127.30			
6.	195.00			
7.	930.70			
8.	325.60			

Duanita and Alisha want to make and sell stuffed kittens. Use the information given in the box to help you find the total production costs for manufacturing the amounts of toys in Exercises 9, 10, and 11.

Fixed Costs		Variable Costs	
Labor	$3.50 per hour	Materials	$2.25 each
	for 30 hours	Packaging	$0.02 each
Advertising	$3.00		
Energy	$3.58		
Transportation	$5.92		

9. 20 kittens   **10.** 40 kittens   **11.** 60 kittens

Duanita and Alisha will sell stuffed kittens for $7.00 each. Use the costs in Exercises 9-11 to find the profit or loss from selling each amount.

12. 20 kittens   **13.** 40 kittens   **14.** 60 kittens

Use your results from Exercises 9–14 to make a spreadsheet that shows the following information for producing and selling 20, 40, and 60 kittens.

	Fixed Cost	Unit Cost	Number Produced	Total Cost	Unit Price	Revenue	Profit (Loss)
15.			20				
16.			40				
17.			60				

For Exercises 18–23, use your results from Exercises 9–17.

18. Write the cost function.

19. Write the revenue function.

20. Write the profit function.

21. Graph the cost function.

22. Graph the revenue function on the same set of axes as the cost function.

23. Find the break-even point algebraically and using graphing techniques.

24. Why is it important to know how much profit (or loss) a business makes?

25. If a company does not make enough money to break even, what kinds of changes should be considered? Write a paragraph explaining the changes you would consider if your company were losing money and what you would do next if each strategy did not work.

26. After one year, the employees in Oscar's company expected to receive a raise in salary. Explain how an employee's thinking about a raise in salary might differ from the owner's thinking. What should Oscar consider before deciding whether to give raises? Why might it be better for the company and for the employees themselves if salary raises are small? When might salary raises be large?

Use the following information to do Exercises 27–30. When Duanita and Alisha were successful in the stuffed kitten business, they expanded and began to make stuffed puppies as well. Their cost for materials for the kittens continued to be $2.27, but the materials for the puppies cost $3.00. They sold the kittens for $7.00 and the puppies for $7.50. To meet the demand, they had to sell more kittens than puppies. They could not obtain materials to make more than 40 kittens per week and more than 20 puppies per week. To pay their expenses, they had to sell a total of more than 50 per week. Let x represent the stuffed kittens and y represent the stuffed puppies.

27. Write the objective function to minimize cost.

28. Write the objective function to maximize revenue.

29. Write the objective function to maximize profit.

30. Write inequalities for the constraints.

31. Graph the constraints and shade the region that satisfies all the inequalities.

32. Find the corners of the region in Exercise 31.

33. Find the quantities that will minimize cost within the constraints.

34. Find the quantities that will maximize revenue within the constraints.

35. Find the quantities that will maximize profit within the constraints.

Find the weekly earnings for each.

1. 42 hours at $4.80 per hour   **2.** 33 hours at $17.25 per hour

Find the weekly take-home pay for each gross pay. Each person is single and claims one withholding allowance (exemption). Tables for federal income tax withholding are in the Reference Section.

3. $201.60   **4.** $569.25   **5.** $202.50   **6.** $844.20

Akira sells stadium cushions. Find the total production cost to manufacture the numbers of cushions indicated in Exercises 7–9.

Fixed Costs		Variable Costs	
Labor	$4,25 per hour	Materials	$1.40 each
	for 30 hours	Packaging	$0.50 each
Advertising	$2.00		
Energy	$2.58		
Transportation	$4.42		

7. 30 cushions   **8.** 60 cushions   **9.** 90 cushions

Find the profit or loss for selling the following amounts at $5.50 each.

10. 30 cushions   **11.** 60 cushions   **12.** 90 cushions

Use your results from the previous exercises to complete the spreadsheet.

	Fixed Cost	Unit Cost	Number Produced	Total Cost	Unit Price	Revenue	Profit (Loss)
13.			30				
14.			60				
15.			90				

16. Refer to Exercises 7–15. Graph the cost and revenue functions.

17. Refer to Exercises 7–15. Find the break-even point algebraically.

Jod and Jessie sell stadium cushions and caps. The cushions cost $1.90; the caps cost $2.25. They sell the cushions for $5.00 and the caps for $6.00. They can obtain no more than 100 cushions and 75 caps per week. To meet demands, they have to sell a total of at least 120 of the two together. They cannot package more than 150 per week. Let x represent the cushions and y represent the caps.

18. Use linear programming to graph the given constraints, and find the points that will give maximum and minimum quantities.

19. What quantities will give the minimum cost?

20. What quantities will give the maximum revenue?

21. What quantities will give the maximum profit?

CUMULATIVE REVIEW

1. Your regular rate of pay is $6.22 per hour. You also receive $1\frac{1}{2}$ times your hourly rate for any hours over 20 in one week. Last week you worked 26 hours. You also received $12.50 in tips. Find your total earnings for the week.

2. You work according to the time schedule shown. You earn $5 per hour. Find your total earnings for the week if you receive $22 in tips that week.

Monday	Tuesday	Wednesday	Thursday	Friday	Saturday
3–6 P.M.	3–7 P.M.	3–6 P.M.	3–6 P.M.	4–9 P.M.	8 A.M.–12 P.M.

3. You earn $9.30 per hour. You work 35 hours a week for 50 weeks and get 2 weeks of paid vacation as well as 18% of your base salary in fringe benefits. Find your yearly earnings including benefits.

4. You are a real-estate agent and sold a house for $110,000. You earn a 4% commission on the first $100,000 of each sale and 6% commission on the amount over $100,000. How much commission did you earn on the sale?

5. You earned $94.00. Your income tax withholding is $7 and your FICA withholding is 7.65% of your earnings. Find your take-home pay.

6. You wrote 33 checks one month. The bank charges $0.0225 for the first 20 checks that you write and $0.08 for each written check over 20. How much are the charges for that month?

7. Your employer made out your paycheck with your first and middle initials and full last name. Write your name as you would to endorse the check.

8. You save $28 per week. You want to buy a stereo system that costs $589. For how many weeks must you save before you can have the stereo system?

9. Felix withdraws the interest from his savings account as soon as the interest is posted twice a year. What is the simple interest earned in $2\frac{1}{2}$ years on his $500 balance? The annual interest rate is 3.5%.

10. Find the value of a $1000 investment after interest has been compounded quarterly for 1 year at an 8.5% annual rate.

11. Find the maximum amount of extra money generated by a deposit of $3000 if the reserve requirement is 12%.

12. The fixed cost of producing hand-painted sweatshirts is $285 per week. The variable costs are $8 per shirt and $1 per shirt for paints. What is the total cost of producing 25 shirts in one week?

13. Use the information in Exercise 12 to find the profit or loss, given that each sweatshirt is sold for $20.

14. Write the equations for the cost function and revenue function of Exercises 12 and 13. Solve these equations to find the break-even point for the sweatshirt-selling operation.

PROJECT 4–1: **Retail Sales**

Work in small groups to set up imaginary companies. Select items that your company would like to sell or services that it would provide. You may use information from the chapter to plan costs or do your own research to find the costs of producing various services or products. Tell how you arrived at these costs for each product or service that your company produces.

1. Decide on the product or service for your company. Describe that product or service. Be sure to include all elements including advertising, energy use, transportation, and packaging.

2. Decide on the number of hours per week each employee (member of the group) will work. Decide the hourly wages for each person. Decide the marital status and number of exemptions for each employee.

3. Complete a payroll register showing each employee's earnings for four consecutive weeks.

4. Furnish a four-week payroll summary for each employee.

5. List your fixed and variable costs. Use a spreadsheet to show fixed costs, variable costs, number produced, total costs, average sale (or sale per unit), total sales, and profit or loss. Your spreadsheet should have entries that result in a loss and others that result in a profit.

6. Write the cost, revenue, and profit functions for your product.

7. Construct a line graph showing the break-even point for the business.

8. List several realistic constraints related to quantities and costs. Use linear programming to minimize costs or maximize sales revenue within the given constraints.

Extensions

1. After you have completed a plan for your company, get together with other groups in your class and try to sell them your product.

2. On the basis of feedback from other groups, redesign your product or service to be more useful or attractive to potential customers.

3. Listen in turn to other groups' presentations, and give them feedback about their product or service. Try to make comments that will help them in redesigning their product or rethinking their service.

Evaluate each expression. Recall that for all real numbers a, $a \neq 0$, and for all integers n:

$$a^{-n} = \frac{1}{a^n}$$

Examples: $\quad 3^{-2} = \frac{1}{3^2} = \frac{1}{9} \qquad (2^{-4})^2 = 2^{-8} = \frac{1}{2^8} = \frac{1}{256}$

1. 4^{-3}
2. 5^{-5}
3. $(3^{-4})^3$

4. $(2^{-3})^2$
5. $(4^{-2})^{-3}$
6. $(3^{-1})^{-3}$

Express with positive exponents. Recall that for all nonzero real numbers a and all integers m and n:

$$\frac{a^m}{a^n} = a^{m-n} \quad \text{if } m \geq n \qquad \frac{a^m}{a^n} = \frac{1}{a^{n-m}} \quad \text{if } m < n \qquad a^0 = 1$$

Example: $\quad$ Express $\frac{x^{-3}y^2z^{-5}}{x^2y^{-4}z^{-5}}$ with positive exponents: $\frac{x^{-3}y^2z^{-5}}{x^2y^{-4}z^{-5}} = \frac{y^6}{x^5}$

7. $\frac{x^{-3}y^{-4}z^{-1}}{x^{-5}y^3z^{-12}}$
8. $\frac{x^0y^3z^{-4}}{x^{-8}y^2z^0}$
9. $\frac{x^9y^{-8}z^{-7}}{x^{-3}y^{-4}z^7}$

Express each percent as a decimal. Recall that *percent* means *per hundred*.

Examples: $\quad 7.5\% = 0.075 \qquad 0.02\% = 0.0002 \qquad 135\% = 1.35$

10. 8.9%
11. 13.6%
12. 0.095%

13. 0.78%
14. 128%
15. 1600%

Evaluate to the nearest hundredth. Use your calculator.

Example: $\quad \dfrac{1500(0.07)(1 + 0.07)^{24}}{(1 + 0.07)^{24} - 1}$

Use the exponent key $\boxed{x^y}$ or $\boxed{\wedge}$ and parentheses as necessary.

1500 $\boxed{\times}$ 0.07 $\boxed{\times}$ $\boxed{(}$ 1 $\boxed{+}$ 0.07 $\boxed{)}$ $\boxed{x^y}$ 24 $\boxed{\div}$
$\boxed{(}$ $\boxed{(}$ 1 $\boxed{+}$ 0.07 $\boxed{)}$ $\boxed{x^y}$ 24 $\boxed{-}$ 1 $\boxed{)}$ $\boxed{\text{ENTER}}$
130.78 $\quad$ To the nearest hundredth

16. $\dfrac{1275(0.005)(1 + 0.005)^{48}}{(1 + 0.005)^{48} - 1}$
17. $\dfrac{1225(0.0025)(1 + 0.0025)^{60}}{(1 + 0.0025)^{60} - 1}$

18. $\dfrac{2400[1 - (1 + 0.015)^{-13}]}{0.015}$
19. $\dfrac{325[1 - (1 + 0.0125)^{-24}]}{0.0125}$

CONSUMER CREDIT

USING SOMEONE ELSE'S MONEY WHILE we wait to acquire our own gives us an inflated sense of wealth. We can now buy almost everything that is for sale without using money that we currently possess.

With an installment loan, we can have a house or a car before we can afford to pay for it. Banks and other lending institutions are willing to finance these purchases if we enter into an installment loan agreement that guarantees repayment of the borrowed money. Banks generate profits in the process because we are willing to repay two or three times as much as we borrow if it enables us to purchase a house.

As Americans, we think of ourselves as honest and hardworking. We have been convinced that owing money and being in debt are fine as long as we can keep up with the payments. Skillful advertisers understand our basic desires for more than we have. Once we buy one item on credit, we find it easier and easier to buy more items on credit. Lending institutions capitalize on our false sense of wealth by offering more and more credit to those of us who can make the installment payments.

Owning something now that we cannot afford to pay for now, except in small monthly installments, is a very attractive and luxurious way to live. However, we can overextend ourselves easily without even realizing it. We may continue to enjoy the use of credit, but as intelligent consumers we should be aware of the hazards. We should know the high price of spending what we do not have and who receives the profits from our use of credit.

In this chapter our high-school students will examine installment buying. They will become aware of the risks of using credit and learn ways to reduce its cost.

MONTHLY PAYMENTS: HOW MUCH CAN YOU AFFORD TO BORROW?

atrick is impressed by Maria's ability to save so much money from her weekly paycheck. He has a job at Paradise Department Store, where he is a cashier in the camera/luggage department. He will graduate from high school this year and would like to attend a community college for at least a year or two.

Patrick and his father have been trying to save money for the state university tuition, but it has been very difficult since Patrick's mother died three years ago. They lost her income, and her illness used up a large portion of their savings. Therefore if Patrick wants to go to college, he will have to help pay the expenses.

Before making a loan, a lending institution must have faith in the borrower's ability to repay it at a future date. Patrick knows that Maria's father is helping her to borrow money to buy a car. Maria will be making monthly payments to the bank to pay back the loan.

Patrick hopes that he too can borrow from the bank for school tuition. He can afford only a certain amount in monthly payments, and he would like to know how much he will be able to borrow.

Maria told Patrick that he will have to pay back more than he borrows over the course of his loan. He is not sure that this is fair, but he wants to go to college, and he knows that the bank will need to make a fair profit as a condition of granting him a loan.

OBJECTIVES: In this lesson, we will help Patrick to:

- *Identify the major functions of credit.*
- *Recognize the features of installment credit.*
- *Calculate monthly payments on a loan.*
- *Compute total payments on a loan.*
- *Determine how much he can afford to borrow.*

NATURE OF CREDIT

Patrick has heard the phrase "Buy now, pay later" in commercial advertising. The use of credit has become a way of life for many consumers, and Patrick thinks that he can handle it too. His economics teacher mentioned that the word *credit* actually means *debt*. People do not like the sound of being in debt, so they use the word *credit* instead.

Credit is a form of debt that occurs whenever cash, goods, or services are provided in exchange for a promise to pay at a future date. Many people cannot afford to pay cash for a large purchase such as a house or an automobile. Therefore they borrow the amount they need for the purchase and pay back that amount plus interest. The total amount paid to the bank is known as the **deferred payment price.** The deferred payment price is considerably more than the original loan amount. The difference between the two amounts is the *interest* or the **finance charge.** It is the profit made by the bank or lending institution. It is also the price paid by the borrower for using someone else's money.

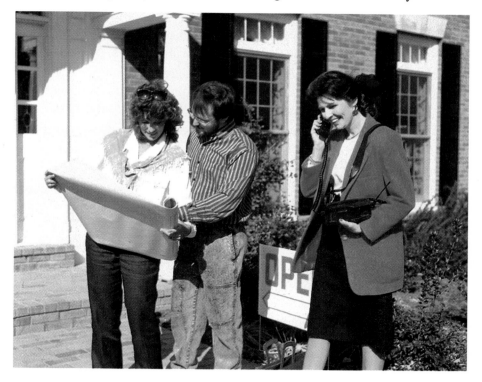

FUNCTIONS OF CREDIT

Patrick has discovered from doing research that the use of credit has the following positive effects:

1. *Credit stabilizes the economy.* It steadies economic activity because it enables individuals and businesses to purchase goods and services even when their incomes are temporarily limited.

2. *Credit promotes business growth.* Many people borrow money to start new businesses or to maintain established ones.

ALGEBRA *REVIEW*

Evaluate. Round answers to the nearest hundredth.

1. $x = \dfrac{2^3}{3^2}$

2. $x = \dfrac{6^4}{3^3}$

3. $x = 4(3 + 2)^3$

4. $x = \dfrac{5}{(5 + 3)^2}$

5. $x = (1 + 0.01)^3$

6. $x = \dfrac{(1 + 0.02)^6}{0.02}$

7. $x = \dfrac{1200(2)^5}{(5 - 3)^3}$

8. $x = \dfrac{600(1 + 5)^3}{(6 - 3)^4}$

9. $x = \dfrac{98(0.005)(1 + 0.005)^{24}}{(1 + 0.005)^{24} - 1}$

10. $x = \dfrac{900(0.0025)(1 + 0.0025)^{36}}{(1 + 0.0025)^{36} - 1}$

Solve for x.

11. $y = x(1 - r)^2$

12. $y = \dfrac{2x(1 + r)}{r^2 z}$

13. $z = \dfrac{3rq^{24}}{x(q + 2)^{24}}$

3. *Credit expands productivity and production.* By borrowing money to attend college, Patrick may increase his earning potential. Similarly, a business that must sell what it produces will have no income until its products are sold. The initial costs of production must be financed by funds that are already in the business or by borrowed funds.

4. *Credit raises the standard of living.* Individuals or couples do not have to wait to build up savings before buying things that make life comfortable, such as a house, a car, furniture, and appliances.

USING INSTALLMENT CREDIT

Using installment credit means making purchases and then making regular payments over a period of months or years. Maria will be getting an **installment loan** when she buys a car. Purchases made on an installment plan often have the following features:

1. A **down payment** is required at the time of the purchase. A down payment is the portion of the purchase that must be paid up front in cash.

2. A substantial finance charge is added to the price.

3. Payments of equal amounts are spread over a specified period of time.

4. To insure that payments on a purchase loan are made as scheduled, protection may be provided to the lender in the form of a security agreement.

Ask Yourself

1. What are the four major functions of credit?

2. What are three features of installment credit?

3. In what must a bank have faith when it approves a loan?

SHARPEN YOUR SKILLS

SKILL 1

Maria can use the monthly payment formula to determine the monthly cost of a car loan. The monthly interest rate is the annual interest rate divided by 12.

> ## Monthly Payment Formula
>
> $$M = \frac{Pr(1 + r)^n}{(1 + r)^n - 1}$$
>
> where M = monthly payment
> P = amount of loan
> r = monthly interest rate
> n = number of payment periods

EXAMPLE 1 Maria finds a used car that she might want to buy. To pay for the car, she must borrow $6500.

QUESTION What is her monthly payment if she borrows $6500 for a 3-year period at an annual interest rate of 9%? How much will she pay for the car in the course of 3 years?

SOLUTION
Use the monthly payment formula and your calculator.

$$M = \frac{Pr(1 + r)^n}{(1 + r)^n - 1}$$

$$M = \frac{6500(0.0075)(1 + 0.0075)^{36}}{(1 + 0.0075)^{36} - 1}$$ $P = 6500; r = 0.09 \div 12 = 0.0075;$
$n = 3 \bullet 12 = 36$

$$M = 206.70$$ To the nearest cent

To input this calculation in your calculator, remember to use the exponent key $\boxed{x^y}$ (sometimes labeled ^ or a^b) and to use parentheses around the denominator as follows:

6500 $\boxed{\times}$ 0.0075 $\boxed{\times}$ $\boxed{(}$ $\boxed{(}$ 1 $\boxed{+}$ 0.0075 $\boxed{)}$ $\boxed{x^y}$ 36 $\boxed{\div}$
$\boxed{(}$ $\boxed{(}$ 1 $\boxed{+}$ 0.0075 $\boxed{)}$ $\boxed{x^y}$ 36 $\boxed{-}$ 1 $\boxed{)}$ $\boxed{\text{ENTER}}$

Maria's monthly payment is $206.70.

The answer to the above calculation is given to many decimal places in your calculator. Use that answer to calculate the deferred payment price by using the answer key $\boxed{\text{ANS}}$ on your calculator.

$\boxed{\text{ANS}}$ $\boxed{\times}$ 36 = 7441.14   Deferred payment price

Notice that 206.70 • 36 = 7441.20. The bank adjusts your last payment so that the deferred payment price is $7441.14.

EXAMPLE 2 If Maria buys a new car instead, she will have to borrow $12,825.

QUESTION What is her monthly payment if she borrows $12,825 for a 4-year period at an annual interest rate of 6%? How much will she pay for the car in the course of 4 years?

SOLUTION
Use the monthly payment formula and your calculator.

$$M = \frac{Pr(1 + r)^n}{(1 + r)^n - 1}$$

$$M = \frac{12,825(0.005)(1 + 0.005)^{48}}{(1 + 0.005)^{48} - 1}$$
$P = 12,825;\ r = 0.06 \div 12 = 0.005;$
$n = 4 \cdot 12 = 48$

$$M = 301.20$$
To the nearest cent

Maria's monthly payment is $301.20. Over four years she will pay [ANS] [×] 48 = $14,457.38, which is the deferred payment price or total payment.

EXAMPLE 3 Maria is also considering five other cars on which she can obtain 8% financing for 3 years, 4 years, or 5 years.

QUESTION What are her monthly and total payments (deferred payment price) on loans of $4200, $5500, $6325, $8275, and $9750?

SOLUTION
Use a spreadsheet program to find the monthly and total payments for the loan amounts at 8% over the periods of 3 years, 4 years, and 5 years. When entering the formula for the monthly payment, use 0.08/12 for the rate. For 3 years at 8% the monthly payment formula in cell C3 is

+A3∗(0.08/12)∗(1+(0.08/12))^36/((1+(0.08/12))^36−1)

For the monthly payment formula for 4 years, use 48 for the number of payment periods. You can use the COPY command of your spreadsheet program to copy the formula of cell C3 to C4. This will automatically change A3 to A4. Then you can use the EDIT command to change 36 to 48.

$$+A4*(0.08/12)*(1+(0.08/12))^{48}/((1+(0.08/12))^{48}-1)$$

The monthly payment formula for 5 years is

$$+A5*(0.08/12)*(1+(0.08/12))^{60}/((1+(0.08/12))^{60}-1)$$

For the total payment, type $12*B3*C3$ in cell D3. Then use the COPY command to copy the formula in the total payment column.

	A	B	C	D	
	Loan	Number	Monthly	Total	
	Amount	of Years	Payment	Payment	
1					
2					
3	4200	3	131.61	4,738.06	12*B3*C3
4	4200	4	102.53	4,921.65	12*B4*C4
5	4200	5	85.16	5,109.65	
6	5500	3	172.35	6,204.60	
7	5500	4	134.27	6,445.01	
8	5500	5	111.52	6,691.21	
9	6325	3	198.20	7,135.29	
10	6325	4	154.41	7,411.76	
11	6325	5	128.25	7,694.89	
12	8275	3	259.31	9,335.10	
13	8275	4	202.02	9,696.81	
14	8275	5	167.79	10,067.23	
15	9750	3	305.53	10,999.06	
16	9750	4	238.03	11,425.25	
17	9750	5	197.69	11,861.69	

If you use $(12*B3)$ instead of 36 in cell C3, you can use the COPY command to complete the entire monthly payment column. The formula is

$$+A3*(0.08/12)*(1+(0.08/12))^{(12*B3)}/((1+(0.08/12))^{(12*B3)}-1)$$

―――― **SKILL 2** ――――

EXAMPLE 4 Patrick is an excellent student and will probably receive a scholarship for college. However, his father explains that Patrick will still have to borrow money.

QUESTION If Patrick can afford monthly payments of $225 and would like to borrow an even multiple of $1000, what is the largest amount of money that he can borrow at 8% for 3 years, 4 years, and 5 years?

SOLUTION

Use a spreadsheet program to determine the monthly payments for $7000 to $12,000 in increments of $1000. The monthly payment formulas are

+A3*(0.08/12)*(1+(0.08/12))^36/((1+(0.08/12))^36−1)	For 3 years
+A3*(0.08/12)*(1+(0.08/12))^48/((1+(0.08/12))^48−1)	For 4 years
+A3*(0.08/12)*(1+(0.08/12))^60/((1+(0.08/12))^60−1)	For 5 years

	A	B	C	D
1	Amount	3 years	4 years	5 years
2	Borrowed			
3	7,000	219.35	170.89	141.93
4	8,000	250.69	195.30	162.21
5	9,000	282.03	219.72	182.49
6	10,000	313.36	244.13	202.76
7	11,000	344.70	268.54	223.04
8	12,000	376.04	292.96	243.32

Patrick can afford to borrow $7000 for 3 years, $9000 for 4 years, or $11,000 for 5 years.

SKILL 3

Patrick solves the monthly payment formula for P, the amount of the loan.

$$M = \frac{Pr(1 + r)^n}{(1 + r)^n - 1}$$

$$M[(1 + r)^n - 1] = Pr(1 + r)^n \qquad \text{Multiply both sides by } [(1 + r)^n - 1].$$

$$\frac{M[(1 + r)^n - 1]}{r(1 + r)^n} = P \qquad \text{Divide both sides by } r(1 + r)^n.$$

He now has a formula to find the exact amount he can borrow.

> ## Amount Formula
>
> $$P = \frac{M[(1 + r)^n - 1]}{r(1 + r)^n}$$
>
> where P = amount of loan
> r = monthly interest rate
> n = number of payment periods
> M = monthly payment

EXAMPLE 5 If Patrick does not restrict himself to borrowing multiples of $1000, he can borrow more money over each time interval.

QUESTION What is the exact amount of money Patrick can borrow at 8% for 3 years if he can afford a monthly payment of $225?

SOLUTION

Use the amount formula. Since the monthly interest rate r is $0.08 \div 12$, which is a repeating decimal, calculate this value, and then use the [ANS] key to enter it in the formula.

$$P = \frac{225((1 + \boxed{\text{ANS}})^{36} - 1)}{\boxed{\text{ANS}}(1 + \boxed{\text{ANS}})^{36}}$$ $M = 225; n = 3 \bullet 12 = 36$

$P = 7180.16$ To the nearest cent

To input this in your calculator use the following keystrokes.

0.08 [÷] 12 [ENTER]
225 [(] [(] 1 [+] [ANS] [)] [xʸ] 36 [−] 1 [)] [÷] [ANS] [(] 1 [+] [ANS] [)]
[xʸ] 36 [ENTER]

Patrick can afford to borrow $7180.16 at 8% for 3 years.

The [ANS] key is a temporary memory that stores the latest calculation. You can also store $0.08 \div 12$ in one of the regular memories of your calculator.

0.08 ÷ 12 [STO →] [ALPHA] [A] [ENTER]

To recall the number stored in memory A, you would enter [ALPHA] [A] in place of [ANS] in the previous calculation.

TRY YOUR SKILLS

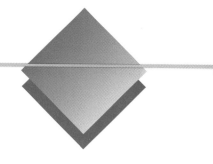

Determine the monthly payment and total payments (deferred payment price) for each car loan. Remember that the deferred payment price is [ANS] times the number of payments.

1. $10,000 at 7% for 5 years
2. $12,000 at 8% for 4 years

Paola can afford a monthly payment of $150 and would like to borrow an even multiple of $1000. Complete the table below to determine the largest amount of money that she can borrow at 10% for 3 years, 4 years, and 5 years.

	Amount Borrowed	3 years	4 years	5 years
3.	$3000			
4.	4000			
5.	5000			
6.	6000			
7.	7000			
8.	8000			

9. Use the amount formula to determine the exact amount of money that Paola can borrow for each time period in Exercise 3. Remember that Paola can borrow (or afford) a payment of $150 a month.

EXERCISE YOUR SKILLS

KEY TERMS

credit
deferred payment
  price
down payment
finance charge
installment loan

1. Why is the term *credit* used instead of *debt*?

2. How does the use of credit raise the standard of living?

3. Why do you think consumers are willing to pay substantial finance charges as a condition of an installment loan?

4. Who receives the revenue from finance charges paid by consumers?

5. What do you think the bank or other lending institution would do if a borrower did not make the payments on an installment loan?

6. What is the deferred payment price?

Determine the monthly payment and total payment (deferred payment price) for each car loan.

7. $13,500 at 7% for 4 years

8. $10,500 at 9% for 5 years

Patrick's friend Ricky is also considering various loans for college tuition. Use a spreadsheet program to find the monthly payment and total payment for the loan amounts listed below at 12.5% over the periods of 3 years, 4 years, and 5 years. (Save your results to use in Lesson 5–2.)

	Loan Amount	Number of Years	Monthly Payment	Total Payment
9.	$ 6,500	3		
10.	6,500	4		
11.	6,500	5		
12.	9,500	3		
13.	9,500	4		
14.	9,500	5		
15.	6,000	3		
16.	6,000	4		
17.	6,000	5		
18.	12,000	3		
19.	12,000	4		
20.	12,000	5		
21.	17,500	3		
22.	17,500	4		
23.	17,500	5		
24.	11,650	3		
25.	11,650	4		
26.	11,650	5		
27.	8,775	3		
28.	8,775	4		
29.	8,775	5		

Kareem can afford a monthly payment of $300 and would like to borrow a multiple of $1000. Complete the table below to determine the largest amount of money that he can borrow at 9% for 3 years, 4 years, and 5 years.

	Amount Borrowed	3 years	4 years	5 years
30.	$ 9,000	$286.20	$223.97	
31.	10,000			
32.	11,000			
33.	12,000			
34.	13,000			
35.	14,000			
36.	15,000			

37. Determine the exact amount of money that Kareem can borrow for each time period in Exercises 30–36.

MIXED REVIEW

1. Matthew's monthly check costs are $0.025 per check for the first 15 checks and $0.12 for each check over 15. If he writes 33 checks this month, how much will the checks cost him?

2. Crystal saves $12.50 per week. At the end of a year, she deposits her savings into a CD that earns 2% for each six-month period. When the CD matures, she withdraws the interest and reinvests the principal. How much interest will she have earned at the end of two years?

3. How much will $6200 be worth at the end of 4 years if it earns 7% interest, compounded semiannually?

4. Olivia earns $9.50 per hour. She works $37\frac{1}{2}$ hours per week for 49 weeks and receives 3 weeks of paid vacation. Her other benefits cost her employer $4000 per year. What is the annual cost to the company of her salary and benefits?

5. Arnold saves $33 per week. He wants to buy a desk that costs $294. How many weeks must he save before he can buy the desk?

6. Alan and Meg are producing and selling sports pennants. The raw materials for each pennant cost $2.13 and the labor is $5.50 an hour. Advertising and other fixed costs are $45. What is the cost of production for 250 pennants, 20 of which can be produced in one hour?

Suppose that the fixed costs of a business are $350 per week and the variable costs are $6.50 per item.

7. What is the cost of manufacturing 92 items in one week?

8. Each item is sold for $10.99. What is the weekly profit?

9. What is the break-even point?

Joan feels that she must warn Patrick and Maria about credit. The first installment loan that her family accepted seemed quite reasonable because it was a mortgage for their house. Then they bought a brand new car on credit.

After a series of violent thunderstorms, the roof in the den started to leak, so the family took out a second mortgage to pay for a new roof and to replace the damaged carpeting. They will be paying on the second mortgage for 7 years.

When Joan's mom and dad were both working, they were easily making all of the installment payments. Four months ago, Joan's mother lost her job. The electronics company at which she was working had been forced to eliminate some jobs in her department because the company lost money last year.

More bills arrive every day. Joan's parents even borrowed money from the credit union to pay some of the bills, but now they are having trouble making the payments to the credit union. The family members are a little stunned by recent events and are wondering how they managed to get into such a mess so quickly.

Joan's mother feels that one of their problems is that they borrowed the money for short time periods and have high monthly payments. She thinks that they should have taken a second mortgage with a payback period of 15 years instead of 7 years. Her father gets upset because he claims that the total payments and total cost are higher when the term of the loan is longer. Joan will be glad when Mom finds another job. Then Mom and Dad will stop being upset, and the family can get back to normal.

OBJECTIVES: *In this lesson, we will help Joan to:*

- *Cite several problems with using credit.*
- *Recognize indications that a family has used too much credit.*
- *Compare the effects of the payback period of a loan on the total payments and interest charges.*

CREDIT: TOOL OR TRAP?

Many Americans borrow money from a bank or other lending institution to obtain the use of merchandise before paying for it. Americans are heavy credit users. Almost everyone uses credit, especially for expensive purchases such as a new car or home repairs. What went wrong for Joan's parents, and how can credit create a problem for millions of others like them? You should always keep the following points about credit in mind.

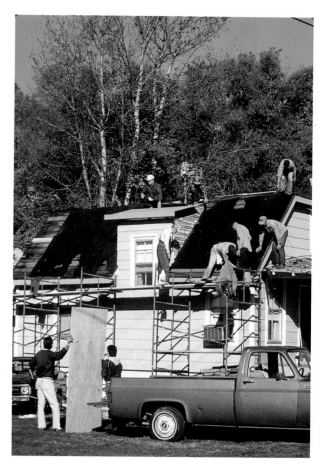

1. *Credit is rented money.* Borrowers pay heavily for the privilege of using credit. Banks and other lending institutions encourage borrowers to indulge in this privilege because the credit that is granted is very profitable for the lenders.

2. *Credit ties up future income.* You are actually spending future income or earnings when you make credit purchases. If you tie up too much future income in making credit payments, you are putting a burden on yourself. If your income happens to disappear as Joan's mother's did, making the payments may become impossible.

3. *Credit makes it easy to overspend.* Businesses with products or services to sell are cashing in on consumers' willingness to use credit for a purchase that they would not consider buying if they had to pay cash. A furniture store owner can sell more bedroom suites by offering "Convenient credit terms—No payments for 6 months."

MANAGING CREDIT

People, businesses, and even governments can be forced into bankruptcy through the unwise use of credit. Excessive debt can undermine job performance, marriage, and health. Installment buying can be helpful if you use it wisely. However, credit can be harmful if you abuse it.

One convenient yardstick for measuring your ability to handle debt is the following: No matter what your income, 10% of take-home pay, excluding a home mortgage, is a comfortable amount to spend making payments on credit

ALGEBRA *REVIEW*

Evaluate. Store the results in your calculator's memory. Give the results to the nearest hundredth.

1. $x = \dfrac{6700(0.002)(1 + 0.002)^{40}}{(1 + 0.002)^{40} - 1}$

2. $y = \dfrac{3300(0.0033)(1 + 0.0033)^{30}}{(1 + 0.0033)^{30} - 1}$

Use the results of Exercises 1 and 2 to determine z and q.

3. $z = 40x$ **4.** $q = 30y$

Use the results of Exercises 3 and 4 to determine s and t.

5. $s = z - 6700$

6. $t = q - 3300$

Evaluate. Store the results in your calculator's memory. Give the results to the nearest hundredth.

7. $x = \dfrac{1800(0.004)(1 + 0.004)^{44}}{(1 + 0.004)^{44} - 1}$

8. $y = \dfrac{6600(0.0066)(1 + 0.0066)^{44}}{(1 + 0.0066)^{44} - 1}$

9. Use the stored results of Exercises 7 and 8 to solve for z if $z = 44x - 44y$.

Determine which expression is smaller.

10. $36(424.70)$ or $60(247.64)$

11. $48(359.28)$ or $36(482.12)$

accounts or installment loans, 15% is a manageable amount, and more than 20% is a dangerous credit overload. Indications that you are suffering from credit overload include the following:

- You must miss some installment payments to make the monthly mortgage payment.
- You seek a new loan before repaying an old one.
- You continue to use your credit cards, even though you can pay only the minimum amount due on your credit card accounts.
- You take out loans to combine debts or ask for extensions on existing loans.
- You receive repeated overdue notices from creditors.
- You have little or no savings or are drawing on savings to pay regular bills that you used to pay out of monthly income.
- You receive telephone calls or letters from creditors demanding payment on overdue bills.

Ask Yourself

1. What are three problems with using credit?

2. What are four indicators that a borrower is experiencing credit overload?

3. Why is it easy to borrow too much money?

SHARPEN YOUR SKILLS

SKILL 1

Maria is learning more about borrowing money and paying it back in monthly amounts. She has discovered that bankers and loan agencies give a name to this "payback" procedure. To make payments on a loan is to **amortize** the loan. The word amortize is derived from the French *à mort* which means "to death." When Maria amortizes her loan, she is gradually "bringing her loan to death."

The cost of a loan is equal to the amount of interest paid over the term of the loan. It is found by subtracting the original loan amount from the total payment.

Total cost = total payment − original loan amount

EXAMPLE 1 In Example 3 of Lesson 5–1, Maria considered loans of $4200, $5500, $6325, $8275, and $9750 at 8%. As the term of a loan gets longer, her monthly payments get smaller, and her total payment gets larger. She wonders how long she should amortize her loan.

QUESTION Does the cost of financing a loan increase or decrease as the term of the loan increases?

SOLUTION
Using a spreadsheet program, add a column to the table in Example 3 of Lesson 5–1 to show the total cost (interest payments) of each loan.

	A	B	C	D	E
1	Loan	Number	Monthly	Total	Total
2	Amount	of Years	Payment	Payment	Cost
3	4200	3	131.61	4,738.06	538.06 +D3−A3
4	4200	4	102.53	4,921.65	721.65
5	4200	5	85.16	5,109.65	909.65
6	5500	3	172.35	6,204.60	704.60
7	5500	4	134.27	6,445.01	945.01
8	5500	5	111.52	6,691.21	1191.21
9	6325	3	198.20	7,135.29	810.29
10	6325	4	154.41	7,411.76	1086.76
11	6325	5	128.25	7,694.89	1369.89
12	8275	3	259.31	9,335.10	1060.10
13	8275	4	202.02	9,696.81	1421.81
14	8275	5	167.79	10,067.23	1792.23
15	9750	3	305.53	10,999.06	1249.06
16	9750	4	238.03	11,425.25	1675.25
17	9750	5	197.69	11,861.69	2111.69

Use the graph function of your spreadsheet program to graph the relationship between amortization periods of 3 years, 4 years, and 5 years and interest payments.

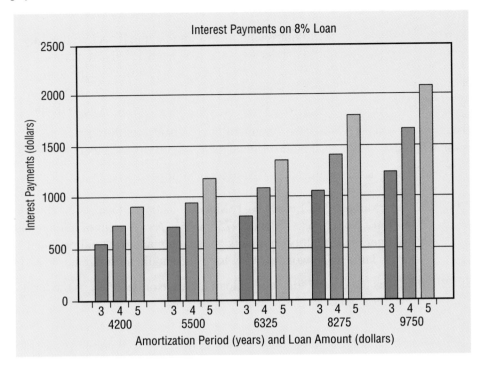

It costs more to finance a specified loan amount at a given interest rate as the amortization period increases.

It is possible to borrow a larger sum if you make the same monthly payments over a longer period of time. Maria must decide whether she prefers a lower monthly payment or a lower total cost for a given loan amount.

SKILL 2

EXAMPLE 2 Patrick points out that Maria can buy a car priced at $9750 or a car priced at $6325 for a monthly payment of approximately $198.

QUESTION Should Maria automatically choose the more expensive car without further investigation, since it appears to cost the same amount of money on a monthly basis? What other factor(s) should she take into account?

SOLUTION
No, she must also consider the total cost of each loan. Since she will be paying for the more expensive car for 5 years instead of 3 years, the total cost of the 5-year loan will be greater—$2111.69 instead of $810.29 for the smaller loan.

EXAMPLE 3 Joan wants to settle the dispute between her parents. Their second mortgage on the house was a loan for $40,000 at 12% for 7 years.

QUESTION Would an amortization period of 15 years, rather than 7 years, have been better for Joan's parents?

SOLUTION

Use your spreadsheet program to create a table showing monthly payments, total payments, and total cost for amortization periods of 7 through 15 years. The monthly rate is 0.12/12 = 0.01, and the loan amount is $40,000. The number of periods for 7 years is 12 • 7 = 84, so the spreadsheet formula for the monthly payment formula for 7 years for cell B6 is

$$40000*(0.01)*(1+0.01)^\wedge 84/((1+0.01)^\wedge 84-1)$$

For the monthly payment formulas for the other years, adjust the number of payments.

	A	B	C	D
1		Amortization Schedule—7 to 15 years		
2		$40,000 loan at 12% interest		
3				
4	Amortization	Monthly	Total	Total
5	Period (years)	Payment	Payment	Cost
6	7	706.11	59,313.18	19,313.18
7	8	650.11	62,410.91	22,410.91
8	9	607.37	65,595.88	25,595.88
9	10	573.88	68,866.06	28,866.06
10	11	547.12	72,219.20	32,219.20
11	12	525.37	75,652.94	35,652.94
12	13	507.47	79,164.77	39,164.77
13	14	492.57	82,752.06	42,752.06
14	15	480.07	86,412.10	46,412.10

Annotations: +C6−40000 (pointing to D6); +B6*(12*A6) (pointing to C6).

In Lesson 5–1, Example 3, you were shown how to modify a spreadsheet formula by replacing a number by a general formula, thus reducing the amount of work required to complete a long table. The same technique can be used for the spreadsheet amortization table above. Instead of 84 in the formula for cell B6, you can write the formula (12*A6). Then you can use the spreadsheet's COPY command to complete all of the Monthly Payment column, column B. For the total payment, use the formula +B6*(12*A6) in cell C6 instead of +B6*84.

Use the graph function of your computer spreadsheet program to graph the relationship between amortization periods of 7–15 years and monthly payments. This graph is shown at the top of the next page.

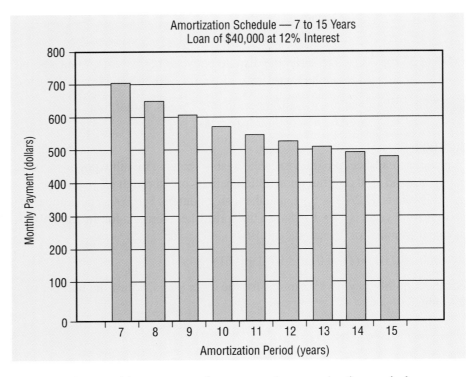

Note that the monthly payments decrease as the amortization period increases. As Joan's mother said, they would have approximately $226 more each month if they had amortized the loan over 15 years instead of 7.

Create another graph that shows the relationship between amortization periods of 7–15 years and total cost.

Note that the total cost increases as the amortization period increases. As Joan's father said, the cost of the loan would be approximately $27,000 more over the course of the loan if they had amortized it over 15 years instead of 7 years.

Joan notes that both her parents have strong support for their arguments. However, she feels that they should have discussed the situation before applying for the second mortgage. Perhaps they should have considered a greater total cost since it would have made their monthly cash flow lower. Maybe they should have compromised on an amortization period of 11 years.

TRY YOUR SKILLS

Use the monthly payment formula and a calculator to complete the table below to determine the total cost of borrowing each sum of money at 10% per year over 3 years, 4 years, and 5 years.

	Loan Amount	Number of Years	Monthly Payment	Total Payment	Total Cost
1.	$ 2,500	3	$80.67	$2904.05	$404.05
2.	2,500	4	63.41		
3.	2,500	5	53.12		
4.	5,000	3			
5.	5,000	4			
6.	5,000	5			
7.	7,500	3			
8.	7,500	4			
9.	7,500	5			
10.	10,000	3			
11.	10,000	4			
12.	10,000	5			

13. In Exercises 1–12, you can see that the payments on a $5000 loan for 3 years and a $7500 loan for 5 years are both approximately $160. Should you automatically select the $7500 loan? Why or why not?

EXERCISE YOUR SKILLS

1. Why should you refrain from taking out a new loan to help make payments on a previous loan if you are having difficulty making these payments?

2. What percent of a family's income can be spent making installment payments, excluding a home mortgage, before the family begins to experience credit overload?

3. How can excessive debt damage an individual's health?

Ricky would like to go backpacking with his friends in Colorado before going off to college. If he did this every summer, he would not be able to earn as much money working at summer jobs. Therefore it would take him longer to repay his loan. He wonders how much it will cost him to amortize his loan over 4 or 5 years instead of 3 years. Using your results from Exercises 9–29 of Lesson 5–1 and a spreadsheet program, add a column to the table to show Ricky the total cost (interest payments) of each loan.

	Loan Amount	Number of Years	Monthly Payment	Total Payment	Total Cost
4.	$ 6,500	3	$217.45	$7828.15	$1328.15
5.	6,500	4	172.77		
6.	6,500	5			
7.	9,500	3			
8.	9,500	4			
9.	9,500	5			
10.	6,000	3			
11.	6,000	4			
12.	6,000	5			
13.	12,000	3			
14.	12,000	4			
15.	12,000	5			
16.	17,500	3			
17.	17,500	4			
18.	17,500	5			
19.	11,650	3			
20.	11,650	4			
21.	11,650	5			
22.	8,775	3			
23.	8,775	4			
24.	8,775	5			

Mary Kim wants to borrow $20,000 at 7.5%. She wants to decide whether to borrow the money for 5, 6, 7, 8, 9, or 10 years.

25. Create a spreadsheet that shows the monthly payments, total payments, and total cost for amortization periods of 5–10 years for the loan.

26. Use the graph function of your computer spreadsheet program to graph the relationship between amortization periods and monthly payments.

27. Create a graph that shows the relationship between the amortization periods and the total costs.

28. Discuss the advantages and disadvantages of borrowing the money for the different time periods.

MIXED REVIEW

1. Lloyd earns $405.98 in gross income each week. How much will be deducted for Social Security and Medicare?

2. If Debbie can afford a monthly payment of $250, how much can she afford to borrow at a yearly interest rate of 8.5% for 4 years?

3. Irma and Louise have decided to crochet ponytail holders for hair. They can prepare 75 hair holders per week. How much will these 75 hair holders cost to make if the costs include 20 hours of labor at $5.00 per hour and the following materials costs:

   Twine at $0.95 per holder
   Beads at $0.75 per holder
   Hair elastics at $0.15 per holder

4. Julio saves $35 from his salary each week. For how many weeks will he have to save to be able to buy a computer word-processing system that costs $500?

5. Use the Rule of 72 to determine how long it will take your $2000 savings account to double in value if it is growing at a rate of 8%.

6. Jen can afford a monthly payment of $250 on her credit card account. How much can she afford to borrow for 4 years at an APR of 8.5%?

CREDIT MANAGEMENT: KEEPING CREDIT COSTS DOWN

5-3

Raul wants to take Joan to the prom this spring, but she is unwilling to discuss the matter. They have been dating for almost two years, so Raul feels that he knows Joan quite well. He also knows that her mother has lost her job, and he imagines that the financial situation at her house is difficult. He has concluded that Joan is reluctant to go to the prom because she cannot afford a new dress.

Raul is thinking about borrowing money to cover his prom expenses, a dress for Joan, and a portable keyboard. However, he wants to avoid making the same mistakes her family made. Therefore he will explore ways to reduce the cost of credit and then determine whether he can afford it.

OBJECTIVES: *In this lesson, we help Raul to:*

- *Study financial advisers' suggestions regarding the wise use of credit.*
- *Determine how much can be saved by understanding the various terms that an installment loan may carry.*

GUIDELINES FOR USING CREDIT WISELY

Credit can be an important tool in money management if it is used wisely. By doing some research, Raul has discovered the following suggestions on the proper use of credit made by financial advisers and credit counselors.

1. *Limit installment debt to 15–20% of take-home pay.* Monthly installment payments, excluding a home mortgage, are likely to be manageable if they are limited to 15–20% of take-home pay. Any amount over 20% may reduce an individual's or family's ability to pay for food, shelter, transportation, clothing, and other essentials. An authority on money matters has recommended that you not owe more than one-third of your discretionary income for the year. **Discretionary income** is the income you have left after paying for basic needs such as food, clothing, and shelter.

2. *Purchase durable products that will outlast the payment period.* If you are going to use credit for items that do not appreciate (increase in value), use it only on the purchase of durable products such as household appliances or cars. Be careful to choose products carefully. For instance, you do not want to make payments on a used car for 3 years if it only lasts for 15 months.

   If you are thinking of borrowing money or opening a charge account, determine the cost, and then decide whether you can afford it. Then shop around for the best credit terms.

WAYS TO REDUCE THE COST OF INSTALLMENT LOANS

Raul has also discovered the following ways to minimize credit costs:

1. *Select a short payment period.* As you discovered in Lesson 5–2, the cost of financing an item is greater with a longer payment period and smaller payments.

2. *Compare interest rates.* The interest rate on a loan can vary dramatically from lender to lender, so it is wise to shop for interest rates just as you would shop for the product itself.

3. *Make a large down payment.* The larger your down payment, the less you will have to borrow.

You can save money in interest charges by paying for a larger portion and financing a smaller portion.

Ask Yourself

1. What are two guidelines for using credit wisely?

2. What should you consider first if you are thinking about borrowing money?

3. What are three ways to reduce the cost of installment loans?

ALGEBRA *REVIEW*

Determine each of the following.

1. 20% of $5500

2. 10% of $7500

3. 30% of $4300

Evaluate. Give the results to the nearest hundredth.

4. $y = 5750 - (0.2)(5750)$

5. $z = 3600 - (0.4)(3600)$

6. Express y as the product of two numbers by factoring Exercise 4.

7. Express z as the product of two numbers by factoring Exercise 5.

SHARPEN YOUR SKILLS

SKILL 1

EXAMPLE 1 Raul is thinking about buying a portable keyboard. He has compared prices and financing costs at various stores. He can purchase it at one store for $650 at an interest rate of 12% and at another store for $650 at an interest rate of 9%. Both loans must be repaid over a period of 2 years.

QUESTION How much money will he save if he purchases the keyboard at the store that offers 9% interest?

SOLUTION
Use the monthly payment formula and your calculator to determine the monthly and total payments at each rate of interest.

$$M = \frac{Pr(1 + r)^n}{(1 + r)^n - 1}$$
$P = 650; n = 2 \cdot 12 = 24$

Monthly Payment at 12%	Total Payment at 12%	Monthly Payment at 9%	Total Payment at 9%	Total Savings
$30.60	$734.35	$29.70	$712.68	$21.67

Remember to use the answer key to find the total payments. However, to find the total savings, use the exact amounts. Why?

EXAMPLE 2 In Example 3 of Lesson 5-1, Maria considered loans of $4200, $5500, $6325, $8275, and $9750 at 8% over periods of 3 years, 4 years, and 5 years. The car dealership is offering 5% financing for one day only.

QUESTION How much can she save on each loan if she finances at a rate of 5% instead of 8%?

SOLUTION
Use a spreadsheet program to find the total payments on each loan at 8% and 5%, and then find the difference (savings). To subtract the exact amount of the total payments, use the function of your spreadsheet that formats a column in exact amounts, such as @round. Remember when you use a display function that a computer or calculator still uses the unrounded amounts in calculations unless you use special formatting functions.

For 5 years at 8%:

@ROUND(+60*A5*(0.08/12)*(1+(0.08/12))^60/((1+(0.08/12))^60-1),2)

For 5 years at 5%:

@ROUND(+60*A5*(0.05/12)*(1+(0.05/12))^60/((1+(0.05/12))^60-1),2)

	A	B	C	D	E
1	Loan	Number	Total	Total	Total
2	Amount	of Years	Payment at 8%	Payment at 5%	Savings
3	4200	3	4,738.06	4,531.60	206.46
4	4200	4	4,921.65	4,642.71	278.94
5	4200	5	5,109.65	4,755.55	354.10
6	5500	3	6,204.60	5,934.24	270.36
7	5500	4	6,445.01	6,079.73	365.28
8	5500	5	6,691.21	6,227.51	463.70
9	6325	3	7,135.29	6,824.37	310.92
10	6325	4	7,411.76	6,991.69	420.07
11	6325	5	7,694.89	7,161.63	533.26
12	8275	3	9,335.10	8,928.33	406.77
13	8275	4	9,696.81	9,147.24	549.57
14	8275	5	10,067.23	9,369.57	697.66
15	9750	3	10,999.06	10,519.78	479.28
16	9750	4	11,425.25	10,777.71	647.54
17	9750	5	11,861.69	11,039.67	822.02

Use the graph function of your computer spreadsheet program to graph the relationship between amortization periods of 3 years, 4 years, and 5 years and total payments for each amount at 8% and 5%.

The total payments, and therefore the total cost, on a loan for a given amount of money over a specified amortization period are always lower when the interest rate is lower.

SKILL 2

Many loans require that you make a *down payment,* which is a percentage of the cost or price. To find the down payment D multiply the cost C by the percent for the down payment r. To find the loan amount P, subtract the down payment D from the cost C. The total amount T is the sum of the down payment D and all the monthly payments nM.

Loans with Down Payment

$D = rC$ where D = down payment
     r = percent for down payment
     C = cost

$P = C - D$ where P = loan amount
     C = cost
     D = down payment

$T = nM + D$ where T = total amount
     n = number of payments
     M = monthly payment
     D = down payment

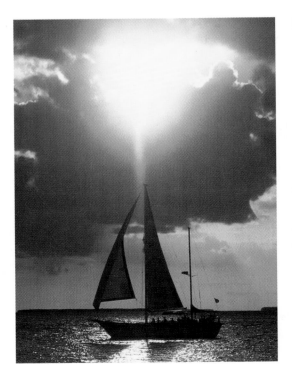

EXAMPLE 3 Joan's aunt and uncle just bought a boat for $14,950 at an interest rate of 12% for 5 years. They are very excited because they were required to make only a 10% down payment. Joan is worried that her aunt and uncle will end up in the same position as her parents did. She wonders whether they could have saved money if they had made a down payment of 20% or 30% instead of 10%.

QUESTION How much money could they have saved if they had made a down payment of 20% or 30%?

SOLUTION
The cost is $14,950 and the interest rate is 12% for 5 years. For 10% down, the down payment is 0.10(14,950) = 1,495 and the loan amount is 14,950 − 1495 = 13,455. Use the monthly payment formula to find the monthly payment and multiply the monthly payment by the number of payments to find the total amount.

Percent Down, r	Down Payment, D	Loan Amount, P	Monthly Payment, M	Total Amount, T	Savings over 10% Down
10%	$1495	$13,455	$299.30	$19,452.94	
20%	2990	11,960	266.04	18,952.62	$ 500.32
30%	4485	10,465	232.79	18,452.29	1000.65

They could have saved $500.32 with 20% down and $1000.65 with 30% down. How much do you think they would have saved with 40% down?

EXAMPLE 4 Maria knows that there is another way to lower the total cost of each loan if she cannot reach a decision on a car in time for the one-day 5% financing offer. She wants to know how much money she can save if she makes a down payment of 20% and therefore borrows only 80% of the cost of the car.

QUESTION How much will Maria save on loans of $4200, $5500, $6325, $8275, and $9750 at 8% over periods of 3 years, 4 years, and 5 years if she makes a down payment of 20% instead of borrowing the entire amount?

SOLUTION
Use a spreadsheet program to find the total amount for the 8% loan without a down payment and with a 20% down payment. Then find the difference (savings).

	A	B	C	D	E	F	G	
1	Loan	Loan Amount	Number	Total Payment	Total Payment	Total Amount	Total	
2	Amount	20% Down	of Years	0% Down	20% Down	20% Down	Savings	
3	4200	3360	3	4,738.06	3790.45	4,630.45	107.61	+A3−0.2*A3
4	4200	3360	4	4,921.65	3937.32	4,777.32	144.33	
5	4200	3360	5	5,109.65	4087.72	4,927.72	181.93	
6	5500	4400	3	6,204.60	4963.68	6,063.68	140.92	
7	5500	4400	4	6,445.01	5156.01	6,256.01	189.00	
8	5500	4400	5	6,691.21	5352.97	6,452.97	238.24	
9	6325	5060	3	7,135.29	5708.23	6,973.23	162.06	
10	6325	5060	4	7,411.76	5929.41	7,194.41	217.35	
11	6325	5060	5	7,694.89	6155.91	7,420.91	273.98	
12	8275	6620	3	9,335.10	7468.08	9,123.08	212.02	
13	8275	6620	4	9,696.81	7757.45	9,412.45	284.36	
14	8275	6620	5	10,067.23	8053.78	9,708.78	358.45	
15	9750	7800	3	10,999.06	8799.25	10,749.25	249.81	
16	9750	7800	4	11,425.25	9140.20	11,090.20	335.05	0.2*A17+E17
17	9750	7800	5	11,861.69	9489.35	11,439.35	422.34	

Use the graph function of your program to show the savings for each amortization period and loan amount.

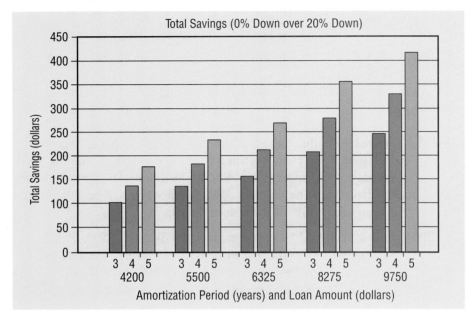

Total Savings (0% Down over 20% Down)

EXAMPLE 5 Bernard is going to buy a stereo for $1200 on which he will make monthly payments. He has collected information about the terms of various loans. These loans have different amortization periods, rates of interest, and down payment requirements. He can amortize the loan over 1 year, 2 years, or 3 years. He can finance it at 9% or 12%. He can make a down payment of 10% or 20%.

QUESTION Which combination of amortization period, interest rate, and down payment yields the lowest total payments? Which yields the highest?

SOLUTION
Use your calculator and the monthly payment formula.

Number of Years	Percent Down	Down Payment	Loan Amount	Total Amount at 9%	Total Amount at 12%
1	10%	$120.00	$1080.00	$1253.37	$1271.48
1	20%	240.00	960.00	**1247.44**	1263.54
2	10%	120.00	1080.00	1304.15	1340.14
2	20%	240.00	960.00	1292.58	1324.57
3	10%	120.00	1080.00	1356.37	**1411.37**
3	20%	240.00	960.00	1339.00	1387.89

The lowest total payment is $1247.44, which occurs with the shorter amortization period, the higher down payment, and the lower interest rate. The highest total payment is $1411.37, which occurs with the longer amortization period, the lower down payment, and the higher interest rate.

TRY YOUR SKILLS

Mindy is going to purchase new furniture. She has compared prices and interest charges at two stores. She can purchase a dining room set for $3300, a living room set for $4200, and a bedroom set for $1800 at both stores. One store offers an interest rate of 15%, and the other store offers an interest rate of 10%. Both stores require that the loan be repaid in 3 years. At neither store will she be required to make a down payment. How much money can she save on each item by financing at 10% instead of 15%? Use the monthly payment formula and your calculator to complete the following chart.

	Cost of Item	Monthly Payment at 15%	Total Payment at 15%	Monthly Payment at 10%	Total Payment at 10%	Total Savings
1.	$3300					
2.	4200					
3.	1800					

Determine the total savings on a loan of $12,450 at 8.5% for 25 years if you make a down payment of 20% or 30% instead of 10%.

	Percent Down	Down Payment, D	Loan Amount, A	Monthly Payment, M	Total Amount, T	Total Savings Over 10%
4.	10%					
5.	20%					
6.	30%					

EXERCISE YOUR SKILLS

1. What is discretionary income?

2. Why does increasing the size of your down payment for a purchase save you money in the long run?

3. Why would a financial institution encourage you to make a smaller down payment and increase the amortization period for an installment loan?

Steve bought the items listed below and amortized the payments at 12% over 3 years with no down payment. Use the monthly payment formula and your calculator to determine how much he could have saved if the interest rate had been 9% instead of 12%.

CD player for $500
Computer for $2500
Stereo cassette deck for $250
TV/VCR for $1000

KEY TERM

discretionary income

	Item Cost	Monthly Payment at 12%	Total Payment at 12%	Monthly Payment at 9%	Total Payment at 9%	Total Savings
4.	$ 500					
5.	2500					
6.	250					
7.	1000					

Use your results from Exercises 4–7 to determine how much Steve would have saved if he had made a down payment of 30% instead of making no down payment. The interest rate remains at 12%.

	Item Cost	Monthly Payment 0% Down	Total Payment 0% Down	Loan Amount 30% Down	Monthly Payment 30% Down	Total Payment 30% Down	Total Savings
8.	$ 500						
9.	2500						
10.	250						
11.	1000						

Daniel is going to buy a new all-terrain vehicle for $42,000 on which he will make monthly payments. He has collected information about the terms of various loans. These loans have different amortization periods, rates of interest, and down payment requirements. He can amortize the loan over 3 years, 4 years, or 5 years. He can finance it at 6% or 8%. He can make a down payment of 10% or 20%. Create a spreadsheet program to find the total payments for each situation.

	Number of Years	Percent Down	Down Payment	Loan Amount	Total Payment at 6%	Total Payment at 8%
12.	3	10%	$4200.00	$37,800	$45,598.17	
13.	3	20%				
14.	4	10%				
15.	4	20%				
16.	5	10%				
17.	5	20%				

18. Which combination of amortization period, interest rate, and down payment yields the lowest total payments?

19. Which combination of amortization period, interest rate, and down payment yields the highest total payments?

20. Use the results of Exercises 12–19 to write a general statement about the effect of amortization period, interest rate, and down payments on total payments.

MIXED REVIEW

1. You are a real-estate agent who has just sold a house for $325,000. Determine your total commission if you earn 5% on the first $150,000 and 8% on any amount over $150,000.

2. You just received a check with your full first name, middle initial, and full last name. Write your name as you would to endorse the check.

3. You have $446.77 in your checking account, and you write a check for $57.49. Show how you would enter this check in your check register.

4. Determine the interest for 6 months if the principal is $5525, the rate is 5%, and the interest is compounded quarterly.

5. You earn $26,000 per year plus 15% of your salary in fringe benefits. Find your yearly earnings, including benefits.

6. What percent of your take-home pay spent on credit payments (excluding mortgage payments) is regarded as a "manageable" amount (safer than "dangerous" but not as safe as "comfortable")?

7. Given a choice, which method would you prefer your bank to use when determining whether a service charge will be imposed; the minimum-balance method or the average-balance method? Explain your answer.

The cost function for a small business is $c = 2.8n + 40$. The revenue function for the business is $r = 6n$.

8. Graph the cost and revenue functions.

9. Find the break-even point.

AMORTIZATION SCHEDULES: SHRINKING INTEREST PAYMENTS

aria knows that she must be careful to choose a loan for which she can afford the monthly payment. Patrick recently told her that he received a raise from Paradise Department Store. Maria hopes that she will receive a pay increase in the next six months too. Her boss also indicated that she might receive a holiday bonus. She plans to save any additional money that she earns.

Maria wonders whether she can pay off her loan early if her financial situation continues to improve. Patrick has read that there can be penalties associated with prepaying a loan.

However, Maria's uncle has just told the family that he saved over $500 by paying the remainder of his car loan 15 months early. She also knows that her mother prepaid her contract on her microwave oven. Therefore Maria is certain that there must be substantial advantages to prepaying a loan.

Maria has a feeling that the money she can save if she prepays her loan is related to the portion of her monthly payment that is interest. Since her monthly loan payment is fixed, she wonders whether her monthly interest payment is constant also.

OBJECTIVES: *In this lesson, we will help Maria to:*

- *Compute the interest due, note reduction, and unpaid balance on her loan on a monthly basis.*
- *Create an amortization schedule for a loan.*
- *Determine prepayment amounts paid on a loan.*
- *Identify the amount saved by prepaying a loan.*

AMORTIZATION SCHEDULES

When Maria receives her car loan, her monthly payment will be identical each month. However, the portion of that payment that is applied as interest changes each month. Banks or other lending institutions provide borrowers with schedules, called **amortization schedules,** that list the interest portion of monthly loan payments. These amortization schedules also list the payment number, the monthly payment, the reduction in unpaid balance (note reduction), and the remaining unpaid balance for each month.

The **payment number** corresponds to the number of months that have elapsed since the money was borrowed. The **interest due** is determined by multiplying the monthly interest rate by the previous month's unpaid balance. The **note reduction** is found by subtracting the interest due from the monthly payment and shows how much of the monthly payment actually goes to paying off the loan. The **unpaid balance** is calculated by subtracting the note reduction from the previous month's unpaid balance and represents how much of the original loan is still unpaid.

PREPAYMENT

Most financial institutions will allow a borrower to prepay the balance of an installment loan. This right of prepayment (paying off the contract before it comes due) saves the borrower interest charges. However, the amount saved by prepaying a loan varies according to each loan agreement. Most loan contracts require the borrower to pay more interest in the early months of the loan period when the lender's risks are greater and less interest near the maturity date when the lender's risks are less.

Many financial institutions use a standard formula to determine the remaining portion of a loan at the time the borrower chooses to prepay it. The formula is based upon the monthly payment the borrower has been making, the monthly interest rate, and the number of remaining payment periods.

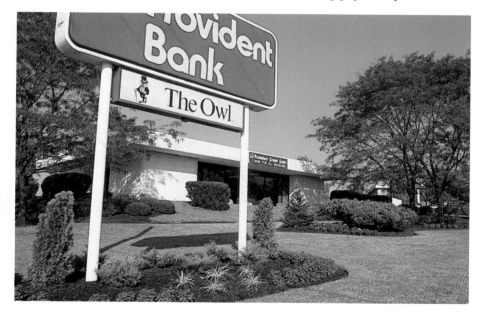

RULE OF 78

Some stores apply the **Rule of 78,** instead of the prepayment formula used by financial institutions, to buyers who purchase merchandise on credit for 1 year and wish to prepay. Since the sum of the numbers from 1 to 12 is 78, the buyer pays a portion of the yearly interest equal to $\frac{12}{78}$ in month 1, $\frac{11}{78}$ in month 2, $\frac{10}{78}$ in month 3, and so on, until $\frac{1}{78}$ is paid in month 12. The largest amount of interest is paid in month 1, and the smallest amount of interest is paid in month 12. By the end of the year, all the interest is paid.

In general, prepaying a loan saves the borrower money. However, some lending institutions charge a **prepayment penalty** to offset a portion of the lost revenues and additional clerical costs that they incur when a borrower prepays a loan. A borrower who is considering prepaying a loan must check the prepayment stipulations stated in the loan contract to determine the costs, if any, of prepayment. This is one of the many reasons that it is very important that the borrower understand the terms of a credit agreement before signing it.

Ask Yourself

1. What does an amortization schedule show?

2. Why do most loan contracts require the borrower to pay more interest in the early months of the loan period?

3. Do you think that you will save money if you prepay a loan?

4. Do you think that it is fair for a bank to charge a prepayment penalty?

ALGEBRA *REVIEW*

1. If the annual interest rate is 15%, what is the monthly interest rate?

2. Express your answer to Exercise 1 as a decimal.

Evaluate. Round to the nearest hundredth.

3. $x = 9 + 8 + 7 + \ldots + 2$

4. $x = 15 + 14 + 13 + \ldots + 5$

5. $x = \dfrac{3}{78} + \dfrac{2}{78}$

6. $x = \dfrac{7}{29} + \dfrac{6}{29} + \dfrac{5}{29}$

7. $x = 3 + 2^4$

8. $x = 3 + 2^{-4}$

9. $x = 1 + (3 + 2)^{-4}$

10. $x = 1 + (3 + 2)^4$

11. $x = 4.5[100 - (6 + 1)^{-12}]$

12. $x = 3.9[98 - (2.30122)^{-3}]$

13. $x = \dfrac{400[1 - (1 + 0.015)^{-10}]}{0.015}$

14. $x = \dfrac{325[1 - (1 + 0.0125)^{-24}]}{0.0125}$

SHARPEN YOUR SKILLS

EXAMPLE 1 Maria is considering a $4500 loan for 2 years at 9%. Her monthly payment is $205.58.

QUESTION What are the interest due, note reduction, and unpaid balance for the first 2 months?

SOLUTION

To determine I_1, the interest due at the end of month 1, Maria multiplied the loan amount L by the monthly interest rate r.

$I_1 = rL$ 　　　 $0.0075 \cdot 4500 = \$33.75$ 　　　 $r = 0.09 \div 12 = 0.0075; L = 4500$

To find R_1, the note reduction in month 1, she subtracted I_1, the interest due at the end of month 1, from the payment P.

$R_1 = P - I_1$ 　　 $205.58 - 33.75 = \$171.83$ 　　 $P = 205.58; I_1 = 33.75$

To calculate B_1, the unpaid balance at the end of month 1, she subtracted the note reduction R_1 from the original loan amount L.

$B_1 = L - R_1$ 　　 $4500 - 171.83 = \$4328.17$ 　　 $L = 4500; R_1 = 171.83$

To determine I_2, the interest due at the end of month 2, she multiplied B_1, the unpaid balance from month 1, by the monthly interest rate r.

$I_2 = rB_1$ 　　 $0.0075 \cdot 4328.17 = \$32.46$ 　　 $r = 0.09 \div 12 = 0.0075; B_1 = 4328.17$

To find the note reduction R_2 in month 2, she subtracted the interest I_2 due from the payment P.

$R_2 = P - I_2$ 　　 $205.58 - 32.46 = \$173.12$ 　　 $P = 205.58; I_2 = 32.46$

To calculate the new unpaid balance B_2, she subtracted the note reduction R_2 from the unpaid balance from month 1.

$B_2 = B_1 - R_2$ 　　 $4328.17 - 173.12 = 4155.05$ 　　 $B_1 = 4328.17; R_2 = 173.12$

Maria was able to see a pattern in her calculations and to use the pattern to predict the following formula for the unpaid balance after 3 months.

$B_3 = B_2 - R_3$ 　　 where $R_3 = P - I_3$ 　and　 $I_3 = rB_2$

Maria realized that the pattern she observed could be extended to cover more than just the first 3 months of the loan. The pattern could be used to create an amortization schedule as long as she wanted.

AMORTIZATION SCHEDULE FORMULAS

$I_1 = rL$	$I_2 = rB_1$	where L = loan amount
$R_1 = P - I_1$	$R_2 = P - I_2$	r = monthly interest rate
$B_1 = L - R_1$	$B_2 = B_1 - R_2$	P = payment amount
		I_1 = interest due at end of month 1
		R_1 = loan reduction at end of month 1
		B_1 = balance at end of month 1

EXAMPLE 2 Maria wants to know what her loan amortization schedule will look like for the first year of her loan for $4500 for 2 years at 9%.

QUESTION What are her interest due, note reduction, and unpaid balance for the first 12 months?

SOLUTION
Maria used her spreadsheet program to create an amortization schedule that shows the payment number, payment amount, interest due, note reduction, and unpaid balance for months 1–12. The rate $r = 0.09 \div 12 = 0.0075$.

	A	B	C	D	E
1	Payment	Payment	Interest	Note	Unpaid
2	Number	Amount	Due	Reduction	Balance
3		@ROUND(E3*0.0075,2)	+B4−C4	+E3−D4	4500.00
4	1	205.58	33.75	171.83	4328.17
5	2	205.58	32.46	173.12	4155.05
6	3	205.58	31.16	174.42	3980.63
7	4	205.58	29.85	175.73	3804.90
8	5	205.58	28.54	177.04	3627.86
9	6	205.58	27.21	178.37	3449.49
10	7	205.58	25.87	179.71	3269.78
11	8	205.58	24.52	181.06	3088.72
12	9	205.58	23.17	182.41	2906.31
13	10	205.58	21.80	183.78	2722.53
14	11	205.58	20.42	185.16	2537.37
15	12	205.58	19.03	186.55	2350.82

Next she used the graph function on her computer spreadsheet program to graph the portion of each monthly payment that is interest.

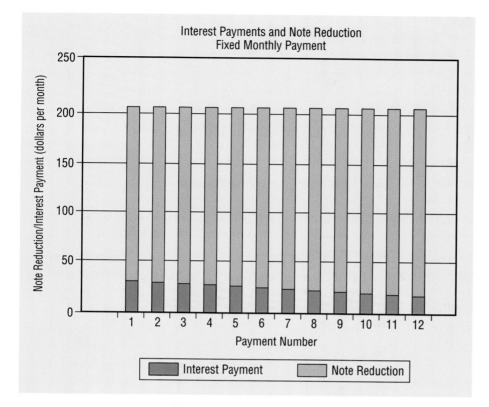

Interest Payments and Note Reduction
Fixed Monthly Payment

Note that the interest portion of the loan decreases each month even though the monthly payment remains constant.

SKILL 2

Maria would like to know how much she will save if she prepays her loan. She can use the prepayment formula to determine the remaining payment if she prepays it at any given time. She can then compare this amount with the amount that she would have paid if she did not prepay the loan.

PREPAYMENT FORMULA

$$A = \frac{M[1 - (1 + r)^{-q}]}{r}$$

where M = monthly payment
r = monthly interest rate
q = number of remaining payment periods
A = prepayment amount

EXAMPLE 3 Maria's uncle decides to prepay his car loan. He has been making payments of $444.89 per month on a 5-year loan of $20,000 at 12%.

QUESTION How much does he owe if he prepays the loan 15 months early? How much money will he save?

SOLUTION

Use the prepayment formula.

$$A = \frac{444.89[1 - (1 + 0.01)^{-15}]}{0.01}$$

$M = 444.89$
$r = 0.12 \div 12 = 0.01$
$q = 15$

$A = \$6168.42$

If he does not prepay the loan, he will pay $15(444.89) = 6673.35$.

$6673.35 - 6168.42 = 504.93$ Subtract.

Therefore, he saves $504.93 by prepaying the loan.

EXAMPLE 4 Some banks charge a prepayment penalty. Maria wonders whether her uncle would have saved money if his bank had charged a prepayment penalty of $250.

QUESTION Would he still have saved money if he had prepaid the loan?

SOLUTION

$504.93 - 250.00 = 254.93$ Subtract the penalty from the original savings.

Yes, he would have saved $254.93.

─────── **SKILL 3** ───────

RULE OF 78

The buyer pays a portion of the yearly interest equal to

$\dfrac{12}{78}$ in month 1

$\dfrac{11}{78}$ in month 2

$\dfrac{10}{78}$ in month 3

⋮

$\dfrac{1}{78}$ in month 12

EXAMPLE 5 Maria's mother purchased a microwave oven for $360, putting $60 down. She signed a contract stating that the balance of $300 was to be paid in 12 equal installments with an interest rate of 21% per year.

QUESTION If she decides to prepay her contract at the time of her ninth payment and the store applies the Rule of 78, how much interest will she save?

SOLUTION

Find the yearly interest.

$0.21 \cdot 300 = 63$ The yearly interest rate is 21%.

After 9 months, Maria's mother has three remaining payments. By prepaying the loan she will not have to pay the portion of yearly interest due in the last three months. Apply the Rule of 78.

$$\frac{3}{78} + \frac{2}{78} + \frac{1}{78} = \frac{6}{78}$$

$\frac{6}{78} \cdot 63 = 4.85$ The yearly interest is $63.

Therefore she saves $4.85.

TRY YOUR SKILLS

Create an amortization schedule for the first 3 months of a $2000 loan at 6% for 4 years for which the monthly payment is $46.97.

	Payment Number	Payment Amount	Interest Due	Note Reduction	Unpaid Balance
1.	1	$46.97	$10.00	$36.97	
2.	2	46.97			
3.	3				

Determine the prepayment savings on a bank loan of $25,000 at 12% for 30 years for which you have been paying $257.15 each month under each condition. Assume that there are no prepayment penalties.

4. It is prepaid 20 months early. Use the prepayment formula.

$A = \dfrac{M[1 - (1 + r)^{-q}]}{r}$ Substitute $M = 257.15$,
$r = 0.12 \div 12 = 0.01$, and $q = 20$.

$A = \dfrac{257.15[1 - (1 + 0.01)^{-20}]}{0.01}$

5. It is prepaid 3 years early.

A $600.00 television purchase is financed at 18% per year with no down payment at a store that applies the Rule of 78. It must be paid in 12 equal installments. Determine the prepayment savings under each condition.

6. It is prepaid after 4 months.

7. It is prepaid after 7 months.

EXERCISE YOUR SKILLS

1. If your bank charges a $200 prepayment penalty, under what condition will you consider prepaying your loan?
2. Why does the interest portion of a monthly payment decrease each month?
3. What is an amortization schedule?

Gregory borrows $9500 from a bank to purchase a car. He finances it at 8% for 4 years, and his monthly payment is $231.92. Use a spreadsheet program to create an amortization schedule that shows the payment number, payment amount, interest due, note reduction, and unpaid balance for the first year.

	Payment Number	Payment Amount	Interest Due	Note Reduction	Unpaid Balance
4.					
5.					
6.					
7.					
8.					
9.					
10.					
11.					
12.					
13.					
14.					
15.					

Determine the prepayment savings on a bank loan of $21,400 at 11% for 25 years for which you have been paying $209.74 each month under each condition. Assume that there are no prepayment penalties.

16. It is prepaid 37 months early.   17. It is prepaid 5 years early.

Determine the prepayment savings on a bank loan of $100,000 at 12.5% for 15 years for which you have been paying $1232.52 each month under each condition. Assume that there is a prepayment penalty of $500.

18. It is prepaid 42 months early.   19. It is prepaid 6 years early.

Use the Rule of 78 to determine how much interest will be saved by prepaying a one-year loan of $1500 at 10% after each time period.

20. 4 months   21. 6 months   22. 10 months

1. Determine which of the following loans has a greater total cost:

   a. $10,000 at 12% for 5 years   **b.** $11,000 at 11% for 3 years

2. During the first week of April, Brielle worked 34 hours and received $44.95 in tips. If her hourly rate of pay is $5.75, how much did she earn that week?

3. Carmen has $8500 in a CD that pays 6% interest, compounded semiannually. How much is her CD worth at the end of 2 years?

4. Angela earns $125 in gross income each week and is single and claims one exemption. How much will her employer withhold for income tax?

5. Rachel receives a commission of 4% on sales up to $10,000 and 6% on all sales over $10,000. How much has she earned in commission if her sales are $18,250?

James Wilcox is a farmer with 160 acres on which to plant. From crop x he can earn $200 per acre; from crop y he can earn $150 per acre. To meet demands, he must plant at least 40 acres of crop x and 60 acres of crop y. Soil conditions will permit him to plant no more than 100 acres of crop x.

6. Write the objective function for his total income.

7. Write inequalities that express each constraint.

8. Graph the inequalities.

9. Find the corners of the region in Exercise 8.

10. What is the number of acres of each crop that James Wilcox should plant to maximize his income?

Maria discovered that there are methods to reduce the cost of credit. Selecting a shorter amortization period, a lower interest rate, and a larger down payment enables the borrower to reduce the total payments on a loan. In addition, prepaying a loan can often save the borrower money. Many car ads mention terms that include cash-back opportunities and significantly lower interest rates than financial institutions are offering.

Maria wonders whether she can save money on the purchase of her car by taking advantage of some of these terms. Patrick warns that merchants can increase their profits by enticing customers to choose credit terms that are favorable to the merchants.

OBJECTIVES: In this lesson, we will help Maria to:

- *Learn how merchants generate credit income and attract customers with special credit plans.*
- *Calculate the total financed price with lower-than-market-rate interest and rebate plans.*
- *Determine the merchants' profits with lower-than-market-rate interest and rebate plans.*
- *Observe the drawbacks associated with rent-to-own plans.*

SPECIAL CREDIT PLANS

1. Since stores must pay credit card companies 4–7% of the purchase price of each item they sell on credit, many customers feel that they are entitled to discounts on their cash purchases. By law, retailers have the right to offer **cash discounts** to customers who pay for their purchases in cash as long as the discounts are offered to all potential customers. However, the law prohibits retailers from adding extra charges to credit card purchases to cover their credit card expenses.

2. The **rent-to-own plan** enables you to purchase expensive items now and pay for them with small payments over a specified period of time. These payments often add up to a larger amount than they would have if you had financed the purchase outright. If you decide to buy the items, then your rental fees are applied to the purchase price. However, if you decide that you cannot afford the items or do not want them, you take them back to the store, which then keeps all the money you have already paid. This is the price you pay for renting. You are spared the embarrassment of breaking an installment loan contract and having your credit rating blemished, but you have lost the money you paid to rent the items.

3. One particularly attractive incentive that is used to promote sales is the extension of credit terms at an interest rate that is lower than the **market rate**, which is the rate currently available on the open market. Automobile companies use this device extensively. Since the dealers can offer to finance your loan through the company's subsidiary finance corporation, they can offer you a lower interest rate than one that might be available through your bank.

4. Some automobile dealers offer an alternative to a low-interest-rate loan called a **rebate** or cash-back plan. When you purchase the car, you receive cash back to be applied to the purchase price, thereby lowering the purchase price.

Ask Yourself

1. Why would a merchant be willing to offer a discount for a cash purchase?

2. What is a rent-to-own plan?

3. What companies frequently offer credit at interest rates that are lower than those offered by banks?

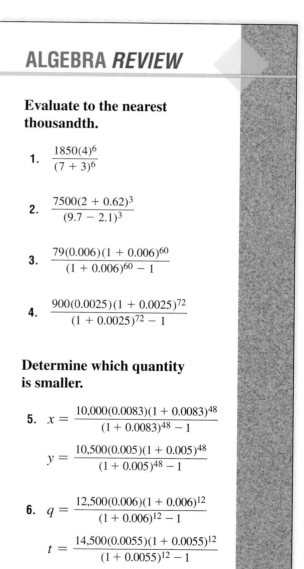

ALGEBRA *REVIEW*

Evaluate to the nearest thousandth.

1. $\dfrac{1850(4)^6}{(7+3)^6}$

2. $\dfrac{7500(2+0.62)^3}{(9.7-2.1)^3}$

3. $\dfrac{79(0.006)(1+0.006)^{60}}{(1+0.006)^{60}-1}$

4. $\dfrac{900(0.0025)(1+0.0025)^{72}}{(1+0.0025)^{72}-1}$

Determine which quantity is smaller.

5. $x = \dfrac{10,000(0.0083)(1+0.0083)^{48}}{(1+0.0083)^{48}-1}$

   $y = \dfrac{10,500(0.005)(1+0.005)^{48}}{(1+0.005)^{48}-1}$

6. $q = \dfrac{12,500(0.006)(1+0.006)^{12}}{(1+0.006)^{12}-1}$

   $t = \dfrac{14,500(0.0055)(1+0.0055)^{12}}{(1+0.0055)^{12}-1}$

7. $48x$ or $48y$

8. $12q$ or $12t$

Maria has just discovered the ad shown for a new luxury model car. She observes that the $1000 cash-back plan offers no financing. Therefore she assumes that she will have to apply for her own bank financing at 13.75% over 48 months if she selects this plan.

New Luxury Model

$1000 CASH BACK

— OR —

$750 CASH BACK Plus

6.9% APR

— OR —

2.9% APR Available

FACTORY AIR

- Wide Vinyl Bodyside Mold
- AM/FM 4 Speaker Stereo
- Tinted Glass
- Power Steering
- Interval Wipers
- Rear Window Defroster

- Instrumentation Group
- Digital Clock w/over Console Light/Security Group
- Dual Elect. Mirrors
- Luxury Wheel Covers
- 1.9L EFI Engine

MSRP	$10,154
FACTORY DISC.	−744
FACTORY REBATE	−1000
OUR DISC.	−1035

YOUR PRICE	**$ 7375**

EXAMPLE 1 Plan 1 offers a purchase price of $7375 and a $1000 rebate with no financing. (13.75% bank financing is available.)

Plan 2 offers a purchase price of $7625 and a $750 rebate at 6.9% over 48 months.

Plan 3 offers a purchase price of $8375 and no rebate at 2.9% over 48 months.

QUESTION   Which plan yields the lowest total financed price to the consumer?

SOLUTION

Use the monthly payment formula and your calculator to find the monthly payment and total financed price for the car under each plan.

	Plan 1	Plan 2	Plan 3
Rebate	$1000	$750	0
Rate	13.75%	6.9%	2.9%
Time	48 months	48 months	48 months
Loan amount	$7375	$7625	$8375
Monthly payment	$200.61	$182.24	$185.01
Total financed price	$9629.24	$8747.35	$8880.24

The lowest total financed price of $8747.35 results from Plan 2 with a rebate of $750 and financing at 6.9%. Maria must carefully consider the details of each financing deal. Looking at the lowest purchase price, the lowest interest rate, or the highest rebate is not enough!

SKILL 2

EXAMPLE 2   Maria realizes that the automobile company and its financing subsidiary would save money if she did not take either rebate. She wonders which plan the company would prefer her to select.

QUESTION   Which plan yields the highest profit to the company?

SOLUTION

Calculate how much the company would save if the customer did not take the rebate and how much interest income it would make for each plan. This represents profit *over the profit built into the price of the car.* The company earns interest income only in Plans 2 and 3, and the amount of that interest is equal to the difference between the total financed price and the amount borrowed.

	Plan 1	Plan 2	Plan 3
Interest	0	$1122.35	$ 505.24
Rebate not paid	0	250.00	1000.00
Profit	0	1372.35	1505.24

In Plan 1, the company gets to keep none of its factory rebate. In Plan 2 it keeps $250 of the rebate, paying out $750. In Plan 3 it keeps all of the $1000 rebate. The most advantageous plan for the company is Plan 3. An automobile company may give its dealers incentives to encourage you to select the plan that is most profitable for the company.

Rent-to-own plans are especially attractive to young people who have not yet established a credit rating and would not qualify for an installment loan. In addition, people who have already overextended their credit and are not eligible for any more may decide to extend themselves even further by using this plan. This plan has the downside risk that all rental money is forfeited if the customer decides not to keep the item.

EXAMPLE 3 Charro wants to buy a TV/VCR, but she cannot afford it at this time. The Houston Home Company offers a rent-to-own credit plan that will allow her to rent it now at $30 per month for 24 months. If she decides to buy it, her rental fees will be applied to the purchase price. If she does not choose the rent-to-own plan, she can purchase the TV/VCR for $600 with an installment loan at 12% interest over 2 years.

QUESTION How much money does she save by obtaining the installment loan rather than using the rent-to-own plan? How much will she lose if she can no longer afford the TV/VCR after 9 months?

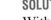

SOLUTION
With the rent-to-own plan, she will pay 24(30) = $720.

To determine her total payments with the installment loan, Charro used her calculator and the monthly payment formula. She will pay $677.86.

This is a savings of 720 − 677.86 = $42.14 over the rent-to-own plan.

She will lose 9(30) = $270 if she can no longer afford the rental payments after 9 months.

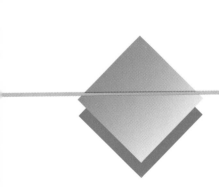

TRY YOUR SKILLS

A car company is offering a new car priced at $34,600. The following financing plans are available. Assume the loans in Plan 1 are from a bank or other lending institution and that the loans in Plans 2 and 3 are from the car manufacturer's financing subsidiary.

	Plan 1	Plan 2	Plan 3
Rebate	$2000	$1000	0
Rate	10.5%	7.5%	3.0%
Time	48 months	48 months	48 months

1. Determine which plan yields the lowest total financed price to the consumer by completing the table below.

	Plan 1	Plan 2	Plan 3
Rebate	$2000	$1000	0
Rate	10.5%	7.5%	3.0%
Time	48 months	48 months	48 months
Loan amount	$32,600		
Monthly payment	$834.67		
Total financed price			

2. Determine which plan yields the highest profit to the company by completing the table below.

	Plan 1	Plan 2	Plan 3
Interest		$5395.73	
Rebate not paid		1000.00	
Profit			

3. How much money would you save if you financed a $1200 item at 8% for 3 years rather than accepting a rent-to-own plan for which you pay $40 per month for 3 years?

1. If you were offered a discount for paying cash, would you still use a credit card for a purchase? Why or why not?

2. Under what circumstances would you use a rent-to-own plan?

3. Why is it important to shop around for loan terms when you are buying a car?

Determine the plan that would provide the lowest total financed price for each car by completing each table in Exercises 4–8. Assume that the loans in Plan 1 are from a bank or other lending institution and that the loans in Plans 2 and 3 are from the car manufacturer's financing subsidiary.

4. Car price: $15,500

	Plan 1	Plan 2	Plan 3
Rebate	$1200	$600	0
Rate	10.25%	7.9%	4.9%
Time	36 months	36 months	36 months
Loan amount			
Monthly payment			
Total financed price			

5. Car price: $12,000

	Plan 1	Plan 2	Plan 3
Rebate	$1500	$800	0
Rate	13%	8.75%	5%
Time	36 months	36 months	36 months
Loan amount			
Monthly payment			
Total financed price			

6. Car price: $9400

	Plan 1	Plan 2	Plan 3
Rebate	$1500	$1200	0
Rate	13%	8.75%	5%
Time	48 months	48 months	48 months
Loan amount			
Monthly payment			
Total financed price			

7. Car price: $7800

	Plan 1	Plan 2	Plan 3
Rebate	$1000	$250	0
Rate	10%	7.75%	6%
Time	48 months	48 months	48 months
Loan amount			
Monthly payment			
Total financed price			

8. Car price: $13,800

	Plan 1	Plan 2	Plan 3
Rebate	$2000	$1000	0
Rate	12%	6.25%	3%
Time	48 months	48 months	48 months
Loan amount			
Monthly payment			
Total financed price			

9. Determine which plan yields the highest profit to the company in Exercise 4 by completing the table below.

	Plan 1	Plan 2	Plan 3
Interest			
Rebate not paid			
Profit			

10. Determine which plan yields the highest profit to the company in Exercise 5 by completing the table below.

	Plan 1	Plan 2	Plan 3
Interest			
Rebate not paid			
Profit			

11. Determine which plan yields the highest profit to the company in Exercise 6 by completing the table below.

	Plan 1	Plan 2	Plan 3
Interest			
Rebate not paid			
Profit			

12. Determine which plan yields the highest profit to the company in Exercise 7 by completing the table below.

	Plan 1	Plan 2	Plan 3
Interest			
Rebate not paid			
Profit			

13. Determine which plan yields the highest profit to the company in Exercise 8 by completing the table below.

	Plan 1	Plan 2	Plan 3
Interest			
Rebate not paid			
Profit			

14. How much money would you save if you financed a $1500 item at 9% for 4 years rather than accepting a rent-to-own plan for which you pay $39.95 per month for 4 years?

15. How much money would you lose if you returned the item in Exercise 14 after 18 months?

16. How much money would you save if you financed a $2500 item at 6.75% for 2 years rather than accepting a rent-to-own plan for which you pay $119.95 per month for 2 years?

17. How much money would you lose if you returned the item in Exercise 16 after 5 months?

MIXED REVIEW

1. What are your weekly earnings if you work 38 hours at $12.75 per hour?

2. Calculate the compound interest for one year on a $775 deposit earning 5.5% if interest is paid quarterly.

3. If you earn $920.00 and your FICA withholding is 7.65% of your earnings, how much do you pay for FICA taxes?

4. If you can afford a monthly payment of $320, how much money can you borrow at 8.25% for 3 years?

5. Calculate the simple interest for 1 year on a $775 deposit from which the interest is withdrawn by the depositor immediately following the posting of the interest at the end of each quarter. The quarterly rate of interest is 1%.

Three students are selling bumper stickers. It takes 20 hours to produce 100 stickers. The cost of labor is $5.35 per hour and the materials for each sticker cost $0.85. Fixed costs such as advertising and energy cost $44.00. Find the cost of producing the given number of stickers.

6. 300 bumper stickers   **7.** 1100 bumper stickers

Suppose that you plan to buy a computer system for $3000 and to finance it by paying it off in monthly installments over 2 years at an annual rate of 12%. You plan to make a down payment of 10%. How much will you save if you change the loan terms as indicated below?

8. Make a down payment of 25% instead of 10%

9. Finance at an annual rate of 9.5% instead of 12%

Some students selling caps for $4.00 at a sports event had these expenses.

Fixed costs: 35 hours of labor at $5.75 per hour
                 Advertising, energy, and transportation: $38.50

Variable costs: caps, at $1.50 each

10. Write an equation for the cost function.

11. Write an equation for the revenue function.

12. Find the break-even point.

Determine the monthly payment on each loan.

1. $25,750 at 9% for 10 years
2. $7200 at 9% for 4 years
3. $1725 at 6.5% for 3 years
4. $99,900 at 8.5% for 30 years
5. Determine the monthly and total payments for a loan of $250,000 at 7% for 25 years.

If you can afford a monthly payment of $475 per month, determine how much money you can borrow for each interest rate and length of loan.

6. 10% for 5 years
7. 8% for 8 years
8. 6% for 25 years

Determine the total cost of each loan.

9. $36,500 at 8.5% for 5 years
10. $88,000 at 7.5% for 15 years

Complete the chart below to determine how much money you can save by borrowing $35,000 at 8% for 3 years or 4 years instead of 5 years.

	Number of Years	Monthly Payment	Total Payment	Savings from 5 Years
11.	3			
12.	4			
13.	5			

Complete the chart below to find the total payments at 10% and 7% and the savings associated with the 7% interest rate for each loan.

	Loan Amount	Number of Years	Total Payment 10%	Total Payment 7%	Total Savings
14.	$2325	3			
15.	5535	5			
16.	7600	4			
17.	9975	6			

18. Graph the results of Exercises 14–17.

Complete the chart below to determine how much money you can save on a loan of $9360 at 11% for 4 years if you make a down payment of 20% or 30% instead of 10%.

	Percent Down	Down Payment	Loan Amount	Total Amount	Savings Over 10% Down
19.	10				
20.	20				
21.	30				

22. Create an amortization schedule to show the interest due, note reduction, and unpaid balance for the first six months on an $8500 loan for 2 years at 7.5% for which the monthly payments are $382.50.

23. How much money will you save by prepaying 13 months early a loan of $37,000 at 9% for 3 years for which the monthly payment is $1176.59? Assume that there are no prepayment penalties.

A $900.00 stereo system purchase is financed at 14% per year with no down payment at a store that applies the Rule of 78. It must be paid in 12 equal installments. Determine the prepayment savings under each condition.

24. It is prepaid after 5 months.   25. It is prepaid after 8 months.

26. Determine the plan that would provide the lowest total financed price for a car priced at $19,500 by completing the table below. Assume that the loan in Plan 1 is from a bank or other lending institution and that the loans in Plans 2 and 3 are from the car manufacturer's financing subsidiary.

	Plan 1	Plan 2	Plan 3
Rebate	$1750.00	$1250.00	0
Rate	10.75%	9.9%	3%
Time	36 months	36 months	36 months
Loan amount			
Monthly payment			
Total financed price			

27. Determine which plan yields the highest profit to the company in Exercise 26 by completing the table below.

	Plan 1	Plan 2	Plan 3
Interest			
Rebate not paid			
Profit			

Use the information that you learned in this chapter to answer each of the following.

28. Name three indicators that a borrower is experiencing credit overload.

29. Is it always better to choose a shorter amortization period? Explain your answer.

30. What are three ways to reduce the cost of installment loans?

31. Why does the interest portion of a monthly payment decrease each month?

32. Will the financing plan with the largest rebate always yield the lowest total financed price? Why or why not?

Find the monthly payment and total payment on each loan.

1. $13,333 at 5% for 4 years **2.** 87,500 at 7.5% for 25 years

3. How much money can you borrow at 5% for 5 years if you can afford a monthly payment of $340?

4. Make a chart with the headings below to determine how much money you can save by borrowing $10,000 at 12% for 3 years or 4 years instead of 5 years.

Number of Years	Monthly Payment	Total Payment	Savings from 5 Years

5. Determine the total cost of a loan for $12,222 at 9.5% for 6 years.

Complete the chart below to find the total payments at 9% and 6% and the savings associated with the 6% interest rate for each loan.

	Loan Amount	Number of Years	Total Payment 9%	Total Payment 6%	Total Savings
6.	$4500	3			
7.	6750	4			

8. Make a chart with the headings below to determine how much money you can save on a loan of $6550 at 7.5% for 3 years if you make a down payment of 20% or 30% instead of 10%.

Percent Down	Down Payment	Loan Amount	Total Amount	Savings Over 10% Down

9. Determine the interest due, note reduction, and unpaid balance for the first two months on a $3600 loan for 3 years at 6% for which the monthly payments are $109.50.

10. How much money will you save by prepaying a loan of $25,000 at 6% for 5 years for which the monthly payment is $483.32 if you prepay it 20 months early? Assume that there are no prepayment penalties.

11. Determine the plan that would provide the lowest total financed price for a car priced at $27,500. Assume the loan in Plan 1 is from a bank and the loans in Plans 2 and 3 are from the car manufacturer's financing subsidiary.

	Plan 1	Plan 2	Plan 3
Rebate	$1950.00	$1000.00	0
Rate	12.75%	9.9%	4.9%
Time	48 months	48 months	48 months

12. Determine which plan yields the highest profit to the company in Exercise 11.

1. Tyrone's monthly check costs are $0.035 per check for the first 20 checks and $0.15 for each check over 20. If he writes 42 checks this month, how much will the checks cost him?

2. Joanie buys a CD for $450 that matures in 1 year, earning 8% simple interest. Every time the CD matures, she withdraws the interest and buys another one-year $450 CD at the same interest rate. If she does this for 3 years in a row, how much interest will she earn in that time?

3. Olga earns $7.75 per hour. She works 35 hours per week for 50 weeks and receives two weeks of paid vacation. Her other benefits cost her employer $3500 per year. What is the annual cost to the company of her salary and benefits?

4. If David can afford a monthly payment of $485, how much can he afford to borrow at a yearly interest rate of 9% for 10 years?

5. Use the Rule of 72 to determine how long it will take your $2300 savings account to double in value if it is growing at an annual rate of 4.5%.

6. You are a real-estate agent who has sold a house for $275,000. Determine your total commission if you earn 6% on the first $100,000 and 7.5% on any amount over $100,000.

7. You have $1546.68 in your checking account, and then you write a check for $597.99. Show how you would enter this check in your check register.

8. Determine the interest for 9 months if the principal is $1775, the rate is 6%, and the interest is compounded quarterly.

9. Determine the prepayment savings on a bank loan of $75,000 at 7.5% for 25 years for which you have been paying $554.24 per month if it is prepaid 5 years early. Assume that there is no prepayment penalty.

10. Determine which of the following loans has a greater total cost:
    a. $20,000 at 10% for 5 years   b. $25,000 at 12% for 3 years

11. Drucilla has $25,500 in a CD that pays 8% interest, compounded quarterly. How much is her CD worth at the end of 2 years?

12. Jacquie earns $1025 in gross income each week. How much will her employer withhold for FICA taxes?

13. Use the Rule of 78 to determine how much interest can be saved by prepaying a one-year loan of $1375 at 8% after 7 months.

14. How much do you save by financing a $65,000 loan at 7% for 5 years instead of 10 years?

15. Explain why making a large down payment decreases the total financed price of a loan.

PROJECT 5–1: Buying a Car

You are about to leave for college, and you need a car. You will have to buy a used car because you cannot afford a new one. Start by looking at newspaper advertisements to find the prices of some of the cars in which you might be interested.

1. Cut out six classified ads for cars that are for sale in your area. Attach the ads in a column along the left-hand side of a large sheet of paper.
2. Divide the remainder of your sheet of paper into columns to record the following information: loan amount, monthly payment, total payments, and total cost.
3. Find the loan amount by subtracting a 10% down payment from the price of each of the six cars you chose.
4. Use the monthly payment formula to find the monthly payment for a loan of 5 years for each of the six cars.
5. Find the total payments made over 5 years for each of the six cars.
6. Find the total cost for each of the six cars you chose.
7. Estimate your possible savings and income by the time you will be ready for college. Will you be able to afford to buy one of these cars?
8. Visit the car dealers that are offering the cars in which you are interested. Compare the values of each of the cars to the sale price. Write a report explaining which of the cars is the best deal.

PROJECT 5–2: Selecting a Loan

Suppose you buy a car for $5000 and have $1000 for a down payment. You must finance $4000. To find the best terms for a loan, you investigate possible sources of credit in your neighborhood. Possible sources of credit are a bank, a finance company, the car dealer, and a credit union. Also, remember that not all banks have the same terms for financing loans, so it is wise to check out several.

1. Select four sources of credit.
2. Complete a chart with the following headings for each of the sources.

Source of Credit	Monthly Payment	Number of Payments	Amount Paid	Finance Charges	Total Cost

3. Compare the total cost of the car for each source.
4. Which source of credit saves you the most money?
5. How much money can you save by using this source of credit?
6. Compare your findings with those of your classmates.

When simplifying expressions, use the following rules for *order of operations:*
- First, perform operations within parentheses or other grouping symbols.
- Next, simplify any exponents.
- Then, multiply and divide in order from left to right.
- Last, add and subtract in order from left to right.

Simplify each expression. Round your answer to the nearest hundredth.

1. $7 + (4 \cdot 2)$ **2.** $(7 + 4) \cdot 2$ **3.** $3 \cdot \left(8 + \dfrac{4}{3}\right)^4$ **4.** $3 \cdot 8^4 + \dfrac{4}{3}$

5. $13^2[(2 \cdot 9)^3(7 \div 9)]$ **6.** $[13^2(2 \cdot 9^3)][7 \div 9]$

7. $2.78(9876 - 500 \div 25)$ **8.** $16[4 - 3.1(42 - 3^4 \div 9) + 7]$

A *relation* is a set of ordered pairs of numbers. The set of ordered pairs $\{(0, 2), (1, 3), (2, 5), (3, 10), (5, 14)\}$ is a relation. The set of replacements for the first variable x is called the *domain* of the relation. The set of replacements for the second variable y is called the *range* of the relation. A *function* is a relation in which each member of the domain is paired with exactly one member of the range. The above relation *is* a function because for each x there is only one y. The relation $\{(2, 3), (3, 4), (2, 6)\}$ is *not* a function because the x value of 2 is paired with y values of 3 *and* 6.

Find the domain and range of each relation.

9. $\{(7, 3), (3, 7), (2, 4)\}$ **10.** $\{(8, 3), (3, 8), (8, 8)\}$

11. Is the relation in Exercise 9 a function? Why or why not?

12. Is the relation in Exercise 10 a function? Why or why not?

Solve for *y* if *x* = 16.

13. $y = x^2 + x$ **14.** $y = x^3$ **15.** $y = x^{\frac{1}{4}}$

16. $y = x^4 - x$ **17.** $y = \sqrt{x}$ **18.** $y = \sqrt[3]{x + 11}$

19. The Berger family spent the following amount of money on food in 1993:

January: $500	February: $610	March: $715
April: $650	May: $509	June: $636
July: $702	August: $653	September: $619
October: $515	November: $598	December: $550

Determine the percent of their 1993 food costs that occurred in each month to the nearest hundredth of a percent.

CREDIT CARDS

THE AMERICAN ECONOMY HAS GRADUALLY shifted away from a system in which individuals and families produced or grew most of what they needed, and offered these products in exchange for goods and services provided by others. In America today, the production and distribution of goods and services is on a much larger scale. Our economy is so specialized that companies exist not only to create products and services, but also to convince us that we should acquire more of these products. The banking industry processes the money exchanged among businesses and offers many services that keep our economy flowing.

One major service offered to consumers by banks, retailers, and other companies is the credit card (or charge card). Bankers and merchants have realized that if they make credit convenient and painless for us, merchants will sell more. The banks can then charge both the consumer and the merchant for the use of their credit cards. We seem to be more willing to hand over a plastic card than money, so we have a tendency to spend more with credit cards.

If we use our credit cards for a period of time and pay small finance charges each month, we may be tempted to use them over and over again until the monthly finance charges become quite large. Our wants become needs. We can no longer afford to buy items that we really need because all of our extra money is spent keeping up the minimum payments for the items that we previously bought on credit.

Fortunately, we do not have to go completely under and declare bankruptcy.

If we watch our spending and begin to repay our debts, we can slowly diminish our credit load. Congress has passed laws to help protect us from ourselves. However, part of the price we pay for living in a free society is the opportunity to make bad choices and the responsibility for coping with the consequences of those choices.

In this chapter our high school students will take a close look at the costs and conveniences of using credit cards. They will learn how individuals qualify for credit cards. They will observe some of the warning signs of credit overload, as well as some of the difficulties we face once we are in over our heads. They will also examine some steps to take to regain financial stability.

6-1 CREDIT CARDS: PLASTIC MONEY

Sylvia Shawn enjoys spending money. She has a bumper sticker on her car that reads "When the Going Gets Tough, the Tough Go Shopping." Her parents are divorced, and she lives with her mother. Since her mother rarely takes her shopping, Sylvia looks forward to shopping with her father during the two weekends that she sees him every month. Paying for everything seems simple for him, but he rarely carries any cash.

Their first stop is usually at a gas station to fill up the tank of his yellow sports car. While there, Sylvia picks up a snack and a soft drink, and her father grabs several packs of baseball cards. He pays for these purchases with his handy oil credit card. Then they are off to Sylvia's favorite shopping mall, where he waits patiently while she tries on eight or ten outfits before selecting two or three. He also lets her buy jewelry and accessories to match her outfits. He always pays with a credit card. It seems as though he has a credit card for every department store!

Since Sylvia will be going away to college next year, her father has been helping her select an elaborate nine-piece set of luggage. For this purchase, he uses a card called a VISA card. He has several VISA cards from different banks across the country. He calls them his "plastic money" cards.

When they finish shopping, Sylvia's dad often takes her to a fancy restaurant for dinner, where he uses his American Express card. He calls it his "entertainment expense account" card, and he uses it for restaurant meals, airline tickets, car rentals, and other travel expenses.

Sylvia does not understand how he got so many cards and why he does not carry real money instead. Her mother has several credit cards, but she rarely uses them. When her parents were still married, they frequently argued about the use of credit cards. Her mother is particularly against using credit cards for cash advances. Sylvia would like to know more about credit cards.

OBJECTIVES: *In this lesson, we will help Sylvia to:*

- *Describe how credit cards are used.*
- *Identify three categories of credit cards and several companies that offer the cards in each category.*
- *Determine the new balance that will be shown on a credit card statement after a purchase is made.*
- *Calculate the effective rate of interest on a credit card purchase.*

WHAT IS A CREDIT CARD AND HOW DOES IT WORK?

A **credit card** is a small plastic card, approximately 2" by $3\frac{1}{2}$", that identifies the holder and extends to the cardholder an unsecured line of credit. Most cards entitle the holder to charge purchases up to a maximum **credit limit.** By signing the application for the card, the cardholder agrees to pay for purchases according to the credit card terms. The **Truth in Lending Act** requires that the credit card company disclose all terms associated with its card, including the annual percentage rate, finance charges, **late payment penalties,** grace period, membership fee, and so on. Examples of companies that issue these cards are local department stores, national retail chains, car rental companies, travel and entertainment companies, airlines, telephone companies, oil companies, hotel/motel chains, and commercial banks.

When a customer uses a credit card in a store, the merchant turns in the charge to the store's bank. The store's bank notifies the bank that issued the card that a purchase has been made. The issuing bank then bills the cardholder for all purchases made that month. The cardholder writes one check for all purchases payable to the issuing bank. Through the bank card system, the card issuer then pays the merchant's bank. The merchant's bank has already paid the merchant.

TYPES OF CARDS

There are four categories of cards: (1) single-purpose credit cards; (2) multipurpose travel, food, and entertainment cards; (3) all-purpose bank credit cards; and (4) debit cards.

Single-purpose credit cards include those from oil companies such as Exxon, Mobil, and Shell and those from stores such as Sears, Montgomery Ward, and Neiman Marcus. Sylvia's father uses an oil company card and cards from the stores in the mall, which are all single-purpose cards. There is no fee for this kind of card, and the card allows him to make purchases or obtain services at the store, station, or company that issued it. The purpose of such a card is to encourage you to buy exclusively from a particular company. If you have a credit card from one specific oil company, you are likely to buy much of your gasoline at that company's stations.

Multipurpose travel and entertainment cards include American Express, Diners Club, and Carte Blanche. There is an annual membership fee for these cards. For this membership fee, these cards provide a variety of services that some credit cards do not provide, such as guaranteed check cashing at hotels and airline counters, free travel insurance, and emergency card replacement. Usually no credit limit is imposed on the cardholder. However, these are *charge cards,* not credit cards, so the cardholder is expected to pay the entire unpaid balance upon receipt of the monthly statement.

All-purpose bank cards include VISA, MasterCard, and Discover cards. These cards usually carry an annual membership fee. They are widely accepted by retailers as well as by restaurants, hotels, airlines, and other companies that wish to cash in on the boost in sales that accompanies the acceptance of credit cards. Some banks offer cardholders special purchase insurance and worldwide travel assistance.

An advantage to using credit cards is that cardholders are covered by Regulation Z, which is the regulation that enforces the Truth in Lending Act passed by Congress. This regulation protects consumers in cases of error, if merchandise is not received, or if services are not rendered. Card users may receive more protection than they would if they paid by check or cash, since it may be easier to get a credit card credit than to get a cash refund.

You can often use these cards to get instant cash up to a limited amount at the bank, credit union, or 24-hour teller facility. Some banks provide special checks with your credit card that you can use to pay for a purchase or to make a deposit into your regular checking or savings account. Using these checks is similar to getting a cash advance.

Debit cards are essentially electronic checks. They enable cardholders to use their automated teller machine (ATM) cards to pay for purchases at local stores. The purchase amount is deducted directly from the cardholder's checking account. Debit cards enable merchants to use credit card companies' existing authorization, draft transmission, and settlement systems.

ALGEBRA *REVIEW*

Express each percent as a decimal.

1. 12.5%
2. 3.6%

3. 119%
4. 125.6%

Express each decimal as a percent.

5. 33.3
6. 122.3

7. 0.76
8. 1.234

Evaluate. Give answer to the nearest hundredth.

9. $\left(1 + \frac{0.125}{12}\right)^{12} - 1$

10. $\left(1 + \frac{0.065}{12}\right)^{11} - 1$

ASK YOURSELF

1. What are three terms that must be disclosed by a bank according to the Truth in Lending Act?

2. What are two companies that offer credit or charge cards in each of the following categories?
   a. Single-purpose cards

   b. Multipurpose travel and entertainment cards

   c. All-purpose bank cards

SHARPEN YOUR SKILLS

SKILL 1

One major difference between an *installment credit* purchase and a *credit card* purchase is the amount of time that elapses before you must pay interest on the purchase. Remember that interest is charged in the first month on an installment loan. However, for a credit card, when you receive the monthly statement, it tells you how much time you have to pay the bill before a finance charge is added. This amount of time is called the **grace period.** Thus you will have the amount of time that elapses between the time of your purchase and the time you receive your statement *plus* the grace period to pay for the purchase before receiving any finance charges. (Usually there is no grace period if you use your credit card to get a cash advance.) Assume here and throughout the lesson that payments on credit card accounts are received on the last day of the month.

EXAMPLE 1 Sylvia wants her father to buy her a sewing machine that costs $550. He charges it on a new VISA card with which he has made no other purchases. He will make monthly payments of $60.

QUESTION What new balance will be shown on his first three monthly statements if his bank applies a 1.5% monthly finance charge and he refrains from using the credit card during this time?

SOLUTION
His first monthly statement will show a previous balance of $0 and a purchase of $550. No interest is charged this month, so the new balance for the first month is $550.

His second monthly statement will show a previous balance of $550, an interest charge of $0.015(550) = $8.25, and a payment of $60, which was received on the last day of the month. Therefore

$$\text{New balance} = \text{old balance} + \text{interest} - \text{payment}$$
$$= 550.00 + 8.25 - 60.00$$
$$= 498.25$$

The new balance for the second month is $498.25.

His third monthly statement will show a previous balance of $498.25, an interest charge of $0.015(498.25) = $7.47, and a payment of $60. Thus

$$\text{New balance} = 498.25 + 7.47 - 60.00 = 445.72$$

The new balance for the third month is $445.72.

EXAMPLE 2 Sylvia noticed that there is a section on the credit card statement that summarizes information each month. For the second month, it appeared as follows:

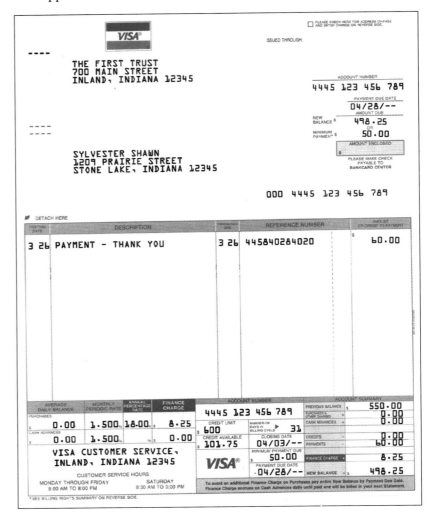

QUESTION What numbers would be included in this section of the statement for the first 8 months of the loan?

SOLUTION
Use a spreadsheet program to calculate the previous balance, new charges, payment received, finance charges, and new balance for each month. Remember that there is a 1.5% monthly finance charge. Be sure to use a rounding function when calculating the finance charges.

	A	B	C	D	E	F	
1	Month	Previous	New	Finance	Payment	New	
2		Balance	Charges	Charges	Received	Balance	
3	1	0.00	550.00	0.00	0.00	550.00	@ROUND(0.015*B4,2)
4	2	550.00	0.00	8.25	60.00	498.25	+B4+C4+D4−E4
5	3	498.25	0.00	7.47	60.00	445.72	
6	4	445.72	0.00	6.69	60.00	392.41	
7	5	392.41	0.00	5.89	60.00	338.30	
8	6	338.30	0.00	5.07	60.00	283.37	
9	7	283.37	0.00	4.25	60.00	227.62	
10	8	227.62	0.00	3.41	60.00	171.03	

SKILL 2

As interest rates in the economy fluctuate, banks may change the interest rates they charge their credit card customers. When interest rates are high, banks often advertise their *monthly* rates. When interest rates are low, they often advertise their *yearly* rates.

When an **annual percentage rate (APR)** of 21% is advertised, the 21% is simple interest. However, credit card interest is charged on a monthly basis. The **monthly interest rate** for an APR of 21% is $0.21 \div 12 = 0.0175$, or 1.75%. Because the interest charge is compounded, you must determine the **effective interest rate** being charged. The following formula can be used to determine the effective interest rate i_{eff} when the APR i is known.

Effective Interest Rate Formula

$$i_{eff} = \left(1 + \frac{i}{12}\right)^{12} - 1$$

where i_{eff} = effective interest rate
i = APR (annual percentage rate)

The effective interest rate on an APR of 21% is

$$i_{eff} = \left(1 + \frac{0.21}{12}\right)^{12} - 1 \qquad i = 21\%$$

$$= 0.2314, \quad \text{or } 23.14\%$$

EXAMPLE 3 Sylvia's mother has a credit card with an APR of 15%. Sylvia claims that the effective interest rate is actually more than 15%. Her mother claims that she pays less than 15% during the first 12 months following a purchase because of the 1-month grace period.

a. What is the effective interest rate on the credit card?

b. What is the effective interest rate for the first year following a purchase?

c. Who is correct, Sylvia or her mother?

d. What is the monthly interest rate?

SOLUTIONS

a. To determine the effective rate on an APR of 15%, substitute $i = 0.15$ into the effective interest rate formula.

$$i_{\text{eff}} = \left(1 + \frac{0.15}{12}\right)^{12} - 1 \;\; = 0.1608$$

The effective interest rate is 16.08%.

b. To determine the effective rate for the first year, note that Sylvia's mother pays interest for only 11 of the 12 months following a purchase because of the grace period. Her effective rate for the first 12 months is

$$i_{\text{eff}} = \left(1 + \frac{0.15}{12}\right)^{11} - 1 \;\; = 0.1464$$

The effective interest rate is 14.64%.

c. Both are correct. An APR of 15% on a credit card produces an effective rate of 16.08%. However, the rate is only 14.64% in the 12 months following a purchase because of the grace period.

d. The monthly interest rate is $0.15 \div 12 = 0.0125$, or 1.25%.

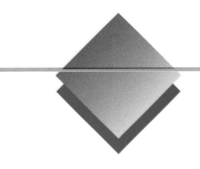

TRY YOUR SKILLS

Using the chart below, determine the new balance shown on your monthly statement for the first 4 months after making a $2300 purchase with your new MasterCard. The card has an APR of 12% and a 1-month grace period. Your monthly payment is $275 and you make no additional purchases.

	Month	Previous Balance	New Charges	Finance Charges	Payment Received	New Balance
1.	1	0.00	$2300.00			
2.	2					
3.	3					
4.	4					

5. Determine the effective interest rate on a credit card with an APR of 12%.

6. Determine the effective interest rate for the credit card in Exercise 5 during the first 12 months following a purchase if the card has a 1-month grace period.

EXERCISE YOUR SKILLS

1. Name one major difference between an installment credit purchase and a credit card purchase.

2. Describe the difference between an annual percentage rate and an effective interest rate.

3. What is the major difference between what you must pay VISA and American Express each month?

Complete the chart below to determine the new balance shown on your monthly statement for the first 3 months after you make a $3000 purchase on your new VISA card that has an APR of 16%. The card carries a 1-month grace period. Assume that you make monthly payments of $325 and that you make no additional purchases.

	Month	Previous Balance	New Charges	Finance Charges	Payment Received	New Balance
4.	1	0.00	$3000.00		0.00	
5.	2					
6.	3					

Use a spreadsheet program to determine the previous balance, new charges, payment received, finance charges, and new balance that would be shown on Ming's first 12 MasterCard statements after a purchase of $3215 if he makes monthly payments of $345. The card has a 1-month grace period and an APR of 13.9%. Assume that he has never used the card and that he makes no further charges on it during this time period.

	Month	Previous Balance	New Charges	Finance Charges	Payment Received	New Balance
7.	1	0.00	$3215.00		0.00	
8.	2					
9.	3					
10.	4					
11.	5					
12.	6					
13.	7					
14.	8					
15.	9					
16.	10					
17.	11					
18.	12					

> ### KEY TERMS
> all-purpose bank cards
> annual percentage rate (APR)
> credit card
> credit limit
> debit cards
> effective interest rate
> grace period
> late payment penalties
> monthly interest rate
> multipurpose travel and entertainment cards
> single-purpose credit cards
> Truth in Lending Act

Determine the effective interest rate and monthly interest rate for each APR. Give answers to the nearest hundredth of a percent.

19. 13.5%   **20.** 11.9%   **21.** 16%   **22.** 19.9%

Determine the effective interest rate on each credit card during the first 12 months following a purchase if each card has a 1-month grace period. Give answers to the nearest hundredth of a percent.

23. 13.5%   **24.** 11.9%   **25.** 16%   **26.** 19.9%

MIXED REVIEW

1. Determine the monthly payment on a loan of $68,500 at 7.9% for 15 years.

2. How much will Sanchez save by prepaying a loan of $5000 at 9% for 5 years for which his monthly payments are $103.79 if he repays it 3 years early?

3. How much will $7500 be worth at the end of 2 years if it earns 4% interest, compounded quarterly?

4. Alberto earns $12.50 per hour. He works 35 hours per week for 48 weeks and receives 4 weeks of paid vacation. His other benefits cost his employer $5200 per year. What is the annual cost to the company of his salary and benefits?

5. Lynn saves $52 per week. She wants to buy a sofa that costs $798. How many weeks must she save before she can buy the sofa?

6. Peter has a loan of $12,000 at an annual rate of 5%. The term of the loan is 7 years and the monthly payments are $169.61. How much will he save by repaying the loan 2 years early?

Suppose that your bank statement shows a closing balance of $325.89. The following transactions occurred but were not shown in the statement: a deposit of $150, an ATM withdrawal of $80, a check in the amount of $200.

7. What amount or amounts must be added to the balance in the statement?

8. What amount or amounts must be subtracted from the statement balance?

9. If no error has occurred, what should your current checkbook balance be?

A company manufactures widgets in two models, plain and fancy. The unit cost for the plain model is $4 and for the fancy model is $6. The selling price is $10 for the plain model and $20 for the fancy model. Let x represent the plain model and y represent the fancy model.

10. What is the objective function for the cost?

11. What is the objective function for the revenue?

12. What is the objective function for the profit?

13. The company can produce 50 widgets a week. Find the weekly profit if it decides to manufacture 30 of the fancy model.

6-2 CREDIT CARD BALANCES: HOW LONG DO THEY LAST?

Trevor grew up next door to Sylvia and her mother. As kids, he and his younger sister Tracey spent a lot of time playing with Sylvia. Sylvia always wanted to pretend that they were a family so that she could take Tracey shopping for new clothes and toys. Trevor's mother gave them some outdated credit cards with which to play. (She cut out a small portion of each one so that they could not be used to make real purchases.) Sylvia would pull out credit cards to pay for all the imaginary purchases that she made for Tracey.

Since Sylvia had always seen her father with his credit cards, she assumed that everyone made all their purchases with credit cards.

Trevor realized even then how much Sylvia enjoyed shopping on credit. His parents have always had their share of credit cards too, but they do not use them as much as Sylvia's father does. Trevor would like to identify the advantages and disadvantages of credit cards, since Sylvia's father and his parents have different views on their use.

OBJECTIVES: *In this lesson, we will help Trevor to:*

- *Identify advantages and disadvantages of credit cards.*
- *Determine how long it takes to pay off a credit card balance if a fixed payment is made each month.*
- *Determine the size of the monthly payment that must be made in order to pay off a credit card balance in a fixed amount of time.*

ADVANTAGES OF CREDIT CARDS

1. Credit cards are very convenient and simple to use when making purchases.
2. You can delay payment for purchases until a predetermined date.
3. A record of your purchases is made automatically.
4. The danger of losing money while shopping is minimized, since you do not need cash at the time of your purchase.
5. It is easy to order merchandise by mail or telephone with a credit card.
6. Salespeople and business owners may come to recognize you if you are a frequent charge customer and may thus provide you with better service.
7. You can make several purchases at one time.
8. Credit cards help you to take advantage of sales.
9. When you have a credit card, you can use it to obtain an immediate cash advance (up to a specified amount) at an automated teller machine (ATM).

DISADVANTAGES OF CREDIT CARDS

1. A credit card encourages people who have a tendency to overspend to charge beyond their income and ability to pay.
2. Using a credit card does not have the same impact as paying cash. You may forget how much you have charged until the bill comes at the end of the month.
3. Since you can make so many charges on the telephone and through the mail, anyone who knows your credit card number, your name, and other information that is easily obtainable can make unauthorized purchases to your account without having your card in their possession.
4. Credit cards can be stolen. If yours are stolen, you should notify the issuing company immediately, or you could be held liable for unauthorized charges. In general, you will not be liable for unauthorized charges once you have notified the company.

ALGEBRA *REVIEW*

Find the domain and range of each relation.

1. $\{(9, 2), (3, 8), (9, 8)\}$

2. $\{(2, 5), (3, 5), (6,6)\}$

3. Is the relation in Exercise 1 a function? Why or why not?

4. Is the relation in Exercise 2 a function? Why or why not?

ASK YOURSELF

1. What are four advantages of credit cards?

2. What are three disadvantages of credit cards?

3. If you lose your credit card, why should you immediately notify the company that issued it?

SHARPEN YOUR SKILLS

SKILL 1

In this lesson you will use the **common logarithm function,** $y = \log_{10} x$. For this function, y represents the exponent to which the number 10 must be raised to obtain x. For example, $\log_{10} 100 = 2$ because $10^2 = 100$, and $\log_{10} 100,000 = 5$ because $10^5 = 100,000$. When writing the common logarithm $y = \log_{10} x$, it is customary to omit the base 10. Thus $\log x = \log_{10} x$.

Your graphing calculator has a special common logarithm key. Evaluate $y = \log 1,000,000$ by pressing the log key on your calculator followed by 1,000,000. Then press the enter key to obtain 6.

A special formula allows you to calculate the length of time that it will take to pay the remaining balance on a credit card if a specified monthly payment M is made. The formula uses the common logarithm function.

Time-to-Pay-Off Formula

$$n = \frac{\log\left(\dfrac{M}{M - Pr}\right)}{\log(1 + r)}$$

where P = amount of loan
r = monthly interest rate
M = monthly payment
n = number of payment periods

EXAMPLE 1 Trevor's father has a balance of $1200 on one of his MasterCard cards that has an APR of 15%. He can afford monthly payments of $150.

QUESTION If he makes no further purchases with this card, how long will it take him to pay off the balance?

SOLUTION

$$n = \frac{\log\left(\dfrac{M}{M - Pr}\right)}{\log(1 + r)}$$

Use the time-to-pay-off formula.

$$n = \frac{\log\left(\dfrac{150}{150 - (1200)(0.0125)}\right)}{\log(1 + 0.0125)}$$

$P = 1200; r = 0.15 \div 12 = 0.0125; M = 150$

$$n = 8.48$$

To the nearest hundredth

Use the following keystrokes: ⦅ LOG ⦅ 150 ÷ ⦅ 150 − 1200 ×
0.0125))) ÷ ⦅ LOG ⦅ 1.0125)) ENTER

It takes 9 months to pay off the balance. In the last month he owes less than $150.

EXAMPLE 2 Trevor's father just received an unexpected raise. As a result, he may be able to make a larger monthly payment to MasterCard.

QUESTION How long will it take him to pay off his bill if he makes monthly payments of $200 instead of $150?

SOLUTION

$$n = \frac{\log\left(\frac{M}{M - Pr}\right)}{\log(1 + r)}$$

$$n = \frac{\log\left(\frac{200}{200 - (1200)(0.0125)}\right)}{\log(1 + 0.0125)} \qquad P = 1200;\ r = 0.15 \div 12 = 0.0125;\ M = 200$$

$$n = 6.28$$

It takes 7 months to pay off the balance if he makes a monthly payment of $200 instead of $150.

SKILL 2

You may wish to pay off a credit card balance before a certain event occurs or before you need to make another purchase. For example, you may want to pay off a balance before you begin college. In this case you will want to know how much you have to pay each month to bring the balance to zero in a specified period of time.

Since the process of paying off a credit card balance is similar to making monthly payments on an installment bank loan, you can apply the monthly payment formula that you studied in Chapter 5. To use the formula for this purpose, let P represent the current balance.

Monthly Payment Formula for Paying Off Balance

$$M = \frac{Pr(1 + r)^n}{(1 + r)^n - 1}$$

where P = current balance
r = monthly interest rate
n = number of payment periods
M = monthly payment

EXAMPLE 3 Trevor's cousin Adam will be going to college in 2 years. Trevor's uncle would like to pay off his VISA card balance before he begins making tuition payments.

QUESTION If his current balance is $4200 and his card carries an annual interest rate of 21%, how large must his monthly payment be if the balance is to be paid off in 2 years?

SOLUTION

$$M = \frac{Pr(1 + r)^n}{(1 + r)^n - 1}$$

$$M = \frac{4200(0.0175)(1 + 0.0175)^{24}}{(1 + 0.0175)^{24} - 1}$$ $P = 4200; r = (0.21 \div 12) = 0.0175;$
$n = 2 \cdot 12 = 24$

$$M = 215.82$$

He must make monthly payments of $215.82 to pay off the balance in 2 years.

TRY YOUR SKILLS

Evaluate to the nearest hundredth.

1. $\log 1000 \div \log 17$
2. $\log 129 \div \log 1.02$
3. $\log (1200 \div 1100) \div \log 1.75$
4. $\log (180 \div 165) \div \log 1.01$

How long will it take to pay off each credit card balance with the given monthly payment at the given APR? Express your answer as a whole number of months.

5. $1500; $170; 12%
6. $3500; $200; 14%
7. $1850; $225; 16%
8. $2000; $210; 15%

Determine the monthly payment that must be made to reduce each MasterCard balance to zero in the specified period of time. Assume that there is an annual finance charge of 15.9%.

9. $28,000 in $2\frac{1}{2}$ years
10. $16,000 in 25 months

EXERCISE YOUR SKILLS

KEY TERM

common logarithm
function

1. If you determine that it will take 7.2 months to pay off a credit balance, why would you round your answer to 8 months instead of 7 months?

2. Why is the process of paying off a credit card balance similar to making monthly payments on an installment bank loan?

Evaluate to the nearest hundredth.

3. $\log 1280 \div \log 25$

4. $\log 277 \div \log 1.15$

5. $\log (1600 \div 1450) \div \log 1.22$

6. $\log (210 \div 195) \div \log 1.0167$

How long will it take to pay off each credit card balance with the given monthly payment at the given APR? Express your answer as a whole number of months.

7. $1500; $110; 12%

8. $170; $15; 12%

9. $1350; $85; 10.9%

10. $180; $36; 10.9%

11. $2750; $180; 16.9%

12. $325; $15; 16.9%

13. $3000; $225; 18%

14. $400; $28; 18%

15. $4500; $220; 12%

16. $525; $42; 12%

17. $5100; $325; 21%

18. $610; $45; 21%

Determine the monthly payment that must be made to reduce each VISA balance to zero in the specified period of time at the given APR.

19. $1900; $1\frac{1}{2}$ years; 15%

20. $2350; 22 months; 16.9%

21. $775; 16 months; 17.5%

22. $895; 15 months; 18%

23. $1675; $3\frac{1}{2}$ years; 9%

24. $1958; $2\frac{1}{2}$ years; 16%

MIXED REVIEW

1. How much money will you save by prepaying a loan of $34,000 at 12.5% for 10 years for which you have been making monthly payments of $497.68 if you prepay it at the end of 7 years?

2. Determine the effective rate of interest to the nearest hundredth of a percent if the APR is 14%.

3. Bevilaqua earns $566.66 in gross income each week. How much will be deducted for Social Security and Medicare?

4. If Marci can afford a monthly payment of $360, how much can she afford to borrow at a yearly interest rate of 9.5% for 5 years?

5. Use the Rule of 72 to determine how long it will take your $5500 savings account to double in value if it is growing at a rate of 3%.

6. If you buy living room furniture for $4100 and finance it at 16% over 3 years, how much will you save by making a down payment of 20% instead of 10%?

7. Suppose that the equation for the cost function of a small business is $c = 1.9n + 118.50$ and the equation for the revenue function is $r = 5n$. Use a graphing calculator to find the break-even point.

6-3 INTEREST CHARGES: THE CHARGES FOR CHARGING

Vernon lives in a suburban area near Dallas. His parents are both college graduates with well-paying jobs. Vernon has played football for his high school team for 3 years. He can run the 100-meter dash in close to 10 seconds. Because of his speed, he usually plays a running back position. He caught two touchdown passes and one 43-yard pass last year in the regional championship game.

When Vernon and his girlfriend Veronica go out to dinner, he often picks up the tab for several other friends who happen to stop by their table to talk football. Vernon carries his father's American Express card for such expenditures.

Vernon has stopped asking his mother to select his clothes for him because she does not always understand the importance of maintaining his image. She has, however, allowed Vernon to use her VISA card to buy his clothes, provided that he does not charge too much in any one month. Last month, after he purchased new snakeskin boots with silver trim, his mother insisted that he not use the card until she had paid down the balance.

Vernon wondered why his parents set limits on their credit card spending. They explained that if he charges more than they can afford to pay when they receive the statement each month, interest payments on his purchases will be added and displayed on the following statement. Vernon wonders about interest charges and whether there are ways to reduce the amount of interest charges.

Recognize the many ways of using credit cards.

Observe which people are likely to use credit each way.

Determine whether and in what ways a consumer should use credit.

Explore the relationship between the size of the monthly payment and the interest charges.

Understand the relationship between APR and interest charges.

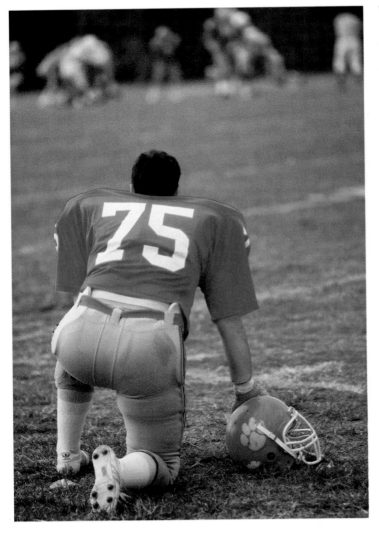

WHO GETS CREDIT AND HOW DO THEY USE IT?

Your level of income determines how much credit you can get. Vernon's family—with its high income—has been offered quite a lot of credit. People with a higher-than-average education are more likely to be offered credit cards than are those with less education. Young families with children are more likely to use cards but are also more likely to incur debt with their cards. Likewise, those who live in the suburbs tend to use their credit cards more than families living in central cities or rural areas.

Some families fall into the category of people who use their cards as a convenience, instead of carrying large amounts of cash or writing many checks. They pay their credit card charges promptly, within the 20- to 25-day grace period, and pay little or no interest.

The other group of card users takes advantage of the revolving installment loan features of their credit cards, almost never completely paying off their debts, making the minimum required payments, and paying interest to the bank with every payment.

Should you use a credit card? According to the experts, the answer to this question is *no* if *any* of the following are true:

- You use the card as an excuse to overspend.
- You frequently buy things impulsively that you do not need.
- You are often late making scheduled payments.
- You have no steady income.

The answer is *yes* if *all* of the following are true:

- You have handled credit responsibly in the past.
- You use the card as a budgeting convenience.
- You recognize the dangers as well as the attraction of using credit cards.

To use credit cards to your best advantage, adhere to the following rules:

1. Keep only the cards that you will use fairly often. For most people in the middle-income bracket, a couple of oil company cards for gasoline and one bank card are enough. The businessperson may wish to use a travel and entertainment card also.

2. Consider every charged purchase as though you were paying cash. Ask yourself: Can I repay the charge promptly and easily?

3. Do not spend more than 20% of your take-home pay on credit payments. As described in earlier chapters, if a family uses more than this amount on credit payments (not including the mortgage payment), it may suffer from credit overload.

4. At the beginning of each month, set a limit on the total amount of charges you will be able to repay easily. Stay within that limit and repay the charges promptly to avoid additional interest charges.

5. Keep your receipts until you receive your statement to verify your spending with the statement.

ASK YOURSELF

1. What are two ways in which consumers use credit cards?

2. Why is it a good idea to charge only what you can pay for each month?

3. What are three reasons not to use a credit card?

ALGEBRA *REVIEW*

Express each percent as a decimal.

1. 13.9%   2. 1.85%

3. 245%   4. 111.1%

Express each decimal as a percent.

5. 456   6. 233.4

7. 0.09   8. 9.876

Find.

9. What is 1.6% of $1600?

10. What is 0.85% of $2100?

11. What is 18% of $4433.22?

12. $1500 is what percent of $7500?

13. $12,750 is what percent of $85,000?

14. $41.25 is what percent of $750?

15. 25 is 100% of what number?

16. $13.75 is 20% of what number?

17. $7800 is 15% of what number?

SKILL 1

Banks and other lending institutions set forth a **minimum monthly payment** that must be made by their credit card holders. However, you can set your own monthly payment as long as it is above the required amount. The amount of your payment affects the length of time it will take you to pay off your balance. It also affects the amount of interest that you will pay.

EXAMPLE 1 The current balance on Vernon's mother's VISA card is $785.00, and her monthly finance charge is 1.5% of the amount owed. Her bank requires a minimum monthly payment of 10% of her unpaid balance rounded to the nearest dollar or $20, whichever is larger. If the amount she owes drops below $20, then she must make a payment equal to the total amount owed.

QUESTION How much interest will she pay in the next year, assuming that she makes a payment of 10% of the amount owed on the last day of the month?

SOLUTION

In the first month she has a balance of $785.00.

$$\begin{aligned} \text{Interest charge} &= \text{balance} \cdot 0.015 \\ &= 785.00 \cdot 0.015 \\ &= 11.78 \qquad \text{To the nearest cent} \end{aligned}$$

$$\begin{aligned} \text{Amount owed} &= \text{balance} + \text{interest} \\ &= 785.00 + 11.78 \\ &= 796.78 \end{aligned}$$

$$\begin{aligned} \text{Payment} &= 10\% \text{ of amount owed} \\ &= 0.10 \cdot 796.78 \\ &= 80 \qquad \text{To the nearest dollar} \end{aligned}$$

In the second month she has a balance of $796.78 - 80 = \$716.78$.

Use a spreadsheet program to work through the first 12 payments, and then total the interest column. To total the interest column use a sum function, such as @SUM(C2..C13) which adds the numbers in cells C2 through C13. In the columns for interest and 10% payment, you will need to use a rounding function. However, the interest is rounded to the nearest cent but the payment is rounded to the nearest dollar. Therefore the formulas are

For cell C2:   @ROUND(0.015*B2,2)

For cell E2:   @ROUND(0.1*D2,0)

	A	B	C	D	E
1	Month	Balance	Interest	Amount Owed	10% Payment
2	1	785.00	11.78	796.78	80.00 — @ROUND(0.1*D2,0)
3	2	716.78	10.75	727.53	73.00 @ROUND(0.015*B2,2)
4	3	654.53	9.82	664.35	66.00
5	4	598.35	8.98	607.33	61.00
6	5	546.33	8.19	554.52	55.00
7	6	499.52	7.49	507.01	51.00
8	7	456.01	6.84	462.85	46.00
9	8	416.85	6.25	423.10	42.00
10	9	381.10	5.72	386.82	39.00
11	10	347.82	5.22	353.04	35.00
12	11	318.04	4.77	322.81	32.00
13	12	290.81	4.36	295.17	30.00
14		Total interest	90.17 — @SUM(C2..C13)		

She pays $90.17 in interest during the first year.

EXAMPLE 2 Vernon wonders whether his mother's interest payments for the first year would increase if her bank allowed her to make smaller monthly payments.

QUESTION How much interest would she pay in the first year if her bank required only that she make a minimum monthly payment of 5% of her unpaid balance (rounded to the nearest dollar) or $20, whichever is larger? How much more interest would she pay in this case?

SOLUTION
In the first month her balance, interest, and amount owed would be the same as in Example 1. However, her payment would be $0.05 \cdot 796.78 = \$40$, rounded to the nearest dollar.

Copy the spreadsheet program from Example 1 and change the payment to 5% of the amount owed as shown on page 264.

Vernon's mother would pay $116.62 in interest during the first year. In the first year, she would pay $116.62 - 90.17 = \$26.45$ more in interest if she made monthly payments of 5% of the amount owed rather than 10% of the amount owed.

	A	B	C	D	E
18	Month	Balance	Interest	Amount Owed	5% Payment
19	1	785.00	11.78	796.78	40.00
20	2	756.78	11.35	768.13	38.00
21	3	730.13	10.95	741.08	37.00
22	4	704.08	10.56	714.64	36.00
23	5	678.64	10.18	688.82	34.00
24	6	654.82	9.82	664.64	33.00
25	7	631.64	9.47	641.11	32.00
26	8	609.11	9.14	618.25	31.00
27	9	587.25	8.81	596.06	30.00
28	10	566.06	8.49	574.55	29.00
29	11	545.55	8.18	553.73	28.00
30	12	525.73	7.89	533.62	27.00
31		Total interest	116.62		

@ROUND(0.05*D19,0)

To compare the interest paid for monthly payments of 5% and 10%, Vernon created a bar graph as shown. He used the months for x values, the interest for 10% payments as the first data range, and the interest for 5% payments as the second data range.

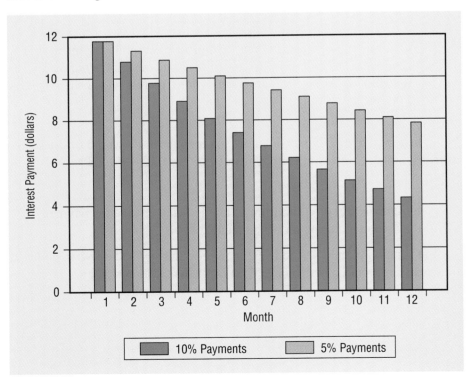

Since Christmas and Hanukkah shopping entices many consumers to buy more than they do at other times of the year, many banks open new MasterCard and VISA accounts for potential shoppers in preparation for the holiday season. The APRs that are offered vary from bank to bank.

EXAMPLE 3 Karl's and Aubrey's parents have two MasterCard accounts. They currently have balances of $450.00 on each account. One account has an APR of 21%, and the other has an APR of 14%.

QUESTION They make monthly payments of 10% of the amount owed (rounded to the nearest dollar) or $20, whichever is larger, on the last day of each month on both accounts. How much will they pay in interest charges during the next year on the account with an APR of 21%? On the account with an APR of 14%? How much will be saved at the lower interest rate?

SOLUTION

Use a computer spreadsheet program to determine the total interest paid on each account during the next year. Remember the monthly interest is the APR divided by 12.

When 10% of the amount owed is less than $20, the payment is $20. Notice that in cell E14 this happens:

$0.10(188.94) = 19$ To the nearest dollar

Therefore in cell E14 and also in cell E15 you must enter 20. The spreadsheet automatically recalculates the remaining cells when you do this.

In working an exercise you may get a negative number in your spreadsheet. This may be because the amount owed is less than $20. Remember that if the amount owed is less than $20, you will only pay the amount owed, so you will need to adjust your spreadsheet.

	A	B	C	D	E
1	MASTERCARD ACOUNT WITH 21% APR				
2	21% APR = 0.0175 per month				
3	Month	Balance	Interest	Amount Owed	Payment
4	1	450.00	7.88	457.88	46.00
5	2	411.88	7.21	419.09	42.00
6	3	377.09	6.60	383.69	38.00
7	4	345.69	6.05	351.74	35.00
8	5	316.74	5.54	322.28	32.00
9	6	290.28	5.08	295.36	30.00
10	7	265.36	4.64	270.00	27.00
11	8	243.00	4.25	247.25	25.00
12	9	222.25	3.89	226.14	23.00
13	10	203.14	3.55	206.69	21.00
14	11	185.69	3.25	188.94	20.00
15	12	168.94	2.96	171.90	20.00
16		Total interest	60.90		
17					
18					
19	MASTERCARD ACOUNT WITH 14% APR				
20	14% APR = 0.0116666 per month				
21	Month	Balance	Interest	Amount Owed	Payment
22	1	450.00	5.25	455.25	46.00
23	2	409.25	4.77	414.02	41.00
24	3	373.02	4.35	377.37	38.00
25	4	339.37	3.96	343.33	34.00
26	5	309.33	3.61	312.94	31.00
27	6	281.94	3.29	285.23	29.00
28	7	256.23	2.99	259.22	26.00
29	8	233.22	2.72	235.94	24.00
30	9	211.94	2.47	214.41	21.00
31	10	193.41	2.26	195.67	20.00
32	11	175.67	2.05	177.72	20.00
33	12	157.72	1.84	159.56	20.00
34		Total interest	39.56		

@ROUND(0.1*D4,0) → (points to E4: 46.00)

change 19.00 to 20.00 → (points to E14: 20.00)

change 18.00 to 20.00 → (points to E32: 20.00)

Karl's and Aubrey's parents pay $60.90 in interest during the year with an APR of 21%. They pay $39.56 in interest during the year with an APR of

14%. This represents a savings of $60.90 - 39.56 = \$21.34$ over the interest charges on the card with an APR of 21%.

To compare the interest payments, create a bar graph as shown.

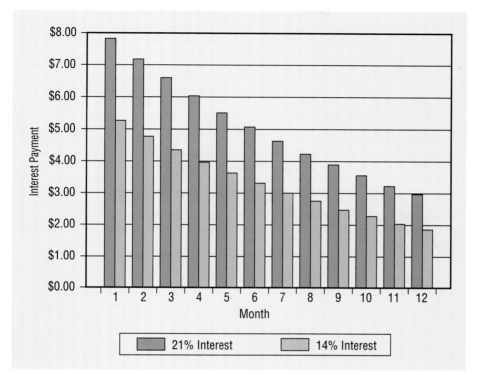

The interest charge is greater when the APR is larger if all other variables are the same.

TRY YOUR SKILLS

Complete the following chart to determine the interest charges over the next four months on a balance of $995.00. Assume that your card has an APR of 15% and you make monthly payments of 5% of the amount due, rounded to the nearest dollar. Assume that you make your payment on the last day of the month.

	Month	Balance	Interest	Amount Owed	Payment
1.	1	$995.00			
2.	2				
3.	3				
4.	4				

5. What is the total interest for the credit card account of Exercises 1–4?

For the credit card account of Exercises 1–5, find the interest cost for the first 4 months if you pay 10% of the amount due instead of 5% of the amount due.

	Month	Balance	Interest	Amount Owed	Payment
6.	1	$995.00			
7.	2				
8.	3				
9.	4				

10. What is the total interest for the credit card account of Exercises 6–9?

11. Compare the results of Exercises 5 and 10 to determine how much more interest you pay when you make payments of 5% of the amount owed instead of 10% of the amount owed.

Complete the chart below to determine how much the interest cost will be if you pay 10% of the amount due each month, rounded to the nearest dollar, and your card carries an APR of 9%. Assume that you make your payment on the last day of the month.

	Month	Balance	Interest	Amount Owed	Payment
12.	1	$995.00			
13.	2				
14.	3				
15.	4				

16. What is the total interest for the credit card account of Exercises 12–15?

17. Compare the results of Exercises 10 and 16 to determine how much more interest you pay if the APR on your card is 15% instead of 9%.

EXERCISE YOUR SKILLS

1. How can you use credit cards to your best advantage?

2. Why is it easier to overspend when you use a credit card?

3. Why do the interest payments decrease when you make larger monthly payments?

The current balance on Pedro's VISA card is $400.00, and his card carries a monthly finance charge of 1.75% of the amount owed. His bank requires a minimum monthly payment of 10% of his unpaid balance rounded to the nearest dollar or $20, whichever is larger. If the amount he owes drops below $20, then he must make a payment equal to the total amount owed. Use a spreadsheet program to determine how much interest he will pay in the next year, assuming that he makes his payment on the last day of the month.

	Month	Balance	Interest	Amount Owed	Payment
4.	1	$400.00			
5.	2				
6.	3				
7.	4				
8.	5				
9.	6				
10.	7				
11.	8				
12.	9				
13.	10				
14.	11				
15.	12				

16. What is the total interest that Pedro must pay?

Complete the chart below to determine his interest cost for the year if Pedro pays 5% of the amount due instead of 10% of the amount due.

	Month	Balance	Interest	Amount Owed	Payment
17.	1	$400.00	$7.00		
18.	2	387.00	6.77		
19.	3				
20.	4				
21.	5				
22.	6				
23.	7				
24.	8				
25.	9				
26.	10				
27.	11				
28.	12				

29. What is the total interest that Pedro must pay?

30. Compare the results of Exercises 16 and 29 to determine how much more interest he pays if he makes payments of 5% of the amount owed instead of 10% of the amount owed.

31. Use the graph function of your computer spreadsheet program to graph Pedro's monthly payments at 10% and 5% of the amount owed.

Complete the chart below to determine how much Pedro's interest cost will be if he pays 10% of the amount due each month and his card carries a monthly finance charge of 1% of the amount owed.

	Month	Balance	Interest	Amount Owed	Payment
32.	1	$400.00	$4.00		
33.	2	364.00	3.64		
34.	3				
35.	4				
36.	5				
37.	6				
38.	7				
39.	8				
40.	9				
41.	10				
42.	11				
43.	12				

44. What is Pedro's total interest?

45. Compare the results of Exercises 16 and 44 to determine how much more interest Pedro pays if the monthly finance charge is 1.75% instead of 1%.

46. Use the graph function of your computer spreadsheet program to graph Pedro's monthly payments with a monthly finance charge of 1.75% and 1%.

MIXED REVIEW

1. Determine which of the following loans has a greater total cost.
   a. $12,000 at 14% for 5 years   **b.** $15,000 at 11% for 4 years

2. During the first week of September, Samantha worked $37\frac{1}{2}$ hours and received $75.75 in tips. If her hourly rate of pay is $6.95, how much did she earn that week?

3. Determine the monthly payment that must be made to reduce a $3333.33 MasterCard balance to zero in 28 months if the card carries an APR of 17.5%

4. Zachary receives a commission of 5.5% on sales up to $12,000 and 7.5% on all sales over $12,000. How much has he earned in commission if his sales are $19,950?

5. How long will it take to pay off a credit card balance of $1800.75 with monthly payments of $200 if the APR is 11%? Express your answer as a whole number of months.

6. Suppose that your family's take-home pay is $4,300 per month. How much can you afford to spend for credit-card payments each month?

6-4 CREDIT CARDS: HOW TO PROTECT THEM

Sylvia Shawn's mother has several credit cards of her own that she obtained after she and Mr. Shawn were divorced. She was able to get credit in her own name after the divorce even though all of the accounts were in Sylvia's father's name while they were still married. The reason that this was possible is that she was protected by the Equal Credit Opportunity Act.

Sylvia remembers how difficult it was for her father when his credit cards were stolen. He had left them in the glove compartment of his car one day at the golf course. Sylvia was with him at the time. As they left the golf course and stopped to buy gasoline, he realized that all his credit cards had been taken. He spent the next three hours making phone calls to all the banks and charge account offices. The thief had spent over $10,000 before the information could be entered into the various computer networks! The thief charged $2650 worth of ski equipment, $3240 worth of video equipment, and a $4500 computer system to Mr. Shawn's now-closed accounts.

For the next three months, Mr. Shawn dealt with the credit card companies. He discovered that under the Fair Credit Reporting Act and the Fair Credit Billing Act, his credit rating would be preserved and he would have to pay only $50 of the unauthorized charges on each card. (This was still a substantial amount because he had so many cards.) He also learned what an inconvenience it is to have all your credit accounts changed and to have to wait for new cards to arrive from every creditor.

From her mother's and father's experiences, Sylvia learned that credit cards are just as valuable as cash.

OBJECTIVES: *In this lesson, we will help Sylvia to:*

⬤ *Discover how consumers are protected by*
- *the Fair Credit Reporting Act.*
- *the Fair Credit Billing Act.*
- *the Equal Credit Opportunity Act.*

⬤ *Recognize ways to protect credit cards.*
⬤ *Calculate credit account interest, payments, and balances.*
⬤ *Calculate the increase in monthly interest caused by new purchases.*

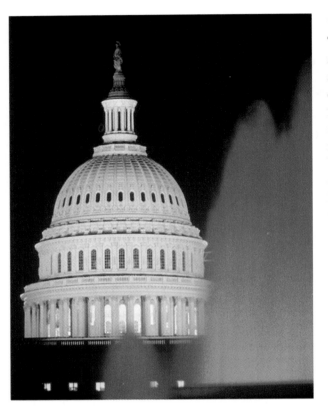

CONSUMER RIGHTS

The **Fair Credit Reporting Act** forbids credit agencies from giving out incorrect credit information about consumers. An individual who has been denied credit, employment, or insurance because of an inaccurate credit agency report may ask to know the source of the report. The individual can then have both incorrect information and any information that cannot be proven removed from the file. The act also allows consumers to examine information in their credit files. If consumers feel that they have been misrepresented, they can add short statements to the files that give their side of the story.

Under the **Fair Credit Billing Act,** consumers can preserve their credit ratings while settling disputes with stores and credit card companies. During this time a **creditor** cannot report you as delinquent to any credit agency for your failure to pay the portion of your bill that is under dispute.

The **Equal Credit Opportunity Act** prevents discrimination on the basis of sex, marital status, race, color, religion, age, or national origin. For instance, women applying for credit must be judged by the same standards as men. If they have a steady income and can qualify in other respects as good credit risks, they are equally entitled to credit.

The law specifically states that creditors cannot:
- Deny credit on the basis of sex, marital status, race, color, religion, national origin, or age (if an individual is old enough to enter into a binding contract).
- Deny credit because an applicant receives any income from a public program.
- Ask questions concerning birth control practices and plans for children, or assume that a female applicant is likely to become pregnant and have an interruption of income.
- Ask about an applicant's marital status (unless a spouse will be contractually liable for the loan, a spouse's income is counted on to repay a loan, or a spouse plans to use the loan).

- Refuse to consider part-time income of a working spouse, alimony, child support, or Social Security payments.
- Cancel a divorced or widowed person's credit when a marriage ends unless the income has dropped so much that the person may not be able to pay.

This act does not entitle you to credit whenever you want it. You must still pass the creditor's tests that indicate your financial ability and willingness to pay.

HOW TO PROTECT YOUR CREDIT CARDS

If your credit cards are stolen, the maximum amount of money that you will have to pay is $50 per card. However, if you carry many cards, like Sylvia's father, you will still have to pay a large sum of money if the thief works quickly before you notice that your cards are gone.

You can take the following precautions to protect yourself:

- Destroy any cards that you do not use. Cut unwanted cards in half and throw them away. Then notify the credit card company to close your account.
- Make a list of all the credit cards that you have (with the account numbers and the name and address of each issuer), and keep this list at home.
- Be sure that you get your card back every time you use it. Dishonest employees of legitimate establishments may retain your card to give to fraudulent users.
- Do not leave your credit cards in the glove compartment of your car. This is one of the first places a credit card thief will look.
- Do not underestimate the value of a credit card. Carrying a credit card is similar to carrying cash. The thief does not underestimate its value!

ASK YOURSELF

1. How does the Fair Credit Reporting Act protect consumers?

2. How does the Fair Credit Billing Act protect consumers?

3. What does the Equal Credit Opportunity Act prevent?

4. What are three ways in which you can protect your credit cards?

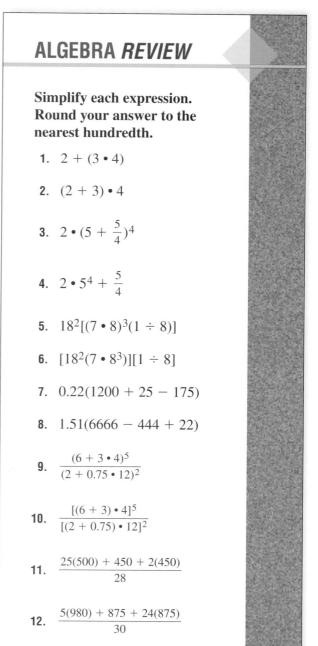

ALGEBRA *REVIEW*

Simplify each expression. Round your answer to the nearest hundredth.

1. $2 + (3 \cdot 4)$

2. $(2 + 3) \cdot 4$

3. $2 \cdot (5 + \frac{5}{4})4$

4. $2 \cdot 5^4 + \frac{5}{4}$

5. $18^2[(7 \cdot 8)^3(1 \div 8)]$

6. $[18^2(7 \cdot 8^3)][1 \div 8]$

7. $0.22(1200 + 25 - 175)$

8. $1.51(6666 - 444 + 22)$

9. $\dfrac{(6 + 3 \cdot 4)^5}{(2 + 0.75 \cdot 12)^2}$

10. $\dfrac{[(6 + 3) \cdot 4]^5}{[(2 + 0.75) \cdot 12]^2}$

11. $\dfrac{25(500) + 450 + 2(450)}{28}$

12. $\dfrac{5(980) + 875 + 24(875)}{30}$

SHARPEN YOUR SKILLS

In Lesson 6–3 you learned that the amount of your monthly payment and the monthly rate of interest influence your credit card balance and the amount of interest that you pay each month. Another factor that influences your balance and interest charges is additional purchases that you make before paying for earlier ones.

EXAMPLE 1 Saul's mother is trying to restrain herself from making any more purchases with her MasterCard until she pays off her current balance. She is quite successful at making no further purchases during some months, but she is unable to resist making more purchases during others. Saul is worried that her balance may be larger at the end of a 12-month period than it was at the beginning of the period. They have therefore decided to follow the account closely for one year.

QUESTION Will she owe more or less at the end of the year if the following hold true?

In month 1 she makes no purchases, and her balance is $1265.80. The monthly interest rate on the card is 1.5%.
She makes monthly payments of 10% of the amount owed, rounded to the nearest dollar.
Purchases are shown on her statements as follows:
  $250.75 in month 3,
  $75.00 in month 5,
  $380.50 in month 8, and
  $45.90 in month 11.

SOLUTION
Use a spreadsheet program to work through the first 12 months. Purchases are added to the amount owed from the previous month and payments are subtracted, so

  Balance = amount owed + purchases − payments

Remember to use the ROUND function for interest in column D and the payment in column F. Interest is rounded to the nearest cent so for cell D2 the formula is

  @ROUND(C2*0.015,2)

The payment is rounded to the nearest dollar so for cell F2 the formula is

  @ROUND(E2*0.1,0)

	A	B	C	D	E	F
1	Month	Purchases	Balance	Interest	Amount Owed	Payment
2	1	0.00	1265.80	18.99	1284.79	128.00
3	2	0.00	1156.79	17.35	1174.14	117.00
4	3	250.75	1307.89	19.62	1327.51	133.00
5	4	0.00	1194.51	17.92	1212.43	121.00
6	5	75.00	1166.43	17.50	1183.93	118.00
7	6	0.00	1065.93	15.99	1081.92	108.00
8	7	0.00	973.92	14.61	988.53	99.00
9	8	380.50	1270.03	19.05	1289.08	129.00
10	9	0.00	1160.08	17.40	1177.48	118.00
11	10	0.00	1059.48	15.89	1075.37	108.00
12	11	45.90	1013.27	15.20	1028.47	103.00
13	12	0.00	925.47	13.88	939.35	94.00
14			Total interest	203.40		

+E3−F3+B4 (formula referencing cell C4)

Despite her purchases, Saul's mother reduced her balance from $1265.80 in month 1 to $925.47 in month 12. She paid a total of $203.40 in interest.

___ **SKILL 2** ___

As you have learned, many bank cards carry an annual fee. If you are late making your scheduled payment, your bank may also charge a late payment penalty.

EXAMPLE 2 In month 9, Saul's mother had to pay her annual fee of $50.00. She was also late making her month 11 payment. Therefore, she received a late penalty fee of $15.00 on her month 12 statement.

QUESTION How do these charges affect her month 12 balance and the interest that she pays for the year?

SOLUTION
Copy the spreadsheet program from Example 1. Insert the annual fee and late penalty payments in the "Purchases" column to recalculate using the new information.

	A	B	C	D	E	F
1	Month	Purchases	Balance	Interest	Amount Owed	Payment
2	1	0.00	1265.80	18.99	1284.79	128.00
3	2	0.00	1156.79	17.35	1174.14	117.00
4	3	250.75	1307.89	19.62	1327.51	133.00
5	4	0.00	1194.51	17.92	1212.43	121.00
6	5	75.00	1166.43	17.50	1183.93	118.00
7	6	0.00	1065.93	15.99	1081.92	108.00
8	7	0.00	973.92	14.61	988.53	99.00
9	8	380.50	1270.03	19.05	1289.08	129.00
10	9	**50.00**	1210.08	18.15	1228.23	123.00
11	10	0.00	1105.23	16.58	1121.81	112.00
12	11	45.90	1055.71	15.84	1071.55	107.00
13	12	**15.00**	979.55	14.69	994.24	99.00
14			Total interest	206.29		

Her month 12 balance is 979.55 − 925.47 = $54.08 larger with the late fee and annual fee. She pays 206.29 − 203.40 = $2.89 more interest.

TRY YOUR SKILLS

Use a spreadsheet program to determine whether Yolanda owes more or less on her VISA account in month 12 than at the beginning of the year. Find her yearly interest payment. Her monthly interest charge is 1.5%, and she makes monthly payments of 10% of the amount owed, rounded to the nearest dollar.

	Month	Purchases	Balance	Interest	Amount Owed	Payment
1.	1	0.00	$554.75	$8.32	$563.07	
2.	2	$76.35				
3.	3	0.00				
4.	4	0.00				
5.	5	75.99				
6.	6	22.80				
7.	7	0.00				
8.	8	97.88				
9.	9	75.00				
10.	10	31.00				
11.	11	66.00				
12.	12	67.00				

13. What is the total interest that Yolanda pays?

14.–26. Recalculate Exercises 1–13 to reflect late fees of $10.00 in months 2 and 7 and an annual fee of $40.00 in month 4.

EXERCISE YOUR SKILLS

1. Why is it important for people to be able to examine information in their credit files?

2. What might happen if consumers were not protected by the Fair Credit Billing Act?

3. How does the Equal Credit Opportunity Act help a woman who becomes divorced or widowed?

KEY TERMS

creditor
Equal Credit
    Opportunity Act
Fair Credit Billing
    Act
Fair Credit
    Reporting Act

Use a spreadsheet program to determine whether Roger owes more or less on his VISA account in month 12 than at the beginning of the year. Find the interest that he pays for the year. His account carries a monthly interest charge of 1.5%, and he makes monthly payments of 10% of the amount owed, rounded to the nearest dollar.

	Month	Purchases	Balance	Interest	Amount Owed	Payment
4.	1	0.00	$875.66	$13.13	$888.79	
5.	2	0.00				
6.	3	0.00				
7.	4	0.00				
8.	5	0.00				
9.	6	$158.80				
10.	7	0.00				
11.	8	85.00				
12.	9	92.00				
13.	10	0.00				
14.	11	0.00				
15.	12	123.00				

16. What is the total interest that Roger pays?

Use a spreadsheet program to determine whether Yory owes more or less on her VISA account in month 12 than at the beginning of the year. Find the interest that she pays for the year. Her account carries a monthly interest charge of 1.5%, and she makes monthly payments of 10% of the amount owed, rounded to the nearest dollar.

	Month	Purchases	Balance	Interest	Amount Owed	Payment
17.	1	0.00	$1450.63	$21.76	$1472.39	
18.	2	0.00				
19.	3	0.00				
20.	4	$234.00				
21.	5	0.00				
22.	6	0.00				
23.	7	0.00				
24.	8	0.00				
25.	9	0.00				
26.	10	314.00				
27.	11	45.00				
28.	12	68.00				

29. What is the total interest that Yory pays?

30.–42. Recalculate Exercises 4–16 to reflect late fees of $20.00 in months 3 and 9 and an annual fee of $40.00 in month 8.

43.–55. Recalculate Exercises 17–29 to reflect late fees of $15.00 in months 2, 4 and 9 and an annual fee of $30.00 in month 10.

MIXED REVIEW

1. If you earn $25,900.00 and your FICA withholding is 7.65% of your earnings, how much do you pay for Social Security and Medicare?

2. If you can afford a monthly payment of $285, how much money can you borrow at 8.75% for 5 years?

3. Determine the monthly payment that must be made to reduce a $2599.90 VISA balance to zero in 42 months if the card carries an APR of 14.5%.

4. Determine the effective rate of interest on a credit card purchase if the APR is 17.9%.

5. How much money can you save by financing a $215,000 mortgage at 7.5% over 15 years rather than 30 years?

6. How much money can you save by financing a $10,500 car at 9% over 3 years rather than 5 years?

A cookie factory makes at least 2000 plain cookies per day and at least 3000 chocolate chip cookies per day. However, the packaging machinery cannot process more than 7500 cookies per day. The profit on each plain cookie is $0.20 and $0.25 on each chocolate chip cookie. Let x and y represent the number of plain and chocolate chip cookies, respectively, that are produced each day.

7. Write the objective function for the profit on the day's production of cookies.

8. Write the inequalities to express the constraints.

9. Graph the inequalities and use the graph to find the number of plain and chocolate chip cookies that gives the greatest profit.

Matilda wants to have a stereo system that costs $700. The store offers to let her purchase the set with an installment loan at 12.5% over 2 years. They also offer her a rent-to-own credit plan that allows her to rent the set for a fixed number of months and to apply all of the rent payments to the purchase price. Find the cost of each of the following rental options.

10. $35 per month for 24 months with no other payment required

11. $24 per month for 36 months with no other payment required

12. $30 per month for 24 months with an additional payment of $50 due at the end of the 24 months

13. Which, if any, of the rental options in Exercises 10–12 is a better deal than the installment loan?

Ursula likes her job as a sales clerk at Joyful Toys. There is only one part of her job that she does not enjoy doing. Each time a customer presents a credit card for a purchase, Ursula must verify that the customer's bank will approve the charge. Sometimes the credit is denied. Ursula does not like to inform a customer of this fact.

One time a customer became angry with Ursula when she told him that his credit was denied. He acted as though she was at fault. He yelled at her and demanded to see the store manager. As he left, he hurled a handful of action figures and a model rocket toward the display case, knocking over a tower of blocks and almost breaking the glass. The store manager did not blame Ursula for the mess or for the fact that several customers left while all this was happening. Nevertheless the incident made her feel very uncomfortable.

The bank that issues a customer's credit card may refuse to pay Joyful Toys if Ursula or another clerk neglects to verify the customer's credit. If that happens, Joyful Toys will lose the entire cost of the items that are purchased. The store already pays a service fee to the bank each time a purchase with a bank card is made.

When Ursula handles a customer's credit card, she must be careful to process the information correctly. After verifying the charge with the customer's bank, she makes an imprint of the card on the sales slip. She circles the expiration date on the sales slip. Ursula then asks the customer to sign the sales slip and compares the customer's signature on the sales slip with the signature on the card. She can then be satisfied that she has done everything possible to verify the customer's right to use the card.

OBJECTIVES: *In this lesson, we will help Ursula to:*

- *Identify costs paid by retailers for selling on credit.*
- *Find the average daily balance of a credit card account for the month.*
- *Explore how the timing of a monthly payment affects the interest charge.*

THE COST TO RETAILERS OF ISSUING THEIR OWN CARDS

Retailers who issue their own cards to sell on credit incur additional costs on every sale. These costs are a result of the following factors:

- The clerical work necessary to record sales and collect payments
- Losses from customers who fail to make payments
- An increased tendency of charge customers to return goods for exchange

Retailers who neglect to investigate a customer's ability to pay are likely to have high losses from unpaid debts. Stores that recklessly advertise generous credit terms to everyone may have to raise their prices to offset their potential losses. Stores that have sound credit policies are likely to have few losses from customers who fail to make payments. Some retailers elect to accept MasterCard, VISA, or American Express instead of (or in addition to) extending their own credit.

Banks that issue all-purpose cards like VISA and MasterCard collect a percentage of every sale made on their cards from retailers. This charge provides some protection for the banks against the money they stand to lose if

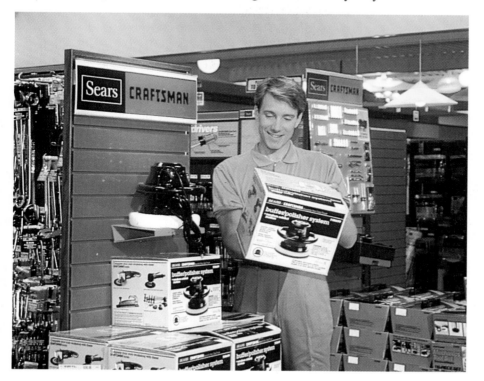

the cardholders fail to pay for the purchases they charged. Even though the salesperson takes every precaution, a few illegitimate uses of cards are bound to slip through.

Banks also pass some of their costs on to their credit card customers by charging interest on unpaid balances. The maximum amount of interest that a bank can charge a customer is set by state laws. Differences in state interest rate limitations caused at least one large bank to move its national credit card operation from New York to South Dakota. In South Dakota the interest rate that a bank can charge is higher than it is in New York.

As was stated in Lesson 6-1, banks often allow customers a grace period before they must pay the bill. If they pay the bill on time, they can avoid interest charges. If cardholders do not wish to pay the entire balance due, they can make a minimum payment (on some types of credit card accounts) and pay interest charges on the remainder.

The amount of interest that a customer must pay is based upon the **average daily balance,** which is found by adding the unpaid balances for each day of the billing cycle and then dividing this number by the number of days in the billing cycle. In this chapter a billing cycle is assumed to be a calendar month. Customers can save money by making their payments as early in the month as is possible.

ASK YOURSELF

1. Why do banks that issue cards such as MasterCard and VISA charge the retailers who accept these cards?

2. How is the average daily balance computed?

3. How is the maximum amount of interest that a bank can charge determined?

ALGEBRA *REVIEW*

When an *average* is to be determined from a number of scores in which some scores are repeated, the number of times a score is repeated is called the *frequency*.

Find the average of the following scores with the indicated frequencies.

Example:

Score	82	75	91
Frequency	3	2	4

$$\text{Average} = \frac{3(82) + 2(75) + 4(91)}{9} = 84.4$$

1.
Score	35	61	78
Frequency	5	3	2

2.
Score	468	721	325
Frequency	1	20	3

3.
Score	10.2	15.5	25.3
Frequency	18	5	2

4.
Score	804	246	760
Frequency	2	28	1

5.
Score	65	945	607
Frequency	20	4	6

SHARPEN YOUR SKILLS

SKILL 1

The average daily balance b is determined by dividing the sum of the daily balances s by the total number of days d in the billing cycle.

Average Daily Balance Formula

$b = \dfrac{s}{d}$ where b = average daily balance

s = sum of the daily balances

d = total number of days in the billing cycle

EXAMPLE 1 Scott's MasterCard account has a monthly interest rate of 1.5%. He had the following daily balances and payments for the month of May:

From 5/1 through 5/16 his daily balance was $385.00.
On 5/17 he made a payment of $80.
From 5/18 through 5/31 his daily balance remained at $305.00.

QUESTION What are the average daily balance, the monthly finance charge, and the ending balance during the month of May?

SOLUTION
Complete the following chart to find the sum of the daily balances.

Dates	Payment	Balance at End of Day	Number of Days	Sum of Daily Balances
5/1–5/16	0.00	$385.00	16	$ 6160.00
5/17	$80.00	305.00	1	305.00
5/18–5/31	0.00	305.00	14	4270.00
			Total 31	$10,735.00

$b = \dfrac{s}{d}$ Use the average daily balance formula,

$b = \dfrac{10,735}{31}$

$b = 346.29$ To the nearest cent

Average daily balance: $10,735 \div 31 = 346.29$
Finance charge (May 17): $0.015(346.29) = 5.19$
Ending balance for May: $305.00 + 5.19 = 310.19$

EXAMPLE 2 Scott always makes his MasterCard payments in the middle of the month when he receives his second bimonthly paycheck. He has often wondered whether the timing of his payment influences the amount he pays in monthly interest.

QUESTION What would his interest payment have been if he paid his May MasterCard bill on the first day of the month instead of on the seventeenth day? On the last day instead of the seventeenth day?

SOLUTION
Set up two charts similar to the chart in Example 1, one with a payment on May 1 and the other with a payment on May 31.

Dates	Payment	Balance at End of Day	Number of Days	Sum of Daily Balances
5/1	$80.00	$305.00	1	$ 305.00
5/2–5/31	0.00	305.00	30	9150.00
			Total 31	$9455.00

Average daily balance: $9455.00 \div 31 = 305.00$
Finance charge (May 1): $0.015(305.00) = 4.58$
Ending balance for May: $305.00 + 4.58 = 309.58$

Dates	Payment	Balance at End of Day	Number of Days	Sum of Daily Balances
5/1–5/30	0.00	$385.00	30	$11,550.00
5/31	$80.00	305.00	1	305.00
			Total 31	$11,855.00

Average daily balance: $11,855.00 \div 31 = 382.42$
Finance charge (May 31): $0.015(382.42) = 5.74$
Ending balance for May: $305.00 + 5.74 = 310.74$

Paying his bill on May 1 instead of May 17 saves: $5.19 - 4.58 = 0.61$
Paying his bill on May 17 instead of May 31 saves: $5.74 - 5.19 = 0.55$
Paying his bill on May 1 instead of May 31 saves: $5.74 - 4.58 = 1.16$

Although the savings in one month do not seem like a lot of money, they will be greater if your balance is larger. The savings accumulate over the months and years. Also, you may have many credit cards with balances. Therefore you can save a substantial amount of money on all of these cards by making credit card payments as early in the month as possible.

When finding the average daily balance be sure to use the correct number of days. Remember that April, June, September and November only have 30 days and February has 28 days unless it is a leap year.

EXAMPLE 3 Consumers often make additional charges on their accounts each month. These additional charges affect the average daily balance in the account.

QUESTION If Scott had purchased a new jacket for $50.00 on 5/11 in Example 1, what would have been his average daily balance, finance charge, and ending balance?

SOLUTION
Use a spreadsheet program.

Dates	Payment	Purchase	Balance at End of Day	Number of Days	Sum of Daily Balances
5/1–5/10	0.00	0.00	$385.00	10	$ 3850.00
5/11	0.00	$50.00	435.00	1	435.00
5/12–5/16	0.00	0.00	435.00	5	2175.00
5/17	$80.00	0.00	355.00	1	355.00
5/18–5/31	0.00	0.00	355.00	14	4970.00
				Total 31	$11,785.00

Average daily balance: $11,785 \div 31 = 380.16$
Finance charge (May 31): $0.015(380.16) = 5.70$
Ending balance for May: $355.00 + 5.70 = 360.70$

TRY YOUR SKILLS

1. Complete the following chart to determine the average daily balance, the finance charge, and the ending balance shown on Jerry's VISA statement for the month of September. Assume that his VISA card carries a monthly interest rate of 1.25%.

Dates	Payment	Balance at End of Day	Number of Days	Sum of Daily Balances
9/1–9/14	0.00	$642.00	14	$8988.00
9/15	$75.00	567.00		
9/16–9/30	0.00	567.00		
			Total	

a. Average daily balance   b. Finance charge

c. Ending balance

2. Complete the chart below to find how much Jerry can save if he makes his September payment on 9/1 instead of 9/15.

Dates	Payment	Balance at End of Day	Number of Days	Sum of Daily Balances
9/1	$75.00	$567.00	1	
9/2–9/30	0.00	567.00		
			Total	

a. Average daily balance   b. Finance charge

c. Ending balance

d. How much can he save if he makes his payment on 9/1 instead of 9/15?

3. Complete the chart below to determine the average daily balance, the finance charge, and the ending balance shown on Savannah's VISA statement for the month of August. Assume that her VISA card carries a monthly interest rate of 1.75%.

Dates	Payment	Purchase	Balance at End of Day	Number of Days	Sum of Daily Balances
8/1–8/12	0.00	0.00	$675.00	12	$8100.00
8/13	$95.00	0.00	580.00		
8/14–8/18	0.00	0.00			
8/19	0.00	$60.00			
8/20–8/31	0.00	0.00			

a. Average daily balance   b. Finance charge

c. Ending balance

EXERCISE YOUR SKILLS

1. Why does a clerk use a computer scanner to verify a customer's credit card before allowing the customer to use it for a purchase?

2. Does the day of the month on which you make a credit card payment influence the interest charge? Why or why not?

Complete each chart. Then determine the average daily balance, the finance charge, and the ending balance that will be shown on each monthly MasterCard statement. Assume that the card carries a monthly finance charge of 1.5%.

3.

Dates	Payment	Balance at End of Day	Number of Days	Sum of Daily Balances
1/1–1/7	0.00	$1948.50	7	$13,639.50
1/8	$98.00	1850.50		
1/9–1/31	0.00	1850.50		

a. Average daily balance
b. Finance charge
c. Ending balance

4.

Dates	Payment	Balance at End of Day	Number of Days	Sum of Daily Balances
2/1–2/10	0.00	$1878.59	10	
2/11	$93.00	1785.59		
2/12–2/28	0.00	1785.59		

a. Average daily balance
b. Finance charge
c. Ending balance

5.

Dates	Payment	Balance at End of Day	Number of Days	Sum of Daily Balances
3/1–3/15	0.00	$1812.87		
3/16	$88.00	1724.87		
3/17–3/31	0.00	1724.87		

a. Average daily balance
b. Finance charge
c. Ending balance

6.

Dates	Payment	Balance at End of Day	Number of Days	Sum of Daily Balances
4/1–4/12	0.00	$1751.38		
4/13	$83.00	1668.38		
4/14–4/30	0.00	1668.38		

a. Average daily balance
b. Finance charge
c. Ending balance

7.

Dates	Payment	Balance at End of Day	Number of Days	Sum of Daily Balances
5/1–5/20	0.00	$1693.90		
5/21	$79.00	1614.90		
5/22–5/31	0.00	1614.90		

a. Average daily balance   **b.** Finance charge

c. Ending balance

8.

Dates	Payment	Balance at End of Day	Number of Days	Sum of Daily Balances
6/1–6/13	0.00	$1639.89		
6/14	$75.00	1564.89		
6/15–6/30	0.00	1564.89		

a. Average daily balance   **b.** Finance charge

c. Ending balance

9.

Dates	Payment	Balance at End of Day	Number of Days	Sum of Daily Balances
7/1–7/18	0.00	$1588.85		
7/19	$71.00	1517.85		
7/20–7/31	0.00	1517.85		

a. Average daily balance   **b.** Finance charge

c. Ending balance

10.

Dates	Payment	Balance at End of Day	Number of Days	Sum of Daily Balances
8/1–8/9	0.00	$1541.24		
8/10	$68.00	1473.24		
8/11–8/31	0.00	1473.24		

a. Average daily balance   **b.** Finance charge

c. Ending balance

11.

Dates	Payment	Balance at End of Day	Number of Days	Sum of Daily Balances
9/1–9/13	0.00	$1495.63		
9/14	$64.00	1431.63		
9/15–9/30	0.00	1431.63		

a. Average daily balance   **b.** Finance charge

c. Ending balance

12.

Dates	Payment	Balance at End of Day	Number of Days	Sum of Daily Balances
10/1–10/21	0.00	$1453.52		
10/22	$61.00	1392.52		
10/23–10/31	0.00	1392.52		

a. Average daily balance   b. Finance charge

c. Ending balance

13. If you pay the bill in Exercise 3 on the first day of the month, how much will you save?

14. If you pay the bill in Exercise 4 on the first day of the month, how much will you save?

15. If you pay the bill in Exercise 5 on the first day of the month, how much will you save?

16. If you pay the bill in Exercise 6 on the first day of the month, how much will you save?

17. If you pay the bill in Exercise 7 on the first day of the month, how much will you save?

18. If you pay the bill in Exercise 8 on the first day of the month, how much will you save?

19. If you pay the bill in Exercise 9 on the first day of the month, how much will you save?

20. If you pay the bill in Exercise 10 on the first day of the month, how much will you save?

21. If you pay the bill in Exercise 11 on the first day of the month, how much will you save?

22. If you pay the bill in Exercise 12 on the first day of the month, how much will you save?

Complete each chart to determine the average daily balance, the finance charge, and the ending balance shown on your VISA statement for the month. Assume that your VISA card carries a monthly interest rate of 1.25%

23.

Dates	Payment	Purchase	Balance at End of Day	Number of Days	Sum of Daily Balances
8/1–8/18	0.00	0.00	$875.00		
8/19	$90.00	0.00	785.00		
8/20–8/25	0.00	0.00			
8/26	0.00	$135.00			
8/27–8/31	0.00	0.00			

a. Average daily balance   b. Finance charge

c. Ending balance

24.

Dates	Payment	Purchase	Balance at End of Day	Number of Days	Sum of Daily Balances
2/1–2/11	0.00	0.00	$500.00		
2/12	0.00	$90.00	590.00		
2/13–2/20	0.00	0.00			
2/21	$65.00	0.00			
2/22–2/28	0.00	0.00			

a. Average daily balance   b. Finance charge

c. Ending balance

25.

Dates	Payment	Purchase	Balance at End of Day	Number of Days	Sum of Daily Balances
4/1–4/8	0.00	0.00	$1275.00		
4/9	0.00	$225.00	1500.00		
4/10–4/17	0.00	0.00			
4/18	$175.00	0.00			
4/19–4/30	0.00	0.00			

a. Average daily balance   b. Finance charge

c. Ending balance

MIXED REVIEW

1. How much money can you save by financing an $8800 car at 10.5% over 3 years rather than 5 years?

2. A $760.00 television purchase is financed at 15% per year with no down payment at a store that applies the Rule of 78. It must be paid in 12 equal installments. Determine the prepayment savings if it is prepaid after 5 months.

3. If you buy a computer system for $3500 and finance it at 18% over 2 years, how much will you save by making a down payment of 25% instead of 10%?

4. What are your weekly earnings if you work 30.5 hours at $14.75 per hour?

5. Calculate the simple interest for one year on a $1200 deposit earning 3.75%.

6. Nancy always withdraws the interest on her savings account as soon as it is posted. Calculate the simple interest for three years on her original $1200 deposit that earns 5% per year.

7. If you buy a used automobile for $7500 and finance it at 11% over 3 years, how much will you save by making a down payment of 25% instead of 10%?

6-6 CREDIT RATINGS: HOW TO DETERMINE YOUR SCORE

Daryl has watched in fascination as his father unfolded his packet of credit cards. He has five oil company cards, six charge accounts at department stores, an American Express card, an ATM cash card, four MasterCards, and six VISA cards. Daryl thinks that carrying that many cards around must be a chore! How in the world can anyone manage to acquire so many cards, much less need them all? There must be a lot of checks to write each month when the bills come. Daryl carries only four cards: his driver's license, his school library card, a credit card for gasoline, and his ID card for the video rental store.

Once when Daryl asked his father about all the credit cards, he explained that his credit rating is excellent. He has lived in the same house for 8 years, he has been a high-salaried employee at the same business for 12 years, he has bought three different cars with installment loans, making prompt payments every month, and he is 41 years old. Creditors are more than willing to give him credit because he has shown he is capable and consistent in making payments. Over a 4-year period, 12 different banks from around the country offered him credit cards. He took most of them up on their offers, realizing how easy it is to use the cards to buy much more than he could afford to pay for with cash. Now he makes the minimum payment each month to each bank and continues to use almost all the credit available to him on each card.

When Daryl watches his father writing all the checks each month, he is not sure carrying 23 credit cards is truly necessary. However, he would like to know how to get at least one general purpose card for himself. Since he has never had a bank card and probably does not even have a credit rating, he is wondering where to begin.

OBJECTIVES: *In this lesson, we will help Daryl to:*

● *Determine an individual's credit rating, using the judgment of the creditor or a credit-scoring table.*
● *Learn how to get credit for the first time.*

HOW CREDIT IS GRANTED

A **credit rating** is an indication of a person's ability to secure goods, services, and money in return for the promise to pay. Daryl knows a favorable credit rating does not come automatically. His father's good rating has developed slowly over time. It has been nurtured, fostered, strengthened, and improved. A good credit rating is an asset of tremendous value to those who develop it over a long period, but it can easily be destroyed. A good credit rating is sensitive to

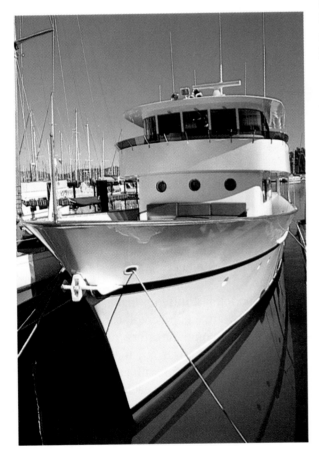

abuse and usually continues only as long as it is justified. Daryl's father has so much credit now that were he to apply for more, he would probably be told he already has sufficient credit.

The decision to grant credit lies with the *judgment of the creditor.* The creditor considers the "three C's" of credit—capacity, character, and collateral.

Capacity Do you have the capacity to repay the loan? How long have you worked, and how much do you earn? Creditors will also ask about your expenses such as rent, mortgage, car payments, and so on.

Character Have you been responsible about meeting financial obligations in the past?

Collateral Are the creditors protected from loss if you fail to pay? Creditors want to know if you have a savings account or other assets (car, home, or other valuable assets) to offer as security for a loan.

Because of laws that prohibit discrimination, large credit-lending institutions have developed **credit-scoring systems** that can be considered more objective than personal judgment. Credit-scoring systems award points for various factors to determine credit worthiness. A scoring system is shown on page 293. The principles underlying a scoring system are as follows:

• In general, you are a better *credit risk* if you are older. People in their mid-30's often face unforeseen expenses due to divorce or other causes of financial stress. Consequently their credit rating points may be lower.

• You are considered more stable the longer you have lived in the same place, and your score will increase if you own your home rather than rent it. People with cars get more points than those without, and owners of newer cars generally earn more points than owners of older cars.

- The length of time you have held the same job is another factor that is considered. The kind of job you have, your income, and your amount of current debt also make a difference.
- It is favorable to have either a savings account or a checking account. If you have both, you score more points.
- If you have one or more credit cards in good standing, you will score higher than those who have none. Department store charge cards, travel and entertainment cards, and oil company cards usually rate somewhat lower than all-purpose bank cards.
- If you have borrowed recently at high interest from a finance company, you will probably lose several points. If you have applied and been turned down for credit more than once in the last 6 months, you may have difficulty getting new credit at all.

Hypothetical Credit-Scoring Table

Fill out your credit profile by answering the nine questions below in Table 1. Circle the one response that applies to you, and then find your total score by adding up the points you got for each response. The points are found in the lower right-hand corner of each box. (For example: if you are 25 years old, you get 5 points.) Once you've totaled your score, look at Table 2 to find out how good a credit "bet" you may be.

1.

#	Question							
1.	age?	under 25 — 12	25–29 — 5	30–34 — 0	35–39 — 1	40–44 — 18	45–49 — 22	50 or over — 31
2.	time at address?	less than 1 yr. — 9	1yr. — 0	2–3 yrs. — 5	4–5 yrs. — 0	6–9 yrs. — 5	10 yrs. or more — 21	
3.	age of auto?	none — 0	0–1 yrs. — 12	2 yrs. — 16	3–4 yrs. — 13	5–7 yrs. — 3	8 yrs. or more — 0	
4.	monthly auto payment?	none — 18	less than $125 — 6	$126–$150 — 1	$151–$199 — 4	$200 or more — 0		
5.	housing cost?	less than $274 — 0	$275–$399 — 10	$400 or more — 12	owns clear — 12	lives with relatives — 24		
6.	checking and savings accounts	both — 15	checking only — 2	savings only — 2	neither — 0			
7.	finance company reference	yes — 0	no — 15					
8.	major credit cards?	none — 0	1 — 5	2 or more — 15				
9.	ratio of debt to income?	no debts — 41	1%–5% — 16	6%–15% — 20	16% or over — 0			

2.

A lender using this scoring table selects a cutoff point from a table like this, which gauges how likely applicants are to repay loans.

Total Score	Probability of Repayment
90	89 in 100
95	91 in 100
100	92 in 100
105	93 in 100
110	94 in 100
115	95 in 100
120	95.5 in 100
125	96 in 100
130	96.25 in 100

Source: Federal Reserve Board. Developed by Fair, Isaac, and Co., Inc. Modified to update.

Daryl would like to know how to obtain *first-time credit.* Credit is easy to get if you have had it before. But what about getting credit for the first time? Below are some suggestions for Daryl to help him establish a good credit reputation. He can do the following:

- Open a charge account at a retail store where his parents have an account.
- Establish checking and savings accounts at the local bank.
- Get one of his parents to **cosign** a small loan. By cosigning, Daryl's father promises to pay if Daryl does not. When Daryl repays on time, he establishes a good credit record.
- Join a credit union.
- Be a responsible employee. Creditors generally ask for length of employment at a job and for personal references.

As Daryl begins to establish his own credit rating, he will be careful not to abuse it. He will treat it as the valuable asset that it is, and continue to strengthen it as time goes on.

ASK YOURSELF

1. What are the "3 Cs" that creditors use to judge a person's acceptability for credit?

2. How does a lending institution determine the credit worthiness of an applicant?

3. How might you establish a good credit reputation?

ALGEBRA *REVIEW*

When you take a test in school, you often receive a score. Mrs. Esposito assigned the following point values to each question on her mathematics test:

> Questions 1–10: 5 points each
> Questions 11–14: 7 points each
> Questions 15–16: 11 points each
> Bonus Question: 10 points

Determine each test score.

1. Sally missed Questions 2 and 12, but answered the others correctly. She also answered the bonus question correctly.

2. Tom missed Questions 13 and 14, but answered the others correctly. He received no bonus points.

3. Janet missed Questions 5, 6 and 11, but answered the others correctly. She received no bonus points.

4.–6. If Mrs. Esposito assigns grades as follows, determine each student's grade on the test.

> $90+ \rightarrow A$
> $80–89 \rightarrow B$
> $70–79 \rightarrow C$
> $65–69 \rightarrow D$
> $0–64 \rightarrow F$

7. Write a general formula for calculating a test score.

SHARPEN YOUR SKILLS

SKILL 1

EXAMPLE 1 Daryl wonders how his father qualifies for so many credit cards. He knows the following facts about his father.

1. He is 41 years old.
2. He has lived at his current address for 8 years.
3. His current car is 2 years old.
4. He has no monthly car payment.
5. His housing cost is $482 per month.
6. He has a checking account and a savings account.
7. He has not been referred to a finance company.
8. He has 23 major credit cards.
9. His ratio of debt to income is 20%.

QUESTION How would a potential creditor rate his credit worthiness? According to the credit-scoring system, what is the probability that he will repay a loan?

SOLUTION

	Points	
1.	18	Use the credit-scoring table to find his credit score.
2.	5	
3.	16	
4.	18	
5.	12	
6.	15	
7.	15	
8.	15	
9.	0	
Score:	114	

Look in part 2 of the credit-scoring table to find the probability that Daryl's father will repay his loan. Since 114 is not listed, use the lower score of 110. The probability that he will repay a loan is 94 in 100, or 94%.

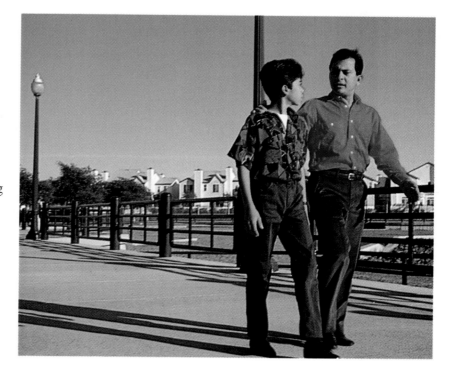

Use the credit-scoring table to find Brooke's credit score for each item.

1. She is 28 years old.
2. She has lived at her current address for 1 year.
3. Her current car is 4 years old.
4. She has no car payment.
5. She lives with relatives.
6. She has a checking account and a savings account.
7. She has not been referred to a finance company.
8. She has no major credit cards.
9. She has no debts.
10. What is Brooke's total score?
11. What is the probability that Brooke will repay a loan?

EXERCISE YOUR SKILLS

1. Why is it important to establish a favorable credit rating?
2. If you were a lender, what is the minimum score on the credit-scoring table that you would accept on an applicant? Why?
3. It has been said that a person's credit rating is one of his or her most valuable assets. Do you agree or disagree? Why?

Use the credit-scoring table on page 293 to determine the credit score for each head of household described below. Then determine the probability that the person will repay a loan.

4. **Family 1**
   Age of the head of the household—35
   Lived at the current address—4 years
   Age of current car—3 years
   Monthly car payment—$165
   Housing cost—$650
   Has a checking account and a savings account.
   Has not been referred to a finance company.
   Has 7 major credit cards.
   The ratio of debt to income is 12%.

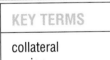

KEY TERMS

collateral
cosign
credit rating
credit-scoring
   systems

5. **Family 2**
   Age of the head of the household—48
   Lived at the current address—12 years
   Age of current car—5 years
   Monthly car payment—$0.00
   Housing cost—$385
   Has a checking account and a savings account.
   Has not been referred to a finance company.
   Has 6 major credit cards.
   The ratio of debt to income is 19%.

6. **Family 3**
   Age of the head of the household—26
   Lived at the current address—2 years
   Age of current car—1 year
   Monthly car payment—$364
   Housing cost—$741
   Has a checking account but not a savings account.
   Has not been referred to a finance company.
   Has 12 major credit cards.
   The ratio of debt to income is 24%.

7. **Family 4**
   Age of the head of the household—43
   Lived at the current address—16 years
   Age of current car—1 year
   Monthly car payment—$0.00
   Housing cost—$640
   Has a checking account and a savings account.
   Has not been referred to a finance company.
   Has 3 major credit cards.
   The ratio of debt to income is 8%.

8. **Family 5**
   Age of the head of the household—32
   Lived at the current address—3 years
   Age of current car—1 year
   Monthly car payment—$237
   Housing cost—$853
   Has a checking account but not a savings account.
   Has been referred to a finance company.
   Has 12 major credit cards.
   The ratio of debt to income is 32%.

9. **Family 6**
   Age of the head of the household—52
   Lived at the current address—22 years
   Age of current car—3 year
   Monthly car payment—$0.00
   Housing cost—$332
   Has a checking account and a savings account.
   Has not been referred to a finance company.
   Has 8 major credit cards.
   The ratio of debt to income is 4%.

10. **Family 7**
    Age of the head of the household—34
    Lived at the current address—6 months
    Age of current car—8 year
    Monthly car payment—$0.00
    Housing cost—$213
    Has a checking account but not a savings account.
    Has not been referred to a finance company.
    Has 2 major credit cards.
    The ratio of debt to income is 8%.

11. **Family 8**
    Age of the head of the household—23
    Lived at the current address—17 years
    Age of current car—1 year
    Monthly car payment—$233
    Housing cost—$150
    Has a checking account and a savings account.
    Has not been referred to a finance company.
    Has 1 major credit card.
    The ratio of debt to income is 3%.

MIXED REVIEW

1. If you borrow $1350.00 for 4 years at 12% to buy furniture, what is the deferred payment price of the furniture?

2. The fixed costs in Nicolette's business are $300 per week. The variable costs are $4.50 per item. What is the total cost if she produces 120 items in one week?

3. You have $778.22 in your checking account and you write a check for $105.89. Show how you would enter this check in your check register.

4. Determine the interest for 9 months if the starting principal is $5525, the annual rate is 5%, and the interest is compounded quarterly.

5. You earn $18,600 per year plus 12% of your salary in fringe benefits. Find your yearly earnings, including benefits.

6-7 REGAINING FINANCIAL STABILITY: SOLUTIONS TO DEBT PROBLEMS

Sylvia and Daryl were talking and began to discuss credit card use. They compared facts about their fathers and their use of credit cards. One credit story led to another and, by the end of the conversation, Sylvia and Daryl were both relieved to know they were not alone in having serious doubts about some of the choices made by their respective fathers. Daryl was especially relieved to know he did not have the only father in town whose stack of credit cards was so thick he appeared to be shuffling a deck of playing cards every time he used one. Once when his father dropped them all on the floor in a restaurant, Daryl was so embarrassed that he rushed out to the parking lot to wait, pretending he did not know the man crawling around on the floor looking for his cards.

Daryl's father was embarrassed, too, but not as much as when he was so far behind in paying all the credit bills that he actually considered trying to pawn some of the new video equipment he had just purchased. At times he has even considered filing for bankruptcy, but he knows that is a drastic step. A person who declares bankruptcy cannot ask for any credit for at least seven years and may be turned down even then.

Sylvia and Daryl are discussing some less radical methods to help their fathers with their finances.

OBJECTIVES: In this lesson, we will help Sylvia and Daryl to:

- *Discover what happens to debtors who cannot pay their debts.*
- *Recognize the signs of carrying too much debt.*
- *Learn methods to safely get out of debt.*
- *Calculate percent of take-home pay that is used for credit payments.*

DEBT COLLECTION PRACTICES

Creditors often employ professional **collection agencies** to collect overdue accounts and repossess articles on which money is due. The collection process has been known to include everything from collection letters, late-night telephone calls, abusive language, and threats of having consumers fired from their jobs because of nonpayment of debt. In order to stop these practices, Congress passed the **Fair Debt Collection Practices Act,** which declares that the following collection actions are illegal:

- Threatening violence, using obscene language, publishing shame lists of debtors, and making harassing phone calls at night.
- Calling a debtor at work or contacting a debtor's employer.
- Claiming the collector is from a state or federal agency or is a government official.
- Revealing the existence of a bad debt to third parties such as neighbors or employers.
- Using false or deceptive means to obtain information about a debtor.

The act also provides that debtors have the right to notify a collection agency in writing that they do not wish to hear from the agency again, except for legal notices, and notices of possible further action. It is important to note that the law applies only to debt collection agencies—not to banks and other financial institutions or stores.

Creditors can use a legal procedure, know as **garnishment,** to withhold a part of a debtor's earnings for the payment of a debt. However, the law also prohibits an employer from firing an employee because of garnishment of wages for indebtedness.

DEBT PROBLEMS

What kinds of families face debt problems? Following is a profile of the typical family that has problems with debt:

- The family is young, with more than the average number of children but only an average income.
- The parent or parents are carefree. They are impulsive shoppers who cave in easily to high-pressure salesmen, and do not postpone buying things when they want them.

- The family does not read much, not even the daily newspaper. Television is their main form of entertainment and their primary source of news and information. Television influences the family's buying decisions.
- The parent or parents tend to blame their situation on "unavoidable circumstances" (such as pregnancy, temporary unemployment, the purchase of a new car) and do not take responsibility for their problems.
- The family moves more often than the average family.
- No one assumes clear responsibility for managing the family finances.
- The family has a single adult at its head (either by choice or as a result of divorce, separation or the death of a partner).

The following are *clear danger signals* of too much debt:

- You continually lengthen the repayment periods on your installment loans and make smaller and smaller down payments.
- The balances on your revolving charge accounts continue to go up.
- The bills for each month begin piling up before you have finished paying the previous month's bills.
- You slowly, but steadily, use a larger part of your income to pay your debts each month.
- You are taking cash advances on your credit card to pay regular monthly bills, such as utilities, rent and food.

SOLUTIONS TO DEBT PROBLEMS

Is there a safe way out? Sylvia and Daryl are convinced there are ways to stop the debt spiral. They have found the following information:

Wage Earner Plan—Chapter XIII   Chapter XIII of the United States Bankruptcy Code is an alternative to declaring bankruptcy. Chapter XIII allows the debtor, creditors, and a judge, acting together, to set a monthly amount for the debtor to pay over an extended repayment period. This is called a *wage earner plan*. The debts are not wiped out, but the court takes a portion of each paycheck and distributes it to the creditors. Chapter XIII does not require the debtor to give up assets such as personal property. Usually, as part of this arrangement, the debtor cannot make additional credit purchases from a creditor until the original debt has been paid.

Bankruptcy   Straight bankruptcy, also called *liquidation bankruptcy,* requires debtors to sell most of their assets at public sale through a trustee in return for a discharge from most, if not all, of their outstanding debts. The concept behind bankruptcy is to wipe out all debts and give the debtor a new start.

The wage earner plan and bankruptcy plan are drastic moves, but not as drastic as the next two that Sylvia and Daryl found.

Pawnbrokers   When individuals or families are not able to get loans from financial institutions such as banks or small finance companies, they sometimes turn to pawnbrokers. This form of borrowing money can be very expensive and should usually be avoided.

For a cash loan, pawnbrokers accept collateral such as jewelry, art objects, watches, and clothing. The pawnbroker actually lends approximately 40% of the resale value of the article at a high interest rate of 36 to 50%. The pawnbroker keeps the collateral for the period of the loan and returns it when the loan has been repaid. Problems occur when people do not save the money to repay the loan and then lose their collateral, which is invariably worth more than the loan.

Loan Sharks   Desperate or unknowing borrowers may turn to loan sharks. They are illegal credit lenders who charge very high rates for money. For example, they will give you $50 now in exchange for your giving them $75 when you get paid. Usually such loans are for one week or one month. Rates are not quoted by these illegal or unethical lenders because they are extremely high—ranging from 100 to 500%. Collection practices often involve threats of harm and violence.

Sylvia and Daryl hope that their fathers will not turn to pawnbrokers or loan sharks. Most states have laws governing small loan practices, but some do not. Some unscrupulous lenders operate without a license in states that do have regulatory laws.

Credit Counseling   Sylvia and Daryl prefer the final method they have found, credit counseling, as the first step toward regaining financial stability. They have recommended it to their fathers. To begin with, they have been told, a person seeking counseling should contact his or her local bank. Many banks are now offering credit counseling to customers who find themselves in difficulty.

Another resource that is available is the National Foundation for Consumer Credit, which is located at 1819 H Street, N.W., Washington, D.C. 20006. The foundation will provide you with the address of one of the consumer credit counseling services near you. These services are nonprofit organizations, backed by local banks, merchants, and educators. They are set up to provide financial counseling to anyone.

Another agency that offers help is the Family Service Agency. Hundreds of Family Service Agencies offer either financial counseling or can refer you to an agency that does. If you do not know which agency offers such help in your area, write to the Family Service Association of America at 44 East Twenty-Third Street, New York, New York 10010.

ASK YOURSELF

1. What action might a creditor use if debtors do not pay their debts?

2. What are three signs of carrying too much debt?

3. What is a safe way to get out of debt?

ALGEBRA *REVIEW*

Solve for x. Round your answers to the nearest hundredth.

1. $x = 12.5\%$ of $435.60

2. $x = 0.2\%$ of $333.99

3. $x = 202.2\%$ of $677.98

4. $32 = x\%$ of $250.88

5. $12 = x\%$ of $1543.00

6. $0.3 = x\%$ of $123.45

7. $75 = 0.2\%$ of x

8. $0.05 = 35\%$ of x

9. $444 = 102\%$ of x

10. The Wolk family saved the following amounts of money:

Jan.	$412	Feb.	$35
Mar.	$95	Apr.	$202
May	$109	Jun.	$333
Jul.	$10	Aug.	$151
Sep.	$1	Oct.	$510
Nov.	$38	Dec.	$0

    Determine the percent of the savings that occurred in each month to the nearest hundredth of a percent.

SHARPEN YOUR SKILLS

SKILL 1

EXAMPLE 1 Daryl and Sylvia have learned that a family that spends more than 20% of their take-home pay on installment payments and credit accounts is in danger of credit overload. They have decided to look at a specific family budget to determine how well the family is handling credit. The family's budget for this month is shown.

Income

Take-home pay	$4550

Expenses

Mortgage Payment	462
Utilities	185
Telephone	45
Cable Television	53
Gasoline and Car Repairs	220
Food	650
Savings	450
*Car Payment	225
*Credit Union Loan	75
*MasterCard	120
*VISA	95
Charge Accounts:	
*Sears	135
*Foley's Garage	25
*American Express	140
Everything Else	1670

*The entries that count as credit payments are marked with an asterisk. (A mortgage is not counted as part of the credit expenses.)

QUESTIONS What percent of the family's take-home pay goes toward paying credit bills? Is this family spending too much of its salary on credit payments?

SOLUTIONS

The family members' total credit payments are

$$225.00 + 75.00 + 120.00 + 95.00 + 135.00 + 25.00 + 140.00 = 815.00$$

They are spending

$$815.00 \div 4550 = 0.179, \text{ or } 17.9\%$$

of their take-home pay on credit payments. They are not spending too much, but they are approaching the 20% limit.

It is often helpful to have a pictorial representation of data.

__EXAMPLE 2__ The pie graph below represents the Taylor family's monthly take-home pay budget.

Everything else (17.1%)
Telephone (0.7%)
VISA (15.2%)
Car/Gas (1.3%)
Car Payment (4.4%)
Savings (6.3%)
Dept. Store (3.8%)
Mortgage (25.4%)
Utilities (5.1%)
Food (7.6%)
Cable TV (0.4%)
Mastercard (12.7%)

__QUESTIONS__ Approximately what percent of their take-home pay are the family members spending on credit? Are they in danger of credit overload?

__SOLUTIONS__

They are making the following credit payments.

VISA	15.2%
Car Payment	4.4%
Dept. Store	3.8%
MasterCard	12.7%
Total	36.1%

They are spending approximately 36.1% of their take-home pay on credit payments. They are significantly over the 20% limit and are in great danger of credit overload.

1. The Irrizarry family's budget for this month follows. To the nearest tenth of a percent, how much of their take-home pay are they spending on credit payments? Are they in danger of credit overload?

Income

Take-home pay	$5300

Expenses

Mortgage Payment	775
Utilities	225
Telephone	85
Cable Television	45
Gasoline and Car Repairs	125
Food	800
Savings	250
*Car Payment	200
*MasterCard	115
*VISA	500
*Neiman Marcus	155
*Macy's	100
*American Express	75
Everything Else	1850

2. The pie graph below represents the Schmidt family's monthly take-home pay budget. To the nearest tenth of a percent, how much of their take-home pay are they spending on credit? Are they in danger of credit overload?

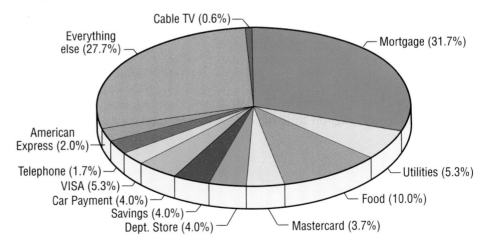

EXERCISE YOUR SKILLS

1. Why do you think the law prohibits an employer from firing an employee because of garnishment of wages for indebtedness?

2. Why is a compulsive shopper at greater risk than a normal shopper of encountering financial problems?

3. Why would you recommend credit counseling as the first step toward regaining financial stability?

The following pie graphs represent the monthly take-home pay budgets of four families. To the nearest tenth of a percent, how much of its take-home pay is each family spending on credit? Are any in danger of credit overload?

4.

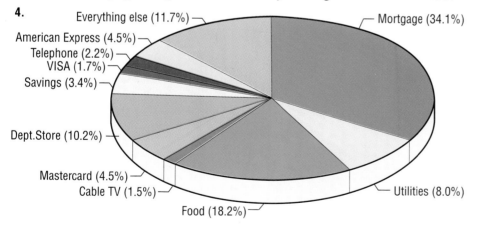

Everything else (11.7%)
American Express (4.5%)
Telephone (2.2%)
VISA (1.7%)
Savings (3.4%)
Dept.Store (10.2%)
Mastercard (4.5%)
Cable TV (1.5%)
Food (18.2%)
Mortgage (34.1%)
Utilities (8.0%)

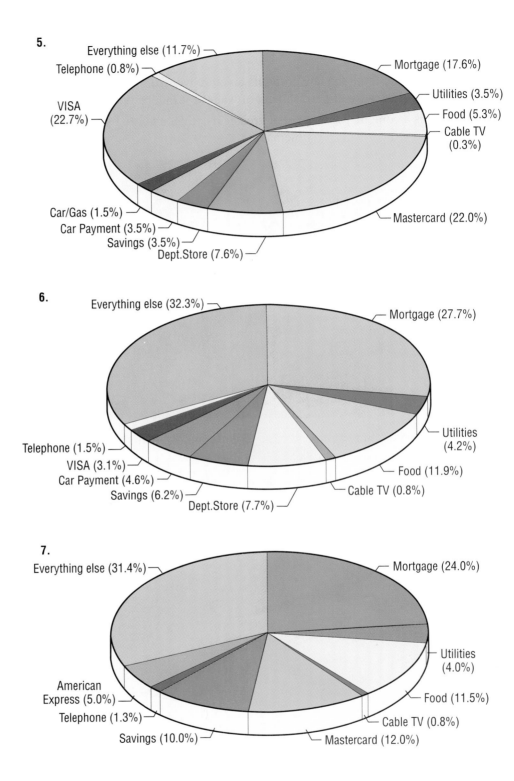

5. Everything else (11.7%) — Telephone (0.8%) — VISA (22.7%) — Car/Gas (1.5%) — Car Payment (3.5%) — Savings (3.5%) — Dept.Store (7.6%) — Mortgage (17.6%) — Utilities (3.5%) — Food (5.3%) — Cable TV (0.3%) — Mastercard (22.0%)

6. Everything else (32.3%) — Telephone (1.5%) — VISA (3.1%) — Car Payment (4.6%) — Savings (6.2%) — Dept.Store (7.7%) — Mortgage (27.7%) — Utilities (4.2%) — Food (11.9%) — Cable TV (0.8%)

7. Everything else (31.4%) — American Express (5.0%) — Telephone (1.3%) — Savings (10.0%) — Mortgage (24.0%) — Utilities (4.0%) — Food (11.5%) — Cable TV (0.8%) — Mastercard (12.0%)

The monthly budgets for four families follow. Their monthly incomes range from $1850.00 (Family D) to $5400.00 (Family B). Determine, to the nearest tenth of a percent, how much each family is spending on credit payments. Then state whether or not each family is in danger of credit overload.

8. **Family A**

Income
Take-Home Pay	$3290.00

Expenses
Mortgage Payment	$650.00
Utilities	148.00
Telephone	43.00
Cable Television	53.00
Gasoline and Car Repairs	183.50
Food	580.00
Savings	330.00
*Car Payment	165.00
*MasterCard	56.00
*American Express	140.00
*VISA	45.00
Charge Accounts:	
*Sears	77.00
Everything Else	819.50

9. **Family B**

Income
Take-Home Pay	$5400.00

Expenses
Rent	$852.00
Utilities	118.70
Telephone	43.60
Sports Club Fee	53.00
Gasoline and Car Repairs	126.00
Food	580.00
Savings	540.00
*Car Payment	237.00
*MasterCard	89.00
*American Express	140.00
*VISA	102.00
Charge Accounts:	
*Hardware Store	34.00
*Penneys	15.00
Everything Else	2469.70

10. **Family C**

Income

Take-Home Pay	$3210.00

Expenses

Mortgage Payment	$640.00
Utilities	235.00
Telephone	74.00
Sports Club Fee	53.00
Gasoline and Car Repairs	54.00
Food	620.00
Savings	321.00
*Car Payment	0.00
*MasterCard	79.00
*American Express	140.00
*VISA	42.00
Charge Accounts:	
*Pharmacy	34.00
*Montgomery Ward	65.00
Everything Else	853.00

11. **Family D**

Income

Take-Home Pay	$1850.00

Expenses

Mortgage Payment	$150.00
Utilities	83.00
Telephone	26.00
Cable Television	33.00
Gasoline and Car Repairs	68.50
Food	225.00
Savings	185.00
*Car Payment	233.00
*MasterCard	26.00
*VISA	35.00
Charge Accounts:	
*Sears	45.00
Everything Else	740.50

1. Debra has a balance of $1325 on her VISA card, which carries an APR of 16%. She makes monthly payments of $150. If she makes no further purchases with this card, how many months will it take her to pay off the balance?

2. Use the Rule of 72 to determine how long it will take Lou's $4400 savings account to double if it is growing at a rate of 3.5%.

3. Determine the effective rate of interest for an APR of 17.2%.

4. Determine Ryan's monthly payment on a loan of $9875.00 if he borrows the money at an annual interest rate of 10% for 15 years.

5. How much will $7200 be worth at the end of 3 years if it earns 5% interest, compounded semiannually?

6. If Sandra works more than 25 hours in one week, she receives $1\frac{1}{2}$ times her regular wage rate of $6.00 per hour for each of the extra hours. One week she worked 33 hours and received $26 in tips. Find her earnings for the week.

7. How much money would you save if you financed an $1800 item at 8% for 3 years rather than accepting a rent-to-own plan for which you pay $60 per month for 3 years?

8. A credit card account has a balance of $1000 and no further purchases are made. The monthly interest charge is 1.5% and $100 of the amount owed is paid off each month. Find the interest paid after two months.

Yumi's parents are purchasing a refrigerator for $420. They finance the purchase with no money down at 15% per year. They pay in 12 equal installments at a store that applies the Rule of 78. Determine the prepayment savings under each condition.

9. It is prepaid after 9 months.

10. It is prepaid after 5 months.

Complete the chart below. Assume that the credit card carries an APR of 11% and that 10% of the amount owed to the nearest dollar is paid off each month.

	Month	Balance	Interest	Amount Owed	Payment
11.	1	$1438.00	$13.18		
12.	2				

A bank offers monthly interest of 0.6% on its checking accounts but charges $0.10 per written check for balances under $500. There is no monthly service charge. Find the amount added or subtracted to the following checking accounts.

13. an account with a balance of $650 and 28 checks written

14. an account with a balance of $480 and 19 checks written

1. Use a spreadsheet program to determine the previous balance, new charges, payment received, finance charges, and new balance that would be shown on Bruce's first 12 MasterCard statements after a purchase of $2315 if he makes monthly payments of $250. The card has a 1-month grace period and an APR of 12.9%. Assume that he has never used the card and that he makes no further charges on it during this time period.

2. Determine the effective rate of interest on a credit card with an APR of 18.5%. Give answer to the nearest hundredth of a percent.

3. How long will it take to pay off a credit balance of $2275 with monthly payments of $225 if the card carries a monthly interest rate of 1.25%?

4. Determine the monthly payment that must be made to pay off a VISA balance of $12,500 in 30 months if there is an APR of 18%.

5. Make a chart with the headings below to determine the interest charges over the next 4 months on a balance of $875.00 if your credit card has an APR of 12% and you make monthly payments of 5% of the amount due, rounded to the nearest whole dollar. Assume that you make your payment on the last day of the month.

Month	Balance	Interest	Amount Owed	Payment

6. What is the total interest in Exercise 5?

7. Make a chart with the headings below to determine the interest charges over the next 4 months on a balance of $875.00 if your credit card has an APR of 12% and you make monthly payments of 10% of the amount due, rounded to the nearest whole dollar. Assume that you make your payment on the last day of the month.

Month	Balance	Interest	Amount Owed	Payment

8. What is the total interest in Exercise 7?

9. Compare your results from Exercises 6 and 8 to determine how much more interest you pay if you pay 5% of the amount owed each month instead of 10% of the amount owed.

10. How much more interest would you have paid in Exercise 8 if the APR on the card had been 21%?

11. Use a spreadsheet program with the following headings to work through the first 8 months of Rebecca's VISA account. The monthly interest rate is 1.5%, and she makes monthly payments of 10% of the amount owed. Her beginning balance was $980.12 and she made purchases of $87.35 in month 2, $25.50 in month 5, $100.00 in month 6 and $53.33 in month 7.

Month	New Charges	Balance	Interest	Amount Owed	Payment

12. In Exercise 11, does Rebecca owe more or less in month 8?

13. Your balance from 9/1 to 9/19 was $400. You made a payment of $50.75 on 9/20. You made no further charges or payments for the remainder of the month. Determine your average daily balance for the month of September. How much did you pay in interest if your card carries a monthly finance charge of 1.75%?

14. How much money would you have saved if you had made your payment on 9/1 instead of on 9/20 in Exercise 13?

15. Your balance from 8/1 to 8/15 was $780. You made a payment of $100 on 8/16. You made a purchase of $220 on 8/25. You made no further charges or payments for the remainder of the month. Determine your average daily balance for August. How much did you pay in interest if your card carries a monthly finance charge of 1.25%?

16. Use the credit scoring system in Lesson 6-6 to find Julie's credit score and the probability that she will repay a loan.

   • She is 32 years old.
   • She has lived at her current address for 11 years.
   • Her current car is 2 years old.
   • Her monthly auto payment is $200.
   • Her monthly housing costs are $525.
   • She has a checking account and a savings account.
   • She has not been referred to a finance company.
   • She has ten major credit cards.
   • Her ratio of debt to income is 15%.

17. The Garcia family's budget for this month follows. To the nearest tenth of a percent, how much of their take-home pay are they spending on credit payments if their monthly income is $6000? Are they in danger of credit overload?

Expenses	
Mortgage Payment	2000.00
Utilities	300.00
Telephone	95.00
Cable Television	50.00
Food	800.00
Savings	350.00
*Car Payment	250.00
*MasterCard	215.00
*VISA	400.00
*Macy's	180.00
*American Express	100.00
Everything Else	1260.00

CHAPTER 6 TEST

1. Determine the effective rate of interest on a credit card with an APR of 16.5%. Give the answer to the nearest hundredth of a percent.

2. How long will it take to pay off a credit balance of $1675 with monthly payments of $180 if the card carries a monthly interest rate of 1.75%?

3. Determine the monthly payment that must be made to pay off a VISA balance of $13,100 in 28 months if there is a finance charge of 12%.

4. Make a chart with the headings below to determine the interest charges over the next 4 months on a balance of $795.00 if your card has an APR of 15% and you make monthly payments of 5% of the amount due, rounded to the nearest whole dollar. Assume you make your payment on the last day of the month.

Month	Balance	Interest	Amount Owed	Payment

5. What is the total interest in Exercise 4?

6. Make a chart with the headings below to determine the interest charges over the next 4 months on a balance of $795.00 if your card has an APR of 15% and you make monthly payments of 10% of the amount due, rounded to the nearest whole dollar. Assume you make your payment on the last day of the month.

Month	Balance	Interest	Amount Owed	Payment

7. What is the total interest in Exercise 6?

8. Compare your results from Exercises 5 and 7 to determine how much more interest you pay if you pay 5% of the amount owed each month instead of 10% of the amount owed.

9. Your balance from 4/1 to 4/18 was $600. You made a payment of $75 on 4/19. You made no further charges or payments for the remainder of the month. Determine your average daily balance for the month of April. How much did you pay in interest if your card carries a monthly finance charge of 1%?

10. How much money would you have saved if you had made your payment on 4/1 instead of on 4/19 in Exercise 9?

11. Your balance from 3/1 to 3/16 was $560. You made a payment of $80 on 3/17. You made a purchase of $160 on 3/22. You made no further charges or payments for the remainder of the month. Determine your average daily balance for March. How much did you pay in interest if your card carries a monthly finance charge of 1.5%?

12. What is a credit rating?

13. Would a family with a monthly income of $3000 and monthly credit payments of $350 be in danger of credit overload? Why or why not?

CUMULATIVE REVIEW

1. You earn $7.25 per hour. You worked 46 hours last week. You receive $1\frac{1}{2}$ times your hourly rate for any hours over 40. Find your total wages for last week.

2. You are a real-estate agent and you sold a house for $210,000. You earn 5% commission on the first $100,000 and 6% commission on any amount over $100,000. How much commission did you earn?

3. You work 35 hours per week for 50 weeks. You get 2 weeks of paid vacation plus 25% of your base salary in fringe benefits. You earn $4500 per month. Determine your yearly earnings, including benefits.

4. The bank charges $0.02 for the first 20 checks you write each month and $0.09 for each check over 20. Determine the charges if you write 25 checks each month.

5. If you write a check for $110.23, how would you write that amount in words?

6. Determine the interest for one year on $600 at 10.5%, compounded semiannually.

7. The fixed costs for your business are $200 per week. The variable costs are $5.10 per item. What is the cost if you produce 80 items in one week?

8. Using the information in Exercise 7, determine your profit or loss for one week if each item sells for $9.50.

9. You borrowed $3400 for 3 years. You pay $111 per month. What is the deferred payment price?

10. In Exercise 9, how much interest did you pay?

11. You have a choice of paying $450 per month for 3 years or $300 per month for 5 years on a $13,500 loan. How much will you save if you choose the 3-year option?

12. Your credit card carries a monthly finance charge of 1.5%. You pay 10% of the amount owed each month. Your beginning balance is $540. Complete a table with the following column heads for 3 months.

Month	Balance	Interest	Amount Owed	Payment

13. Determine the plan that would provide the lowest total financed price for a $16,750 car. Assume the loan in Plan 1 is from a bank or other lending institution and the loans in Plans 2 and 3 are from the car manufacturer's financing subsidiary.

	Plan 1	Plan 2	Plan 3
Rebate	$1200.00	$600.00	0
Rate/Time	10.75%/36 months	7.5%/36 months	3.9%/36 months

14. Determine which plan yields the highest profit to the company in Exercise 13.

PROJECT 6–1: Shopping on Credit

1. Call or visit 25 different merchants in your area, including at least 3 department stores.
2. Find out which credit cards are accepted by these stores. Try to find at least 5 places that accept credit cards other than MasterCard or VISA.
3. Make a poster presenting the information you gathered, including the name and address of each store and the hours they are open.
4. Write a 1-page report on what you learned through this project. Include comments about the following:
   - Why do so many merchants accept MasterCard, VISA, and Discover cards?
   - What, if any, are the differences among MasterCard, VISA and Discover?
   - What advantage do merchants gain from accepting credit cards?
   - What are the disadvantages of credit cards for merchants and consumers?
   - What department stores issue their own credit cards? What are the advantages and disadvantages of these cards to the merchants and the consumers?

PROJECT 6–2: Audio-Visual Presentations

Use the kinds of techniques you see in advertising to point out some of the disadvantages or advantages to the use of credit cards. First, discuss with a small group of your classmates how you feel about using credit cards. Then choose one of the following activities to complete independently or with a group of classmates.

1. Create a poster presenting one or more of the following:
   - Advantages of using credit cards.
   - Disadvantages of using credit cards.
   - A collage of available credit cards—include a wide variety, such as diner's cards, oil company cards, department store cards, and bank cards.

   Illustrate the poster and make it eye-catching or humorous.
2. Draw a cartoon illustrating some negative aspect or disadvantage of credit buying. Exaggeration is permissible in cartoons to emphasize your point.
3. Write a humorous skit to illustrate the negative aspects of using credit cards and perform your skit for your class.

A *random experiment* is the occurrence of something that has an uncertain result. For example, this is the set of all possible results, or *outcomes* of the random experiment of tossing a die:

{1, 2, 3, 4, 5, 6}

This list of all possible outcomes is the *sample space* of the experiment. An *event* is a subset of a sample space. For example, "rolling an even number" is {2, 4, 6} since 2, 4, and 6 are the only outcomes that are even.

List each event. The sample space is the set of six possible outcomes of rolling a die.

Example   Rolling a prime number

Solution   {2, 3, 5}

1. Rolling an odd number

2. Rolling a perfect square

3. Rolling a number less than 5

4. Rolling a number greater than π

The *probability* of an event is the ratio

$$P = \frac{\text{number of outcomes in the event}}{\text{number of outcomes in the sample space}}$$

Find the probability of obtaining the given outcome on the roll of a die. The sample space is the set of all possible rolls.

Example   Rolling a perfect square

Solution $P = \dfrac{m}{n}$ $\dfrac{\text{number of outcomes in the event}}{\text{number of outcomes in the sample space}}$

$\qquad\quad = \dfrac{2}{6}$ $\dfrac{\text{2 perfect squares, 1 and 4}}{\text{6 die faces}}$

$\qquad\quad = 0.33333\ldots,$

5. Rolling an odd number

6. Rolling an even number

7. Rolling a number less than 5

8. Rolling a multiple of 3

The following are useful properties of probability.

   The probability of an impossible event is 0.
   The probability of a certain event is 1.
   The probability of any event is a number between 0 and 1, inclusive.

Tell whether the probability of the event is 0, 1, or a number between 0 and 1.

9. Rolling a 7 on a die toss

10. Winning money in a lottery

11. Living to the age of 98

12. Tossing a coin heads or tails

CHAPTER 7

Planning

FOR TEENAGERS THE FUTURE IS NOW. They often consider it a waste of time to think about matters such as life insurance or retirement planning that have nothing to do with today. Indeed, it is true that a young person without family responsibilities rarely needs life insurance. It is also understandable that for someone with more immediate financial concerns, such as how to pay the rent, retirement is something to worry about later–much later. However, this attitude is probably a mistake. Young people should prepare themselves at an early age for the time down the road when such matters will be very important.

A young person who does devote some time to learning about life insurance and retirement faces some difficult decisions. For example, a young person's life insurance premium is much lower than the premium for someone who is much older, but is that enough of a reason to buy the insurance? Some families have purchased cash-value life insurance so that they can borrow on the policy when it comes time to pay for college. Is that the best way to accomplish this goal?

The earlier retirement is planned for, the more secure and comfortable the retirement years are likely to be. Unfortunately, people from earlier generations have not always learned this lesson. Many older citizens have found themselves trying to get by on fixed-income pensions that do not adequately cover expenses. As a result, they sometimes have had to rely on their grown children for support at a time when those children need to conserve their financial resources to raise and educate their own children.

If the teenagers in this chapter take a closer look at the financial decisions that their parents are making, they may become a bit more understanding when those parents have to turn down a request that would strain the family budget. Who knows? The teenagers might even learn some pointers from the chapter that they can pass on to their parents to assist them in their retirement planning.

7–1 Life Insurance: Who Needs It?

7–2 Spreading the Risk: How Insurance Works

7–3 Value for the Future

LIFE INSURANCE: WHO NEEDS IT?

Lily never used to care about life insurance. As a topic of conversation she ranked it just below her grandmother's cataract surgery and slightly above her cousin's discussion of his recent vacation illustrated with 243 slides. Her mother tried once to discuss the subject of insurance with her, but Lily didn't really pay close attention.

Lily's attitude about life insurance quickly changed last year after the father of her very close friend, Annika, was killed in an automobile accident. Of course, Annika and her mother found it very difficult to deal with the shock of the unexpected death. Then, as if the death itself were not bad enough, their adjustment to their new circumstance became complicated by an unpleasant fact: Annika's father had not purchased enough life insurance

to help support the family after his death. Annika's mother had to struggle to pay the bills with the income from her job.

She had hoped to find a better-paying job without having to move, since she knew how much it would mean to Annika to stay with her friends and finish her senior year at the same high school. After eight months, however, Annika and her mother finally had to move to another city, where a higher-paying job finally did become available.

As a result of all this, Lily began to wonder what it was about life insurance that could so dramatically affect the quality of a person's life. Her friend Manuel has offered to help her learn more about life insurance— the various types of policies and how much they cost.

OBJECTIVES: *In this lesson, we will help Lily and Manuel to:*

- *Understand what life insurance is.*
- *Decide who needs life insurance.*
- *Determine how much life insurance a family should buy.*
- *Examine the kinds of life insurance that are available.*
- *Calculate the cost of various amounts of life insurance.*

LIFE INSURANCE

Life insurance is a contract to pay a specified amount of money to a designated person upon the death of the policyowner. If Annika's father had purchased adequate life insurance, his wife would have received enough money to replace the income lost when he died.

How Does Insurance Work? All types of insurance are based on two ideas: risk sharing and statistical probability. Every person faces the possibility of financial disaster caused by an unpredictable event such as an accident, fire, flood, illness, or the death of the principal wage earner. These risks can be lessened if they are shared by a large group of people paying money into a central pool. Then those who have contributed to the pool have the right to call on that reserve when they suffer a loss. Insurance companies manage such pooled money, called **premiums,** that they have collected from their policyowners. In the event of the of the policyowner's death, the company disburses money from the pool to the **beneficiaries,** those people named in the policy as recipients of the benefits.

Who Needs Life Insurance? When a family has only one provider, it is placing its financial security in the hands of that person. If that person should die, then the family may face serious economic difficulty. Although Annika's father was not the sole provider (her mother also worked), his family did depend upon his income. He should have bought a life insurance policy that was large enough to replace all or most of the income that his **dependents** (his wife and children) needed to survive after he died.

In today's families, women are increasingly responsible for providing all or part of the family income. For that reason, women as well as men must decide how much life insurance they need. If a wife and mother were to die, her husband and children might have to not only replace her income but also pay someone to provide child care and housework.

Remember that life insurance is purchased to protect the dependent members of a family. Therefore it is usually a mistake to buy life insurance to cover the life of anyone other than the principal breadwinners. In particular, the money that might be spent on a child's life insurance policy can probably be put to better use, even though the child's premiums are very low. If you are a high school or college student and have no dependents, you probably do not need life insurance.

How Much Insurance Is Needed? An insurance company will generally sell any amount of insurance that a person wants to buy. However, there are some general guidelines that may be helpful in determining how much life insurance a family should purchase.

An ideal goal is to have the insurance provide enough income for the family to continue its current standard of living. According to one major financial services company, this can usually be done without maintaining 100% of current gross pay. Instead, you should attempt to replace your family's usual *net income,* that is, the amount that remains after income and FICA taxes have been deducted. You can probably accomplish this goal by purchasing insurance that provides 75% of the previous gross income. If the premiums for doing this are too high for the family budget, you should at least aim at a replacement income of 60% of the current gross income to avoid a serious lowering of your family's standard of living.

LIFE INSURANCE: TWO TYPES

Manuel's older brother, Martin, is married to Rachel, his childhood sweetheart. The couple has a young child. Manuel's research will be useful to Martin and Rachel as they decide what kind of insurance they need. Manuel has learned that they can choose from two basic types of insurance. The first type, **term insurance,** offers pure protection. The second, called **cash-value insurance,** offers less protection for the same money. However, it also builds up a cash value that the policyowner can draw on in later years.

Term Insurance If Martin and Rachel choose term insurance, they will be protected for a fixed period of time, or *term.* The term is usually five or ten years. If the insured person dies during the term, the beneficiary will receive the **face value of the policy.** If Annika's father had had $100,000 worth of term insurance, the beneficiary, her mother, would have received $100,000 on his death. If the insured person does not die during the term, the policy either expires or is renewed for another term. Term insurance generally has the lowest premiums of any insurance available until approximately the age of 50. Because the amount of coverage can be increased or decreased at the end of each term, term insurance is a very flexible option. Manuel discovered three common types of term insurance.

Group life insurance, which people can buy through their place of employment, provides term insurance for a large number of people under a single policy without the need for a medical examination.

A policy of the second type, **renewable convertible term insurance** covers a person for a period of time such as one, five, or ten years and can be renewed without a medical examination. In addition to being renewable, renewable convertible term insurance is convertible into a different form of insurance.

The third alternative is **decreasing term insurance.** Under this option the amount of benefits that an insurance company will award the beneficiaries decreases over time as the children grow up and cease being dependents. After this happens, the parents no longer need as much insurance as they did in earlier years. The appeal of decreasing term insurance is that the premiums are lower than those for renewable term insurance. Often decreasing term insurance is used to provide cash to pay off a mortgage. Policies have been designed so that the face value declines exactly with a person's outstanding mortgage balance.

Cash-Value Insurance   If Martin and Rachel choose cash-value insurance, they will be buying protection plus savings. As they pay the premiums, the policy will build up cash value much like a savings account.

Manuel learned about three traditional types of cash-value insurance. The first is **whole life insurance,** also called *straight life* or *ordinary life*. With this kind of insurance policy, you pay premiums for your entire life that are more than enough for the insurance coverage you need. The remaining part of your premium is invested for you by the insurance company, building up the cash value of the policy. You always have the option of cashing in the policy, that is, taking out the money and the interest it has earned. In addition, since the policy has a cash value, you can borrow part of the cash value. If you should die with an outstanding loan, the amount of benefit due to your beneficiaries is reduced. However, the rate of return tends to be low in comparison to the return that you would get investing on your own.

Limited payment life insurance has a fixed number of premium payments that stop after a certain number of years; for example, 20 years. The purchaser pays higher premiums, but the policy is paid up to its cash value in fewer years than is the case for whole life insurance.

If the couple wants a policy in which the cash value buildup is accomplished especially quickly, they might choose an **endowment policy.** An endowment policy, like other cash value insurance, is protection plus savings, but the emphasis is on the savings. Because of this rapid accumulation over a short period of time, the premiums on endowment policies are very high.

Universal life insurance is a relatively new type of cash value insurance. As with whole life insurance, it gives both protection and savings. Two basic types of universal life differ in the death benefit. In Type A the death benefit remains constant; in Type B the death benefit varies with the cash value so that as the cash value increases, the death benefit also increases. Usually, the insurer guarantees a minimum rate of interest throughout the duration of the policy. As with whole life, you can borrow or withdraw a part of the cash value.

The types of life insurance and their characteristics are summarized in the following chart.

LIFE INSURANCE	
Term Insurance	
Group life	Purchased through an employer; low rates
Renewable convertible term	One-, five-, or ten-year terms; renewable and convertible
Decreasing term	Like renewable convertible term but less expensive and with a gradually decreasing death benefit
Cash-Value Insurance	
Whole life	Combines savings with insurance; can be cashed in for its accumulated cash value
Limited payment	Higher premiums for a fixed number of years, such as 20; builds up cash value more quickly than whole life
Endowment	Emphasis on high cash buildup; very high premiums
Universal life	Combines savings with insurance; can be cashed in for its accumulated cash value

ALGEBRA *REVIEW*

A thrown dart may land anywhere on its target. So the probability of landing in the top half of the square is, $P = 0.5$.

Determine the probability of a dart landing in each region.

1. Lower rectangle

2. Upper-right square

3. Upper-left triangle

Determine the probability of a dart landing in each region.

4. Inner circle

5. Inner diamond

6. Lower-left triangle

What Does it Cost? The cost of life insurance varies widely. The range between the most expensive and the least expensive can be as much as 100%. To find the best price for both term insurance and cash-value life insurance, you should shop around and compare the prices that different insurance companies offer.

How About Taxes? The proceeds paid to a beneficiary from some kinds of life insurance policies are generally free from federal income taxes but may be subject to other taxes, such as an estate tax. Tax laws change from time to time, so you should obtain up-to-date information from a tax adviser.

Ask Yourself

1. What is life insurance?

2. Which member(s) of a family should have life insurance?

3. How much life insurance should a family assume?

4. What are the two basic kinds of life insurance?

5. What is a beneficiary?

6. What advantage does decreasing term insurance have over regular term insurance?

SHARPEN YOUR SKILLS

SKILL 1

To determine how much insurance to buy to replace either 75% or 60% of gross income, you can use the following chart and formula. This chart is also in the Reference Section.

MULTIPLES-OF-SALARY CHART								
	Current Age							
	25 Years		35 Years		45 Years		55 Years	
Current Gross Earnings	75%	60%	75%	60%	75%	60%	75%	60%
$ 7,500	4.0	3.0	5.5	4.0	7.5	5.5	6.5	4.5
9,000	4.0	3.0	5.5	4.0	7.5	5.5	6.5	4.5
15,000	4.5	3.0	6.5	4.5	8.0	6.0	7.0	5.5
23,500	6.5	4.5	8.0	5.5	8.5	6.5	7.5	5.5
30,000	7.5	5.0	8.0	6.0	8.5	6.5	7.0	5.5
40,000	7.5	5.0	8.0	6.0	8.0	6.0	7.0	5.5
65,000	7.5	5.5	7.5	6.0	7.5	6.0	6.5	5.0

> ## Replacement Life Insurance Formula
> $R = mS$ where R = required replacement insurance
> m = multiple from the table
> S = original gross salary or wage

EXAMPLE 1 Annika's father was earning $30,000 per year when he was 35 years old.

QUESTION How much insurance should he have bought to replace 75% of his regular gross income?

SOLUTION

Look at the Multiples-of-Salary Chart. Find the income level of $30,000 in the left-hand column. Then read across to the column for 35 years and 75%. The number 8.0 appears in that space. Use the replacement life insurance formula.

$R = mS$ The multiple of salary m is 8.0.
  $= 8 \cdot 30,000$
  $= 240,000$

Annika's father should have purchased $240,000 worth of life insurance at age 35 to replace 75% of his gross income.

The following table shows the comparative premium rates for two of the major types of insurance. This table is also in the Reference Section. *Note:* To use the table to find the cost of $50,000 worth of life insurance, divide the $100,000 premium by 2. To find the cost of $200,000 worth, multiply by 2.

COMPARISON TABLE FOR TERM AND WHOLE LIFE PREMIUMS			
Policy face value is $100,000			
Age	Five-Year Renewable Term	Whole Life	First-Year Difference
20	$205	$ 775	$ 570
25	207	918	711
30	218	1112	894
35	254	1374	1120
40	363	1729	1366
45	562	2127	1565
50	878	2689	1811

EXAMPLE 2 Ramón is a 30-year-old father who is comparing the premiums for different types of life insurance.

QUESTIONS
1. How much would five-year term insurance for $200,000 cost Ramón per year?

2. How much would the same amount of whole life insurance cost him?

3. In one year, how much would he save by buying term insurance instead of whole life?

SOLUTIONS

Use the Comparison Table for Term and Whole Life Premiums.

1. For age 30 the number in the "Five-Year Renewable Term" column is $218. Ramón's annual premium would be 2 • 218, or $436.

2. For age 30, the number in the "Whole Life" column is $1112. Ramón's annual premium would be 2 • 1112, or $2224.

3. The difference is 2224 − 436, or $1788. Ramón would save $1788 by buying term insurance.

TRY YOUR SKILLS

Use the Multiples-of-Salary Chart and the Replacement Life Insurance Formula in Skill 1 to find the amount of life insurance each of these income earners should buy under the conditions described.

	Current Earnings	Age	Income Replacement	Amount of Insurance
1.	$65,000	35	75%	
2.	65,000	45	60%	
3.	40,000	55	75%	

Use the Comparison Table for Term and Whole Life Premiums in Skill 2 to find the yearly premium for the amounts of insurance listed below at the ages shown.

	Amount of Insurance	Age	Type	Premium
4.	$100,000	25	Term	
5.	100,000	25	Whole life	
6.	200,000	35	Term	
7.	200,000	35	Whole life	
8.	300,000	45	Term	
9.	300,000	45	Whole life	

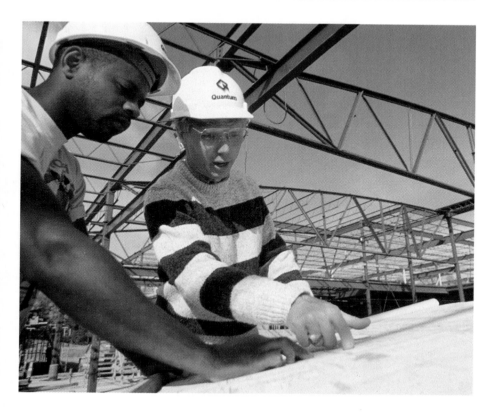

EXERCISE YOUR SKILLS

1. Why do people need life insurance?

2. When would it be a good idea for both a husband and a wife to have a life insurance policy?

3. If an aged grandmother were living with a family, would it be a good idea to insure that grandmother's life? Why or why not?

KEY TERMS

beneficiaries
cash-value
    insurance
decreasing term
    insurance
dependents
endowment policy
face value of the
    policy
group life insurance
life insurance
limited payment life
    insurance
premiums
renewable
    convertible term
    insurance
term insurance
universal life
    insurance
whole life insurance

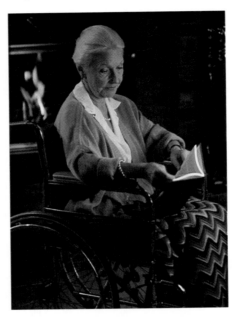

4. Why would "death insurance" be a more accurate term than "life insurance" to describe the kind of policy that has been discussed in this lesson?

5. Why is group life insurance cheaper than life insurance bought as an individual policy?

6. What is the advantage of limited payment life insurance?

Use the Multiples-of-Salary Chart to find the amount of life insurance each of these income earners should buy under the conditions described.

	Current Earnings	Age	Income Replacement	Amount of Insurance
7.	$40,000	25	60%	
8.	30,000	35	75%	
9.	30,000	45	60%	
10.	23,500	55	75%	
11.	23,500	25	60%	
12.	15,000	35	75%	
13.	15,000	45	60%	

Use the Comparison Table for Term and Whole Life Insurance to find the yearly premium for the amounts of insurance listed below at the ages shown.

	Amount of Insurance	Age	Type	Premium
14.	$ 50,000	30	Term	
15.	50,000	30	Whole life	
16.	150,000	40	Term	
17.	150,000	40	Whole life	
18.	250,000	50	Term	
19.	250,000	50	Whole life	

20.–22. For each of the three pairs of policies (term and whole life) in Exercises 14–19, which is the less expensive? How much cheaper is the less expensive choice?

MIXED REVIEW

Assume that a person receives $1\frac{1}{2}$ times the usual hourly rate for each hour or part of an hour beyond 40 hours per week. Find the total wages earned in one week in each case.

1. Hourly wage rate: $8.25; total hours worked: 50

2. Hourly wage rate: $11.00; total hours worked: $47\frac{1}{2}$

3. Suppose that you can pay off a $12,500 loan either by paying $395 per month for 3 years or by paying $260 per month for 5 years. How much will you save if you pay the loan off in 3 years?

Complete the chart below to determine how much the interest cost will be if you pay 10% of the amount due to the nearest dollar each month and your credit card carries an APR of 9%.

	Month	Balance	Interest	Amount Owed	Payment
4.	1	$880.00			
5.	2				

6. What is the total interest for the two months?

7. Tom's weekly gross pay is $115. The amount withheld for taxes is $2 and 7.65% of the gross pay is withheld for FICA taxes. Find Tom's take-home pay for the week.

8. Write an inequality for each of the following 3 constraints: you want to work no more than 15 hours per week, you want to earn at least $100 a week, and you won't get a job paying more than $9 an hour. Let x represent the number of hours you work in 1 week and y represent your weekly salary.

Martin and Rachel finally decided to purchase term insurance on Martin's life rather than the other types of life insurance that were available. Since Martin is still in his twenties, he was able to get a policy with a relatively low premium. The oldest brother, Benjamin, has also started a family recently. Because he is in his thirties, the term insurance that Benjamin purchased is almost $50 more per year than Martin has to pay for the same $100,000 policy. Manuel knew that insurance premiums increase as the age of the insured person increases, but he wondered why the amount of the increase was $50. Why not a $5 increase—or $500?

Lily has recently had a similar experience. Her father told her that he had rejected an insurance salesman's argument that Lily's life should be insured "because her premiums would be so low." She and her father both know that she doesn't need life insurance because no one depends on her for support. Nevertheless, like Martin, she wants to know why her premiums would be so low.

Manuel decided to talk with his cousin Roberto, who works for a major insurance company. One of Roberto's duties is to help prepare new tables that show the premiums that people have to pay for insurance. Manuel hoped that Roberto could explain to him and Lily how the numbers in the tables are computed.

OBJECTIVES: *In this lesson, we will help Lily and Manuel to:*

- *Understand how life-expectation tables are used to estimate the probability that an individual will die within one year.*
- *Learn how an insurance company determines its premium schedule to make a reasonable profit.*

STATISTICAL TABLES

Manuel and Lily visited Roberto at his office. As they were talking, Manuel noticed that on Roberto's desk was a statistical table entitled Expected Deaths per 100,000 Alive at Specified Age. Part of the table is shown below. This table is also in the Reference Section.

	EXPECTED DEATHS PER 100,000 ALIVE AT SPECIFIED AGE	
Age	Expected Deaths Within 1 Year	Expected to be Alive in 1 Year
15	63	99,937
16	79	99,921
17	91	99,909
18	99	99,901
19	103	99,897
20	106	99,894
21	110	99,890
22	113	99,887
23	115	99,885
24	117	99,883
25	118	99,882
26	120	99,880
27	123	99,877
28	127	99,873
29	132	99,868

Roberto explained that information in the table enabled him to calculate the probability that a person of a given age will die or not die sometime during the next 12-month period. This information helps him to decide how large a life insurance premium should be for each age.

For example, the probability that a 16-year-old person will still be alive 1 year from today is found by using the formula for the **probability of an event.** This probability is written as $P(E)$, which is read "P of E."

Probability of an Event

$P(E) = \frac{m}{n}$ where $P(E)$ = the probability of an event E

m = the number of times the event occurs

n = the number of all possible outcomes

Recall that the probability of an event is always a number between 0 and 1, inclusive.

From the chart, Roberto is able to determine the probability that a 16-year-old person will be alive 1 year from today.

$$P(E) = \frac{99,921}{100,000}$$

Number of 16-year-old people alive 1 year later
Total number of 16-year-old people

$$= 0.99921$$

The probability that a 16-year-old will be alive 1 year from today is 0.99921, that is, almost 1. The event E is practically certain.

The probability that a 16-year-old will die within one year is

$$P(E') = \frac{79}{100,000}$$

$$= 0.00079$$

This probability is 0.00079, or almost zero. The event E' ("E prime") is very unlikely.

In the above formula, E' is the event "E does not occur," or "not E." For this reason it is called the **complement** of E. Notice that

$$P(E) + P(E') = 0.99921 + 0.00079$$
$$= 1.00000$$

In words, *the sum of the probabilities of an event and its complement is 1*. For example, it is certain (a probability of 1) that a 16-year-old will either be dead or alive 1 year from today.

Manuel found the discussion about probabilities very interesting, but he still did not have an answer to his question: How is the size of an insurance premium determined?

Ask Yourself

1. How do you think insurance companies can obtain information about life expectancy?

2. If a person's life expectancy is very high, does that mean that he or she can safely drive without a seat belt at high speeds? Explain your answer.

3. In earlier centuries, life expectancy for infants and very young children was much lower than it is today. Why do you think that the life expectancy for children has improved in recent years?

SHARPEN YOUR SKILLS

SKILL 1

The **expected value** is the amount of money to be won or lost in the long run. If an event can assume two values, then the *expected value* of the event is the sum of the product of each value and its probability.

> **Expected Value Formula**
>
> $E = P_1v_1 + P_2v_2$ where v_1 and v_2 are values and P_1 and P_2 are the corresponding probabilities

EXAMPLE 1 In a certain game of chance, you win $4 when a coin shows heads and $1 when it shows tails.

QUESTION What is the expected value of money that you could win?

SOLUTION
The probability of obtaining heads and the probability of obtaining tails are each 0.5.

$E = P_1v_1 + P_2v_2$ Expected value formula
$E = 0.5(4) + 0.5(1)$ $v_1 = 4,\ v_2 = 1,\ P_1 = 0.5,\ P_2 = 0.5$
$ = 2.50$

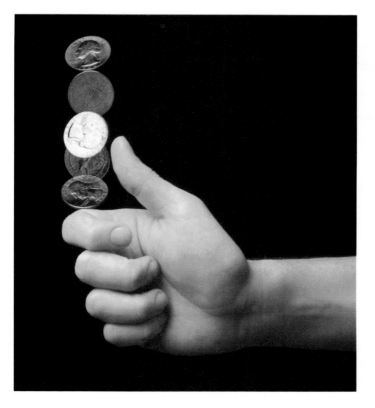

For the "privilege" of tossing the coin you should pay no more than $2.50. Notice that on a single toss you will win either $1 or $4, never $2.50. For a large number of tosses you can expect to gain about $2.50.

SKILL 2

Roberto's insurance company must make a reasonable profit on the insurance policies that it writes. Otherwise, it will not be able to pay benefits on the policies or even to survive as a company. Roberto uses the idea of expected value to find the premium that gives the **break-even value** for the insurance company; that is, the value of the premium that gives zero profit after paying for all expenses.

EXAMPLE 2   The direct and indirect expenses for each policy that the insurance company writes are about $20 per policy.

QUESTION   How can Roberto use probability to determine the proper premium for a $50,000 insurance policy on a 28-year-old person?

SOLUTION
The possible outcomes in the next year for a 28-year-old person are as shown in the table on page 331. The variable x represents the break-even value for the premium. Notice that the probability that a person dies is 1 minus the probability that a person lives.

COMPANY'S GAIN OR LOSS ON ONE POLICY (before deducting expenses)		
Possible Outcome	**Probability of Outcome**	**One-year Gain/Loss**
The person lives.	$\frac{99,873}{100,000}$, or 0.99873	x (gain)
The person dies.	$\frac{127}{100,000}$, or 0.00127	$x - 50,000$ (loss)

$$E = P_1 v_1 + P_2 v_2$$ Use the expected value formula.
$$E = 0.99873x + 0.00127(x - 50,000)$$
$$0 + 20 = 0.99873x + 0.00127x - 63.5$$ Profit: 0; expenses: 20
$$83.5 = 1.00000x$$
$$x = 83.5$$

The break-even premium for one year of life insurance on a 28-year-old person is $83.50. The company will have to charge more than $83.50 to make a reasonable profit.

Roberto explained to Manuel and Lisa that this example applies to a *one-year term* life insurance policy, not to a whole life policy. A whole life policy has additional savings features that would raise the premiums significantly. He also mentioned that most term policies are for five or ten years, not one year, and that five-year and ten-year life expectation tables can be used to prepare premium schedules for such policies.

The profit that an insurance company makes on policies is determined by the amount of revenue (money) that is received as premiums minus any benefits paid out and any other expenses such as overhead.

Profit on Insurance

$P = R - B - C$ where P = profit
R = revenue received as premiums
B = benefits paid out
C = costs or expenses

EXAMPLE 3   Roberto told Lisa and Manuel that his company charges a 20-year-old person $110 for a one-year term $50,000 life insurance policy.

QUESTION   What profit does the company expect to make on each of 100,000 such policies if the expenses for each policy are $20?

SOLUTION
Let x represent the profit for each policy. Then $100,000x$ represents the profit for 100,000 policies. Use the table of expected deaths on page 331. There are 106 expected deaths for 20-year-old people. The death benefit is $50,000. So, the benefits paid out are $B = 106(50,000)$. The cost is $20 per policy, so $C = 20(100,000)$. The revenue is $110 per policy, so $R = 110(100,000)$.

$$P = R - B - C \qquad \text{Use the profit on insurance formula.}$$
$$100,000x = (110)100,000 - 106(50,000) - 20(100,000)$$
$$100,000x = 11,000,000 - 5,300,000 - 2,000,000$$
$$100,000x = 3,700,000$$
$$x = 37 \qquad \text{Divide each side by 100,000.}$$

The company will make $37 profit on each policy and 37(100,000) or $3,700,000 on 100,000 policies.

EXAMPLE 4   To use the table of expected deaths on page 331 for another amount of people, write a proportion.

QUESTION   What is the number of expected deaths for 5000 20-year-olds?

SOLUTION
From the table on page 331 the expected deaths for 100,000 20-year-olds is 106. Let d represent the expected deaths for 5000 20-year-olds. Then

$$\frac{106}{100,000} = \frac{d}{5000} \qquad \text{Multiply both sides by 5000.}$$

$$\frac{106(5000)}{100,000} = d$$

$$5.3 = d$$

The expected deaths for 5000 20-year-olds is 5.3.

Find the expected value.

1. The expected gain from a coin toss that pays $3 for heads and $2 for tails

2. The expected gain from a roll of a die that pays $10 for a 6 or a 2 and $1 for any other result

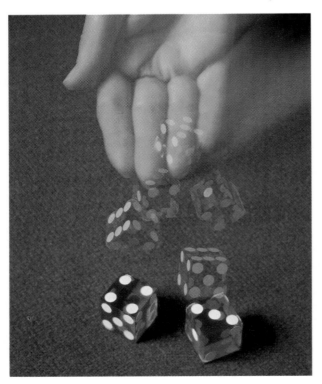

A person's life expectancy for one year is shown in the following table. She is contemplating the purchase of a one-year term policy with a face value of $60,000. Refer to the table to answer Exercises 3–5.

COMPANY'S GAIN OR LOSS ON ONE POLICY (before deducting expenses)		
Possible Outcome	Probability of Outcome	One-Year Gain/Loss
The person lives.	0.99906	x
The person dies.	?	$? - ?$

3. Find the probability that the person will die within one year.

4. What is the algebraic expression that represents the company's one-year gain or loss (before expenses) if the person dies?

5. What is the algebraic expression that represents the company's expected gain or loss (before expenses) on the policy?

6. The company has direct and indirect expenses of $25 for each policy that it issues. Find the premium that the company must charge to break even; that is, neither to make nor to lose money on this policy.

Use the formula for profit on insurance $P = R - B - C$ and the table of expected deaths to calculate the profit that a company makes on one-year term life insurance policies for the policies and populations described in Exercises 7–10. Begin by letting x represent the money made on each policy to cover profit. The cost for each policy is $25.

7. 1000 19-year-olds; face value: $100,000; annual premium: $200

8. 1000 23-year-olds; face value: $40,000; annual premium: $100

9. 5000 28-year-olds; face value: $135,000; annual premium: $350

10. 50,000 25-year-olds; face value: $45,000; annual premium: $125

EXERCISE YOUR SKILLS

1. Why do you think that the probability of a 16-year-old being alive one year from now is higher than that of a 25-year-old?

2. Why don't insurance companies charge the same life insurance premium for everyone regardless of his or her age?

3. Is there any age group for which an insurance company could not issue a life insurance policy? Explain your answer.

4. Suppose that two insurance companies both sell five-year term life insurance policies but that one of the companies charges an annual premium that is $30 higher than the other company's. What are some possible explanations for this?

5. From time to time, insurance companies update the tables that they use to help them set premium schedules. Why do you think that they have to update the tables?

6. Because they deal with large populations of people, life insurance companies can use principles of probability to help them set profitable premium schedules. Does this mean that an insurance company can never suffer a loss from selling insurance policies? Explain your answer.

KEY TERMS
break-even value
complement
expected value
probability of an event

A winning lottery ticket pays $200. Find the expected value of the lottery for each number of sold tickets.

7. 50 tickets   8. 100 tickets   9. 500 tickets

10. In a certain store the probability that a person makes one purchase is 0.7; the probability that a person makes two purchases is 0.3. Find the expected number of purchases.

Find the break-even premium for a one-year term insurance policy for each indicated individual. In each case, assume that the direct and indirect expenses for issuing one policy are $30. Use the table of expected deaths on page 331.

11. a $50,000 policy for a man or woman of 24

12. a $100,000 policy for a man or woman of 29

13. Find the break-even premium for a one-year term insurance policy for $200,000 for a man or woman of 35. The insurance company expects 99.825% of all 35-year-old people to live at least one more year.

14. Find the break-even premium for a one-year term insurance policy for $150,000 for a man or woman of 38. The insurance company expects 99.765% of all 38-year-old people to live at least one more year.

15. Find the break-even premium for a one-year-term insurance policy for $250,000 for a man or woman of 39. The insurance company expects 99.725% of all 38-year-old people to live at least one more year.

What profit does an insurance company expect to make for each one-year term insurance policy? Assume that the direct and indirect expenses for each policy are $25.

16. Face value: $50,000; age of insured: 25; annual premium: $135

17. Face value: $150,000; age of insured: 22; annual premium: $390

18. Face value: $290,000; age of insured: 26; annual premium: $800

19. Face value: $250,000; age of insured: 29; annual premium: $650

20 Face value: $75,000; age of insured: 24; annual premium: $200

21. Face value: $125,000; age of insured: 27; annual premium: $325

An insurance company employee has the responsibility of determining the company's schedule of life insurance premiums. She is reviewing the following portion of a life-expectancy table.

EXPECTED DEATHS PER 100,000 ALIVE AT SPECIFIED AGE		
Age	Expected Deaths Within 1 Year	Expected to be Alive in 1 Year
45	315	99,685
46	341	99,659
47	371	99,629
48	405	99,595
49	443	99,557

The company's marketing department has told her that the public loses interest in buying life insurance whenever the annual premium is greater than $800. The company's current expenses for each policy issued are $35 and it aims at making a profit of $50 on each policy.

22. At what age will a person first be inclined to resist purchasing one-year term life insurance with a face value of $200,000?

23. To appeal to the person of Exercise 22, the company manages to reduce its expenses to $20 per policy. Will this be enough to enable the employee to lower the premium to $799 without cutting into the company's profits? If not, how much of a reduction in profits would the company have to accept for the employee to be able to lower the premium to $799?

24. The company wants to sell a one-year term policy to a 49-year-old person and keep the premium below $800. Assume that the expenses have been reduced to $20 per policy and that the profit on the policy has been raised to $55. What would be the largest possible face value of such a policy?

25. The company wants to sell a one-year term policy to a 45-year-old person and keep the premium below $500. Asume that the expenses are $20 per policy and that the profit on the policy is $60. What would be the largest possible face value of such a policy?

Suppose that you write a check for the amount shown. Write the amount in words.

1. $500

2. $10,100.10

3. Which approach will save you more money; to make a 20% down payment on a purchase, or to save up until you can afford a 30% down payment? Explain your answer.

4. Determine the effective rate of interest if the APR is 14.75%.

Complete the following chart to determine how much the interest cost will be if you pay 5% of the amount due to the nearest dollar each month and your credit card carries an APR of 9%.

	Month	Balance	Interest	Amount Owed	Payment
5.	1	$1000.00			
6.	2				

7. What is the total interest for the two months?

8. What happens to the nation's money supply when the Federal Reserve System increases the percent of required reserves that banks must hold?

9. You earn $7.25 per hour. You worked 46 hours last week. You receive $1\frac{1}{2}$ times your hourly rate for any hours over 40. Find your total wages for last week.

10. Last month the Wilson family spent $205 on the car payment, $506 on credit card payments, $850 on utilities and rent, $560 on food, and $2,000 on everything else. To the nearest tenth of a percent, what percent of their take-home pay did the family spend on credit payments?

11. Use the Time-to-Pay-Off formula to find the number of months required to pay off a credit card balance of $1,850. The APR is 15% and the monthly payment is $120.

12. You have a loan of $25,000 at 10.5% for 5 years for which you have been making monthly payments of $537.35. How much money will you save if you prepay the loan at the end of 40 months?

13. Suppose that a bank's reserve requirement is 25%. How much new money can be created from a deposit of $2000?

Some students purchased a number of damaged tote bags for $1.00 each and repaired them for $0.50 each. They plan to sell the repaired tote bags.

14. What is the unit cost for purchasing and repairing the tote bags?

15. Find the total cost for 56 tote bags.

16. The fixed costs are $62. What are the total costs?

While Lily and Manuel are taking a look at life insurance, their friend Eleanor has other concerns—the financial well-being of her grandparents. Her grandmother, who retired earlier this year, has had to cut back on some expenses to help her family make ends meet.

Eleanor's grandfather retired three years ago. Since then, he has had some costly health problems that are not completely covered by Medicare. In addition to dealing with these past cost burdens, both grandparents have had to purchase supplementary health insurance to cover some of their future medical expenses that will not be covered by Medicare.

Eleanor is not certain of the details, but she does know that the retirement pensions that her grandparents receive are just enough to keep them comfortable. She also knows that her grandparents have had a little difficulty adjusting to the fact that they do not have as much income as they did when they were both working. Eleanor's mother is ready to help her parents in case of a financial emergency but hopes that she will not have to step in. Their circumstances have caused her to begin reevaluating her own retirement plans, and even Eleanor has suddenly become interested in topics such as annuities and Individual Retirement Arrangements.

Examine reasons for investing in retirement plans that are tax deferred, such as annuities and Individual Retirement Arrangements (IRAs).

Calculate the future value of regular payments invested at compound interest.

Compare the future value of cash-value life insurance with the future value of the same amount invested at compound interest.

Compare the difference in accumulated cash value between investing directly and saving indirectly through whole life insurance.

SOCIAL SECURITY IS NOT ENOUGH

Everyone faces the fear of outliving one's pension. Eleanor's grandparents have some degree of financial security because they planned ahead for their retirement. They knew that they could not rely on their monthly Social Security checks, since that system was never intended to provide one's sole source of retirement income, even though the system is adjusted upward periodically for inflation. Instead, Eleanor's grandparents learned to take advantage of supplementary sources of retirement income such as annuities and Individual Retirement Arrangements.

ANNUITIES

Years ago, Eleanor's grandmother placed some of her income from college teaching in a tax-deferred annuity. An **annuity** is an investment plan that provides income upon retirement. Annuities, which are usually purchased through insurance companies, offer two advantages—forced savings and tax deferral. Grandmother chose to make payments into her annuity through a payroll deduction. The amount that she requested was automatically deducted from her paycheck each month.

Now that she has retired, Grandmother withdraws the money that she originally contributed as well as interest on that money. She doesn't pay taxes on the returned contributions, but she does pay taxes on the interest income from the annuity.

Grandmother's annuity has provided her with two major tax advantages. First, she has never had to pay any taxes on any of the interest that has been accumulating for all those years—until now. During this long period of time, that interest has been earning more interest. She would never have received all that additional money if her tax obligation had not been delayed until her retirement years. Second, even though she now has to pay the taxes, they are less than they would have been if she had paid them in earlier years. The reason? Grandmother's tax *rate* on her pension income is lower than it was on her much higher working income. The lower your income, the lower your tax rate.

INDIVIDUAL RETIREMENT ARRANGEMENTS (IRAS)

An **Individual Retirement Arrangement (IRA),** sometimes called an individual retirement *account,* has the same **tax shelter** advantages just described for annuities. It allows employees (and some self-employed people) to put up to $2000 a year into such a plan to shelter the accumulated interest from taxes. If the employee's income is less than a certain amount, then the contribution can also be fully deducted from the adjusted-gross-income line on the federal tax return. The maximum income for a full deduction is, at this writing, $40,000 for a married couple and $25,000 for an individual. (A partial deduction is allowed on up to $10,000 above these levels.) Employees who have no employer pension plan are exempt from these income maximums.

You may withdraw money from an IRA, but you will pay a tax penalty and possibly an interest penalty unless you are either disabled or over the age of $59\frac{1}{2}$. You may choose the kind of institution that you want to handle your IRA, for example, a bank, a brokerage firm, or a mutual fund group. You may also set up more than one IRA among a variety of institutions. It is important never to combine your IRA account with non-IRA funds.

EMPLOYER PENSION PLANS

Some employers offer pension plans for their employees. If such a plan is approved by the Internal Revenue Service (IRS) as a *qualified plan* or as a **401(k) plan,** then an employer will make tax-sheltered contributions to the plan that are larger than an employee could get through a regular IRA. In some such plans, the employee may also contribute to the plan, but whether or not those contributions are tax deductible depends on the exact nature of the plan. Unlike its rules for IRAs, there are no IRS-imposed income maximums on participants in these employer plans. A self-employed person may set up a similar tax-sheltered plan, called a **Keogh plan.**

These employer and Keogh plans have the disadvantage that, unlike IRAs, they are somewhat complicated to set up and to administer. To avoid these complications, some employers or self-employed people prefer to establish a **Simplified Employee Pension Plan (SEP)** that allows a person to contribute to an IRA without all of the usual IRA rules and limitations.

BALANCING INSURANCE AND INVESTMENTS

Eleanor is especially curious about cash-value life insurance as an investment vehicle compared to a direct investment such as a mutual fund.

The CVLI/BTID Controversy   A controversy has arisen in the last few years over the merits of cash-value life insurance compared to term insurance. Consumers have assumed, perhaps because of persuasive insurance salespeople, that cash-value insurance is a good way to combine adequate life insurance coverage with a means for saving money. However, some people in the insurance industry have begun to advise their customers that they can save much more money by buying term insurance and investing the money that they save in instruments such as certificates of deposit, mutual funds or even stocks or bonds. These opposing viewpoints are expressed in the code phrase CVLI/BTID, which stands for *Cash-Value Life Insurance/Buy Term, Invest the Difference.*

If your employer offers you the opportunity to buy life insurance at work, you may be able to save significantly on the annual premium. Another way of saving on the cost of the premium is to find a company that sells **low-load insurance policies,** policies with a very low commission. Over a long period the annual savings can be very important. One insurance expert has concluded that over 20 years, a full-commission cash-value insurance policy would have to give a yield that was $1\frac{1}{2}$ percentage points higher than a low-commission insurance policy to match the cash value of the low-load policy. This could amount to a difference of several thousand dollars over 20 years.

The most difficult choice in your planning may be the selection of interest rates that you can reasonably expect to earn over the next 30 years. Since even professional economists cannot accurately predict future interest rates, your best course may be to start with the current rates for government bonds, both short-term (six-month Treasury bills) and long-term (30-year bonds) and make your own projection about rates in the future. Since government bonds are the safest investments of all, you have a right to expect higher returns from any investment that is not as safe as they are.

Do not have unrealistic expectations about the 30-year performance of your own investments. If you can obtain a long-term average of five percentage points more than the yearly return on a 30-year government bond, then you will be doing reasonably well. You will be doing very well indeed if you consistently average ten points a year above such bonds.

On the basis of your research, you can decide whether it is better for you to save through an insurance company or to use the insurance company solely for protecting your family's earning power and do your investing directly. Careful decision making on your part may lead to high long-term yields. Even our teenagers, who see retirement as something many years in the future, know the difference between having $20,000 at retirement and having $200,000!

Ask Yourself

1. What are tax-deferred annuities?

2. What are Individual Retirement Arrangements?

3. What is the earliest age at which you can normally withdraw money from an IRA without penalty?

ALGEBRA *REVIEW*

Find the value of A or B in the following formulas for the values given.

$$B = \frac{(1.01)^n - 1}{0.01}$$

$$A = \frac{P[(1 + r)^n - 1]}{r}$$

1. $B; n = 3$ 2. $B; n = 30$

3. $A; n = 4, r = 0.03, P = 1000$

4. $A; n = 10, r = 0.065,$ $P = 1500$

5. $A; n = 20, r = 0.09,$ $P = 2000$

6. $A; n = 40, r = 0.10, P = 750$

7. $A; n = 20, r = 0.15,$ $P = 2000$

SHARPEN YOUR SKILLS

SKILL 1

To determine the future value of money that is deposited into an interest-bearing account on a regular basis, use the formula for the **future value of a periodic investment.**

> ### Future Value of a Periodic Investment Formula
>
> $$A = \frac{p[(1 + r)^n - 1]}{r}$$
>
> where A = the future value of the investment
> p = the investment made at the end of each period
> r = the interest rate for the period
> n = the number of periods

EXAMPLE 1 Eleanor's father and mother are both 43. They have just opened IRA accounts that they hope will supply them with extra money when they retire 22 years from now at the age of 65. Each year, they will deposit $2000 dollars into each account, which they are assuming will pay 8% interest, compounded annually.

QUESTION By the time they are 65, how much money will be in each parent's account?

SOLUTION

$$A = \frac{p[(1 + r)^n - 1]}{r} \qquad \text{Future value of a periodic investment}$$

$$= \frac{2000[(1 + 0.08)^{22} - 1]}{0.08} \qquad p = 2000,\ r = 0.08,\ n = 22$$

$$= 110{,}913.51$$

When they retire, Eleanor's parents will each have $110,913.51 in their accounts. This is $66,913.51 more than the $44,000 that they each contributed to their accounts over the 22 years.

SKILL 2

EXAMPLE 2 Suppose Eleanor's grandfather purchased $50,000 worth of whole life insurance when he was 25.

QUESTIONS 1. How much would his annual premium be?
2. How much would he pay for the insurance over 40 years?
3. If he had invested the premium money every year in an annuity that paid $8\frac{1}{2}\%$ interest, compounded annually, how much would his savings be worth at the end of 40 years?

SOLUTIONS

1. See the Comparison Table for Term and Whole Life Premiums in the Reference Section. The premium for $50,000 of whole life insurance is half of the premium for $100,000; that is, $918 ÷ 2, or $459.

2. The total paid in 40 years would be

   $459 \cdot 40$, or $18,360

   He would have paid $18,360 in premiums over 40 years.

3. Use the formula for the future value of a periodic investment.

$$A = \frac{p[(1 + r)^n - 1]}{r}$$

$$= \frac{459\,[(1 + 0.085)^{40} - 1]}{0.085} \qquad p = 459,\ r = 0.085,\ n = 40$$

$$= 135{,}718.28 \qquad \text{To the nearest cent}$$

His annuity would be worth about $135,718, or about $85,000 more than the cash value of the insurance policy after 40 years. This money would remain untaxed until he began to receive his periodic payments during retirement. If Eleanor's grandfather had chosen to use the $18,360 to buy the annuity, then he would have been without the $50,000 insurance for 40 years.

--- **SKILL 3** ---

The following table shows the cash value (accumulation of savings) of a whole life insurance policy of $100,000 over the first 20 years. This table is also in the Reference Section.

ACCUMULATED CASH VALUE OF $100,000 WHOLE LIFE POLICY
AGE OF ISSUE: 25

Year	Person's Age	Cash Value	Year	Person's Age	Cash Value
1	25	$ 0	11	35	$10,187
2	26	700	12	36	11,501
3	27	1500	13	37	12,860
4	28	2300	14	38	14,246
5	29	3100	15	39	15,667
6	30	4020	16	40	17,094
7	31	5158	17	41	18,555
8	32	6349	18	42	20,014
9	33	7538	19	43	21,563
10	34	8898	20	44	23,197

EXAMPLE 3 Eleanor was interested in knowing whether her grandfather could have been better off financially by buying term insurance instead of whole life at age 25 and investing the money saved in an IRA at $8\frac{1}{2}\%$.

1. If Eleanor's grandfather had bought $50,000 worth of whole life cash-value insurance at age 25, what would the cash value have been when he was 44?
   2. How much money could a 25-year-old man accumulate between the ages of 25 and 30 by buying a $50,000 term policy instead of a whole life policy and investing the difference in the two premiums in an IRA at $8\frac{1}{2}\%$?
   3. How does the money accumulated in Question 1 compare with the cash value that would have accumulated in the whole life policy in the same five-year period?

SOLUTIONS

1. Looking at the table, you can see that the value at age 44 would have been $23,197 ÷ 2, or $11,598.50.

2. Eleanor used the Comparison Table for Term and Whole Life Premiums in the Reference Section to find the difference between the premiums for the two kinds of policies at age 25. To obtain the premiums for a $50,000 policy, she divided each table entry by 2.

   Whole life premium − term premium = annual saving
   $$459 \qquad - \qquad 103.50 \qquad = 355.50$$

   Next, Eleanor found how much an annual investment of $355.50 will grow to in five years compounded annually at $8\frac{1}{2}\%$. She used the formula for the future value of a periodic investment.

   $$A = \frac{p[(1 + r)^n - 1]}{r}$$

   $$A = \frac{355.50[(1.085)^5 - 1]}{0.085} \qquad p = 355.50, r = 0.085, n = 5$$

   $$A = 2106.47$$

   The annual payments will grow to about $2106.

3. From the table entitled Accumulated Cash Value of $100,000 Whole Life Policy, Eleanor found that the cash value of the policy would be worth about $3100 ÷ 2, or $1550 at the end of 5 years. By buying a term policy instead of a whole life policy and investing the difference at $8\frac{1}{2}\%$, Eleanor's grandfather would have accumulated

   $$2106 - 1550$$

   or an extra $556 at the age of 30 to reinvest tax-sheltered for another 30 or 35 years; that is, until he retired. When he renewed his term policy at age 30, 35, 40, and so on, his premium would gradually rise. This would reduce the size of the premium savings that he realized in the early years.

   If Eleanor's grandfather had bought term insurance instead of whole life, he would have been better off *provided* that he could be reasonably sure of averaging an $8\frac{1}{2}\%$ return or better. Otherwise, the whole life option might have been a better choice after all.

TRY YOUR SKILLS

Use the formula for the future value of a periodic investment to answer Exercises 1–4.

1. If you begin to contribute $2000 a year at age 45 to an IRA account that you expect to pay 8% per year, how much will you have in your account at age 65?

2. If you begin contributing the same amount at age 35, how much will you have at age 60?

3. Suppose you purchase $100,000 worth of insurance when you become 25. Use the Comparison Table for Term and Whole Life Insurance in the Reference Section to find the premium per year. Then calculate how much you will pay in premiums in 40 years.

4. If you invested the amount of the insurance premium in Exercise 3 in an annuity at a return of $8\frac{1}{2}\%$, how much would your annuity be worth at the end of 40 years?

Use the table in this lesson entitled Accumulated Cash Value of $100,000 Whole Life Policy Issued at Age 25 to find the cash value of a $100,000 policy at the following ages. The policy is bought at age 25.

5. 30   6. 35   7. 40   8. 42

Use the Comparison Table for Term and Whole Life Premiums in the Reference Section, the table entitled Accumulated Cash Value of $100,000 Whole Life Policy in the Reference Section, and the formula for the future value of a periodic investment to answer Exercises 9–14.

9. What is the annual premium for a $100,000 five-year renewable term life insurance policy for a 30-year-old person?

10. What is the annual premium for a $100,000 whole life insurance policy for a 30-year-old person?

11. How much money is saved each year if a 30-year-old person buys the term policy instead of the whole life policy?

12. Use the formula for the future value of a periodic investment to find how much tax-deferred money a 30-year-old person could accumulate between the ages of 30 and 35 by buying a $100,000 term policy instead of a whole life policy and then investing the money saved because of the lower premium in an IRA that gives an annual return of $7\frac{1}{2}\%$.

13. What would be the cash value of the whole life insurance policy for a 30-year-old person after 5 years?

14. Compare the cash value of the insurance policy with the value of the IRA at the end of 5 years. Which is the more profitable investment?

EXERCISE YOUR SKILLS

KEY TERMS

annuity
401(k) plan
future value of a
   periodic
   investment
Individual
   Retirement
   Arrangement
   (IRA)
Keogh plan
low-load insurance
   policies
Simplified
   Employee
   Pension Plan
   (SEP)
tax shelter

1. Why should people invest in retirement plans such as tax-deferred annuities and Individual Retirement Arrangements?

2. What are two advantages of purchasing tax-deferred annuities through payroll deductions?

3. Under what circumstances may a person not defer the taxes on his or her IRA contribution?

Use the formula for the future value of a periodic investment to answer Exercises 4–6.

4. Suppose that at the age of 35 you begin contributing $2000 a year to an IRA account that you think will pay an average of 10% a year. How much can you expect to have in your account at age 65?

5. If you begin contributing the same amount at age 35, how much more will you have at age 65 than if you had begun at age 45?

6. If you begin contributing the same amount at age 35, how much more will you have at age 65 than if you had begun at age 40?

7. Suppose you purchase $100,000 worth of whole life insurance when you become 30. Use the Comparison Table for Term and Whole Life Insurance in the Reference Section to find the premium per year. Then calculate how much you will pay in premiums over 35 years.

Suppose that you wanted to know how well you would do at the end of 35 years if you had invested the insurance premium in Exercise 7 in an annuity at a return of 9% per year.

8. Use the formula for the future value of a periodic investment to find the value of the annuity in 35 years.

9. Explain why the results of Exercise 8 do not necessarily imply that the annuity investment is the better choice.

Use the Comparison Table for Term and Whole Life Premiums in the Reference Section, the table entitled Accumulated Cash Value of $100,000 Whole Life Policy in the Reference Section, and the formula for the future value of a periodic investment to answer Exercises 10–13.

10. How much money is saved each year if a 35-year-old person buys a term policy instead of a whole life policy?

11. Determine how much tax-deferred money a 35-year-old person could expect to accumulate between the ages of 35 and 40 by buying a $100,000 term policy instead of a whole life policy and investing the money that is saved because of the lower premium in an IRA that will earn $7\frac{3}{4}\%$ per year.

12. What would be the cash value of the whole life insurance policy for a 35-year-old person after 5 years?

13. Compare the cash value of the insurance policy with the value that the person expects the IRA to have at the end of 5 years. Which seems to be the more profitable investment?

14. Under what circumstances would it be better for a young person to choose a term policy instead of a whole life policy and then use the money saved because of the lower premium to invest into an annuity or IRA?

15. Under what circumstances would it be better for a young man or woman to purchase a whole life policy rather than a term policy, even though the premium money that would be saved might be more profitably invested in an annuity or IRA?

MIXED REVIEW

Each of two real-estate salespeople sold a house for $180,000. Find the commission earned by each person.

1. Rafael gets a straight 5% commission on the entire sales price.

2. Sally gets a $4\frac{1}{2}$% commission on the first $100,000 and a 6% commission on any amount over $100,000.

3. Benjamin Franklin once remarked, "Money makes money, and the money that money makes makes more money." In two words or fewer, what money concept was Franklin referring to?

4. For an all-purpose credit account, set up a chart to show the first 3 monthly payments, and find the total interest paid over 3 months at 1.5% interest per month. The beginning balance is $720, and you make monthly payments of 10% of the amount owed to the nearest dollar. During the second month you make an additional purchase of $125.

5. In the Larsen family, $190 is spent on the car payment, $400 on credit card payments, $980 on utilities and rent, $760 on food, and $1,170 on everything else. To the nearest tenth of a percent, what percent of their take-home pay is the family spending on credit payments?

6. The probability that a person will be alive in 1 year is 0.99894. What is the probability that the person will die within one year?

7. Ralph is considering renting a computer on a rent-to-own credit plan that allows him to rent the computer for $90 per month for 24 months. If he decides to buy the computer, the rental fees will be applied to cover the entire purchase price. How much will the computer cost him under this plan?

8. Suppose that Ralph purchases the computer in Exercise 7 for $1,800 with an installment loan with monthly payments at 12% annual interest over 2 years. Is this a better deal than the rent-to-own plan? Explain your answer.

1. Why is it important for people to have adequate life insurance?

2. How does a person decide whether to purchase term insurance or cash-value insurance?

3. Why might young families be wise to purchase decreasing term insurance?

4. Why do people need to supplement their Social Security pensions with their own retirement plans?

Use the Multiples-of-Salary Chart in the Reference Section to find the amount of life insurance that each of these income earners should buy.

5. Juanita is 35 years of age and has gross earnings of $40,000. She needs 60% income replacement.

6. Joshua is 45 years of age and has gross earnings of $30,000. He needs 75% income replacement.

Use the Comparison Table for Term and Whole Life Premiums in the Reference Section to find the yearly premiums for the amounts of insurance given below.

7. $50,000 whole life insurance at 25 years of age

8. $150,000 term insurance at 45 years of age

9. $200,000 term insurance at 40 years of age

10. $250,000 whole life insurance at 35 years of age

John Andrews is considering the purchase of a one-year term policy with a face value of $90,000. The partially completed table below shows how an insurance company evaluates John's ability to remain alive for one year.

COMPANY'S GAIN OR LOSS ON ONE POLICY (before deducting expenses)		
Possible Outcome	Probability of Outcome	One-Year Gain/Loss
John Andrews lives.	0.99906	x
John Andrews dies.	?	$x - 90,000$

11. Find the probability that John Andrews will die in the next year.

12. The company has direct and indirect expenses of $20 for each policy it issues. Find the premium that the company must charge in order to break even; that is, neither to make nor to lose money on this policy.

Use the table entitled Expected Deaths per 100,000 Alive at Specified Age on page 331 to find the break-even premium for a one-year term insurance policy for each indicated individual. In each case, assume that the direct and indirect expenses for issuing one policy are $20.

13. An $80,000 policy for a man or woman of 28

14. A $250,000 policy for a man or woman of 20

Use the table entitled Expected Deaths per 100,000 Alive at Specified Age on page 331. What profit does an insurance company expect to make for each one-year term insurance policy? Assume that the direct and indirect expenses for each policy are $20.

15. Face value: $60,000; age of insured: 21; annual premium: $120

16. Face value: $175,000; age of insured: 19; annual premium: $310

Use the formula for the future value of a periodic investment to answer Exercises 17–18.

17. At the age of 30, André begins to contribute $2000 a year to an IRA account that he expects to pay 11% a year. How much does he expect to have in the account at the age of 42?

18. If André had begun his IRA contributions at the age of 20, how much more would he expect to have in the account at age 42 than he actually expects because of his beginning at the age of 30?

In Exercises 19–20, use the Comparison Table for Term and Whole Life Premiums on page 326, and the formula for the future value of a periodic investment.

19. Miguel purchases $150,000 worth of whole life insurance when he is 35 years old. How much will he pay in premiums over 30 years?

20. If Miguel had invested the amount of the insurance premium in Exercise 19 in an annuity that gave him an annual compounded return of 7% interest, how much would the account be worth at the end of 30 years?

Use the Comparison Table for Term and Whole Life Premiums in the Reference Section, the table entitled Accumulated Cash Value of $100,000 Whole Life Policy in the Reference Section, and the formula for the future value of a periodic investment in Lesson 7–3 to answer Exercises 21–24.

21. How much money does a 40-year-old person save each year if he buys the term policy instead of the whole-life policy?

22. Determine how much tax-deferred money a 40-year-old person could accumulate between the ages of 40 and 45 by buying a $100,000 term policy instead of a whole life policy and investing the difference between the two premiums in an IRA account that is expected to return $7\frac{1}{2}$% per year.

23. What would be the cash value of the whole life insurance policy for a 40-year-old person after 5 years?

24. Compare the cash value of the insurance policy with the value of the IRA at the end of 5 years. Which is the more profitable investment?

1. Melanie is 45 years old and has gross earnings of $65,000. Use the Multiples-of-Salary Chart for Net Income Replacement on page 325 to find the amount of life insurance that she should buy. She needs 60% income replacement.

2. Use the Comparison Table for Term and Whole Life Premiums on page 326 to find the yearly premium for a $100,000 whole life insurance policy purchased at the age of 20.

3. Use the table entitled Expected Deaths per 100,000 Alive at Specified Age on page 331 of Lesson 7–2 to find the break-even premium for a one-year $150,000 term insurance policy for a man or woman of 18. Assume that the direct and indirect expenses for issuing one policy are $20.

Use the table entitled Expected Deaths per 100,000 Alive at Specified Age on page 331. What profit does an insurance company expect to make for each one-year term insurance policy? Assume that the direct and indirect expenses for each policy are $20.

4. Face value: $90,000
   Age of insured: 20
   Annual premium: $170

5. Face value: $350,000
   Age of insured: 29
   Annual premium: $980

Use the formula for the future value of a periodic investment to answer Exercises 6 and 7.

6. At the age of 25, Marie began contributing $1500 a year to an annuity paying 10.5% annually. How much will she have in the account at the age of 40?

7. If Marie had begun her annuity contributions at the age of 20, how much more would have been in the account at age 40 than was there because she began at the age of 25?

Use the Comparison Table for Term and Whole Life Premiums in the Reference Section, the table entitled Accumulated Cash Value of $100,000 Whole Life Policy in the Reference Section, and the formula for the future value of a periodic investment to answer Exercises 8–11.

8. How much money is saved each year if a 45-year-old person buys the term policy instead of the whole life policy?

9. Find how much tax-deferred money a 45-year-old person could accumulate between the ages of 45 and 50 by buying a $100,000 term policy instead of a whole life policy and investing the difference between the two premiums in an IRA at 8% per year.

10. What would be the cash value of the whole life insurance policy for a 45-year-old person after 5 years?

11. Compare the cash value of the insurance policy with the value of the IRA at the end of 5 years. Which is the more profitable investment?

CUMULATIVE REVIEW

A company has the following fringe benefits policy:

Employee	Salary	Fringe Benefits Policy
Bank vice-president	$51,000	7.65% of gross for FICA taxes
		3% of gross for retirement
		$160 per month for life and health insurance
		25 nonworking days

1. Find the amount that the company must pay annually for extra taxes, retirement, life and health insurance, and other indicated fringe benefits.

2. Find how much salary is paid for the indicated number of nonworking days in a total of 240 operating days.

3. Frank's earnings as a real-estate salesman are based on a graduated commission. He receives 3.5% on each sale up to $150,000 and 5% of the amount above $150,000. How much would he earn for selling a $215,000 house?

4. Suppose you wrote 28 checks in June. Your bank charges $0.03 per check for the first 20 checks and $0.15 for each check over 20. How much were your June bank charges?

5. What would be the interest on $950 at 9.5% for one year on an investment that is compounded semiannually?

6. Find the total financed price, the total interest, and the dealer's profit for each car-sale plan for a car that is priced at $12,900.

	Plan A	Plan B
Rebate	0	$1500
Rate	4.5%	10.5%
Time	48 months	48 months
Monthly payment		
Total financed price		
Total interest		
Dealer's profit		

7. Suppose that your credit card charges 1.75% interest per month. You pay 10% of the balance to the nearest dollar each month. Complete a table using the following columns for 3 months. Your beginning balance is $690.

Month	Balance	Interest	Amount Owed	Payment

8. Use the Comparison Table for Term and Whole Life Premiums in the Reference Section to find the yearly premium for $200,000 of whole life insurance purchased at the age of 35.

PROJECT 7-1: Comparing Types of Available Insurance

Work with a group of your classmates to complete this activity.

1. Each member of the group should choose two insurance companies that sell policies in your area.
2. Find out what types of life insurance policies each company sells.
3. Obtain a list of the premiums for each type of policy.
4. Create a bulletin board to display your findings.
5. Discuss your findings with the other members of your group.
6. Evaluate the various policies and premiums. Decide which company and which policy provide the best investment at the lowest cost.
7. Prepare a brochure stating which policy is the best and why. Make the brochure colorful and eye-catching so it might be used as an advertisement for the policy.

PROJECT 7-2: How are Insurance Premiums Determined?

1. Contact various insurance companies to find out what statistics they use to determine the premiums for different types of life insurance.
2. Contact people (actuaries) who compile the statistics used by insurance companies to determine premiums. Find out how they compile the statistics.
3. Write a report summarizing your findings.
4. Use reference books such as a world almanac to study the changes in the life expectancy of men and women in the last century.
5. Compare the changes in life expectancies with the changes in life insurance premiums over the years for which you can find figures. How do they compare?

A *scatter plot* is a graph of several points. Four scatter plots are shown below. A *line of best fit* is superimposed on each graph.

A *line of best fit* is the straight line that is closer to all the points than is any other line you might draw. The *correlation coefficient r* is a number between −1 and 1 that indicates how good the fit is. If *r* is 1 or −1, then the fit is perfect. If *r* is close to 1 or −1, the fit is good. If *r* is 0 or near 0, then the fit is poor.

Graph the table of values at the right as a scatter plot. Then graph the equations of Exercises 1–3 over the scatter plot and tell whether you think the fit of each line is *poor, good,* or *excellent.*

x	0	1	3	5	7
y	0	2	4	6	7

1. $y = 0.5x + 2$ **2.** $y = -0.2x - 2.5$ **3.** $y = x + 0.5$

A graphing calculator can be used to find an equation of a line that best fits a scatter plot. The calculator will give you values for *a* and *b* in the *linear-regression equation* $y = a + bx$. It will also give you the correlation coefficient for the line.

Write an equation for a linear-regression equation that corresponds to the given calculator display. Round the constants to the nearest hundredth.

Example $a = -0.136986$ **Solution** $y = 3.26x - 0.14$
               $b = 3.260274$
               $r = 0.973955$

4. $a = 0.486765$ **5.** $a = 1.186441$
     $b = 2.242647$ $b = -3.067797$
     $r = 0.837326$ $r = -0.993112$

6. $a = 0.612805$ **7.** $a = 4.254237$
     $b = 0.484756$ $b = 1.271186$
     $r = 0.474731$ $r = 0.159854$

8. In Exercises 4–7, which line is the best fit for its scatter plot? Which line is the poorest fit?

9. Use a graphing calculator to find the linear-regression equation for the scatter plot of Exercises 1–3.

Investments

IN STOCKS AND BONDS

WALL STREET IS MORE THAN A PLACE, FOR MANY OF us, Wall Street represents wealth. It is an arena in which millions of dollars circulate every day. Portions of companies, even whole companies, are bought and sold to other companies and to wealthy investors, and the money flashes by as numbers on computer screens.

Many companies, however, are also owned by thousands of individual stockholders, including wage-earning investors. Companies with over 100,000 stockholders include Westinghouse, Atlantic Richfield, Eastman Kodak, Texaco, Chrysler, Occidental Petroleum, Mobil, Ford, Sears, and Southwestern Bell, to name a few.

Some of the students whom we have met in earlier chapters will investigate various aspects of stock investments. They will study the differences among stocks, bonds, and mutual funds and how these investments compare with money that is kept in a savings account. They will also examine some of the ways to avoid risky investment schemes.

Jeff owns some stock that his grandfather bought for him when Jeff was three years old. Granddad retired seven years ago from a large engineering firm that offered stock ownership in the company as part of the employee benefit plan. Granddad had also been able to buy some of the shares for Jeff as a way of saving toward his college costs.

Jeff didn't know much about stocks when Granddad gave him the stock certificates on his 16th birthday, but now that the time is approaching for him to go to college, he has become more interested. Four times a year, the engineering company sends him payments that his father told him are called *dividend checks*. Although the amounts have varied over the years, Jeff notices that they have added significantly to his new checking account.

Stocks are not Jeff's only investment. Ever since he was five years old, Jeff's parents have bought him a $50 U.S. savings bond every month. Jeff is not sure that these investments will cover the high costs of his college education. He is asking himself questions such as the following.

Will he need to continue working this summer tutoring algebra?

Should he find a job that pays more than tutoring?

Should he cash in the savings bonds and invest the money in mutual funds?

Jeff has become interested in mutual funds because his father owns some shares in a mutual fund that he buys through an employee benefit plan. Dad says he is more than willing to let the mutual fund manage his investments, since the daily and weekly changes in the world of business and finance are so varied and complex that he prefers not to be bothered with them.

OBJECTIVES: *In this lesson, we will help Jeff to:*

- *Examine several types of investments, including stocks, bonds, and mutual funds.*
- *Calculate the number of bonds or shares of stock an investor can buy at a specified price.*
- *Read and understand daily stock and bond transactions in the newspaper.*

STOCKS

Jeff's stock certificates, called simply **shares** or **stock,** represent ownership in a **corporation,** a form of business organization that people beginning a business very often choose because it limits the liability of its owners, the **shareholders.** If the corporation is successful, Jeff will take part in its success through **dividends** (the periodic payments that a corporation distributes to its stockholders). The dividends from Granddad's company are declared four times a year.

Granddad and Jeff may also make money on their stock through an increase in the market value of the shares of stock. Such an increase in value, called a **capital gain,** will normally occur if a company increases its profits. On the other hand, if the corporation is not successful, it may have to reduce or eliminate its dividends, and the value of the stock may decrease. Then all of the shareholders such as Granddad and Jeff could even lose part or all of their investment.

A corporation always issues *common stock* (like Jeff's) and may also issue *preferred stock*.

Common Stock The owners of **common stock** of a corporation are entitled to elect a board of directors and to receive any dividends that the board declares. However, a corporation's board can choose not to declare a dividend. This may occur if a corporation has poor earnings. Sometimes, even a profitable corporation will pay little or no dividends in order to have funds for research or expansion. In that case, shareholders will expect to see the value of their shares increase to compensate for the lack of dividends.

Preferred Stock Like owners of common stock, owners of **preferred stock** have a share in the ownership of a corporation. However, they usually do not have a right to vote in the election of corporate directors. Unlike the holders of common stock, they are entitled to a specified, fixed dividend. If the company encounters financial difficulty, all preferred shareholders must be paid before any dividends are declared on the common stock.

Newspaper Stock Listings Newspapers print closing stock and bond prices for the previous trading day along with other information about the stock or bond, such as the volume of trading. A typical line from a stock table is shown below in simplified form. It shows how the stock price for Growth Company of America performed the previous day.

52-Week							
High	Low	Stock	Div	High	Low	Last	Change
$35\frac{3}{4}$	$19\frac{3}{4}$	GrowCo	1.10	$22\frac{1}{8}$	$21\frac{5}{8}$	22	$+\frac{1}{4}$

The abbreviated name of this company, GrowCo, appears in column three. The first two columns show the highest and lowest prices at which GrowCo traded during the previous 52 weeks ($35.75 and $19.75, respectively). After the name of the stock comes the annual per-share dividend of $1.10. Yesterday's high and low price are shown next. Finally, the day's closing price appears along with the price change for the day. In this case the stock price for GrowCo varied between $21\frac{5}{8}$ and $22\frac{1}{8}$ ($21.625 and $22.125), ending the trading session at $22; this was up $\frac{1}{4}$, or 25 cents, from the previous day's closing price of $21.75.

BONDS

When you buy a **bond,** you are lending money to either a corporation or a government body, such as a city, state, county, or independent agency. The **face value of a bond** is the original loan amount that appears on the bond, usually a multiple of $1000. In return for your money you get three things:

1. The bond issuer's promise to pay you back the entire face value
2. A maturity date, which is the earliest date when the bond will be redeemed
3. A promise to pay interest at a specified interest rate on the face value

For example, if you buy a "$1000 Joy $8\frac{1}{8}$ 14" 20-year bond from Joy Motor Company in 1994, you will be paid an annual interest rate of $8\frac{1}{8}\%$, or $81.25 per $1000, until the year 2014. On the date of maturity (sometime in 2014), Joy will pay back the $1000. If you decide to sell the bond before the maturity date, it will probably be worth either more or less than the face value, depending on the prevailing interest rate for all similar bonds at the time you choose to sell.

Corporate Bonds   Private companies issue **corporate bonds** to raise money for plant expansion or other company purposes. If a corporation's earnings are not great enough to cover its obligations to both stockholders and bondholders, then the bondholders have priority. All bond interest must be paid before any dividends can be paid to the stockholders.

Municipal Bonds   States, cities, counties, school districts, and other governmental bodies issue **municipal bonds** to raise money for schools, hospitals, streets, and so forth. They are usually partially or totally exempt from local and federal income taxes.

U.S. Savings Bonds   Jeff's parents have purchased over $6000 in **U.S. savings bonds** for him. Like many other Americans, they take advantage of the payroll savings plan that allows individuals to buy a bond each month through a payroll deduction.

U.S. savings bonds are backed by the power of the federal government to tax and the country's faith in the government to pay its debts. However, such bonds are sometimes criticized because bondholders have to pay for their exceptional safety by accepting a relatively low rate of interest. If a savings bond is lost, stolen, or destroyed, the bondholder may be able to have it replaced by the government if he or she can supply evidence of actually owning the bond.

U.S. savings bonds have one additional important advantage; there are no transaction costs such as commissions to a broker. Unless you are purchasing a large number of bonds that you plan to hold for many years, the large commission that you pay may severely reduce or even wipe out the interest that you get from a bond that is not a savings bond.

MUTUAL FUNDS

The company at which Jeff's father works assists its employees in buying mutual funds. A **mutual fund** is a means of pooling funds with thousands of other people to acquire a wide variety of stocks, bonds, and other types of investments. This way of investing appeals particularly to small investors, such as Jeff's father, who do not have enough money to **diversify** (buy a well-balanced assortment of investments) or enough knowledge to make sound stock and bond investment decisions.

When investors buy shares from a mutual fund company, the company uses that money to purchase stocks, bonds, and similar **securities.** Thus, instead of investing $1000 in a particular corporation, Jeff's father can buy $1000 worth of shares in a mutual fund and have a stake in perhaps 100 or more different corporations.

BEFORE YOU INVEST . . .

Jeff wants to invest in stocks either directly or indirectly through a mutual fund. However, he also remembers something his grandfather told him: "Buying stocks or mutual fund shares should be part of your long-term investment program, but first be sure that you have a savings account or a reputable money-market account that equals at least six months of your living expenses." Jeff has a savings account at his local bank but was unfamiliar with money-market funds. Jeff's father told him that a **money-market account** was similar to a NOW account that combines the features of a savings and checking account.

A money-market account can be opened with a mutual fund group or with a bank. The interest received on a money-market account is higher than on a savings account, but the account usually has certain restrictions such as requiring any withdrawal to be at least $500 or limiting the number of withdrawals that may be made in a month. If you open such an account at a bank insured by the Federal Deposit Insurance Corporation (FDIC), your account will be insured up to $100,000. However, the interest will be less than you get from a mutual fund group because the money-market fund at the mutual-fund group is not insured by the federal government. These uninsured money-market funds are generally regarded as reasonably safe, but you should investigate the background of the mutual fund group before sending it any money.

Jeff decided to follow his Granddad's advice. In addition to the savings account he already has, he opened a money-market account at the bank and will bring his combined savings in the two accounts to a half year of living expenses before he invests in stocks or a mutual fund.

Ask Yourself

1. What is your relationship to a company when you buy its stock?

2. What is your relationship to a company when you buy its bond?

3. What is the main difference between buying a mutual fund and buying a stock?

ALGEBRA *REVIEW*

Solve for x.

1. $2650 = 42.50x$

2. $19.875x = 1700$

3. $25.375x = 8764$

4. $112\frac{1}{2} = x + 108\frac{3}{4}$

5. $x - \frac{7}{8} = 5\frac{1}{4}$

6. $x + 34\frac{5}{8} = 35\frac{1}{2}$

SHARPEN YOUR SKILLS

Stock prices are quoted in fractional parts of a dollar. This means that

$$\$0.25 = \frac{1}{4} \qquad \$0.50 = \frac{1}{2} \qquad \$0.75 = \frac{3}{4}$$

To determine what $\frac{1}{8}$, $\frac{3}{8}$, $\frac{5}{8}$, and $\frac{7}{8}$ are equivalent to in dollars, divide the numerator by the denominator.

$$\frac{1}{8} = 0.125 \qquad \frac{3}{8} = 0.375 \qquad \frac{5}{8} = 0.625 \qquad \frac{7}{8} = 0.875$$

EXAMPLE 1 Jeff would like to cash in his U.S. savings bonds. He has $2200 in savings bonds that have matured.

QUESTION How many shares could he buy of a stock that is selling for $33\frac{3}{8}$ dollars per share? Disregard transaction costs such as commissions.

SOLUTION
The price of one share of the stock is $33\frac{3}{8}$, or $33.375. To find how many shares of stock Jeff can buy, divide the amount that he has ($2220) by the price per share ($33.375).

$$2200 \div 33.375 = 65.917603$$

Jeff knows that he cannot buy a part of a share of stock, so he drops the decimal portion of the number. Jeff can buy 65 shares of the stock (before transaction costs). The shares will cost $65 \cdot 33.375 = \$2169.38$. Notice that you must use the fractional portion of a cent to find the cost that is rounded to the nearest cent.

Bond prices that appear in the newspaper bond tables are somewhat misleading, since they show the true price *divided by 10*. Thus a price of $104\frac{1}{2}$, or 104.5, actually refers to a bond that has a market price of $1045.

EXAMPLE 2 The bond table in the newspaper shows that a high quality corporate bond is selling at $104\frac{1}{2}$.

QUESTION How many corporate bonds could Jeff buy with the $2200 from his U.S. savings bonds? Disregard transaction costs such as commissions.

SOLUTION

To find how many bonds Jeff can buy, divide the amount that he has by the price of one bond.

$$2200 \div 1045 = 2.105 \qquad 104\tfrac{1}{2} \cdot 10 = 1045$$

Jeff cannot buy a part of a bond, but he can buy two bonds. The two bonds will cost $2 \cdot 1045 = \$2090$. Jeff probably would not purchase the bond because of the high commission cost.

――――――――― **SKILL 3** ―――――――――

EXAMPLE 3 Jeff has been reading the financial section of the newspaper to learn how the prices of stocks are reported. Following is a partial listing of stock transactions for one day.

52-Week							
High	**Low**	**Stock**	**Div**	**High**	**Low**	**Last**	**Change**
$53\tfrac{1}{4}$	$39\tfrac{1}{2}$	BlcInc	1.25	$41\tfrac{1}{2}$	$40\tfrac{3}{4}$	41	$-\tfrac{1}{4}$
163	$135\tfrac{1}{4}$	BxCo	2.00	$137\tfrac{1}{4}$	$136\tfrac{3}{4}$	$137\tfrac{1}{4}$	$+\tfrac{3}{4}$
$4\tfrac{3}{4}$	$2\tfrac{1}{4}$	Maldmer		$2\tfrac{5}{8}$	$2\tfrac{1}{2}$	$2\tfrac{5}{8}$	$+\tfrac{1}{8}$
$12\tfrac{3}{4}$	$7\tfrac{5}{8}$	TemCont	0.75	$11\tfrac{1}{8}$	11	$11\tfrac{1}{8}$	$+\tfrac{1}{4}$

QUESTIONS

1. What was the closing price for TemCont?
2. What was the highest price at which BlcInc traded?
3. How much did BxCo stock vary during trading?
4. How does BxCo's closing price compare with its closing price for the previous trading session?
5. What was the closing price of BxCo at the end of the previous trading session?
6. What is the annual dividend per share paid by BlcInc?
7. What would your annual dividend income be from 100 shares of BlcInc?
8. Which companies closed at their high prices for the day? Which did not?
9. What is the highest price at which Maldmer has traded in the previous 52 weeks? What is the lowest price?

SOLUTIONS

1. $11\frac{1}{8}$ (or $11.125)

2. $41\frac{1}{2}$ (or $41.50)

3. $137\frac{1}{4} - 136\frac{3}{4} = \frac{1}{2}$

4. $\frac{3}{4}$ of a point (75 cents) higher

5. $137\frac{1}{4} - \frac{3}{4} = 136\frac{1}{2}$

6. $1.25

7. $100 \cdot 1.25 = \$125$

8. TemCont, BxCo, and Maldmer closed at their high for the day; BlcInc did not.

9. $4\frac{3}{4}$ is the highest price at which Maldmer has traded in the previous 52 weeks; $2\frac{1}{4}$ is the lowest.

In this lesson, the effect of transaction costs has been ignored. These include transfer fees, which are usually very small, and the commission that you pay the broker when you buy or sell. The effect of commissions on your stock price is considered in the next lesson.

TRY YOUR SKILLS

Use the following prices per share to find the number of shares of stock you can afford to purchase if you have $50,000 to invest in each of the following companies. Also, find the total cost of the shares. Remember that you cannot purchase part of a share of stock. Ignore transaction costs such as brokerage commissions.

	Company	Price per Share	Number of Shares	Total Cost
1.	AdoreMin	$6\frac{1}{2}$	7692	
2.	ATR	$24\frac{7}{8}$		$49,998.75
3.	AcnInc	$9\frac{1}{8}$		
4.	AcmeAl	$10\frac{3}{4}$		

5. How many bonds can you buy with $50,000 if the price of a bond is $97\frac{1}{2}$?

6. AdoreMin's closing price for the day is $6\frac{5}{8}$, $\frac{1}{2}$ point lower than yesterday. What was yesterday's closing price?

7. The annual dividend for ATR is $1.33 per share. What is the dividend income on 490 shares?

EXERCISE YOUR SKILLS

KEY TERMS

bond
capital gain
common stock
corporate bonds
corporation
diversify
dividends
face value of a bond
money-market
   account
municipal bonds
mutual fund
preferred stock
securities
shares
shareholders
stock
U.S. savings bonds

1. What is the risk involved in buying stocks?

2. Who is more likely to realize a capital gain, the owners of common stock or the owners of preferred stock? Why?

3. For what purposes might your city or town raise money by issuing municipal bonds for sale?

Use the following prices per share to find the number of shares of stock you can afford to purchase if you have $50,000 to invest in each of the following companies. Also find the total cost of the shares. Ignore transaction costs such as brokerage commissions.

	Company	Price per Share	Number of Shares	Total Cost
4.	Aber&Son	56		
5.	AceText	$8\frac{7}{8}$		
6.	AmerScCo	$10\frac{5}{8}$		
7.	AmerTool	90		
8.	AyerHrd	$3\frac{1}{4}$		
9.	Babbage	$74\frac{1}{2}$		
10.	BLTInc	$53\frac{5}{8}$		

11. Jeff has $3400 that he can invest in either a growth stock that is trading at $70 per share or a high-quality bond that pays annual interest of 5.8% and matures in the year 2010. Which would be a more suitable investment for him? Explain your answer.

Use the following portion of a table of stock transactions to answer Exercises 12–20.

52-Week							
High	Low	Stock	Div	High	Low	Last	Change
$57\frac{3}{4}$	$40\frac{3}{4}$	LollPp	1.42	$41\frac{3}{8}$	41	$41\frac{1}{4}$	$-\frac{1}{8}$
$49\frac{7}{8}$	$29\frac{3}{8}$	LukInd	0.97	30	$29\frac{5}{8}$	$29\frac{5}{8}$	——
$91\frac{1}{8}$	67	MamaMi	2.24	$81\frac{3}{4}$	$80\frac{1}{4}$	$81\frac{1}{4}$	$-\frac{1}{4}$
24	$15\frac{5}{8}$	NajiCo	0.46	$17\frac{3}{4}$	$17\frac{3}{8}$	$17\frac{5}{8}$	$-\frac{1}{8}$
$36\frac{3}{8}$	22	NolInc	0.64	$24\frac{1}{4}$	$23\frac{7}{8}$	$24\frac{1}{4}$	$+\frac{1}{4}$

12. Which stock finished up for the day?

13. Which stock closed at the same price at which it closed at the previous trading session?

14. How much did NajiCo vary during the day?

15. Which stock varied by exactly $1\frac{1}{2}$ during the day?

16. What is the difference between the closing price of NajiCo for the day and the closing price for the previous trading session?

17. By how much does LollPp's closing price differ from its high for the last 52 weeks?

18. Which stock closed exactly 2 points above its low for the last 52 weeks?

19. If you had 100 shares of NolInc, how much would you receive in dividends in 1 year?

20. Which stock pays the highest dividend?

MIXED REVIEW

A restaurant is advertising for a cashier in the local newspaper. The job is for an 8-hour day, 5 days per week. There is a 2-week paid vacation each year. Find the weekly and yearly earnings for the given hourly rate.

1. $7.50   2. $8.25   3. $9.00

4. Find the simple interest earned in 3 years on $890.00 at the annual rate of 7%.

Use a spreadsheet program to fill in the missing entries from the table below that shows the activity in Victor's VISA account over a period of 3 months. The account carries a monthly interest charge of 1.5% and Victor makes monthly payments of 10% of the amount owed to the nearest dollar.

	Month	Purchases	Balance	Interest	Amount Owed	Payment
5.	1	0.00	$624.50			
6.	2	$92.00				
7.	3	0.00				

8. Find the total interest paid in the 3-month period.

9. Alice is considering renting a piano on a rent-to-own credit plan that allows her to rent the piano for $200 per month for 24 months. If she decides to buy the piano, the rental fees will be applied to cover the entire purchase price. How much will the piano cost under this plan?

10. Suppose that Alice purchases the piano of Exercise 9 for $4,000 on an installment loan with monthly payments at 10% annual interest over 2 years. Is this a better deal than the rent-to-own plan? Explain.

Use the Income Tax Withholding Table in the Reference Section to find the take-home pay for the given wages and withholding allowances. Assume that the person is single. The FICA withholding is 7.65% of gross pay.

	Monthly Salary	Withholding Allowances	Take-home Pay
11.	$1540	0	
12.	2750	1	
13.	4130	3	

Maria has continued to save money. Recently, she has been wondering whether she should invest in the stock market or a mutual fund to make more money than she can get in a savings account. Maria has a good friend, Cal, whose father, Clarence Sr., has owned thousands of shares of stock in various companies.

Sometimes Clarence Sr. has done really well in the market, and sometimes he hasn't. Cal still talks about October 19, 1987, known as Black Monday in business circles. The Dow Jones average dropped 508 points that day, losing 22.6% of its value. Cal's father lost money in that crash. The market did rebound eventually and had record highs within a couple of years. But that was too late for Clarence Sr.

As a result of financial difficulties caused by the crash, Cal's family had to sell its large home in the expensive part of town and buy a much smaller house in a suburb farther away from the city. Clarence Sr. also sold his luxury car and bought a used one inexpensively from the car dealership that he owned. Cal himself had to give up his vision of having his own car when he reached 17.

Maria wonders how investing in stocks can cause such an upheaval in a family. She is aware of how stock prices and mutual fund prices can vary but believes that she could choose some fine companies or a good fund to invest in if she had the money to invest. Maria will find out how stocks and mutual funds are bought and sold as well as other things that she needs to know if she is considering investing real money.

* *Discover what factors must be considered before deciding to buy stocks or a mutual fund.*
* *Understand the process that takes place when stocks are traded.*
* *Calculate profit or loss from buying and selling shares of stock.*
* *Understand how transaction costs affect the value of stock or mutual fund investments.*

SHOULD I INVEST IN THE STOCK MARKET?

Clarence Sr. watched his acquaintances acquire large sums of money by "playing the market." Unfortunately for him, somewhere along the line he crossed over the line between *investing* in stocks and *gambling* with money his family needed to secure its financial independence.

Cal's family always knew that stock ownership involved some risk. Perhaps Clarence Sr. was careful at first with his choice of stocks. Then, as those choices paid well and he was able to make bigger profits, he was swept along by the tide of his increasing fortune, and he began to take bigger risks. When the market dropped so suddenly in October 1987, some of the riskier companies lost so much of their value that he was forced to sell many of his holdings at a loss.

It is hard to define the line between investing safely and gambling with money needed for essentials; this line is defined differently in every family. You should be aware that carefully conducted studies have shown that very few people successfully "play the market," that is, very few people can *consistently* buy stocks for a few weeks or months and then sell the stocks at a profit.

On the other hand, if you carefully choose a group of stocks for the fundamental value of the underlying companies or the ability of those companies to sustain long-term growth, you may, *if you hold such stocks over several years,* be able to build up your family's wealth, especially if the overall economy is healthy. In the past, stocks in well-managed companies have also been a good buffer against inflation when held for many decades. Moreover, stocks, unlike some other forms of investment, are easy to liquidate if you need cash.

Clarence Sr. was lucky that he could still afford to buy a house and car. He would say that you should consider all of the following points before you invest in stocks.

1. Be sure you can afford to lose what you risk.
2. Determine specific long-term and short-term investment goals.
3. Ask yourself if you have the emotional temperament suited to handling the ups and downs of investing.
4. Plan to spend time as well as money on the companies you invest in.
5. Choose an experienced, reputable broker to give you guidance and advice.
6. Do not expect too much too soon.
7. After you have set your goals and objectives, stick to them.
8. Be aware that you are buying part of a particular company, not a lottery ticket on the next stock market move.
9. Have an overall family investment plan to protect you from falling into hit-or-miss investing.

THE PROCESS OF BUYING AND SELLING

If Maria decides that she can follow the above guidelines, then she may open an account with a **broker,** that is, a salesperson who specializes in buying and selling stocks and bonds. When Maria buys or sells a stock, she must pay the broker a **commission.** Commission rates vary according to the brokerage house and the size of the trade.

If her brokerage firm is a **full-service firm,** then it will issue reports on individual companies, and its brokers will offer opinions on individual stocks. If she has an account with such a firm, she can get its reports and study them. However, the quality of advice in these reports and from individual brokers will vary from very good to poor. To find a competent broker, she should do what she would do to find a competent physician or lawyer: obtain some names from a person whose advice she trusts. In this case, Maria decides to contact an older friend who has years of investing experience.

If Maria had preferred to do her own research, she could have chosen to open an account with a **discount broker,** who charges much lower commissions than a full-service broker but provides no investment advice at all.

Many investors purchase a hundred shares of a particular stock, which is called a **round lot.** A trade of fewer than 100 shares is called an **odd lot.** Because odd lots can be inconvenient to trade, Maria may have to pay a higher broker commission for an odd lot than for a round lot.

Stocks are traded on national and regional stock exchanges. The largest of these is the *New York Stock Exchange,* where over 84 percent of all listed securities are bought and sold. The *American Stock Exchange* is the other national exchange in the United States. Many stocks trade not on an exchange but rather in the **over-the-counter market,** a national network of dealers and brokers who trade among themselves by telephone, telegraph, or teletype.

Stocks are bought and sold on a security exchange by the auction method. A stock may be offered for sale at a certain price, or someone may bid for the same stock at a lower price. A sale is made when either the buyer or the seller meets the other's price. Maria can instruct her broker to purchase or sell shares of stock for her with a limit on the price that she will accept.

Ask Yourself

1. How does investing in stocks differ from "playing the market"?

2. What are some guidelines that you should follow before you decide to invest in stocks? Which guideline do you think is the most important for you?

3. Why should investors be careful about following the advice of a stockbroker?

ALGEBRA *REVIEW*

One way to think about **absolute value** is to look at the number line. The distance from zero to any point on the number line is positive.

Find each distance.

Examples: From 0 to A is 4
           From 0 to B is 4

1. From 0 to C

2. From 0 to D

3. From 0 to E

4. From 0 to F

Find the absolute value of each change in price.

5. Old price: $100
   New price: $85

6. Old price: $85
   New price: $100

7. Old price: $65x$
   New price: $95x$

8. Old price: $5a$
   New price: $1.5a$

SKILL 1

Commission rates are charged on a percent-of-sale basis. In other words, in addition to the cost of the stocks or bonds, you must pay a percentage of the cost for commission. To determine the cost of a commission, you multiply the number of shares, the price of one share, and the commission rate.

Commission Cost Formula

$c = npr$ where c = cost of commission
n = number of shares
p = price of one share
r = commission rate

EXAMPLE 1   Maria would like to buy 250 shares of International Enterprises, Inc. at $17\frac{3}{4}$ per share. She wants to compare the commissions charged by Osgood and Company, a full-service broker with a commission rate of 2%, and Willcox and Company, a discount broker with a commission rate of 1%.

QUESTION   How much commission will each broker charge, and what will the total cost be?

SOLUTION

Osgood and Company   *Willcox and Company*

$c = npr$ $c = npr$
$c = 250 \cdot 17.75 \cdot 0.02$ $c = 250 \cdot 17.75 \cdot 0.01$
$c = 88.75$ $c = 44.375$

The commission is $88.75.   The commission is $44.38.

The purchase price of the stock is $250 \cdot 17.75 = 4437.50$. Add the commission to $4437.50 to find the total cost.

$88.75 + 4437.50 = 4526.25$ $44.38 + 4437.50 = 4481.88$

The total cost at Osgood is $4526.25, and the total cost at Willcox is $4481.88. Maria will pay $88.75 - 44.38 = \$44.37$ more at Osgood.

If you decide to sell a stock, you will pay a brokerage commission, just as you did when you bought the stock. However, the selling commission and buying commission will normally not be equal since the selling price usually differs from the purchase price.

When you sell stock that you have owned for a period of time, your stock may have increased in value. This increase is called a *capital gain*. Usually, you will have to pay income taxes on the capital gain on a stock that you have sold. If the value of your stock decreases and you sell, you will have a loss.

EXAMPLE 2 Maria would like to know what the capital gain would be on the 250 shares of International Enterprises. Assume that she purchased the stock from the full-service broker.

QUESTION What would be Maria's capital gain if the market price of the stock increases to $72\frac{1}{4}$ over the next 10 years?

SOLUTION
Multiply the new price of one share by the number of shares.

$$72.25 \cdot 250 = 18,062.50$$

The capital gain is found by subtracting the total purchase price from the total value of the stock after 10 years.

$$18,062.50 - 4526.25 = 13,536.25$$

Maria's capital gain after 10 years is $13,536.25. If she does not sell, her capital gain may increase or decrease in value. If she does decide to sell, then the amount that will be subject to tax is the capital gain reduced by both her buying commision and her selling commission.

The rate of increase or decrease is the *change in price* divided by the original price. A formula for the rate of increase or decrease involves *absolute value*. To find the percent of increase or decrease, divide the rate by 100.

Rate-of-Change Formula

$$r = \frac{|P_n - P_o|}{P_o}$$

where r = rate of increase or decrease
P_o = original price
P_n = new price

EXAMPLE 3 Maria would like to know the percent of decrease or increase when the price of a share of stock varies.

QUESTIONS
1. If the market price of a share of stock drops from $45 to $40, what is the percent of decrease?
2. If the market price of a share of stock increases from $45 to $60, what is the percent of increase?

SOLUTIONS

1. $r = \dfrac{|P_n - P_o|}{P_o}$ Use the rate-of-change formula.

$ = \dfrac{|40 - 45|}{45}$ $P_n = 40, P_o = 45$

$ = \dfrac{5}{45}$ $|40 - 45| = |-5| = 5$

$ = 0.11$

The percent of decrease is 11%.

2. $r = \dfrac{|P_n - P_o|}{P_o}$ Use the rate-of-change formula.

$ = \dfrac{|60 - 45|}{45}$ $P_n = 60, P_o = 45$

$ = \dfrac{15}{45}$ $|60 - 45| = |15| = 15$

$ = 0.33$

The percent of increase is 33%.

SKILL 4

Maria has learned that commission rates can vary widely on the purchase of stocks or mutual funds. Some brokers charge as little as 1%, while mutual funds have commissions that vary from 0 to over 8%.

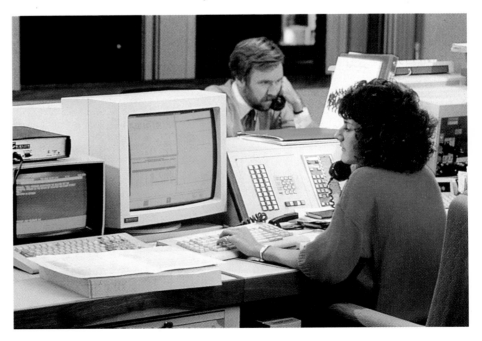

EXAMPLE 4 Maria decided to compare how various commission rates affect an investment of $1000.

QUESTION How does the commission rate affect the purchase price of a stock or mutual fund?

SOLUTION
She wrote an equation for the relationship between the total cost T of a stock or mutual fund purchase and the commission rate r.

Total price = price + commission
$$T = 1000 + 1000r$$

Next, she used a spreadsheet program to calculate the total price T for commission rates r ranging between 0 (no commission) and 8%.

r	0	0.005	0.01	0.02	0.03	0.04	0.05	0.06	0.07	0.08
T	1000	1005	1010	1020	1030	1040	1050	1060	1070	1080

Using the spreadsheet's graphing program, Maria graphed the values with r on the horizontal axis and T on the vertical axis. The graph showed clearly that the commission rate and total purchase price had a *linear* relationship. Whenever the commission rate increases by 1 percentage point, her cost also increases, by $10.

Effect of Commission on Total Price

TRY YOUR SKILLS

Find the cost of these shares of stock before commission.

1. 300 shares at $21\frac{1}{8}$

2. 225 shares at $63\frac{1}{4}$

3. 45 shares at $32\frac{1}{2}$

4. 40 shares at $16\frac{5}{8}$

5.–7. For Exercises 1–3, find the commission and the total cost if the commission rate is 1.5%.

Suppose that the 40 shares of Exercise 4 are held for 5 years. Find the capital gain or capital loss in each case. Ignore the effect of commissions.

8. Market price is $\$42\frac{7}{8}$

9. Market price is $14

10.–11. Find the percent gain or loss in market price for the stocks of Exercises 8 and 9.

Use a spreadsheet graphing program to show how the total cost of a mutual fund purchase is related to the commission rate for commission rates of 0 to 8% when the amount invested is as given.

12. $500

13. $2000

EXERCISE YOUR SKILLS

1. How does temperament affect the way a person handles investments?

2. How should an investor determine how much money to invest in stocks?

3. Why might it be wiser for some people to invest in a completely safe security, such as U.S. savings bonds, rather than stocks?

Find the purchase price of these shares of stock at the prices shown. Ignore the effect of commissions. Find the capital gain or capital loss at the current market value one year later by subtracting the purchase price from the current market value. Note that if the difference is a negative number, then there is a loss. Calculate the percent of increase or decrease. Put parentheses around a decrease.

	Company	Number of Shares	Original Purchase Price/Share	Total Price	Current Market Price/Share	Total Price	Gain or Loss	Percent of Change
4.	AP&P	306	$41\frac{7}{8}$	$12,813.75	$43\frac{3}{8}$		$459 gain	3.6%
5.	Zola	215	$10\frac{1}{4}$		$16\frac{1}{2}$			
6.	ABM	400	$42\frac{1}{2}$		$65\frac{7}{8}$			
7.	Lands	6000	20		$18\frac{1}{4}$			
8.	Dr. Pop	150	$29\frac{3}{8}$		$26\frac{3}{8}$			
9.	Doledo	310	$9\frac{7}{8}$		$17\frac{1}{4}$			
10.	Binbury	55	$65\frac{5}{8}$		$74\frac{1}{4}$			

For the following companies from Exercises 4–10, find the commission cost and the total original purchase cost given the commission rates.

11. 306 AP&P, 2%

12. 215 Zola, 1%

13. 400 ABM, 1.5%

14. 6000 Lands, 0.5%

Maria is reviewing the prospectuses (informational booklets) of two mutual funds: a "load" fund and a "no-load" fund. The load fund charges a commission of 8%; the no-load fund charges no commission. The following table assumes that a $1000 investment in each fund will vary in value as shown over the next 10 years and end up with a value of $4050.

| | | | | Percent Return | |
Elapsed Time	Cost of No-Load Fund	Cost of Load Fund	Current Value of Fund	No-Load Fund	Load Fund
0 years	$1000	$1080	$1000		
$\frac{1}{2}$ year	1000		1100		
1 year	1000		950		
2 years	1000		1300		
4 years	1000		1750		
6 years	1000		2300		
8 years	1000		3200		
10 years	1000		4050		

15. Find the missing table values. Round percents to the nearest whole percent.

16. Draw a line graph that shows the 10-year performance of the no-load fund. Use the x axis for the number of years from 0 to 10, and use the y axis for the percent of gain or loss from 10% loss to 310% gain.

17. Draw a graph that shows the ten-year performance of the load fund on the same x and y axes as the graph of Exercise 16.

18. Which fund had the better 10-year performance? How much better did that fund perform in 10 years?

Find the amount that each company must pay each year for the indicated fringe benefits.

1. Salary: $38,000
   7.65% for FICA taxes
   4% of gross pay for retirement
   $1800 per year for training

2. Salary: $25,000
   7.65% for FICA taxes
   $75 per month for health insurance
   $80 per month for life insurance

Find the new balance B for the indicated investments. Use the compound interest formula $B = P(1 + r)^n$, where P is the initial principal, r is the interest rate for the period, and n is the number of periods.

3. Starting principal: $1000; annual compounding at 8% for 5 years

4. Starting principal: $12,500; semiannual compounding at 7% per year for 12 years

5. Use the monthly payment formula $M = \dfrac{Pr(1 + r)^n}{(1 + r)^n - 1}$ to find the monthly payment for a car loan of $8000 at 9% a year for 4 years. In the formula, r represents the monthly interest rate and n represents the total number of payments.

6. How long will it take to pay off a credit card balance of $2500 with monthly payments of $150 if the APR is 9%? Express your answer as a whole number of months.

The following chart shows the payments in a credit card account during the month of April. The monthly interest rate is 1.5%.

Dates	Payment	Balance at End of Day	Number of Days	Sum of Daily Balances
4/1–4/29	0.00	$406.00	29	$11,774.00
4/30	$95.00	311.00	1	311.00
		Total	30	$12,085.00

7. Determine the average daily balance.

8. Determine the finance charge on April 30.

9. Determine the ending balance for April.

10. Your balance from 1/1 to 1/15 was $450. Then you made a payment of $100 on 1/16. You made no further charges or payments for the remainder of the month. Determine your average daily balance for January. How much did you pay in interest if your card carries a monthly finance charge of 1.75%?

8-3 INVESTMENT PROS AND CONS

During the many months that Maria was saving up for her car, Nelson kept track of her savings account and noticed how steadily it was growing. After hearing his explanation of the effects of compound interest, she became as intrigued as he was by the idea that compound interest can make invested money appear to take on a life of its own and seemingly grow without any outside help.

Nelson knows that Maria is learning about the stock market and mutual funds and that she is thinking about the risks involved in investing real money in one of these areas. They have spent many hours discussing the problem Clarence Sr. has been having since the value of his investments took such a beating on Black Monday.

Having watched what happened to Cal and his family, Nelson and Maria understand that a wise investor must not view investments as a gamble and must not jump into any investment scheme that promises large, fast profits. If such an "opportunity" seems too good to be true, it probably is!

OBJECTIVES: *In this lesson, we will help Nelson and Maria to:*

- *Examine ways to avoid risky investment schemes.*
- *Graph the change in the market value of an investment over a long period of time.*
- *Use a line of best fit to predict the approximate future value of an investment.*

INVESTMENT BOOBY TRAPS

Nelson has seen ads such as these in the newspaper:

"Worried about inflation? Diamonds are an investor's best friend."

"Make quick money! For only $1000, you can earn $16,000."

"Let me show you how I can quadruple the value of your stock portfolio in 18 months!"

Investment come-ons such as these operate on the principle that greed will overcome common sense. They often separate the investor from his or her money without anything of value being provided in return. However, in the past the situation was far worse. For example, unlike today, stock prices were frequently manipulated by those in a position to distribute misleading information about stocks.

After the stock market crash of 1929 the U.S. Congress established the **Securities and Exchange Commission (SEC),** an independent federal agency whose purpose is to prevent the return of the unsound stock selling practices and schemes that were common before the crash. The commission has two basic responsibilities.

- To require companies that offer securities for sale in more than one state to file with the commission and to make complete and accurate information about the company available to investors
- To protect investors against misrepresentation and fraud in the issuance and sale of securities

The SEC requires companies, including mutual fund companies, to disclose facts that are essential for an investor to make an informed investment decision. However, this agency cannot protect the public from making poor investment decisions. To become an investor who is not taken advantage of, you should take to heart the following tips from the SEC.

- Do not deal with security firms or salespeople with whom you are not personally familiar. Consult people whose knowledge about such matters you trust; perhaps your local banker can play this role.
- Be sure you understand the risk that you are assuming when you buy stock or mutual fund shares. Risk cannot be eliminated, but it can be managed; be sure that the level of risk is not too great for the gain you expect to achieve.
- Tell any salesperson who is promising a specific investment result to put the recommendations and expectations in writing.
- Give at least as much consideration to your investments as you would to any other valuable asset or property.
- Do not play the market, that is, do not buy a stock with the intention of selling with a profit in only a few weeks or months.

- Do not listen to high-pressure sales talk.
- Beware of tips, rumors, and promises of spectacular profits. A deal that sounds too good to be true probably *is* too good to be true.

DOES THE BROKER HAVE A DEAL FOR YOU!

While the great majority of brokers are very honest, you may occasionally run into a con artist who is interested only in taking your money. Be very wary of a salesperson who does not have a reputation that you can check on or who does any one of the following.

- Plugs one certain stock or mutual fund and refuses to sell you anything else
- Promises a quick, sure profit
- Claims to have inside information
- Urges you to hurry "before the price goes up"

If you wind up falling for a slick operator's con game at some time during your life, take comfort from the fact that almost everyone gets stung in that way at least once. Finally, when you do decide to invest, never risk more than you could feel comfortable losing. Otherwise, you may lose money even from a well-chosen investment. This might very well happen if a temporary (but severe) drop in its price were to frighten you into selling a basically sound investment at a loss.

Ask Yourself

1. What is the Securities and Exchange Commission?

2. What are the basic responsibilities of the U.S. Securities and Exchange Commission?

Select the equation that *best fits* the points in each of the following graphs. Explain why the equation fits best and the others do not fit. Remember that none of the equations will fit perfectly.

1. a. $y = x + 1$
    b. $y = 2x$
    c. $y = x + 2$

2. a. $y = x$
    b. $y = 0.5x - 1$
    c. $y = x - 1$

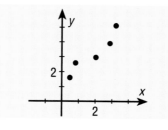

3. a. $y = -x$
    b. $y = x$
    c. $y = -x + 1$

SHARPEN YOUR SKILLS

EXAMPLE 1 Nelson has consulted a mutual funds reference manual at his neighborhood library. From the manual he recorded the 5-year price performance of the Vantage Balanced Mutual Fund, a fund recommended by Maria's investor friend. The prices were taken at 6-month intervals from July 1, 1988, to July 1, 1993, as shown.

Date	Years After Start	Price of Vantage Balanced Mutual Fund
July 1, 1988	0.0	$10.00
Jan. 1, 1989	0.5	10.60
July 1, 1989	1.0	11.80
Jan. 1, 1990	1.5	11.60
July 1, 1990	2.0	12.20
Jan. 1, 1991	2.5	12.40
July 1, 1991	3.0	13.60
Jan. 1, 1992	3.5	14.80
July 1, 1992	4.0	15.50
Jan. 1, 1993	4.5	16.20
July 1, 1993	5.0	16.80

QUESTION How can Nelson graph this price information?

SOLUTION

Nelson enters the data into his graphing calculator with the years after the start as x values and the prices as y values (11 ordered pairs of numbers starting with $x_1 = 0$, $y_1 = 10.0$). He chooses a range.

Xmin: -0.5 Ymin: -2
Xmax: 6   Ymax: 20
Xscl: 0.5   Yscl: 2

He plots all of the price/date information first as a **scatter plot** (left below), then as a **broken line graph** (right below). The broken line graph shows how the prices of the Vantage Balanced Mutual Fund changed over 5 years.

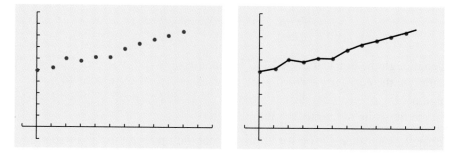

SKILL 2

EXAMPLE 2 Nelson wants to find an equation that will fit the points he has graphed.

QUESTION How can Nelson use a linear regression equation to find a line of best fit?

SOLUTION

Nelson decides to use his graphing calculator to find the *line of best fit* for the scatter plot. A linear equation of the form

$$y = a + bx$$

is called the **linear regression equation.** To obtain a and b, Nelson chooses "linear regression equation" from a menu on the graphing calculator. (Some graphing calculators use a code word for "linear regression equation," for example "LinReg.") As soon as Nelson chooses "linear regression equation," the calculator displays the following:

$a = 9.804545455$
$b = 1.369090909$
$r = 0.9850588294$

Using the values for a and b rounded to the nearest hundredth, Nelson writes an equation for the line of best fit

$y = 9.80 + 1.37x$

Nelson can graph the linear regression equation on the same set of axes as the original scatter plot as shown below.

Even though the line is the "best fit," it contains only some of the 11 points. The quantity r, the *correlation coefficient,* is a measure of how good the fit is. A value of $r = +1$ or -1 indicates a perfect fit. In this case $r = 0.985059$ shows that the fit, while not perfect, is rather good.

Nelson has another reason for knowing that the linear regression equation is not perfect, since if $x = 0$ then

$y = 9.80 + 1.37(0)$
$y = 9.80$

Nelson already knows that when $x =$ "0 years after the start," the actual price of the mutual fund is $10.00, not $9.80. He concludes that if he uses the equation to predict future prices of the mutual fund, then he cannot expect the predictions to be completely accurate.

EXAMPLE 3   Nelson wants to predict future prices of the Vantage Balanced Mutual Fund.

QUESTION   What price does Nelson's linear regression equation predict for the mutual fund in July 1994?

SOLUTION
Use Nelson's linear regression equation. July 1, 1994, is 6.0 years after the start.

$y = 1.37x + 9.80$
$y = 1.37(6.0) + 9.80$ Substitute 6.0 for x.
$y = 18.02$

The predicted price for July 1994 is a bit more than $18. As a check, Nelson uses the trace function on his graphing calculator and notices that the point that corresponds to the ordered pair (6, 18) is almost on his line of best fit, so the algebraic solution and the graphical solution agree closely.

Nelson is happy with the results that he has obtained using the linear regression model. He believes that the model, while not perfect, is good enough to help him understand how the mutual fund may perform in the future. The model will provide a basis for advising Maria on whether she should invest in the fund.

TRY YOUR SKILLS

Nelson's friend Maria is interested in taking a bit more risk than Nelson to have a greater chance of creating a higher value for her investment. So instead of looking at the Vantage Balanced Fund, she decides to study the 5-year performance of the Vantage Maximum Capital Gain Fund, shown below.

Date	Years After Start	Price of Vantage Maximum Capital Gain Fund
July 1, 1988	0	$10.00
Jan. 1, 1989	0.5	10.40
July 1, 1989	1.0	12.40
Jan. 1, 1990	1.5	11.50
July 1, 1990	2.0	12.60
Jan. 1, 1991	2.5	11.50
July 1, 1991	3.0	13.70
Jan. 1, 1992	3.5	16.20
July 1, 1992	4.0	15.30
Jan. 1, 1993	4.5	17.50
July 1, 1993	5.0	18.00

1. Make a scatter plot for the above data beginning with July 1, 1988.
2. Make a broken line graph for the data.
3. To find a line of best fit using a graphing calculator, what should be your values for x_1 and y_1?
4. For the graph of the line of best fit, what would be a good range?
5. Which of the following equations could represent a linear equation for the line of best fit?

   a. $y = bx^2 + a$ b. $y = ax^b$ c. $y = a + bx$ d. $y = ab^x$

6. Use a graphing calculator to find the linear regression equation that is the line of best fit for the data. How good is the fit?
7. Use the equation that you found in Exercise 6 to predict a possible value for the Vantage Maximum Capital Gain Fund in July 1995.

EXERCISE YOUR SKILLS

KEY TERMS

broken line graph
linear regression
  equation
scatter plot
Securities and
  Exchange
  Commission
  (SEC)

1. Why was the Securities and Exchange Commission established?

2. Why do some people lose money on investments despite the protection provided by the Securities and Exchange Commission?

3. What signals should alert you to suspect that a broker might not be acting in your best interest?

Maria's sister, Sarita, is interested in taking a bit more risk than she would have to assume in a balanced mutual fund but not as much as in a fund that tries to attain a maximum of capital gains. She decides to study the 5-year performance of the Vantage Growth and Current Income Fund, shown below.

Date	Years After Start	Price of Vantage Growth and Current Income Fund
July 1, 1988	0	$20.00
Jan. 1, 1989	0.5	21.60
July 1, 1989	1.0	24.80
Jan. 1, 1990	1.5	23.80
July 1, 1990	2.0	25.60
Jan. 1, 1991	2.5	25.20
July 1, 1991	3.0	28.40
Jan. 1, 1992	3.5	31.20
July 1, 1992	4.0	32.10
Jan. 1, 1993	4.5	34.00
July 1, 1993	5.0	35.00

4. Make a scatter plot for the above data beginning with July 1, 1988.

5. Make a broken line graph for the data.

6. Which of the following equations could represent a linear regression equation for the line of best fit?

  a. $y = x^b + a$ b. $y - bx = a$ c. $y = b^x + a$ d. $axy = b$

7. Use a graphing calculator to find the linear regression equation that is the line of best fit for the data. For your first data points, use $x_1 = 0$ and $y_1 = 20.00$. How good is the fit?

Use the equation that you found in Exercise 7 to predict a possible value for the Vantage Growth and Current Income Fund on the following dates.

8. January 1, 1995   9. January 1, 1998

10. January 1, 1999   11. July 1, 1999

12. How well does the equation of Exercise 7 predict the actual value of the Vantage Growth and Income Fund on July 1, 1988?

For the 5-year performance of Vantage Growth and Current Income Fund, find an equation for the line of best fit for the first ten data points, omitting the data for July 1, 1993.

13. What price does the equation predict for the value of the fund on July 1, 1993?

14. How does the predicted price compare with the actual price?

15. What do you conclude from this result?

MIXED REVIEW

1. Find the total cost of 100 shares of Daring Ventures, Inc. at $63\frac{1}{2}$. The commission cost is 1.5%.

Determine the total savings on a loan of $49,800 at 8.5% for 25 years if you make a down payment of 20% or 30% instead of 10%.

	Percent Down	Down Payment, D	Loan Amount, A	Monthly Payment, M	Total Amount, T	Savings Over 10% Down
2.	10%	$4,980	$44,820		$113,250.83	
3.	20%					
4.	30%					

5. Why do some large lending institutions use a credit scoring system such as the one that appears on page 293 of Lesson 6–6?

6. Use the formula for the Future Value of a Periodic Investment on page 344 of Lesson 7–3 to find the amount that will be in an IRA account after 25 years if the account consistently pays a compounded rate of 7.5% per year and $1,500 is contributed to the IRA every year.

7. Suppose that your name is Charles Lewis and your uncle sends you a check made out to "Chuck Lewis." Show how you would endorse that check.

8. A certain insurance company expects 99.830% of all 34-year-old people to live at least one more year. What is the break-even premium for a $50,000 policy issued to such a person by the company if the direct and indirect expenses for issuing the policy amount to $15?

9. If you earn $65,000 per year, how much do you pay in Social Security and Medicare taxes?

10. If you earn $159,000 per year, how much do you pay in Social Security and Medicare taxes?

11. Suppose that your family's take-home pay is $3500 per month. How much can you afford to spend for credit card payments each month?

STOCK PRICES AND INFLATION

Maria read an article in the newspaper that referred to some investors who feared that a "bear market" might be developing. She was not sure what bears had to do with stock prices. Nelson told her that in a **bear market,** most stocks go down and that in a **bull market,** most stocks go up. He also told her that not all the stocks go up or down at the same time.

Maria also remembers from her study of the Federal Reserve System that the U.S. economy is affected by inflation. She also knows that though the rates of inflation have

varied from year to year, inflation has affected salaries and prices for a number of years. One means of measuring the effects of this inflation is the *Consumer Price Index.*

Maria has often heard her parents speak of nickel postage stamps and 10-cent newspapers as though their childhood was not that long ago. To Maria, 35 years might as well be 100, but she does know that most adults talk that way. She decides that when she is an adult, she will remember what she thought when she was a teenager and not tell her children about how low prices were when she was a kid.

OBJECTIVES: *In this lesson, we will help Maria to:*

⬤ *Understand the effects of inflation as charted by the Consumer Price Index.*
⬤ *Use the Consumer Price Index to calculate changes in prices of products.*
⬤ *Use a regression model to illustrate changes in the Consumer Price Index.*
⬤ *Use the concept of probability to estimate whether to invest money in a mutual fund or stocks.*

THE CONSUMER PRICE INDEX

Inflation affects the cost of every product and service that we purchase because it cheapens the value of the dollar. Even though salaries may also have increased, the purchasing power of those salaries may be less than the salaries of 10 years ago. The **Consumer Price Index (CPI)** is an economic yardstick that can help Maria to judge the changes in the buying power of the dollar. Economists use the CPI to measure **inflation** (a general increase in prices) or **deflation** (a general decrease in prices). Periods of deflation are usually accompanied by high unemployment.

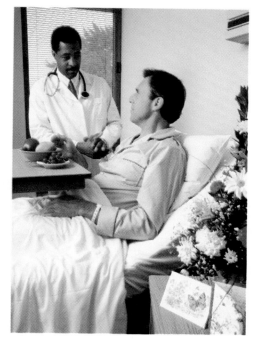

The Bureau of Labor Statistics compiles and publishes two CPIs. The CPI-W represents the consumption experience of urban wage earners and clerical workers; this represents approximately 40% of the population. The CPI-U is a broader index that covers all urban consumers, about 80% of the population. The CPI-U includes the consumption habits of diverse groups such as salaried workers, the unemployed, the retired, and the self-employed.

The following chart shows the breakdown of the CPI for all urban consumers in various categories. The categories listed are very general; more detailed categories are available from the Bureau of Labor Statistics and other sources. The chart illustrates the changes in the cost of all items and also the costs for four major categories for the years 1960–1992. These special categories are food, shelter, apparel, and medical care. The year 1984 is the base year, the year in which a **market basket** of items is assigned an arbitrary value of $100. Each separate category, such as Shelter or Apparel, is also assigned a value of $100 in 1984. As a result, all years that precede or follow 1984 have CPI

CONSUMER PRICE INDEX (1960–1992)					
Year	All Items	Food	Shelter	Apparel	Medical Care
1960	29.6	30	25.2	45.7	22.3
1970	36.8	39.2	35.5	59.2	34
1975	53.8	59.8	48.8	72.5	47.5
1980	82.4	86.8	81	90.9	74.9
1984	100	100	100	100	100
1985	107.6	105.6	107.7	105	113.5
1986	109.6	109	115.8	105.9	122
1987	113.6	113.5	121.3	110.6	130.1
1988	118.3	118.2	127	115.4	138.6
1989	124	125.1	132.8	118.6	149.3
1990	130.7	132.4	140	124.1	162.8
1991	136.2	135.8	146.3	128.8	177
1992	140.7	138.3	152.2	131	189.4

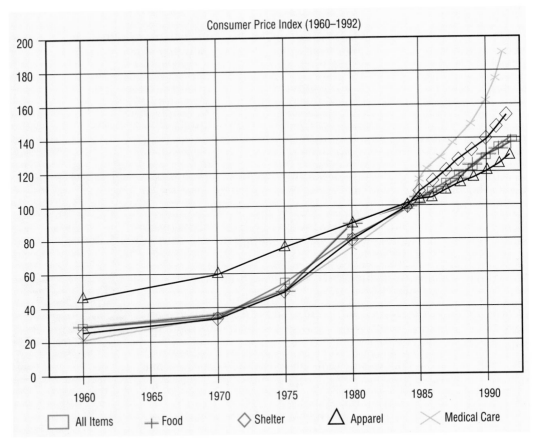

Consumer Price Index (1960–1992)

☐ All Items + Food ◇ Shelter △ Apparel ✕ Medical Care

values that are proportionally related to the base year. Notice that in the combined category, "All Items," a basket of goods that cost $100 in 1984 would have cost $29.60 in 1960 and $140.70 in 1992.

Maria wanted to see the numbers displayed in a graph for easy comparison among the categories. She entered the numbers for each category in a spreadsheet and created the graph shown above. From the graph, Maria could see that the rate of price increase was not the same for every category. For example, the rate of increase for the price of shelter was greater than the rate for all items.

Ask Yourself

1. What is a bear market?

2. What is a bull market?

3. What is the Consumer Price Index?

4. Why do you think a period of deflation is usually accompanied by unemployment?

ALGEBRA *REVIEW*

Solve each proportion.

1. $\dfrac{x}{3} = \dfrac{8}{2}$

2. $\dfrac{x+1}{5} = \dfrac{x-1}{4}$

3. $\dfrac{2.5}{a} = \dfrac{1}{2}$

4. $\dfrac{3}{15} = \dfrac{b-1}{10}$

5. $\dfrac{7}{4x+2} = \dfrac{1}{2x-4}$

SHARPEN YOUR SKILLS

SKILL 1

EXAMPLE 1 Maria's research revealed to her that inflation can be measured by keeping track of the annual change in the CPI. For example, notice the "all-items CPI" between 1988 and 1989:

CPI (all items) for 1988: 118.3
CPI (all items) for 1989: 124

QUESTION What is the percent of increase in the CPI between 1988 and 1989?

SOLUTION
Use the rate-of-change formula from Lesson 8–2. The rate of change in the CPI is the change divided by the CPI for the earlier year.

$$r = \frac{|P_n - P_o|}{P_o} \qquad \text{Rate-of-change formula}$$

$$r = \frac{|124 - 118.3|}{118.3}$$

$$= 0.04818$$

Since the CPI increased, the rate of inflation is 4.8%.

SKILL 2

EXAMPLE 2 Since 1984 is the base year for the CPI, Maria would like to compare her parents' spending power in 1984 with what they make now.

QUESTION Maria's parents spent $20,000 in 1984. How much would they have needed in 1992 just to buy the same things?

SOLUTION
Write a proportion. Let x represent the unknown 1992 expenditure.

$$\frac{1992 \text{ expenditure}}{1992 \text{ CPI}} = \frac{1984 \text{ expenditure}}{1984 \text{ CPI}}$$

$$\frac{x}{140.70} = \frac{20,000}{100} \qquad \text{Next, use the } \textit{Rule of Proportions.}$$

$$x \cdot 100 = 140.70 \cdot 20,000 \qquad \text{If } \frac{a}{b} = \frac{c}{d}, \text{ then } ad = bc.$$
$$100x = 2,814,000$$
$$x = 28,140$$

In 1992 it would have taken about $28,000 to buy the same market basket of goods that $20,000 would have bought in 1984.

EXAMPLE 3   Maria would like to estimate the cost of living for the year 2000.

QUESTION   How can Maria use the CPI table for 1960–1992 to estimate the cost of living for the year 2000?

SOLUTION

Maria knew from talking with Nelson that she can use a graphing calculator to find a line of best fit. She decides to use the data points from the "All Items" column of the 1960–1992 CPI table. To make the data easier to work with, she simplifies the values for x that she will have to enter. She lets x represent the number of years after 1960 and y represent the CPI for each year in the table beginning with 1960 as shown below.

x	0	10	15	20	24	25	26	27	28	29	30	31	32
y	29.6	36.8	53.8	82.4	100	107.6	109.6	113.6	118.3	124	130.7	136.2	140.7

Maria enters the data in a graphing calculator and creates a scatter plot using the following range values.

Xmin: −2   Ymin: −10
Xmax: 33   Ymax: 150
Xscl: 2   Yscl: 20

She knows that Nelson used a straight line as a model for his investigation of mutual funds. She decides to investigate the three other curves that the calculator provides as a model for data.

Two of the other three curves cannot use her data (they will not take 0 as a value for x). The two models that do accept all of her data are the *linear regression* equation that Nelson used and another curve called the **exponential regression equation.** The graphing calculator provides a correlation coefficient r for each model and values for two constants, a and $b,$ that appear in the formulas for the models.

Regression Model	Formula	Equation	Correlation Coefficient
Linear	$y = a + bx$	$y = 10.73788254 + 3.850867094x$	$r = 0.973223596$
Exponential	$y = ab^x$	$y = 26.19388895(1.05527341)^x$	$r = 0.9876350425$

Maria uses her graphing calculator to graph each model on top of the scatter plot.

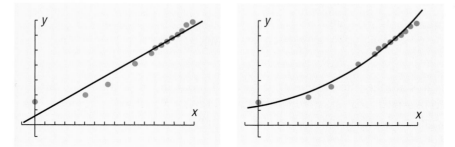

She decides to use the exponential model instead of the linear model for two reasons.

1. The exponential graph seems to fit the scatter points better.

2. She knows that the closer the correlation coefficient is to 1, the better the fit. The correlation coefficient for the exponential curve is closer to 1 than is the correlation coefficient for the straight line.

To estimate the cost of living for the year 2000, Maria notes the year 2000 is 40 years after 1960. Maria uses the ZOOM and TRACE features of the graphing calculator to find that y is about 225 when x is close to 40. To check, Maria substitutes into the exponential equation.

$y = 26.19388895(1.05527341)^{40}$ Substitute 40 for x.
$y = 225$ To the nearest dollar

According to the exponential model, the CPI will be about 225 in the year 2000.

EXAMPLE 4 Maria would like to predict the rate of inflation for the year 2000.

QUESTION If the exponential model is correct, what will be the rate of inflation?

SOLUTION
Maria notices that the exponential equation is similar to the compound interest formula (see page 107 in Lesson 3–2).

$B = p(1 + r)^n$ where p = initial investment
                         r = interest rate for the period
                         n = number of periods

She compares the two formulas

$$B = p(1 + r)^n \qquad\qquad y = 26.19(1 + 0.055)^x$$

and notices that the compound interest formula is also an exponential model. She writes 1.055 as $1 + 0.055$ to sharpen the comparison and correctly concludes that the predicted annual increase in the CPI (the rate of inflation) is 0.055, or about 5.5%. She also notices that the model is not perfect, since in 1960 when x was 0,

$$
\begin{aligned}
y &= 26.19(1.055)^0 \\
&= 26.19 \cdot 1 \qquad\qquad \text{Recall } b^0 = 1 \text{ when } b \neq 0. \\
&= 26.19
\end{aligned}
$$

which is not the true value for the CPI in 1960. (The true value was 29.6.) Maria suspects that the predicted inflation rate, 5.5%, might also be inaccurate. In a few more years she will know.

SKILL 4

Maria asks her friend who knows about investments what she should expect her fund to be worth in 5 years if inflation continues at a low rate and what her fund might be worth if deflation occurs at a low rate. Her friend's opinion is that her $1000 might drop to about $700 if deflation were to occur. Her friend agrees with Maria that if a mild inflation occurs, the fund will probably be worth at least $1800. Her friend also believes that the Federal Reserve will do whatever it can to avoid a deflation, since a deflation is usually accompanied by a recession and above-average unemployment. However, a deflation might occur anyway. The investor friend thinks that the probability of a deflation is only about 1 in 5; that is, about 0.20.

EXAMPLE 5   How can Maria use her new knowledge about inflation to help her with her decision about investing in a mutual fund?

QUESTION   What is the expected value of her portfolio in 5 years?

SOLUTION
Maria decides to calculate the expected value of her portfolio in 5 years. If the probability of a deflation is 0.20 and inflation is the only other choice, then the probability of inflation can be found by solving the following probability formula for $P(\text{inflation})$.

$$
\begin{aligned}
P(\text{inflation}) + P(\text{deflation}) &= 1 \\
P(\text{inflation}) &= 1 - P(\text{deflation}) \\
&= 1 - 0.20 \\
&= 0.80
\end{aligned}
$$

The expected value of her fund is found by using the expectation formula.

$$E = a_1P_1 + a_2P_2$$
$$= 1800 \cdot 0.80 + 700 \cdot 0.20$$
$$= 1440 + 140$$
$$= 1580$$

The expected value of Maria's fund in 5 years is $1580. She knows that this value is useful as a guide in deciding whether to invest in a fund or to stick to a savings account. She also knows that the actual value of the fund is likely to be around either $1800 (if there is some inflation) or $700 (if there is deflation), not around $1580. In deciding whether to invest in the fund, Maria has to face the possibility that the fund will drop in value to $700 in 5 years.

TRY YOUR SKILLS

Use the table of the Consumer Price Index (1960–1992) on page 389 to find the percent of increase in the CPI (all items) for the following time periods. Round your answers to the nearest tenth of a percent.

1. From 1989 to 1990

2. From 1984 to 1985

Use the table of the Consumer Price Index (1960–1992) on page 389 to write and solve a proportion that can be used to find how much money a family would need to buy a market basket of goods in the given year that was worth $10,000 in 1987.

3. 1984

4. 1990

Suppose that an exponential regression model gives the following equations as a description of the cost of living over a five-year period. What would be the expected rate of inflation?

5. $y = 100(1 + 0.04)^x$

6. $y = 53.6(1.062)^x$

7. Suppose that the probability of low inflation is 0.15 and the only other possibility is moderate deflation. Find the probability of moderate deflation.

8. Suppose that there are three possibilities: low inflation with a probability of 0.3, severe deflation with a probability of 0.1, and perfect price stability (neither inflation or deflation). Find the probability of perfect price stability.

1. How do economists use the Consumer Price Index?

2. In the table for the CPI, which category had the largest increase between 1984 and 1992? Which had the least increase?

3. Which category in the CPI table most nearly matched the "all items" column? Why do you think this happened?

4. The "all items" column in the table for the CPI consists of a market basket of 400 items. Why is this a better indication of how prices are behaving than the cost of a few selected items would be?

The prices given are for 1984. Find the cost of the indicated items in 1992. Use the table of the Consumer Price Index (1960–1992) on page 389.

5. $500 worth of food

6. $1000 in medical care

7. A $25 market basket

Find the percent of change in the CPI for the indicated item.

8. Food between 1984 and 1985

9. Shelter between 1989 and 1990

10. Apparel between 1987 and 1988

11. Medical care between 1985 and 1986

Use a graphing calculator to make an exponential regression model for the cost of medical care from 1960 to 1992 as shown in the table on page 389. Then answer Exercises 12–16.

12. What is the equation for the model?

13. How accurately does the model agree with the actual cost of medical care in 1960?

14. What value does the model suggest for the average annual rate of increase in the cost of medical care between 1960 and 1992?

15. Use the graphing calculator to make a scatter plot of the data, and on the same calculator screen, graph the exponential regression equation that you found.

16. Do you think that a linear regression model would have given a more accurate picture than the exponential regression model for the increase in the cost of medical care from 1960 to 1992? Explain your answer.

Maria's sister, Sarita, has become interested in a mutual fund that stresses both growth and current income. She believes that if there is inflation in the next 5 years, then her $1000 investment will grow to about $1750 but that if there is a deflation, then the fund may drop to $800. She thinks that the next 5 years will be either inflationary or deflationary with a 20% probability that the economy will be deflationary.

17. What is the probability that the economy will be inflationary?

18. What is the expected value of Sarita's fund in 5 years?

MIXED REVIEW

The following table shows cost and sales figures for a company that sells board games. Use a computer spreadsheet to fill in the missing information.

	Unit Cost	Number Produced	Fixed Cost	Total Cost	Unit Price	Revenue	Profit (Loss)
1.	$8.93	15	$310.00		$25.00		
2.	7.50	125	310.00		25.00		

Use the following table for Exercises 3–5.

Comparison Table for Term and Whole Life Annual Premiums for a Policy with a Face Value of $100,000			
Age	5-year Renewable Term	Whole Life	First-year Difference
20	$205	$ 775	$570
25	207	918	711
30	218	1112	894

3. What is the annual premium for a $100,000 whole life policy for a 20-year-old person?

4. What is the annual premium for a $200,000 term policy for a 30-year-old person?

5. Determine the difference that a 25-year-old person will pay in annual premiums the first year if he or she takes a term policy with a face value of $250,000 instead of a whole life policy with the same face value.

Enter the following transactions in a check register form. Make up your own form with the following headings: Check Number, Date, Checks/Deposits, Amount, and Balance. Under "Checks/Deposits" have two lines for each entry, the top one labeled "To:" and the bottom one labeled "For:." Find the new balance after each transaction. The starting balance is $450.

6. Check 201, September 27, to Aaron Jones, $158.50 for car repair

7. Deposit on September 29 of $130

8. Check 202, September 29, to Superior Supermarket, $87.63 for groceries

9. The probability that a person will be alive in 1 year is 0.98329. What is the probability that the person will die within 1 year?

10. Last month, the Antonelli family spent $241 on the car payment, $452 on credit card payments, $841 on the mortgage, $86 on utilities, $475 on food, and $1500 on everything else. To the nearest tenth of a percent, what percent of their take-home pay did the family spend on credit payments?

1. How can you make money from stocks?

2. How is buying bonds the same as lending money to a municipality or a corporation?

3. Why are mutual funds suitable for small investors?

4. What do you think is the most important factor in deciding whether or not to buy stocks?

5. What method is used by brokers to buy and sell stocks on a security exchange?

6. How does the Securities and Exchange Commission help to protect investors?

7. Why is it important to adopt a long-term point of view when buying stocks whether directly in the stock market or indirectly through a mutual fund?

Use the prices per share shown below to find the number of whole shares of stock that you can afford to purchase if you have $20,000 to invest in each of the following companies. Then find the total cost of the shares. Ignore brokerage commissions.

	Company	Price per Share	Number of Whole Shares	Total Cost
8.	NunnInc	42		
9.	OctOfAm	$7\frac{5}{8}$		
10.	OscarsPies	$18\frac{3}{4}$		
11.	OwlInd	150		

Find the capital gain or loss from buying 2500 shares at the given purchase price and selling them at the current market value. Ignore the effect of commissions.

	Company	Buy	Total Value	Sell	Total Value	Gain/Loss
12.	Disnel	$63\frac{1}{2}$		$68\frac{1}{2}$		
13.	AP & P	$18\frac{1}{2}$		$30\frac{1}{4}$		
14.	Banter	$24\frac{3}{8}$		$19\frac{5}{8}$		
15.	Zola	$43\frac{1}{2}$		89		

The 5-year performance of the Vantage Long-Term Growth Fund is shown below.

Date	Number of Years After Start	Price of Vantage Long-Term Growth Fund
July 1, 1988	0	$10.00
Jan. 1, 1989	0.5	10.70
July 1, 1989	1.0	12.90
Jan. 1, 1990	1.5	11.80
July 1, 1990	2.0	13.10
Jan. 1, 1991	2.5	12.50
July 1, 1991	3.0	14.30
Jan. 1, 1992	3.5	16.10
July 1, 1992	4.0	17.60
Jan. 1, 1993	4.5	18.90
July 1, 1993	5.0	20.10

16. Make a scatter plot for the above data beginning with July 1, 1988.

17. Make a broken line graph for the data.

18. Use a graphing calculator to find the linear regression equation that is the line of best fit for the data.

19. Use the equation that you found in Exercise 18 to predict a possible value for the Vantage Long-Term Growth Fund in July 1995.

Use a graphing calculator to make an exponential regression model for the cost of food from 1960 to 1992 as shown in the table on page 389. Then answer Exercises 20–22.

20. What is the equation for the model?

21. How accurately does the model agree with actual cost of food in 1960?

22. What value does the model suggest for the average annual rate of increase in the cost of food between 1960 and 1992?

CHAPTER 8 TEST

For the companies listed below, find the number of whole shares that you can buy with $100,000. Ignore commissions.

1. Z Mart at $65\frac{1}{4}$ per share

2. Disnel at $38\frac{3}{4}$ per share

3. Zaxta at $87\frac{1}{2}$ per share

4. Algin at $25\frac{7}{8}$ per share

5.–8. Find the total value of the shares of stock in Exercises 1–4.

For the following companies, find the commission cost and the total purchase cost (price plus commission) for the given information.

9. 306 Cts&Dgs; $20/share; 2%

10. 215 Elmo; $33.50/share; 1%

11. 400 SuprCo; $100.375/share; 1.5%

12. 6000 Langly; $7/share; 0.5%

The value of the Vantage Conservative Growth Fund grew annually for 5 years as follows, beginning on January 1, 1988: $15.00 (starting value), $15.70, $16.80, $17.80, $19.00, $20.60.

13. Make a scatter plot for the above data beginning with January 1, 1988.

14. Make a broken line graph for the data.

15. Use a graphing calculator to find the linear-regression equation that is the line of best fit for the data.

16. Use the equation that you found in Exercise 15 to predict a possible value for the Vantage Conservative Growth Fund in January 1996.

Use the Consumer Price Index (1960–1992) on page 389 in Lesson 8–4 to find the approximate cost of each of the following items in 1989. Choose from the categories Food, Shelter, and Apparel. For each item the 1984 price is given.

17. Woman's jacket; $50.75

18. House; $77,000

19. Rent; $435

20. Pizza slice; $0.70

An investor believes that if there is inflation in the next 5 years, then her $10,000 investment will grow to about $18,000 but that if there is a deflation, then the investment may drop to $7500. She thinks that the next 5 years will be either inflationary or deflationary with a 30% probability that the economy will be deflationary.

21. What is the probability that the economy will be inflationary?

22. What is the expected value of the investment in 5 years?

CUMULATIVE REVIEW

1. If Phyllis works more than 25 hours in one week, she receives $1\frac{1}{2}$ times her regular wage rate of $4.80 for each of the extra hours. One week she worked 32 hours and received $68.75 in tips. How much money did Phyllis earn that week?

2. Angela earns a weekly salary of $150 at the real estate agency where she works and gets a 4% commission for each house that she sells. In a 4-week period her sales were $100,000. What were her total earnings?

3. Suppose you wrote 30 checks in June. Your bank charges $0.040 per check for the first 25 checks and $0.10 for each check over 25. How much were your June bank charges?

Andy can save $35 a week. He wants to buy a 25-inch remote stereo color TV that sells regularly for $419.99 but will be on sale for 10 weeks at $369.99.

4. Find how much he can save between now and the end of the sale. Will he be able to buy the set on sale? Explain your answer.

5. How long will he have to save in order to be able to afford the TV at the regular price?

Use a calculator to find the amount of interest and the new balance that will accumulate over 2 years on the following principal amounts at the given interest rate compounded as shown.

	Principal	Interest Rate	How Often Compounded	Interest Earned	New Balance
6.	$5000	8%	Annually	1st period: _____	
				2nd period: _____	
7.	$9000	9.5%	Semiannually	1st period: _____	
				2nd period: _____	

The following table shows cost and sales amounts for a company that sells sports equipment. Use a computer spreadsheet to fill in the missing information.

	Unit Cost	Number Produced	Fixed Cost	Total Cost	Unit Price	Revenue	Profit (Loss)
8.	$6.50	20	$550.00		$86.00		
9.	6.50	295	550.00		86.00		

10. Use the monthly payment formula $M = \dfrac{Pr(1 + r)^n}{(1 + r)^n - 1}$ to determine how much money will be saved by purchasing a $950 computer with monthly payments at an interest rate of 12% rather than 18%. The loan is to be repaid in 2 years.

11. Determine the effective interest rate on a credit card with an APR of 14%.

1. Find several companies of medium or large size that make high-quality products with which you are familiar.

2. In the financial section of your local public library, find your chosen companies' names in a standard reference manual such as *Standard & Poor's*. Make a note of the the telephone numbers of the four companies that appeal to you the most.

3. Phone each company. Ask the person responsible for shareholder relations to send you a copy of the most recent annual report and any other available information about the company. The vast majority of companies will respond favorably to your request.

4. The material from the companies may tell you whether any of the companies has a local branch. Find out what you can about that branch, such as the role it plays in its community and its employees' attitudes.

5. Track your companies' stock prices for several weeks in the newspaper.

6. Share with the rest of the class all of the information that you have about your companies.

Extension

On the basis of what you have learned, select the company that you think would be the best investment for at least the next 5 years. Share your reasons for this choice with the rest of the class.

A *number line* assigns each real number a point on the line. A number line is helpful in comparing numbers and examining some of the relationships between numbers, especially inequalities.

Graph each number on a number line and describe the position of each number.

Example   2.6

The number 2.6 is a little more than half way between 2 and 3.

1. -1.5 **2.** $6\frac{1}{3}$ **3.** $-\frac{8}{3}$ **4.** π **5.** $\sqrt{2}$ **6.** $(-2.5)^2$

Inequalities can be represented on the number line. A *solid circle* at an end point means that the point is included in the set. An *open circle* means the number is excluded.

Graph each inequality on a number line.

Example $x \leq 7$

7. $x > 1$ **8.** $x < -2$ **9.** $-2 \leq x$ **10.** $5 \geq x$

A *compound inequality* satisfies more than one condition. When two conditions are linked by "and," the solution set is the *intersection* of the sets satisfying the conditions. A number in both sets is in the solution set. Sometimes "and" is not written but it is clear that two conditions are given. When two conditions are linked by "or," the solution is the *union* of two sets. A number is in the solution set if it satisfies either one or the other condition or both conditions.

Graph the solution set on the number line.

Example: **a.** $-3 \leq x \leq 5$ **b.** $x < -1$ or $x \geq 2$

a. $-3 \leq x \leq 5$ means $-3 \leq x$ and $x \leq 5$

b. $x < -1$ or $x \geq 2$

11. $1 \leq x < 10$ **12.** $x \leq -2$ or $x > 3$ **13.** $x \leq 5$ and $x > 0$

14. $1 \geq x$ or $1 \leq x$ **15.** $-5 < x < 0$ **16.** $0 < x$ and $x < 1$

Federal

Paying the Price

WHETHER OR NOT WE AGREE WITH HOW our tax dollars are spent, most of us would not want to do without at least some of the services that our federal, state, and local governments provide.

Taxes are one of the largest items in the family budget. The average taxpayer works every year from January 1 to around early May just to cover tax obligations. However, most of the tax that we pay never goes through our hands; our employer withholds some of our pay for taxes and sends that withheld money directly to the Internal Revenue Service (IRS). In most communities, part of the withheld money goes also to the state and local governments for its taxes. If it were not for the discipline of having the government regularly withholding part of our pay, many of us might find it difficult to meet our tax obligation at all.

Even so, before every April 15, each of us must calculate our income for the previous year, and if it is above a certain minimum amount, we must file a tax form with the IRS. Since there are so many kinds of income, the IRS has devised forms to accommodate the various possibilities. These forms do take some getting used to.

From time to time, attempts have been made to simplify the forms for the benefit of the taxpayer, especially the taxpayer who does not have a complicated tax situation.

Income Tax:

Three of our high school students will examine the three tax forms that individuals most frequently encounter. The students will demonstrate how, for many individuals, the forms are fairly easy to complete. They will also come in contact with sources of assistance in filling out the forms. Unfortunately, these sources do not extend their assistance to making the tax payments themselves; for that, we are on our own!

9–1 Finding Your Income Tax

9–2 Forms and More Forms: 1040EZ and 1040A

9–3 Form 1040, Schedule A, and Help!

9-1 FINDING YOUR INCOME TAX

Luis had big plans for his first paycheck. His car needed some tires, and the amount that he expected to receive for his first week's work should have been just enough. For that reason, he was quite surprised when he received the check—$11.89 was missing! He examined the stub more closely and noticed that deductions had been made from his salary.

The largest deduction caught his attention: "Federal Income Tax Withheld." He remembered the mass of papers that his father and mother shuffle and sort once a year during the first week of April. The activity is usually accompanied by intense discussions about depreciation, Schedule A deductions, and other equally strange notions. In addition, there are invariably many unkind comments about wasteful government spending. Now it is Luis's turn to deal with income taxes.

He is determined to figure out what it all means. He received his W-2 form from his employer and knows that he must file a federal income tax return. He hopes to receive a refund but wonders whether the amount withheld from his paycheck will *really* be enough to cover his taxes. He plans to compare the withheld amount with the amount of the tax given in the federal tax table. If he gets back as much as he thinks, he will be able to buy a whole new set of tires for his car.

OBJECTIVES: *In this lesson, we will help Luis to:*

Realize how much of the year a person works to pay income taxes.

Discover some of the characteristics of the federal income tax system.

Find how much income tax will be owed in specific cases.

Compare the tax owed with the amount withheld by the employer.

THE PRICE TAG OF GOVERNMENT

Taxes are what we pay for civilized society.
Oliver Wendell Holmes, Jr.

In this world nothing is certain but death and taxes.
Benjamin Franklin

Since Luis first realized that all his earnings were being taxed by the federal government, he has noticed newspaper and magazine articles about taxes. One that recently caught his attention was entitled "Tax-Freedom Day." As he read, he discovered that the Tax Foundation Incorporated, a private organization, calculates an annual Tax Freedom Day, the day each year when the average American will stop working to pay for government services. To Luis this meant that if he were working full time and receiving a full salary, the amount of money that he would pay the government in taxes for the year would be equivalent to what he would earn from the beginning of January to May 1.

The table that follows shows how the cost of government as measured by the number of working days has increased considerably between 1930 and the present. The Tax Foundation estimates that typical workers pay approximately one third of their earnings into local, state, and federal taxes. Luis certainly hopes that these governments are using his money wisely.

TAX FREEDOM DAYS	
1930	February 13
1940	March 8
1950	April 3
1955	April 9
1960	April 17
1965	April 15
1970	April 28
1975	April 28
1980	April 28
1985	May 1
1990	May 5
1991	May 8

HOW DID THIS HAPPEN?

Luis wondered whether governments in our country have always needed so much money. As he looked into the matter, he discovered that for its first 100 years the United States covered its expenses by imposing taxes on certain goods manufactured here and on products brought in from other countries. Then our country changed from a society that was mainly rural and agricultural to one that was mainly industrial. This change was accompanied by a growth in the role played by the federal government. As a consequence, the federal government's need for income increased. The 16th Amendment, adopted in 1913, gave Congress the power to pass income tax laws. We now have taxes on income as well as on sales and property.

UNITED STATES INDIVIDUAL INCOME TAX

Luis also discovered some features of the elaborate process that we go through to supply funds to our government. The **individual income tax** is imposed on an individual's earnings from wages, salary, tips, interest, rents, dividends, and capital gains. It is the largest revenue-producing tax for the federal government. Each year, on or before April 15, U.S. citizens and other legal residents determine their tax liability and usually either pay the amount that is due or request a refund. The only other possibility is to allow the government to apply the refund toward next year's tax liability.

Here are some of the characteristics of our income tax system.

Ability to Pay   The federal income tax rules are designed to match one's ability to pay. To accomplish this goal, the IRS has established a **progressive tax schedule,** one that takes an increasingly higher share of any additional income that you earn. That explains why Luis's parents will pay a higher percent of their income for taxes than he does. A person with a very low income may even owe no federal income tax at all.

Voluntary Compliance Another characteristic of the federal personal income tax is voluntary compliance. The Internal Revenue Service relies on Luis and the rest of us to meet our tax responsibilities faithfully. However, any U.S. citizen or other legal resident who does not pay taxes voluntarily can be fined or even imprisoned; tax evasion is illegal.

Pay-as-You-Earn Luis especially appreciates a feature of the federal tax system known as **pay-as-you-earn.** As we receive types of income that are subject to a tax, money is held back to pay the tax. These types of income include tips, pensions, gambling winnings, and, in Luis's case, wages. This system helps prevent Luis and other taxpayers from owing large amounts of taxes on April 15 of the following year. Luis's employer is responsible for withholding tax from Luis's paycheck and depositing it with the IRS. Luis learned how much money had been withheld from his paycheck during the year when he received his Form W-2 from his employer early in the following year.

1 Control number	22222	For Official Use Only ▶ OMB No. 1545-0008		
2 Employer's name, address, and ZIP code		6 Statutory employee ☐ Deceased ☐ Pension plan ☐ Legal rep. ☐ 942 emp. ☐ Subtotal ☐ Deferred compensation ☐ Void ☐		
		7 Allocated tips	8 Advance EIC payment	
		9 Federal income tax withheld	10 Wages, tips, other compensation	
3 Employer's identification number	4 Employer's state I.D. number	11 Social security tax withheld	12 Social security wages	
5 Employee's social security number		13 Social security tips	14 Medicare wages and tips	
19a Employee's name (first, middle initial, last)		15 Medicare tax withheld	16 Nonqualified plans	
		17 See Instrs. for Form W-2	18 Other	
19b Employee's address and ZIP code				
20	21	22 Dependent care benefits	23 Benefits included in Box 10	
24 State income tax	25 State wages, tips, etc.	26 Name of state	27 Local income tax	28 Local wages, tips, etc.	29 Name of locality

Copy A For Social Security Administration   Cat. No. 10134D   Department of the Treasury—Internal Revenue Service

Form **W-2 Wage and Tax Statement 19--**

For Paperwork Reduction Act Notice and instructions for completing this form, see separate instructions.

FORM W-2

Every employer for whom Luis works during the year must provide him with a W-2 form for the period of time during which he worked. This form provides a record of wages earned as well as federal, state, and other taxes withheld during the year. When Luis files his tax return, he must submit a copy of each of his W-2 forms at the same time.

TAX TABLES

You can find your **tax liability,** the amount you owe, in the **tax tables,** which you can obtain from the IRS. (A typical set of these tables are in the Reference Section at the back of the book.) Before you can use the tables, however, you need to find your *taxable income.* **Taxable income** is the amount of money that you actually make during the year *minus* certain adjustments, deductions, and exemptions that the government allows.

After you know your taxable income, you use the column in the table that is appropriate for your filing status. A portion of a tax table is shown in Skill 1. There are columns for single people, married people filing jointly, married people filing separately, and heads of households. Read down the income column until you find your taxable income. Then read across to the column that applies to you. The amount shown is your tax. (For married couples, filing a **joint return** often results in lower taxes, especially if one income is lower than the other.)

ALGEBRA *REVIEW*

Evaluate each expression for the indicated value of *I*.

1. $0.15I$
   for $I = 5,600$

2. $1200 + 0.25\,(I - 7800)$
   for $I = 8000$

3. $0.15I$
   for $I = 21,430$

4. $3217.50 + 0.28(I - 21,450)$
   for $I = 21,470$

5. $3217.50 + 0.28(I - 21,450)$
   for $I = 51,890$

6. $11,743.50 + 0.31(I - 51,900)$
   for $I = 51,910$

7. $11,743.50 + 0.31(I - 51,900)$
   for $I = 75,750$

8. $11,743.50 + 0.31(I - 51,900)$
   for $I = 105,380$

Ask Yourself

1. What is an income tax?

2. How long does the average person have to work in a year to earn the amount of money that he or she pays to the government in taxes?

3. What are two characteristics of our income tax system?

SKILL 1

EXAMPLE 1 Elizabeth's taxable income for the past year was $2,275.

QUESTION How much tax does Elizabeth, who is single, owe?

SOLUTION

Use the portion of the Tax Table shown below. Read down the income column headed "If line 5 (Form 1040EZ), line 22 (Form 1040A), or line 37 (Form 1040) is—" to find 2,275.

Notice that 2,275 is listed twice, once under the column marked "At least" and once under "But less than." Use the line where 2,275 is shown under the "At least" column. Read across to the column headed "Single." The amount in that column is 343.

Elizabeth owes $343.00 in income tax for the year.

Tax Table

If line 5 (Form 1040EZ), line 22 (Form 1040A), or line 37 (Form 1040) is–		And you are—				If line 5 (Form 1040EZ), line 22 (Form 1040A), or line 37 (Form 1040) is–		And you are—			
At least	But less than	Single	Married filing jointly *	Married filing separately *	Head of a household	At least	But less than	Single	Married filing jointly *	Married filing separately *	Head of a household
		Your tax is—						Your tax is—			
1,900	1,925	287	287	287	287	3,400	3,450	514	514	514	514
1,925	1,950	291	291	291	291	3,450	3,500	521	521	521	521
1,950	1,975	294	294	294	294	3,500	3,550	529	529	529	529
1,975	2,000	298	298	298	298	3,550	3,600	536	536	536	536
2,000						3,600	3,650	544	544	544	544
						3,650	3,700	551	551	551	551
2,000	2,025	302	302	302	302	3,700	3,750	559	559	559	559
2,025	2,050	306	306	306	306	3,750	3,800	566	566	566	566
2,050	2,075	309	309	309	309	3,800	3,850	574	574	574	574
2,075	2,100	313	313	313	313	3,850	3,900	581	581	581	581
2,100	2,125	317	317	317	317	3,900	3,950	589	589	589	589
2,125	2,150	321	321	321	321	3,950	4,000	596	596	596	596
2,150	2,175	324	324	324	324	**4,000**					
2,175	2,200	328	328	328	328						
2,200	2,225	332	332	332	332	4,000	4,050	604	604	604	604
2,225	2,250	336	336	336	336	4,050	4,100	611	611	611	611
2,250	2,275	339	339	339	339	4,100	4,150	619	619	619	619
2,275	2,300	343	343	343	343	4,150	4,200	626	626	626	626
2,300	2,325	347	347	347	347	4,200	4,250	634	634	634	634
2,325	2,350	351	351	351	351	4,250	4,300	641	641	641	641
2,350	2,375	354	354	354	354	4,300	4,350	649	649	649	649
2,375	2,400	358	358	358	358	4,350	4,400	656	656	656	656

Taxpayers whose taxable income is over $100,000 do not use tax tables. Instead, they use the **Tax Rate Schedules** X, Y-1, Y-2, or Z. Schedule X, for filers who are single, is shown below.

Tax Rate Schedules

Caution: Use **only** if your taxable income (Form 1040, line 37) is $100,000 or more. If less, use the **Tax Table**. Even though you cannot use the tax rate schedules below if your taxable income is less than $100,000, all levels of taxable income are shown so taxpayers can see the tax rate that applies to each level.

Schedule X—Use if your filing status is **Single**

If the amount on Form 1040, line 37, is: Over—	But not over—	Enter on Form 1040, line 38	of the amount over—
$0	$21,450	.... 15%	$0
21,450	51,900	$3,217.50 + 28%	21,450
51,900	....	11,743.50 + 31%	51,900

You can construct a formula for any tax that is calculated using the Tax Rate Schedules. (Recall that the symbol < means "is less than" and the symbol ≤ means "is less than or equal to.")

Formulas for Tax Rate Schedule X (for *single* taxpayers)

i.	$t = 0.15I$	if $I \le 21{,}450$
ii.	$t = 3{,}217.50 + 0.28(I - 21{,}450)$	if $21{,}450 < I \le 51{,}900$
iii.	$t = 11{,}743.50 + 0.31(I - 51{,}900)$	if $51{,}900 < I$

where I = taxable income
t = tax on the income

Although only taxpayers with income over $100,000 use the Tax Rate Schedules, the above formulas can be used by *any* single taxpayer as a rough check of the tax obtained from the Tax Table. The difference between the two methods of finding the tax may be about 5 or 10 dollars, although for someone with a high income the difference can be almost 20 dollars. The Tax Table, the Tax Rate Schedules, and other tax-related forms and schedules are those that were in effect for 1992. From time to time, tax laws are changed so that the particular details shown here may differ from those currently in effect. However, the overall principles and procedures affecting your tax obligations have not significantly changed.

EXAMPLE 2 Marilyn's taxable income is $102,000.

QUESTION How much does Marilyn, who is single, pay in taxes?

SOLUTION
Since $51,900 < I$, use the third Schedule X formula.

$t = 11,743.50 + 0.31(102,000 - 51,900)$
$t = 27,274.50$

Marilyn's tax is $27,274.50.

―――――― **SKILL 3** ――――――

To find taxable income I, you must reduce the total income T by any *deductions D* and any *exemptions E*.

Taxable income = total income − (deductions + exemptions)
$$I = T - (D + E)$$

Deductions and exemptions will be explained more fully in later lessons.

EXAMPLE 3 Luis's monthly income for this year was $675. He had a deduction of $3600 and an exemption of $2300. He has one withholding allowance, and he is single.

QUESTIONS 1. What was Luis's total income for the year?
2. What was his taxable income?
3. How much income tax does he owe?
4. How much was withheld from his paychecks?
5. Does he have a refund? If so, how much is it?

SOLUTIONS

1. His total yearly income was

   675 • 12, or $8100

2. To find Luis's taxable income, you must reduce his total income by a *deduction* of $3600 and an *exemption* of $2300. Luis's taxable income is

   $I = T - (D + E)$
   $I = 8100 - (3600 + 2300)$
   $I = 2200$

   Luis's taxable income is $2200.

3. Use the Tax Table in the Reference Section or the abbreviated table in this section. The tax on $2200 is $332.

4. To find the amount that would have been withheld from Luis's check, look at the withholding tables for Single Persons—Monthly Payroll Period. A portion of that table is shown below.

SINGLE Persons—MONTHLY Payroll Period

If the wages are—		And the number of withholding allowances claimed is—					
		0	1	2	3	4	5
At least	But less than	The amount of income tax to be withheld is—					
500	520	45	16	0	0	0	0
520	540	48	19	0	0	0	0
540	560	51	22	0	0	0	0
560	580	54	25	0	0	0	0
580	600	57	28	0	0	0	0
600	640	61	32	3	0	0	0
640	680	67	38	9	0	0	0
680	720	73	44	15	0	0	0
720	760	79	50	21	0	0	0
760	800	85	56	27	0	0	0
800	840	91	62	33	3	0	0
840	880	97	68	39	9	0	0
880	920	103	74	45	15	0	0
920	960	109	80	51	21	0	0
960	1,000	115	86	57	27	0	0

   In the table, locate Luis's monthly income of $675. It falls on the line that reads "At least 640, But less than 680." Use the column for one withholding allowance, the number that Luis chose when he completed his Form W-4 (see Lesson 1–4). Under that column, find the amount withheld, $38 per month. To find the yearly withholding, multiply the monthly amount by 12: 38 • 12 = 456. The withheld amount was $456.

5. The amount owed is less than the amount already withheld. The difference is 456 − 332, or 124. Luis can expect a refund of $124.

TRY YOUR SKILLS

Use the Tax Table in this lesson or in the Reference Section to find the taxes owed on each of the following incomes. (In Exercise 4 a "head of household" could be a single parent or a person caring for an elderly parent.)

	Filing Status	Taxable Income	Tax Owed
1.	Single	$ 2,050	_____
2.	Married, filing jointly	2,200	_____
3.	Married, filing separately	2,268	_____
4.	Head of household	2,132	_____
5.	Single	11,000	_____
6.	Married, filing jointly	27,500	_____
7.	Head of household	42,000	_____
8.	Head of household	52,600	_____

People with taxable incomes over $100,000 must use the Tax Rate Schedules, but those with taxable incomes of less than $100,000 can also use the Tax Rate Schedules as a rough check on the tax that they obtain from the Tax Table. Use the appropriate Tax Rate Schedule formula to find the tax owed on each of the following incomes.

	Filing Status	Taxable Income	Formula	Tax Owed
9.	Single	$ 2,245	_____	_____
10.	Single	23,000	_____	_____
11.	Single	104,000	_____	_____
12.	Single	151,900	_____	_____
13.	Single	132,000	_____	_____
14.	Single	185,000	_____	_____
15.	Single	200,000	_____	_____

16. The taxpayers of Exercises 9 and 10 are not permitted to use the Tax Rate Schedule. Why not? If they were, which taxpayer would have a lower tax by using the Tax Rate Schedule than by using the Tax Table?

EXERCISE YOUR SKILLS

1. Why is it necessary for people to pay federal income taxes?
2. What is the date by which personal income tax returns must be filed for the previous year?
3. Does the voluntary compliance nature of our income tax system mean that you legally have the right not to pay if you wish? If not, what does it mean?

Use the Tax Table in the Reference Section of the book to find the taxes owed on each of the following incomes.

	Filing Status	Taxable Income	Tax Owed
4.	Married, filing jointly	$ 8,870	_____
5.	Married, filing separately	47,850	_____
6.	Head of household	7,280	_____
7.	Single	10,170	_____

Use Tax Rate Schedule X or Y-1 in the Reference Section to find the tax owed on each of the incomes given in Exercises 8–12. (For taxpayers with taxable incomes less than $100,000, this method serves only as a rough check on the actual tax found in the tax table.) Use the two sets of formulas shown below, where I = taxable income and t = tax on the income.

Formulas for Tax Rate Schedule X (for single taxpayers)

i. $t = 0.15I$ if $I \leq 21,450$
ii. $t = 3,217.50 + 0.28(I - 21,450)$ if $21,450 < I \leq 51,900$
iii. $t = 11,743.50 + 0.31(I - 51,900)$ if $51,900 < I$

Formulas for Tax Rate Schedule Y-1 (married filing jointly or qualifying widow or widower)

i. $t = 0.15I$ if $I \leq 35,800$
ii. $t = 5,370.00 + 0.28(I - 35,800)$ if $35,800 < I \leq 86,500$
iii. $t = 19,566.00 + 0.31(I - 86,500)$ if $86,500 < I$

	Filing Status	Taxable Income	Tax Owed
8.	Single	$107,000	_____
9.	Married, filing jointly	138,500	_____
10.	Qualifying widow	30,750	_____
11.	Single	42,654	_____
12.	Qualifying widow	68,756	_____

For the taxpayers of Exercises 10–12, find their actual tax using the Tax Table in the Reference Section. Then answer Exercises 13–15.

13. Which of the three taxpayers has the greatest dollar difference between the two methods of calculating the tax?

14. Which of the three differences is the greatest when expressed as a percent of the tax found by using the Tax Table?

15. Which, if any, of the taxpayers of Exercises 10–12 would pay a lower tax if he or she were allowed actually to use the Tax Rate Schedule?

Find the yearly income, the taxable income, the taxes owed, and the amounts to be withheld annually for the incomes described below. Use a deduction of $3600 for a single person, $6000 for a married couple, and $5250 for a head of household. Each exemption is $2300. Assume that the married couples are filing a joint return and that the head of household is single. Use the Tax Table where possible. Otherwise, use the Tax Rate Schedules.

	Filing Status	Monthly/ Yearly Income	Number of Exemptions	Taxable Income	Tax Owed	Withholding Allowances Number	Withholding Allowances Annual Amount	Refund (+) or Money Owed (−)
16.	Single	$ 825/9900	1	$ 4,000		1	$744	+$140
17.	Single	1555/	1			1		
18.	Single	785/	1			0		
19.	Single	2378/	0			0		
20.	Single	4634/	0			1		
21.	Married	3500/	4	26,800	4,024	4		
22.	Married	3777/	5			6		
23.	Married	5300/	2			2		
24.	Married	4200/	5			4		
25.	Head of household	1780/	3			3		
26.	Head of household	2116/	2			2		
27.	Head of household	3209/	4			6		

28. Write a set of algebraic formulas for Tax Rate Schedule Y-2 (for married couples filing separately), which is in the Reference Section.

29. Because of unusually large medical expenses, a married woman finds that it is to the family's advantage for her and her husband to file separately one year rather than jointly. Her taxable income that year is $111,500. Use one of the formulas that you wrote in Exercise 28 to calculate her income tax.

MIXED REVIEW

1. Daniel has a small business grooming dogs. He charges $20 for each dog that he grooms but gives a 10% discount to each customer who brings him a new customer. One week he had 27 customers, six of whom brought in a new customer. What were Daniel's gross earnings that week?

Use the table for weekly withholding amounts for single persons in the Reference Section and the fact that 7.65% of gross pay is withheld for FICA taxes to find the missing entries in the payroll summary below. Assume that the number of claimed withholding allowances is 1.

	Week Ending	Gross Pay	Withholding	FICA	Take-home Pay
2.	5/17	$115.00			
3.	5/17	$287.00			

4. Your credit card balance from 4/1–4/12 was $750. You made a payment of $125 on 4/13. You made no further charges or payments for the remainder of the month. The monthly finance charge is 1.25% and is calculated on the last day of the month. Determine your average daily balance for April.

Make a table to show the multiplier effect on the amounts below if the Federal Reserve requirement is 20%. Show the first five levels and find the total extra money that is generated at the fifth level.

5. Initial deposit: $1,200

6. Initial deposit: $10,000

7.–8. For the initial deposits of Exercises 5–6, find the total amount of new money created by the multiplier effect.

9. You have a loan of $17,500 at 9.5% for 5 years. Your monthly payments are $367.53. Determine how much money you will save if you prepay the loan at the end of 30 months.

Some students purchased some canvas bags for $2.00 each and painted them for $1.50 each. They plan to sell the painted canvas bags.

10. What is the unit cost for purchasing and painting the canvas bags?

11. Find the cost for purchasing and painting 56 canvas bags.

12. The fixed costs are $75. What are the total costs?

FORMS AND MORE FORMS: 1040EZ AND 1040A

Larry learned his lesson about automated teller machines the hard way. He still remembers the Saturday afternoon when he took so much cash from half a dozen different ATMs around town that he had to rush to the bank on Monday to put some of the money back in. He even had to borrow $100 from his mother to cover some of the checks that he had already written on his checking account. Since that weekend, he has been working to pay her back. She has hired him to do everything from mowing the lawn to doing the laundry.

Larry also has an outside job as a clerk at a sporting goods store. When he received his W-2 form in January, Larry was amazed that his earnings had been so high. But where had all that money gone? He knew that he had spent quite a lot on his car. Then, of course, he and Lorrie Anne had gone out most weekends. Larry had bought a few gifts for her, too. Even as he was spending most of his paycheck, Larry knew that he was supposed to be saving for college.

With all of his failure to control his spending, Larry has one consolation; the income taxes on his earnings have been withheld throughout the year. He would certainly be in debt to his mother if he had to borrow $309 to pay the taxes. As it is, Larry will actually get a refund. All he has to do now is fill out the proper form. Larry isn't sure which form to use, the 1040EZ or the 1040A. He wonders whether his choice of a form will affect the time he will have to wait for his refund.

OBJECTIVES: *In this lesson, we will help Larry and his parents to:*

- *Choose one of the two forms 1040EZ or 1040A.*
- *Fill out Form 1040EZ for Larry.*
- *Fill out Form 1040A for Larry's parents.*

WHICH FORM IS THE RIGHT ONE?

Larry would like to have his refund as soon as possible. For one thing, he has not saved very much of the income from his job to use at college next year. He imagines that if he puts his tax refund into a savings account, his mother will notice that he is at least making an effort to save! On the other hand, he really could use a jacket to go with the new skis that he just bought.

Larry knows that the IRS will not send his refund at all if he does not send in a properly completed form. The question is, which form? He can choose from the regular 1040, the 1040A, or the 1040EZ. Larry hopes that he can use the 1040EZ, since it sounds "E-Z" to fill out.

Form 1040EZ   As Larry will see when he reads the instructions, if he uses **Form 1040EZ,** he must meet all the following requirements:

1. His filing status is "single."
2. He does not claim any dependents.

3. He was under 65 for the entire year and not blind at the end of the year.

4. His taxable income was less than $50,000.

5. His income included only wages, salaries, tips, taxable scholarship or fellowship grants, and taxable interest of $400 or less.

6. He did not receive any advance earned-income credit payments.

7. He was not a nonresident alien at any time during the year.

8. His wages were not over $55,000 if he had more than one employer.

If he does not meet these requirements, he must use Form 1040 if he has deductions to itemize (see Lesson 9-3) or Form 1040A if he has none.

Form 1040A   If Larry chooses Form 1040A, he can still get his refund. The **Form 1040A** is longer and more complex than Form 1040EZ because it can be used by people who have a more complicated tax situation. It is the form that his parents will use. To use it, you must know a little more about *exemptions, dependents,* and *deductions.*

The filer is the person whose name first appears at the beginning of a joint return. For a nonjoint return, it is the only person whose name appears at the beginning of the return. The IRS allows you to subtract $2300 from your gross income for each exemption that you can claim. An **exemption** may normally be claimed for each family member, including the filer. However, except for the filer's spouse, a family member may be counted as an exemption only if he/she meets the IRS's tests for being a **dependent.** If you are a dependent of your parents (or of someone else), then you cannot claim an exemption for yourself.

The qualifications for being a dependent are fairly complicated. For a complete account of the rules, you should obtain an IRS publication such as *Your Federal Income Tax, for Individuals* (Publication 17). (All IRS publications are free.) However, the basic requirements, all of which must be satisfied, are that a dependent must:

- Be a relative of the filer or a member of the filer's household
- Be a citizen or resident of the United States
- Have a gross income (before adjustments) of less than $2300
    unless he or she is under 19 and is a child of the filer, or
        unless he or she is under 24 and qualifies as a student and is a child of the filer
- Have received more than half of his or her support from the filer

Every person filing a return may subtract a number called a **deduction.** People with expenses such as large medical bills, mortgage interest, and charitable gifts may be able to list, or itemize, these expenses and subtract them from their income as **itemized deductions** on Schedule 1040, as described in Lesson 9–3. The IRS allows every filer to choose the larger of his or her itemized deductions or a fixed number called a **standard deduction.** Most people choose the standard deduction, since it is usually larger than the itemized deduction.

People who use Form 1040A can find the standard deduction that they are allowed by referring to the standard deduction charts in the booklet that accompanies the form. They contain the same information as shown in the following charts, which have been adapted from IRS Publication 17.

Standard Deduction Tables

Standard Deduction Chart for Most People*

If Your Filing Status is:	Your Standard Deduction Is:
Single	$3,600
Married filing joint return or Qualifying widow(er) with dependent child	6,000
Married filing separate return	3,000
Head of household	5,250

* DO NOT use this chart if you were 65 or older or blind, OR if someone can claim you as a dependent.

Standard Deduction Chart for People Age 65 or Older or Blind*

Check the correct number of boxes below. Then go to the chart.

You   65 or older ☐   Blind ☐

Your spouse, if claiming spouse's exemption   65 or older ☐   Blind ☐

Total number of boxes you checked ☐

If Your Filing Status is:	And the Number in the Box Above is:	Your Standard Deduction is:
Single	1	$4,500
	2	5,400
Married filing joint return or Qualifying widow(er) with dependent child	1	6,700
	2	7,400
	3	8,100
	4	8,800
Married filing separate return	1	3,700
	2	4,400
	3	5,100
	4	5,800
Head of household	1	6,150
	2	7,050

* If someone can claim you as a dependent, use the worksheet in Table 20-3, instead.

Caution: If you are married filing a separate return and your spouse itemizes deductions, or if you are a dual-status alien, you cannot take the standard deduction even if you were 65 or older or blind.

Standard Deduction Worksheet for Dependents*

If you were 65 or older or blind, check the correct number of boxes below. Then go to the worksheet.

You	65 or older ☐	Blind ☐	
Your spouse, if claiming spouse's exemption	65 or older ☐	Blind ☐	

Total number of boxes you checked ☐

1. Enter your **earned income** (defined below). If none, go on to line 3	1. _____
2. Minimum amount	2.　$600
3. Compare the amounts on lines 1 and 2. Enter the **larger** of the two amounts here	3. _____
4. Enter on line 4 the amount shown below for your filing status. • Single, enter $3,600 • Married filing separate return, enter $3,000 • Married filing jointly or Qualifying widow(er) with dependent child, enter $6,000 • Head of household, enter $5,250	4. _____
5. Standard deduction. a. Compare the amounts on lines 3 and 4. Enter the **smaller** of the two amounts here. If under 65 and not blind, stop here. This is your standard deduction. Otherwise, go on to line 5b	5a. _____
b. If 65 or older or blind, multiply $900 ($700 if married or qualifying widow(er) with dependent child) by the number in the box above. Enter the result	5b. _____
c. Add lines 5a and 5b. This is your standard deduction for 1992.	5c. _____

Earned income *includes wages, salaries, tips, professional fees, and other compensation received for personal services you performed. It also includes any amount received as a scholarship that you must include in your income.*

* Use this worksheet ONLY if someone can claim you as a dependent.

Ask Yourself

1. What is the maximum taxable income that a person may have and still be allowed to use Form 1040EZ?

2. What basic requirements must a person meet to be claimed as a dependent?

3. What kinds of income besides salaries must be reported on federal tax forms?

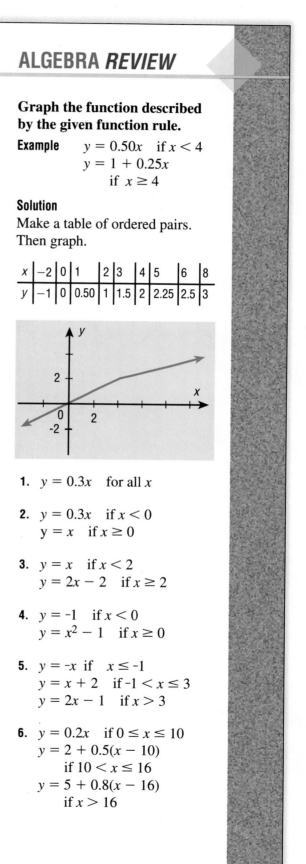

ALGEBRA *REVIEW*

Graph the function described by the given function rule.

Example　$y = 0.50x$　if $x < 4$
$y = 1 + 0.25x$
　　if $x \geq 4$

Solution
Make a table of ordered pairs. Then graph.

x	−2	0	1	2	3	4	5	6	8
y	−1	0	0.50	1	1.5	2	2.25	2.5	3

1. $y = 0.3x$　for all x

2. $y = 0.3x$　if $x < 0$
$y = x$　if $x \geq 0$

3. $y = x$　if $x < 2$
$y = 2x - 2$　if $x \geq 2$

4. $y = -1$　if $x < 0$
$y = x^2 - 1$　if $x \geq 0$

5. $y = -x$ if　$x \leq -1$
$y = x + 2$　if $-1 < x \leq 3$
$y = 2x - 1$　if $x > 3$

6. $y = 0.2x$　if $0 \leq x \leq 10$
$y = 2 + 0.5(x - 10)$
　　if $10 < x \leq 16$
$y = 5 + 0.8(x - 16)$
　　if $x > 16$

SHARPEN YOUR SKILLS

EXAMPLE 1 Larry meets the requirements for using Form 1040EZ.

QUESTION How will he fill out the form to determine the tax he owes?

SOLUTION
Larry completed the form as shown on page 426. He used the Tax Table and the instructions on the back of the form. Then he followed these steps.

Step 1   Larry filled in his name, address, and Social Security number.

Step 2   Larry reported all of his wages on line 1 and the interest from his savings account on line 2. He then added lines 1 and 2 to get his **adjusted gross income** on line 3.

Step 3   Larry used a worksheet to enter "Yes" and "3600" on line 4 of the form. The worksheet is shown below.

Standard Deduction Worksheet for Dependents*

If you were 65 or older or blind, check the correct number of boxes below. Then go to the worksheet.

You	65 or older ☐	Blind ☐
Your spouse, if claiming spouse's exemption	65 or older ☐	Blind ☐

Total number of boxes you checked ☐

1. Enter your **earned income** (defined below). If none, go on to line 3	1. *5,460*
2. Minimum amount	2. $600
3. Compare the amounts on lines 1 and 2. Enter the **larger** of the two amounts here	3. *5,460*
4. Enter on line 4 the amount shown below for your filing status. • Single, enter $3,600 • Married filing separate return, enter $3,000 • Married filing jointly or Qualifying widow(er) with dependent child, enter $6,000 • Head of household, enter $5,250	4. *3,600*
5. Standard deduction. **a.** Compare the amounts on lines 3 and 4. Enter the **smaller** of the two amounts here. If under 65 and not blind, stop here. This is your standard deduction. Otherwise, go on to line 5b	5a. *3,600*
b. If 65 or older or blind, multiply $900 ($700 if married or qualifying widow(er) with dependent child) by the number in the box above. Enter the result	5b._____
c. Add lines 5a and 5b. This is your standard deduction for 1992.	5c._____

Earned income *includes wages, salaries, tips, professional fees, and other compensation received for personal services you performed. It also includes any amount received as a scholarship that you must include in your income.*

* Use this worksheet ONLY if someone can claim you as a dependent.

He then subtracted line 4 from line 3 to get line 5, his **taxable income.**

Step 4   On line 6, Larry entered the federal income tax withheld shown on his W-2 form. Then, to find his tax on line 7, Larry first found the portion of the Tax Table shown on page 411 of Lesson 9–1. Since his taxable income was between $2,050 and $2,075, he selected the tax that he found to the right of these numbers under the column headed "Single." He entered this tax, $309, on line 7 of the form.

Step 5   Since the tax is less than the $478 that was withheld from his paycheck, Larry was entitled to a refund. He subtracted line 7 from line 6: 478 − 309, or 169. His refund was $169.

To verify his work, Larry used the checklist on the back of Form 1040EZ. The checklist is similar to the following.

Checklist for Form 1040EZ

1. Did you check your computations (additions, subtractions, and so on), especially when calculating your taxable income, federal income tax withheld, and your refund or the amount you owe?

2. Did you check the "Yes" box on line 4 if your parents (or someone else) can claim you as a dependent on their return, even if they chose not to claim you? If no one can claim you as a dependent, did you check the "No" box?

3. Did you enter an amount on line 4? If you checked the "Yes" box on line 4, did you fill in the worksheet to figure the amount to enter? If you checked the "No" box, did you enter 5,900?

4. Did you use the amount from line 5 to find your tax in the Tax Table? Did you enter the correct tax on line 7?

5. If you didn't get a label, did you enter your name, address (including ZIP Code), and Social Security number in the spaces provided on Form 1040EZ?

6. If you got a label, does it show your correct name, address, and Social Security number? If not, did you enter the correct information?

7. Did you attach your W-2 form(s) to the left margin of your return? And did you sign and date Form 1040EZ and enter your occupation?

Notice that Larry used whole dollar amounts when filling out his income tax return. In Example 2, Larry's parents use dollars and cents when filling out their income tax return. The IRS allows you to round to the nearest dollar, if you prefer.

Form **1040EZ**

Department of the Treasury—Internal Revenue Service

Income Tax Return for
Single Filers With No Dependents   **19--**

OMB No. 1545-0675

Name & address

Use the IRS label (see page 10). If you don't have one, please print.

Please print your numbers like this:

$$9\ 8\ 7\ 6\ 5\ 4\ 3\ 2\ 1\ 0$$

L A B E L H E R E	Print your name (first, initial, last)	
	LARRY L. LENDER	
	Home address (number and street). If you have a P.O. box, see page 10.	Apt. no.
	1040 LESSON ST.	
	City, town or post office, state, and ZIP code. If you have a foreign address, see page 10.	
	OMAHA, INDIANA 45533	

Your social security number

3 4 5   4 5   6 7 8 9

Please see instructions on the back. Also, see the Form 1040EZ booklet.

Presidential Election Campaign (See page 10.)
Do you want $1 to go to this fund?

Note: *Checking "Yes" will not change your tax or reduce your refund.* ▶

✓ | Dollars | Cents

Report your income

Attach Copy B of Form(s) W-2 here. Attach tax payment on top of Form(s) W-2.

Note: *You* **must** *check Yes or No.*

1   Total wages, salaries, and tips. This should be shown in box 10 of your W-2 form(s). Attach your W-2 form(s).   1

5 460 . 00

2   Taxable interest income of $400 or less. If the total is more than $400, you cannot use Form 1040EZ.   2

200 . 00

3   Add lines 1 and 2. This is your **adjusted gross income.**   3

5 660 . 00

4   Can your parents (or someone else) claim you on their return?
☒ **Yes.** Do worksheet on back; enter amount from line E here.
☐ **No.** Enter 5,900.00. This is the total of your standard deduction and personal exemption.   4

3 600 . 00

5   Subtract line 4 from line 3. If line 4 is larger than line 3, enter 0. This is your **taxable income.**   5

2 060 . 00

Figure your tax

6   Enter your Federal income tax withheld from box 9 of your W-2 form(s).   6

478 . 00

7   **Tax.** Look at line 5 above. Use the amount on **line 5** to find your tax in the tax table on pages 22-24 of the booklet. Then, enter the tax from the table on this line.   7

309 . 00

Refund or amount you owe

8   If line 6 is larger than line 7, subtract line 7 from line 6. This is your **refund.**   8

169 . 00

9   If line 7 is larger than line 6, subtract line 6 from line 7. This is the **amount you owe.** Attach your payment for full amount payable to the "Internal Revenue Service." Write your name, address, social security number, daytime phone number, and "1992 Form 1040EZ" on it.   9

Sign your return

I have read this return. Under penalties of perjury, I declare that to the best of my knowledge and belief, the return is true, correct, and complete.

Keep a copy of this form for your records.

Your signature	Date
X *Larry Lender*	4-12--
	Your occupation

For IRS Use Only — Please do not write in boxes below.

For Privacy Act and Paperwork Reduction Act Notice, see page 4 in the booklet. Cat. No. 11329W Form 1040EZ (19--)

EXAMPLE 2 Larry's parents must complete the more complicated Form 1040A.

QUESTION What steps must they take to complete the form?

SOLUTION

The Lenders' completed Form 1040A appears on pages 428–429. An explanation of some of the lines follows.

Step 1 The Lenders have filled in their names, address, and Social Security numbers.

Step 2 Mr. and Mrs. Lender are married filing a joint return.

Step 3 On lines 6a and 6b the Lenders claim three exemptions, one for each of them, and one for Larry. On line 6c Larry is claimed as a dependent.

Step 4 Line 7 is the total of Mr. and Mrs. Lender's income from their salaries. These two amounts were shown on their W-2 forms:

> Mr. Lender earned $17,600.
> Mrs. Lender earned $31,000.

Line 8a shows the interest income received on a savings account that they have at their credit union.

Lines 8b, 9, 10, 11, 12, and 13 are left blank. If the Lenders had received any dividends from stock or money-market investments, they would have included these payments on line 9.

Line 14, the **total income,** is the sum of lines 7, 8a, and 9–13.

Step 5 Since the Lenders have no IRA deduction, line 15 is left blank. For the Lenders, line 16, the **adjusted gross income,** is the same as line 14.

Step 6 Line 17 at the top of page 2 of the form is the same as line 16. None of lines 18a, 18b, or 18c applies to the Lenders, so they are left blank.

Line 19, the **standard deduction,** is $6,000, since the Lenders are married and filing jointly.

Line 20 is the adjusted gross income reduced by the standard deduction.

In line 21 the Lenders multiply $2,300 by 3, the number of exemptions, to get their total exemption allowance of $6,900.

After subtracting their exemptions from line 20, the Lenders arrive at their **taxable income** in line 22.

Step 7 The Lenders find their tax on lines 23 and 27 by using the Tax Table in the Reference Section.

Lines 24, 25, and 26 are left blank.

On lines 28a and 28d the Lenders write the sum of the amounts shown as withheld on their two W-2 forms. This is their **total payment.**

Lines 28b and 28c are left blank.

Step 8 The Lenders complete the form by subtracting their total tax (line 27) from their total payments (line 28d) to get line 30, the amount of their **refund,** $1,783.80. They both sign and date the form and mail it to the IRS.

Form **1040A**	Department of the Treasury—Internal Revenue Service **U.S. Individual Income Tax Return** (0) **19--**		IRS Use Only—Do not write or staple in this space.

Label
(See page 14.)

Use the IRS label. Otherwise, please print or type.

L A B E L   H E R E

Your first name and initial — *NATHAN S.*   Last name — *LENDER*

If a joint return, spouse's first name and initial — *MAUREEN T.*   Last name — *LENDER*

Home address (number and street). If you have a P.O. box, see page 15. — *1040 LESSON ST.*   Apt. no.

City, town or post office, state, and ZIP code. If you have a foreign address, see page 15. — *OMAHA, INDIANA 45533*

OMB No. 1545-0085

Your social security number — *432 : 12 : 7777*

Spouse's social security number — *459 : 38 : 4141*

For Privacy Act and Paperwork Reduction Act Notice, see page 4.

Presidential Election Campaign Fund (See page 15.)
Do you want $1 to go to this fund?   **Yes** ☐ **No** ☐
If a joint return, does your spouse want $1 to go to this fund?   ✔

Note: *Checking "Yes" will not change your tax or reduce your refund.*

Check the box for your filing status
(See page 15.)

Check only one box.

1 ☐ Single
2 ☑ Married filing joint return (even if only one had income)
3 ☐ Married filing separate return. Enter spouse's social security number above and full name here. ▶ _____
4 ☐ Head of household (with qualifying person). (See page 16.) If the qualifying person is a child but not your dependent, enter this child's name here. ▶ _____
5 ☐ Qualifying widow(er) with dependent child (year spouse died ▶ 19 ____). (See page 17.)

Figure your exemptions
(See page 18.)

If more than seven dependents, see page 21.

6a ☑ **Yourself.** If your parent (or someone else) can claim you as a dependent on his or her tax return, **do not** check box 6a. But be sure to check the box on line 18b on page 2.

b ☑ **Spouse**

c **Dependents:** (1) Name (first, initial, and last name)	(2) Check if under age 1	(3) If age 1 or older, dependent's social security number	(4) Dependent's relationship to you	(5) No. of months lived in your home in 1992
LARRY L. LENDER		*345 45 6789*	*SON*	*12*

d If your child didn't live with you but is claimed as your dependent under a pre-1985 agreement, check here . . . . ▶ ☐

e Total number of exemptions claimed.

No. of boxes checked on 6a and 6b — *2*

No. of your children on 6c who:
• lived with you — *1*
• didn't live with you due to divorce or separation (see page 21)

No. of other dependents on 6c

Add numbers entered on lines above — *3*

Figure your total income

Attach Copy B of your Forms W-2 and 1099-R here.

If you didn't get a W-2, see page 22.

Attach check or money order on top of any Forms W-2 or 1099-R.

7	Wages, salaries, tips, etc. This should be shown in box 10 of your W-2 form(s). Attach Form(s) W-2.	7	*48,600*	*00*
8a	**Taxable** interest income (see page 24). If over $400, also complete and attach Schedule 1, Part I.	8a	*300*	*00*
b	**Tax-exempt** interest. DO NOT include on line 8a. 8b			
9	Dividends. If over $400, also complete and attach Schedule 1, Part II.	9		
10a	Total IRA distributions. 10a	10b Taxable amount (see page 25). 10b		
11a	Total pensions and annuities. 11a	11b Taxable amount (see page 25). 11b		
12	Unemployment compensation (see page 29).	12		
13a	Social security benefits. 13a	13b Taxable amount (see page 29). 13b		
14	Add lines 7 through 13b (far right column). This is your **total income.** ▶	14	*48,900*	*00*

Figure your adjusted gross income

15a	Your IRA deduction from applicable worksheet. 15a			
b	Spouse's IRA deduction from applicable worksheet. **Note:** *Rules for IRAs begin on page 31.* 15b			
c	Add lines 15a and 15b. These are your **total adjustments.**	15c		
16	Subtract line 15c from line 14. This is your **adjusted gross income.** If less than $22,370, see "Earned income credit" on page 39. ▶	16	*48,900*	*00*

Cat. No. 11327A   19-- **Form 1040A** page 1

Name(s) shown on page 1	Your social security number
	432 12 7777

Figure your standard deduction, exemption amount, and taxable income

17	Enter the amount from line 16.	17	48,900	00

18a Check if: { ☐ **You** were 65 or older   ☐ Blind } **Enter number of**
{ ☐ **Spouse** was 65 or older ☐ Blind } **boxes checked ▶** 18a ☐

b If your parent (or someone else) can claim you as a dependent, check here . . . . ▶ 18b ☐

c If you are married filing separately and your spouse files Form 1040 and itemizes deductions, see page 35 and check here ▶ 18c ☐

19 Enter the **standard deduction** shown below for your filing status. **But if you checked any box on line 18a or b,** go to page 35 to find your standard deduction. **If you checked box 18c,** enter -0-.
- Single—$3,600   • Head of household—$5,250
- Married filing jointly or Qualifying widow(er)—$6,000
- Married filing separately—$3,000

19	6,000	00

20	Subtract line 19 from line 17. (If line 19 is more than line 17, enter -0-.)	20	42,900	00
21	Multiply $2,300 by the total number of exemptions claimed on line 6e.	21	6,900	00
22	Subtract line 21 from line 20. (If line 21 is more than line 20, enter -0-.) This is your **taxable income.** ▶	22	36,000	00

Figure your tax, credits, and payments

If you want the IRS to figure your tax, see the instructions for line 22 on page 36.

23	Find the tax on the amount on line 22. Check if from: ☒ Tax Table (pages 48–53) or ☐ Form 8615 (see page 37).	23	5,433	00

24a Credit for child and dependent care expenses. Complete and attach Schedule 2. 24a

b Credit for the elderly or the disabled. Complete and attach Schedule 3. 24b

c Add lines 24a and 24b. These are your **total credits.** 24c

25	Subtract line 24c from line 23. (If line 24c is more than line 23, enter -0-.)	25		
26	Advance earned income credit payments from Form W-2.	26		
27	Add lines 25 and 26. This is your **total tax.** ▶	27	5,433	00

28a Total Federal income tax withheld. If any tax is from Form(s) 1099, check here. ▶ ☐ 28a 7,216 80

b 1992 estimated tax payments and amount applied from 1991 return. 28b

c **Earned income credit.** Complete and attach Schedule EIC. 28c

d Add lines 28a, 28b, and 28c. These are your **total payments.** ▶ 28d 7,216 80

Figure your refund or amount you owe

Attach check or money order on top of Form(s) W-2, etc., on page 1.

29	If line 28d is more than line 27, subtract line 27 from line 28d. This is the amount you **overpaid.**	29	1,783	80
30	Amount of line 29 you want **refunded to you.**	30	1,783	80
31	Amount of line 29 you want **applied to your 1993 estimated tax.** 31			
32	If line 27 is more than line 28d, subtract line 28d from line 27. This is the **amount you owe.** Attach check or money order for full amount payable to the "Internal Revenue Service". Write your name, address, social security number, daytime phone number, and "1992 Form 1040A" on it.	32		
33	Estimated tax penalty (see page 41). 33			

Sign your return

Keep a copy of this return for your records.

Under penalties of perjury, I declare that I have examined this return and accompanying schedules and statements, and to the best of my knowledge and belief, they are true, correct, and complete. Declaration of preparer (other than the taxpayer) is based on all information of which the preparer has any knowledge.

Your signature, *Nathan Lender*   Date 4-13-   Your occupation ACCOUNTANT

Spouse's signature. If joint return, BOTH must sign. *Maureen Lender*   Date 4-13-   Spouse's occupation SYSTEMS ANALYST

Paid preparer's use only

Preparer's signature ▶   Date   Check if self-employed ☐   Preparer's social security no.

Firm's name (or yours if self-employed) and address ▶   E.I. No.   ZIP code

°U.S. Government Printing Office: 1992 — 315-155

TRY YOUR SKILLS

Fill out a 1040EZ form for each of the following incomes to determine the amount owed or refund due to the nearest dollar. Assume there is no taxable interest. All of the people are single; make up names and addresses. You may contribute to the election campaign fund; this choice will not affect anything else. The standard deduction amount for line 4 will be $3,600 unless the adjusted gross income from line 3 is less than that.

	Income	Amount Withheld	Can Be Claimed as a Dependent by Someone?
1.	$ 2,400	$ 225	Yes
2.	9,020	1,200	No
3.	3,800	420	Yes
4.	46,000	8,748	No

Fill out a 1040A form for each of the following couples to determine the amount owed or refund due to the nearest dollar. Their Social Security numbers and those of their dependents are shown in parentheses.

5. (689-78-6523) (658-31-4353)
Fred and Fawn Turner
10 Springfield Ave.
Austin, Texas 78749
Filing a joint return
Wages: Fawn, $9,700
        Fred, $18,000

Dependent:
Angela, age 7 (632-29-4168)

Withheld: from Fawn, $0
              from Fred, $1,200
Interest income: $350
Dividend income: $298

6. (652-81-3099) (618-39-4788)
Sten and Greta Larson
9500 Crosswicks Drive
Orlando, Florida 32819
Filing a joint return
Wages: Sten, $37,500
        Greta, $29,320

Dependents:
Karl, age 10   (399-68-9915)
Kristina, age 8 (015-33-2751)
Withheld: from Sten, $4,512
             from Greta, $1,504
Interest income: $287
Dividend income: $155

EXERCISE YOUR SKILLS

1. Why do you think the government does not require everyone to file the same income tax form?

2. Why do you think most married people file joint returns?

3. Why do you think people are allowed to subtract standard deductions to find their taxable income?

Fill out a 1040EZ form for each of the following incomes to determine the amount owed or refund due to the nearest dollar. For Exercises 4–8, make up a name and address. For Exercises 9–12, use the given information. All of the people are single. You may choose whether or not to contribute to the election campaign fund; this choice will not affect anything else you fill out. The standard deduction amount for line 4 will be $3,600 unless the adjusted gross income from line 3 is less than that.

	Wages, Salaries, and Tips	Interest	Amount Withheld	Can Be Claimed as a Dependent by Someone?
4.	$ 1,900	$100	$ 340	No
5.	4,200	50	600	Yes
6.	8,500	0	1,400	Yes
7.	18,600	350	2,040	No
8.	Two jobs: $15,400	200	3,700	No
	$ 2,200		0	

9. Daryl Hawkins
   1234 Sesame Street
   Phoenix, Arizona 85442
   S.S. No. 551-76-3333
   Wages: $20,650
   Withheld: $2,400
   Interest income: $50.00
   Daryl will be claimed as a
   dependent on another's return.

10. Brent Poulson
    1415 16th Ave.
    Dallas, Texas 75277
    S.S. No. 432-55-7777
    Wages: $6,200
    Withheld: $192
    Interest income: $246.80
    Brent will be claimed as a
    dependent on another's return.

11. Yazmin Rogers
    2020 Twentieth Ave.
    Seattle, Washington 98043
    S.S. No. 456-78-4321
    Wages: $39,400
    Withheld: $7,008
    Interest income: $150
    Yazmin will not be claimed
    as a dependent elsewhere.

12. Renata Taylor
    2323 Ranger Street
    Cincinnati, Ohio 45208
    S.S. No. 543-77-2222
    Wages: $35,170
    Withheld: $5,796
    Interest income: $310
    Renata will not be claimed
    as a dependent elsewhere.

Fill out a 1040A form for each of the following couples or individuals to determine the amount owed or refund due to the nearest dollar. Their Social Security numbers are shown in parentheses.

13. (421-32-4587) (476-58-9023)
Flora and Ernest Jones
10101 Barley Street
Anaheim, California 98043
Filing a joint return
Wages: Flora, $19,400
          Ernest, $21,620
Withheld: from Flora, $2,350
            from Ernest, $1,836

Dependents:
Bea, age 6   (452-79-5644)
Phillip, age 14   (751-44-9086)
Earl, age 17   (555-44-9922)

Interest income: $200
Dividend income: $180

14. (458-29-7011) (492-71-6322)
Emilio and Rosa Ortiz
13422 Euclid Avenue
Des Moines, Iowa 55105
Filing a joint return
Wages: Emilio, $48,750
          Rosa, $53,790
Withheld: from Emilio, $9,212
            from Rosa, $8,268

Dependents:
Rosita, age 4   (711-42-1195)
Julio, age 8   (329-44-7890)
Carlos, age 12   (666-21-4434)

Interest income: $280
Dividend income: $195

15. (463-44-9292)
Janice Parker
9286 Spring Blvd.
Two Egg, Florida 32304
Filing as head of household
Wages: $38,450
Withheld: $3,800

Dependents:
Dimitri, age 12 (428-77-4083)
Newton, age 14 (523-97-6702)
Sophia, age 16 (452-95-2965)

Interest income: $40
Dividend income: $300

16. (716-43-9026)
John Madison, Jr.
1409 S. Alta Vista
San Remo, Texas 77566
Filing as head of household
Wages: $47,060
Withheld: $8,952

Dependent:
John Madison, Sr., age 87
(490-62-6710)

Interest income: $390
Dividend income: $309

MIXED REVIEW

Determine the monthly payment and total payments (deferred payment price) for each car loan.

1. $15,500 at $6\frac{1}{2}$% for 5 years

2. $9,000 at 10% for 4 years

3. Use the Comparison Table for Term and Whole Life Premiums on page 326 to find the yearly premiums for $100,000 of whole life insurance and of 5-year term insurance, each purchased at the age of 45. Which is the less expensive insurance and what is the first-year saving if the less expensive insurance is purchased?

4. How many whole shares of Truffles Company stock can you buy if you have $37,665 to invest and the stock is selling for $102\frac{1}{4}$ per share? Ignore commission costs.

Use the credit scoring table on page 293 of Lesson 6–6 to determine the credit score for each family described below. Then determine the probability that the head of the household will repay the loan.

5. Family 1
- Age of the head of household: 40
- Lived at current address: 6 years
- Age of their current car: 2 years
- Monthly car payment: $190
- Housing cost: $800 per month
- They have a checking account and a savings account
- They have not been referred to a finance company
- They have 4 major credit cards
- Their ratio of debt to income is 15%

6. Family 2
- Age of the head of household: 24
- Lived at current address: 3 years
- Age of their current car: 4 years
- Monthly car payment: $90
- Housing cost: $350 per month
- They have a checking account but no savings account
- They have not been referred to a finance company
- They have 8 major credit cards
- Their ratio of debt to income is 15%

Suppose that an exponential regression model gives the following equations as a description of the cost of living over a 5-year period. What would be the expected rate of inflation?

7. $y = 150(1 + 0.06)^x$

8. $y = 83.5(1.027)^x$

FORM 1040, SCHEDULE A, AND HELP!

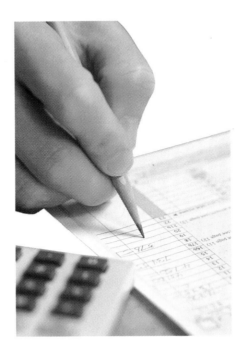

Evelyn went into business with her friends selling T-shirts and sweatshirts decorated with hand-painted designs. After a few months the partners had their production process fairly well under control, and the enterprise began to show a profit. Now it is February; the business has been in operation since August. Evelyn knows that she must file an income tax return for the income she has received from the business.

She wants her tax return to be completed very carefully. She heard some of her friends talking about the problems of the father of one of their classmates. They were discussing several of his visits to the IRS office. Apparently, the IRS examined some of his tax returns from earlier years. The way Evelyn understood the matter, he had to pay over $12,000 in back taxes that the IRS said he owed.

Evelyn knew she couldn't begin to scrape together $12,000 to pay the government, especially not at any one time. Her parents assure her that unless her profits are over $60,000 in one year, she will not have to pay that much. They have had years when their tax bill was that high, however, so Evelyn knows that the federal income tax is not an insignificant amount.

If Evelyn decides that she cannot fill out the proper forms and needs some help, several sources of assistance are available. Evelyn will explore some of these sources.

OBJECTIVES: *In this lesson, we will help Evelyn to:*

Find where a taxpayer can go for help in filing a tax return.
Complete Form 1040 and Schedule A.

TAX HELP: WHERE TO GET IT

When Evelyn first looked at the array of different tax forms that are available at the local IRS office, she was not sure which ones to pick up. She finally settled on Form 1040, a booklet of instructions for Form 1040, and a few other IRS publications that looked as if they might have something to do with business.

Evelyn's parents have been filing the 1040 form every year since they bought their first house. The interest that they pay on the mortgage is deductible. However, to take advantage of this deduction, they must use the 1040 form.

Evelyn, who cannot tell a deduction from a dependent, looks at the 1040 form and wonders whether it is truly written in English. She recognizes the words, but making sense out of what they say is a formidable task. Fortunately for Evelyn, there are people who make their livings interpreting all the rules set up by the IRS.

In fact, by the middle of the 1980s, over one third of all taxpayers sought help from the IRS, and millions of others paid from ten dollars to thousands of dollars for tax help from other sources. For Evelyn and millions like her, four major sources of tax help are available: the IRS itself, professional tax preparers including mass-market tax preparation services, CPA tax specialists, and tax lawyers. The first three of these sources are described below.

Internal Revenue Service   The IRS has a toll-free telephone answering service and nearly 1000 tax assistance centers located throughout the United States. Under certain circumstances the IRS will figure your tax for you on Form 1040, Form 1040A, or Form 1040EZ. Of course, you must provide complete information.

Relying on an IRS employee's opinion about a tax issue is not always a good idea. The IRS has been known to give taxpayers wrong forms, answer questions incorrectly, make mathematical errors, and, very rarely, dispense misleading or incorrect advice in its publications. Moreover, if you make a mistake because of incorrect advice from the IRS, you will probably have little success in using this fact as an excuse for the mistake. In other words, claiming that you were misinformed will probably not help if the IRS selects your return for examination.

Professional Tax Preparation Services There are tax preparation services that will prepare your 1040A forms for a fee, sometimes as low as $10 to $25. However, since you have to find and provide all the essential records, which is most of the work, you may be able to do the whole job yourself. The IRS has exerted enormous effort to simplify the forms so that you can fill them out without help and thus save the money you would otherwise pay a tax preparer. If you feel that a preparation service is necessary, however, you should talk personally with the person who works on the return to find out something about the preparer's background.

Certified Public Accountants For complicated tax work, a certified public accountant (CPA) who specializes in taxes is the best source of tax help.

Do-It-Yourself Being in business for herself with her friends has given Evelyn more confidence, especially since the business has started to grow and make a profit. Evelyn found in the early stages of the business that visits to her local bookstore and library were very beneficial. She did notice in the business section of the bookstore a whole shelf of books offering advice on how to fill out tax returns. Evelyn's parents have read a few of these tax guides, and every year they study the new changes in the tax law.

Another source of help that has gained acceptance in recent years is **tax preparation software** that can be used by any person who has a personal computer. Programs for both federal taxes and state taxes are available. These programs may cost over $50 (less for state tax programs) and cannot normally be reused for the following year's taxes. However, after your initial purchase, you can usually buy the tax program for later years at a discount from the original price.

ALGEBRA *REVIEW*

Evaluate each expression for the indicated values of x and y.

1. $x + 0.025y$
   $x = 80, y = 10,000$

2. $x(1 + 0.025)y$
   $x = 80, y = 10,000$

3. $x - 0.075y$
   $x = 10,000, y = 30,000$

4. $x - 0.075y$
   $x = 2250, y = 30,000$

5. $x - 0.075y$
   $x = 4250, y = 44,000$

Graph each inequality on a number line.

6. $x \geq -2$ 7. $0 \leq x$

8. $x < 5$ 9. $x > -4$

10. $2 \leq x < 10$

11. $x < -1$ or $3 < x$

12. $-3 < x \leq 5$

13. $x > -2$ and $x < 1$

14. $x < 0$ or $x \geq 2$

Ask Yourself

1. What are four major sources of help in completing income tax forms?

2. What is the best source of tax help for someone who has a complicated return?

3. What sources of help are available for someone who wants to prepare his or her own tax return?

SHARPEN YOUR SKILLS

EXAMPLE 1 Mr. and Mrs. Enterprise file the 1040 form each year mainly because of the mortgage interest deduction, which requires filling out Schedule A. Many parts of the 1040 form are similar to those on the 1040A form. Quite a few of the lines on the 1040 form do not apply to the Enterprises.

QUESTION How should the Enterprises complete page 1 of Form 1040?

SOLUTION
To fill out Schedule A, the Enterprises first have to find their adjusted gross income on line 31 of Form 1040. This line is at the bottom of the first page of the form, which is shown on page 438. Here is an explanation of some of the lines of page 1.

Lines 1–5	Filing status of taxpayer; similar to 1040A
Line 6	Exemptions; similar to 1040A
Lines 7–9	Income from wages, salaries, interest, and dividends; similar to 1040A
Lines 10–22	Other kinds of income that an individual or family might have
Line 23	The family's total income
Lines 24a, 24b	The IRA contribution of $2000 from each of the Enterprises' two salaries. Neither of Evelyn's parents is covered at work by a retirement plan, so each is allowed a full deduction of the $2000 contribution.
Lines 25–29	Other adjustments to income that do not apply to the Enterprises
Line 30	The total adjustments. For the Enterprises this is just the IRA contributions.
Line 31	The adjusted gross income

1040

Department of the Treasury—Internal Revenue Service

U.S. Individual Income Tax Return

19-- (0)   IRS Use Only—Do not write or staple in this space.

For the year Jan. 1–Dec. 31, 1992, or other tax year beginning _____ , 1992, ending _____ , 19 __   OMB No. 1545-0074

Label

(See instructions on page 10.)

Use the IRS label. Otherwise, please print or type.

L A B E L   H E R E

Your first name and initial	Last name	Your social security number
FREEMONT O. ENTERPRISE		526 31 7055
If a joint return, spouse's first name and initial Last name		Spouse's social security number
ELAINE P. ENTERPRISE		475 42 9266

Home address (number and street). If you have a P.O. box, see page 10.   Apt. no.

300 MAIN STREET

City, town or post office, state, and ZIP code. If you have a foreign address, see page 10.

MUNCIE , INDIANA 47304

For Privacy Act and Paperwork Reduction Act Notice, see page 4.

Presidential Election Campaign ▶
(See page 10.)

Do you want $1 to go to this fund? ✔ Yes ☐ No

If a joint return, does your spouse want $1 to go to this fund? ✔ Yes ☐ No

Note: *Checking "Yes" will not change your tax or reduce your refund.*

Filing Status

(See page 10.)

Check only one box.

1 ☐ Single

2 ✔ Married filing joint return (even if only one had income)

3 ☐ Married filing separate return. Enter spouse's social security no. above and full name here. ▶ _____

4 ☐ Head of household (with qualifying person). (See page 11.) If the qualifying person is a child but not your dependent, enter this child's name here. ▶ _____

5 ☐ Qualifying widow(er) with dependent child (year spouse died ▶ 19 __). (See page 11.)

Exemptions

(See page 11.)

6a ☒ **Yourself.** If your parent (or someone else) can claim you as a dependent on his or her tax return, **do not** check box 6a. But be sure to check the box on line 33b on page 2

b ☒ **Spouse**

c **Dependents:**

(1) Name (first, initial, and last name)	(2) Check if under age 1	(3) If age 1 or older, dependent's social security number	(4) Dependent's relationship to you	(5) No. of months lived in your home in 1992
EVELYN ENTERPRISE		237 11 8642	DAUGHTER	12

If more than six dependents, see page 12.

d If your child didn't live with you but is claimed as your dependent under a pre-1985 agreement, check here ▶ ☐

e Total number of exemptions claimed

No. of boxes checked on 6a and 6b   **2**

No. of your children on 6c who:
• lived with you **1**
• didn't live with you due to divorce or separation (see page 13) _____

No. of other dependents on 6c _____

Add numbers entered on lines above ▶ **3**

Income

Attach Copy B of your Forms W-2, W-2G, and 1099-R here.

If you did not get a W-2, see page 9.

Attach check or money order on top of any Forms W-2, W-2G, or 1099-R.

7	Wages, salaries, tips, etc. Attach Form(s) W-2		7	47,650 00
8a	**Taxable** interest income. Attach Schedule B if over $400		8a	350 00
b	**Tax-exempt** interest income (see page 15). DON'T include on line 8a	8b		
9	Dividend income. Attach Schedule B if over $400		9	
10	Taxable refunds, credits, or offsets of state and local income taxes from worksheet on page 16		10	
11	Alimony received		11	
12	Business income or (loss). Attach Schedule C or C-EZ		12	
13	Capital gain or (loss). Attach Schedule D		13	
14	Capital gain distributions not reported on line 13 (see page 15)		14	
15	Other gains or (losses). Attach Form 4797		15	
16a	Total IRA distributions . 16a	b Taxable amount (see page 16)	16b	
17a	Total pensions and annuities 17a	b Taxable amount (see page 16)	17b	
18	Rents, royalties, partnerships, estates, trusts, etc. Attach Schedule E		18	
19	Farm income or (loss). Attach Schedule F		19	
20	Unemployment compensation (see page 17)		20	
21a	Social security benefits 21a	b Taxable amount (see page 17)	21b	
22	Other income. List type and amount—see page 18		22	
23	Add the amounts in the far right column for lines 7 through 22. This is your **total income** ▶		23	48,000 00

Adjustments to Income

(See page 18.)

24a	Your IRA deduction from applicable worksheet on page 19 or 20	24a 4,000 00		
b	Spouse's IRA deduction from applicable worksheet on page 19 or 20	24b		
25	One-half of self-employment tax (see page 20)	25		
26	Self-employed health insurance deduction (see page 20)	26		
27	Keogh retirement plan and self-employed SEP deduction	27		
28	Penalty on early withdrawal of savings	28		
29	Alimony paid. Recipient's SSN ▶ _____	29		
30	Add lines 24a through 29. These are your **total adjustments** ▶		30	4,000 00

Adjusted Gross Income

31 Subtract line 30 from line 23. This is your **adjusted gross income.** *If this amount is less than $22,370 and a child lived with you, see page EIC-1 to find out if you can claim the "Earned Income Credit" on line 56* ▶ | 31 | 44,000 00 |

Cat. No. 11320B   Form **1040** (19--)

Tax Compu-tation (See page 22.)	32	Amount from line 31 (adjusted gross income)	32	44,000 00
	33a	Check if: ☐ **You** were 65 or older, ☐ Blind; ☐ **Spouse** was 65 or older, ☐ Blind. Add the number of boxes checked above and enter the total here ▶ 33a		
	b	If your parent (or someone else) can claim you as a dependent, check here ▶ 33b ☐		
	c	If you are married filing separately and your spouse itemizes deductions or you are a dual-status alien, see page 22 and check here ▶ 33c ☐		
	34	Enter the **larger** of your: { **Itemized deductions** from Schedule A, line 26, **OR** **Standard deduction** shown below for your filing status. **But if you checked any box on line 33a or b,** go to page 22 to find your standard deduction. **If you checked box 33c,** your standard deduction is zero. • Single—$3,600 • Head of household—$5,250 • Married filing jointly or Qualifying widow(er)—$6,000 • Married filing separately—$3,000 }	34	9,210 00
	35	Subtract line 34 from line 32	35	34,790 00
	36	If line 32 is $78,950 or less, multiply $2,300 by the total number of exemptions claimed on line 6e. If line 32 is over $78,950, see the worksheet on page 23 for the amount to enter	36	6,900 00
If you want the IRS to figure your tax, see page 23.	37	**Taxable income.** Subtract line 36 from line 35. If line 36 is more than line 35, enter -0-	37	27,890 00
	38	Enter tax. Check if from **a** ☒ Tax Table, **b** ☐ Tax Rate Schedules, **c** ☐ Schedule D, or **d** ☐ Form 8615 (see page 23). Amount, if any, from Form(s) 8814 ▶ **e** _____	38	4,181 00
	39	Additional taxes (see page 23). Check if from **a** ☐ Form 4970 **b** ☐ Form 4972	39	
	40	Add lines 38 and 39 ▶	40	4,181 00
Credits (See page 23.)	41	Credit for child and dependent care expenses. Attach Form 2441	41	
	42	Credit for the elderly or the disabled. Attach Schedule R	42	
	43	Foreign tax credit. Attach Form 1116	43	
	44	Other credits (see page 24). Check if from **a** ☐ Form 3800 **b** ☐ Form 8396 **c** ☐ Form 8801 **d** ☐ Form (specify)_____	44	
	45	Add lines 41 through 44	45	0 00
	46	Subtract line 45 from line 40. If line 45 is more than line 40, enter -0- ▶	46	4,181 00
Other Taxes	47	Self-employment tax. Attach Schedule SE. Also, see line 25	47	
	48	Alternative minimum tax. Attach Form 6251	48	
	49	Recapture taxes (see page 25). Check if from **a** ☐ Form 4255 **b** ☐ Form 8611 **c** ☐ Form 8828	49	
	50	Social security and Medicare tax on tip income not reported to employer. Attach Form 4137	50	
	51	Tax on qualified retirement plans, including IRAs. Attach Form 5329	51	
	52	Advance earned income credit payments from Form W-2	52	
	53	Add lines 46 through 52. This is your **total tax** ▶	53	4,181 00
Payments Attach Forms W-2, W-2G, and 1099-R on the front.	54	Federal income tax withheld. If any is from Form(s) 1099, check ▶ ☐ [54] 5,100 00		
	55	1992 estimated tax payments and amount applied from 1991 return [55]		
	56	**Earned income credit.** Attach Schedule EIC [56]		
	57	Amount paid with Form 4868 (extension request) [57]		
	58	Excess social security, Medicare, and RRTA tax withheld (see page 26) [58]		
	59	Other payments (see page 26). Check if from **a** ☐ Form 2439 **b** ☐ Form 4136 [59]		
	60	Add lines 54 through 59. These are your **total payments** ▶	60	5,100 00
Refund or Amount You Owe Attach check or money order on top of Form(s) W-2, etc., on the front.	61	If line 60 is more than line 53, subtract line 53 from line 60. This is the amount you **OVERPAID** ▶	61	919 00
	62	Amount of line 61 you want **REFUNDED TO YOU** ▶	62	919 00
	63	Amount of line 61 you want **APPLIED TO YOUR 1993 ESTIMATED TAX** ▶ [63]		
	64	If line 53 is more than line 60, subtract line 60 from line 53. This is the **AMOUNT YOU OWE.** Attach check or money order for full amount payable to "Internal Revenue Service." Write your name, address, social security number, daytime phone number, and "1992 Form 1040" on it	64	
	65	Estimated tax penalty (see page 27). Also include on line 64 [65]		

Sign Here
Keep a copy of this return for your records.

Under penalties of perjury, I declare that I have examined this return and accompanying schedules and statements, and to the best of my knowledge and belief, they are true, correct, and complete. Declaration of preparer (other than taxpayer) is based on all information of which preparer has any knowledge.

Your signature	Date	Your occupation
▶ *Freemont Enterprise*	4/12/--	*Production Manager*
Spouse's signature. If a joint return, BOTH must sign. ▶ *Elaine Enterprise*	4-12--	Spouse's occupation *Travel Agent*

Paid Preparer's Use Only

Preparer's signature ▶	Date	Check if self-employed ☐	Preparer's social security no.
Firm's name (or yours if self-employed) and address ▶		E.I. No.	
		ZIP code	

*U.S. Government Printing Office: 19-- — 315-148

EXAMPLE 2 With page 1 of Form 1040 now completed, the Enterprises can now fill out Schedule A.

QUESTION How should the Enterprises use Schedule A to itemize their deductions?

SOLUTION

Mr. and Mrs. Enterprise have paid $4500 in mortgage interest on their home and also have several other deductible expenses, such as state income taxes and charitable contributions. Here is an explanation of the Enterprises' Schedule A, which appears on page 442.

Line 1	Medical expenses: $4250.00
Lines 2–4	This amount from line 31 (or line 32) of Form 1040 is used to reduce the amount of the medical deduction from $4250 to $950.
Line 5	During the tax year the Enterprises paid $1320 to the state of Indiana in income taxes. Of this, $1200 was money withheld by the two employers during the tax year. The balance was paid by Mr. Enterprise himself in April of the year when he sent the State of Indiana a check for $120 to complete his income tax payment for the year before that.
Line 6	Real-estate taxes on their home: $1240
Line 8	Total deductible taxes: $2560
Lines 9a–12	Interest payments paid on their home mortgage: $4500
Lines 13–16	Contributions to a local charity: $1200
Line 26	Totals from lines 4, 8, 12, and 16: $9210. The Enterprises enter this amount on line 34 of Form 1040 shown on page 439.

EXAMPLE 3 The Enterprises can now complete Form 1040 shown on page 439.

QUESTION How will the Enterprises use Schedule A to complete Form 1040?

SOLUTION

Line 32	This is the same as line 31, the adjusted gross income.
Lines 33–34	These lines determine your deduction. They are similar to the lines of Form 1040A except that here there is a choice of taking an itemized deduction based on Schedule A.
Line 35	The gross adjusted income reduced by the itemized deductions: $44,000 - 9210 = \$34,790$.

Line 36	The exemptions: 3 exemptions • $2300 = $6900.
Line 37	The taxable income: line 35 − line 36, that is, 34,790 − 6900, or $27,890.
Lines 38–40	Since the taxable income is less than $100,000, the Enterprises use the Tax Table to find their tax. To the right of the line that reads "27,850 27,900" they find their tax, $4,181, in the second column. If the income had been greater than $100,000, they would have used the Tax Rate Schedule instead.
Lines 41–52	The Enterprises have no credits to subtract from their tax on line 40 or other taxes to pay.
Line 53	The total tax. Since there are no credits or other taxes, the total tax is the same as line 40 above.
Lines 54–60	Payments already made on the tax. For the Enterprises this is just line 54, the payments withheld on the W-2 forms.
Lines 61–62	Since line 60 is greater than line 53, the Enterprises have a refund: $5100 − $4181 = $919.

The Enterprises both sign their Form 1040. Then, before the filing deadline, they attach the W-2 forms to the form and mail it together with Schedule A to the IRS in an envelope that the IRS has provided.

If line 5 (Form 1040EZ), line 22 (Form 1040A), or line 37 (Form 1040) is—		And you are—			
At least	But less than	Single	Married filing jointly *	Married filing separately	Head of a house-hold
			Your tax is—		
26,000					
26,000	26,050	4,499	3,904	4,960	3,904
26,050	26,100	4,513	3,911	4,974	3,911
26,100	26,150	4,527	3,919	4,988	3,919
26,150	26,200	4,541	3,926	5,002	3,926
26,200	26,250	4,555	3,934	5,016	3,934
26,250	26,300	4,569	3,941	5,030	3,941
26,300	26,350	4,583	3,949	5,044	3,949
26,350	26,400	4,597	3,956	5,058	3,956
26,400	26,450	4,611	3,964	5,072	3,964
26,450	26,500	4,625	3,971	5,086	3,971
26,500	26,550	4,639	3,979	5,100	3,979
26,550	26,600	4,653	3,986	5,114	3,986
26,600	26,650	4,667	3,994	5,128	3,994
26,650	26,700	4,681	4,001	5,142	4,001
26,700	26,750	4,695	4,009	5,156	4,009
26,750	26,800	4,709	4,016	5,170	4,016
26,800	26,850	4,723	4,024	5,184	4,024
26,850	26,900	4,737	4,031	5,198	4,031
26,900	26,950	4,751	4,039	5,212	4,039
26,950	27,000	4,765	4,046	5,226	4,046
27,000					
27,000	27,050	4,779	4,054	5,240	4,054
27,050	27,100	4,793	4,061	5,254	4,061
27,100	27,150	4,807	4,069	5,268	4,069
27,150	27,200	4,821	4,076	5,282	4,076
27,200	27,250	4,835	4,084	5,296	4,084
27,250	27,300	4,849	4,091	5,310	4,091
27,300	27,350	4,863	4,099	5,324	4,099
27,350	27,400	4,877	4,106	5,338	4,106
27,400	27,450	4,891	4,114	5,352	4,114
27,450	27,500	4,905	4,121	5,366	4,121
27,500	27,550	4,919	4,129	5,380	4,129
27,550	27,600	4,933	4,136	5,394	4,136
27,600	27,650	4,947	4,144	5,408	4,144
27,650	27,700	4,961	4,151	5,422	4,151
27,700	27,750	4,975	4,159	5,436	4,159
27,750	27,800	4,989	4,166	5,450	4,166
27,800	27,850	5,003	4,174	5,464	4,174
27,850	27,900	5,017	4,181	5,478	4,181
27,900	27,950	5,031	4,189	5,492	4,189
27,950	28,000	5,045	4,196	5,506	4,196

Schedule A—Itemized Deductions

(Schedule B is on back)

► **Attach to Form 1040.** ► **See Instructions for Schedules A and B (Form 1040).**

OMB No. 1545-0074

19--

Attachment
Sequence No. **07**

Name(s) shown on Form 1040

FREEMONT AND ELAINE ENTERPRISE

Your social security number

526 31 7055

Medical and Dental Expenses		**Caution:** *Do not include expenses reimbursed or paid by others.*			
	1	Medical and dental expenses (see page A-1)	1	4,250 00	
	2	Enter amount from Form 1040, line 32. 2 44,000			
	3	Multiply line 2 above by 7.5% (.075)	3	3,300 00	
	4	Subtract line 3 from line 1. If zero or less, enter -0-. ►	4		950 00
Taxes You Paid	5	State and local income taxes	5	1,320 00	
	6	Real estate taxes (see page A-2)	6	1,240 00	
(See page A-1.)	7	Other taxes. List—include personal property taxes . ►	7		
	8	Add lines 5 through 7 . . . . ►	8		2,560 00
Interest You Paid	9a	Home mortgage interest and points reported to you on Form 1098	9a	4,500 00	
	b	Home mortgage interest not reported to you on Form 1098. If paid to an individual, show that person's name and address. ►			
(See page A-2.)		....			
		....			
Note: Personal interest is not deductible.		....	9b		
	10	Points not reported to you on Form 1098. See page A-3 for special rules	10		
	11	Investment interest. If required, attach Form 4952. (See page A-3.)	11		
	12	Add lines 9a through 11 . . . . ►	12		4,500 00
Gifts to Charity		**Caution:** *If you made a charitable contribution and received a benefit in return, see page A-3.*			
	13	Contributions by cash or check	13	1,200 00	
(See page A-3.)	14	Other than by cash or check. If over $500, you **MUST** attach Form 8283	14		
	15	Carryover from prior year	15		
	16	Add lines 13 through 15 . . . . ►	16		1,200 00
Casualty and Theft Losses	17	Casualty or theft loss(es). Attach Form 4684. (See page A-4.) . . . . ►	17		
Moving Expenses	18	Moving expenses. Attach Form 3903 or 3903F. (See page A-4.). . . . . . ►	18		
Job Expenses and Most Other Miscellaneous Deductions	19	Unreimbursed employee expenses—job travel, union dues, job education, etc. If required, you **MUST** attach Form 2106. (See page A-4.) ►	19		
	20	Other expenses—investment, tax preparation, safe deposit box, etc. List type and amount ►			
(See page A-5 for expenses to deduct here.)		....	20		
	21	Add lines 19 and 20	21		
	22	Enter amount from Form 1040, line 32. 22			
	23	Multiply line 22 above by 2% (.02)	23		
	24	Subtract line 23 from line 21. If zero or less, enter -0-. . . . . ►	24		
Other Miscellaneous Deductions	25	Other—from list on page A-5. List type and amount ► .... ►	25		
Total Itemized Deductions	26	Is the amount on Form 1040, line 32, more than $105,250 (more than $52,625 if married filing separately)? • **NO.** Your deduction is not limited. Add lines 4, 8, 12, 16, 17, 18, 24, and 25. • **YES.** Your deduction may be limited. See page A-5 for the amount to enter. } ►	26		9,210 00
		Caution: *Be sure to enter on Form 1040, line 34, the **LARGER** of the amount on line 26 above or your standard deduction.*			

For Paperwork Reduction Act Notice, see Form 1040 instructions.   Cat. No. 11330X   Schedule A (Form 1040) 19--

TRY YOUR SKILLS

Suppose that the Enterprises had made no contribution to their IRAs. Find what effect, if any, that would have had on each of the following.

1. Their adjusted gross income
2. Their medical deduction
3. Their mortgage deduction
4. Their total deductions
5. Their taxable income
6. Their tax
7. Their total payments (line 60)
8. Their refund

EXERCISE YOUR SKILLS

1. Why should taxpayers be informed about income tax laws even when they choose to have someone fill out their forms for them?
2. Why should taxpayers with complicated tax work use a certified public accountant who specializes in taxes?
3. When should taxpayers file a Schedule A instead of claiming the standard deduction?

Use the information provided for the two families below to answer Exercises 4–16. The numbers in parentheses are Social Security numbers.

<table>
<tr><td>(465–88–1033) (543–66–7121)
Howard and Florence Charisma
4565 Gulfstream
Panama Beach, Florida 32233
Wages: Howard, $37,800
        Florence, $21,866
Withheld: from Howard, $4020
        from Florence, $1992
Interest income: $250
Dividend income: $300
IRA for Florence: $2000</td>
<td>Filing Status: Married filing jointly
Dependents:
Loren, age 5   (305–67–7868)
Charlie, age 18 (690–13–9870)

Itemized deductions:
Medical expenses: $6500
Real-estate taxes: $1200
Mortgage interest on home: $6420
Contributions: $1300</td></tr>
</table>

<table>
<tr><td>(143–68–2569) (390–22–1378)
Macon and Milly Logan
12322 Commonwealth Avenue
Boston, Massachusetts 02108
Wages: Macon, $44,000
        Milly, $31,700
Withheld: from Macon, $5440
       from Milly, $4200
Interest income: $375
Dividend income: $200
IRA for Macon and Milly: $4000
(Neither Macon nor Milly has a pension plan at work.)</td>
<td>Filing Status: Married filing jointly
Dependents:
Molly, age 17   (531–22–3657)
Mark, age 17   (531–22–3656)

Itemized deductions:
Medical expenses: $700
Real-estate taxes: $2253
Local income taxes: $4000
Mortgage interest on home: $7564
Contributions: $1600</td></tr>
</table>

4. Complete the first page of Form 1040 for Howard and Florence Charisma to find their adjusted gross income to the nearest dollar.

5. Complete Schedule A for the Charismas to find their total itemized deductions to the nearest dollar.

6. Complete page 2 of Form 1040 for the Charismas to find their tax owed or refund due to the nearest dollar.

7. Complete the first page of Form 1040 for Macon and Milly Logan to find their adjusted gross income to the nearest dollar.

8. Complete Schedule A for the Logans to find their total itemized deductions to the nearest dollar.

9. Complete page 2 of Form 1040 for the Logans to find their tax owed or refund due to the nearest dollar.

Mr. Charisma expects to be self-employed next year, earning the same amount of money as this year. He must fill out a Schedule C form for self-employed persons. However, you are not expected to fill out Schedule C to answer questions 10–13.

10. If he earns $37,800 next year, on which line of Form 1040 will he enter his business income as a self-employed person?

11. If Mr. Charisma becomes self-employed, he will be able to start a Keogh plan (self-employment retirement plan) by contributing 20% of his earnings to the plan each year. On which line of Form 1040 will he enter his contribution?

12. Prepare a Form 1040 and a Schedule A for the Charisma family to show their adjusted gross income, their total itemized deductions, and their tax owed or refund due to the nearest dollar next year when Mr. Charisma is self-employed with a Keogh plan. Assume that all of the earnings and expenses are the same as for this year. Note that Mr. Charisma is both an employer and an employee, so his FICA, or self-employment tax (line 47) is calculated at the rate of $2 \cdot 7.65\%$, or 15.3%. One-half of his self-employment tax (line 25) is half of the line 47 entry. You may assume that Mr. Charisma's estimated tax payments (line 55) are the same as what his withheld taxes had been in Exercises 4–6.

13. How much money will the Charisma family save on taxes next year because of the Keogh plan?

14. Suppose that Mr. and Mrs. Logan had not contributed to their IRAs this year. How much more would they have paid in taxes?

15. Suppose that instead of owning their own home, the Logans were renting a house and paying $9817 in rent. (This is the total of their current real-estate taxes and mortgage interest.) Would they find it worthwhile to use Schedule A?

16. Use the assumptions of Exercise 15, with the IRA deduction, to find out how much more or how much less the Logans would pay in taxes if they were renting instead of owning.

1. At the beginning of a monthly billing period, Albert's Mastercard balance was $385. The monthly finance charge was 1.7% of the amount owed. His bank required him to pay at least 15% of his unpaid balance or $25, whichever was larger. If he always paid the minimum required amount and made no additional purchases, in how many months did Albert reduce his balance to below $200?

Find the break-even premium for a 1-year term insurance policy for each indicated individual. Assume that the direct and indirect expenses for issuing one policy are $20.

2. A $100,000 policy for a person with a life expectancy of 99.750%

3. A $200,000 policy for a person with a life expectancy of 98.950%

Use the prices per share shown below to find the number of shares of stock you can afford to purchase if you have $50,000 to invest in each of the following companies. Also, find the total cost of the shares in each company. Round all money amounts to the nearest penny. Remember that you cannot purchase part of a share of stock. Ignore transaction costs such as brokerage commissions.

	Company	Price per Share	Number of Shares	Total Cost
4.	BullyPulp	$ 97		
5.	ElbGrs, Inc.	$18\frac{3}{4}$		
6.	Strm&Drng	$210\frac{3}{8}$		

Use the Tax Table in the Reference Section to find the taxes owed on each of the following incomes.

	Filing Status	Taxable Income	Tax Owed
7.	Married, Filing Jointly	$68,500	_____
8.	Single	17,800	_____

9. Suppose that your family's take-home pay is $4100 per month. How much can you afford to spend for credit-card payments each month?

10. The company wants to sell a 1-year term policy to a 45-year-old person and to keep the premium below $600. Assume that the expenses are $30 per policy and that the profit on the policy is $75. What would be the largest possible face value of such a policy?

11. If you earn $83,500 per year, how much do you pay in Social Security and Medicare taxes?

12. If you earn $142,500 per year, how much do you pay in Social Security and Medicare taxes?

CHAPTER 9 REVIEW

1. What does the term *filing status* mean?

2. May a person with taxable income over $100,000 use the tax table? If not, how does such a person find out how much tax he or she owes?

3. How does having dependents affect the amount of taxes that people pay?

Use the Tax Table in the Reference Section to find the tax owed each of the following incomes.

	Filing Status	Taxable Income	Tax Owed
4.	Single	$26,450	
5.	Married, filing jointly	87,000	
6.	Married, filing separately	8,490	
7.	Head of household	38,400	
8.	Single	1,256	
9.	Married, filing jointly	19,500	

Use Tax Rate Schedules X and Y-1 in the Reference Section and the formulas on page 416 of Lesson 9–1 to find the tax owed on each of the following incomes.

	Filing Status	Taxable Income	Tax Owed
10.	Single	$150,000	
11.	Married, filing jointly	108,000	
12.	Married, filing jointly	121,347	
13.	Single	106,734	

Find the yearly income, the taxable income, the taxes owed, and the amounts to be withheld annually for the incomes described below. Use a deduction of $3600 for a single person, $6000 for a married couple, and $5250 for a head of household. Each exemption is $2300. Assume that the married couple is filing a joint return and that the head of household is single. Use the Tax Table in the Reference Section.

	Filing Status	Monthly/ Yearly Income	Number of Exemptions	Taxable Income	Tax Owed	Withholding Allowances Number	Withholding Allowances Annual Amount	Refund (+) or Money Owed (−)
14.	Single	$ 655/	1	$1960		1		
15.	Married	$ 940/	3			2		
16.	Single	$1000/	1			2		
17.	Head of Household	$4450/	4			4		

Fill out a 1040EZ form for each of the following incomes to determine the amount owed or refund due to the nearest dollar. Assume that there is no taxable interest. For each exercise, make up a name and address. All of the people are single. You may choose whether or not to contribute to the election campaign fund; this choice will not affect anything else you fill out. The standard deduction amount for line 4 will be $3600 unless the adjusted gross income from line 3 is less than that.

	Wages, Salaries, and Tips	Interest	Amount Withheld	Can Be Claimed as a Dependent by Someone?
18.	$22,400	$350	$2460	No
19.	5,671	97	823	No
20.	12,500	290	2528	Yes
21.	18,600	350	3440	Yes

22. Fill out a 1040A Form for the following married couple filing jointly to determine the amount owed or refund due to the nearest dollar. Their Social Security numbers are shown in parentheses.

(321–45–3660) (544–36–1565)
Frederick and Patricia Moore
222 Beachfront Lane
Memphis, Tennessee 38132
Wages: Frederick, $23,026.75
        Patricia, $22,617.16
Withheld: from Frederick, $4217
        from Patricia, $4431

Dependents:
William, age 3 (534–56–8567)
Arthur, age 12 (574–24–6149)
David, age 15 (422–22–7469)

Interest income: $254
Dividend income: $195

23. Fill out Form 1040 for the following married couple filing jointly to determine their adjusted gross income, their total itemized deductions, and their tax owed or refund due to the nearest dollar. The numbers in parentheses are Social Security numbers. Use Schedule A for itemized deductions. Assume no tax penalty is owed.

(284–11–6997) (492–22–6451)
Alex and Jennifer Robertson
496 Jefferson Ave.
Longmont, Colorado 80502
Wages: Alex, $43,400
       Jennifer, $22,000
Withheld: from Alex, $5020
       from Jennifer, $2392
Interest income: $250
Dividend income: $385
IRA for Alex: $2000

Dependents:
Dorry, age 7 (551–33–1092)
Lorry, age 9 (493–44–0122)

Itemized Deductions:
Medical expenses: $3500
Real-estate taxes: $2900
Mortgage interest
  on home: $7595
Contributions: $980

CHAPTER 9 TEST

Use the Tax Table in the Reference Section to find the tax owed by each of the following taxpayers.

1. A single person with taxable income of $25,600

2. A married couple, filing jointly, with taxable income of $36,450

3. A married person, filing separately, with taxable income of $38,725

4. A head of a household with taxable income of $2600

Use the Tax Rate Schedules in the Reference Section to find the tax owed by each of the following taxpayers.

5. A single person with taxable income of $122,400

6. A married couple, filing jointly, with taxable income of $178,650

7. Write a set of algebraic formulas for Tax Rate Schedule Z (for heads of households), which is found in the Reference Section.

8. Use one of the formulas from Exercise 7 to find the tax on a single parent with two young children. The family's total taxable income is $107,550.

9. Fill out a Form 1040EZ for the taxpayer described below to determine the amount owed or refund due to the nearest dollar. He cannot be claimed as a dependent on another taxpayer's return.

   (354–20–8238)
   Nick Piocci   Salary: $16,780 per year
   3821 Oak Avenue   Interest income: $235
   Berwyn, Illinois 60402   Withheld taxes: $1822

10. Fill out 1040A form for the following married couple who file jointly and have four dependents to determine the amount owed or refund due to the nearest dollar. Fiala is blind; neither Miguel nor Fiala is over 65. To make a decision concerning the deduction on line 34, use the chart on page 422.

    (102–00–9812) (324–57–3214)
    Miguel and Fiala Ferguson
    2325 Ski Slope Drive   Interest income: $250
    Red River, New Mexico 84222   Dividend income: $320
    Miguel earns $34,222   Withheld taxes: $3564
    Fiala does not work.   IRA deduction (Miguel): $1800

11. Suppose that in Exercise 10, the Fergusons have the following expenses.

    Medical expenses: $8240   State income tax payments: $2400
    Real estate taxes: $1500   Charitable contributions: $750

    Find the Fergusons' adjusted gross income, total itemized deductions, and tax owed or refund due to the nearest dollar using Form 1040 and Schedule A. To compute the standard deduction on line 19, use the chart on page 422. (You may ignore the directions concerning page 35 of the tax booklet.) Does the family benefit by using Form 1040 instead of Form 1040A? If so, by how much?

CUMULATIVE REVIEW

Molly is considering each of the following employment opportunities.

Company	A	B	C
Salary	$42,500	$48,000	$38,800
Retirement benefits	4.0% of gross pay	3% of gross pay	4.5% of gross pay
Medical insurance	$60/month	$50/month	$56/month
Educational expense	$450/year	$900/year	$1400/year
Travel allowance	$1900/year	$200/month	$450/month

1. Calculate the total amount of money being offered, including fringe benefits for each position.

2. Determine which of the three positions offers the most money.

Three students plan to sell souvenir umbrellas at the school football game. Their cost for 30 umbrellas is as follows.

Labor	25 hours at $4.50 per hour
Materials	umbrellas at $5.50 each
Advertising, energy, and transportation	$36.50

3. Find the cost of producing the 30 umbrellas.

4. What price should they put on the umbrellas in order to make $100 in profit?

Complete the chart below that shows one month's activity in a MasterCard account. Then find the quantities asked for in Exercises 5–7. Assume that the monthly finance charge is 1.5%.

Dates	Payment	Balance at End of Day	Number of Days	Sum of Daily Balances
1/1–1/8	0.00	$1549.50	8	$12,396.00
1/9	$183.00	1366.50		
1/10–1/31	0.00	1366.50		

5. Average daily balance 6. Finance charge   7. Ending balance

Use the formula for the Future Value of a Periodic Investment on page 344 of Lesson 7–3 to answer Exercises 8–9.

8. If you contribute $2000 to an IRA account that pays 11% a year beginning at the age of 20, how much will you have in your account at age 60?

9. If you begin contributing to your IRA at the age of 30, how much less will you have in your account at the age of 60 than you would have if you had begun your contributions at the age of 20?

Use the Tax Table or the Tax Rate Schedule in the Reference Section to find the taxes for the following taxpayers.

10. A single taxpayer with a taxable income of $151,000

11. A head of household with a taxable income of $92,500

PROJECT 9–1: State Income Taxes

1. Use reference books (such as a world almanac) to find the methods different states use to determine how much state income tax their residents pay.
2. Evaluate the methods of determining state income taxes to find out in which states people pay the highest taxes.
3. Find out in which states people pay the lowest taxes.
4. Devise a way to report your findings in an interesting way. Show the five highest, the five lowest, the most unusual, and so on. Or make a bar graph comparing the income tax rates in different states.

Extensions

1. Make up some problems based on your findings to calculate the state income taxes for people making the same salaries but living in different states. Have your classmates solve some of the problems.
2. Find which cities have income taxes and what their rates of taxation are.
3. Discuss with your classmates some reasons why some states have higher income taxes than others.

PROJECT 9–2: The History of Income Taxes

Work with a group of classmates to complete this activity. Use reference books (such as encyclopedias, history books, and political science books) to find out when federal income taxes were first introduced in the United States.

1. What is the name of the law that made federal income taxes legal?
2. Why and how was it passed?
3. When was the Internal Revenue Service established?
4. Who administers the Internal Revenue Service?
5. How are changes in tax laws made?
6. Do income taxes in fact strictly conform to the "earn more, pay more" principle?

Extensions

1. Organize a debate with your classmates to argue the pros and cons of federal income taxes. Also include how the taxes could be made more equitable.
2. Investigate income tax laws of other countries in the world.
3. Calculate what percent of income the average American pays in taxes— include federal, state, city, county, and social security taxes.

Very often we want to summarize a set of numbers or scores with just one number. The mean, median, and mode are *statistics* used to do this.

The *mean* is the sum of the scores divided by the total number of scores.

The *median* is the score that lies in the middle when the scores are ordered from least to greatest. If there are an even number of scores then the median is the average of the two middle scores.

The *mode* is the score that occurs the most times. If no score occurs most frequently then the set of numbers has no mode.

Find the mean, median, and mode. Round to the nearest tenth.

Example   {70, 70, 70, 74, 75, 76, 82, 91, 97}

Mean: $\dfrac{70 + 70 + 70 + 74 + 75 + 76 + 82 + 91 + 97}{9} = 78.3$ The mean is 78.3.

Median: 70, 70, 70, 74, **75**, 76, 82, 91, 97   The median is 75.
Mode: **70, 70, 70,** 74, 75, 76, 82, 91, 97   The mode is 70.

1. {60, 68, 68, 78, 78, 78, 89, 90, 90}

2. {56, 89, 23, 48, 56, 91, 184, 56}

3. {200, 220, 220, 220, 240, 400, 800, 1000, 1000, 2000}

4. {4.1, 4.5, 4.5, 5.3, 5.3, 5.3, 6.5, 6.5, 6.9, 8.2, 8.3}

Sometimes when many scores are repeated, frequencies are used to express the results. Frequency means how many times a score is repeated.

Find the mean, median, and mode. Round to the nearest tenth.

Example   Thirty-two students took a test. The scores are shown in the table.

Score	65	76	82	83	84	87
Frequency	1	3	4	8	9	7

Mean: $\dfrac{65 + 3(76) + 4(82) + 8(83) + 9(84) + 7(87)}{32} = 82.8$ The mean is 82.8.

Median: There are 32 scores. The first score is 65. The 16th score is 83 and the 17th score is 84. The median is $(83 + 84) \div 2 = 83.5$.

Mode: The score that occurs most often is 84. The mode is 84.

5.

Score	60	65	74	75	82	85	86	90	92
Frequency	1	3	7	10	9	7	8	4	2

6.

Score	205	214	254	301	420	475	500
Frequency	6	9	14	60	34	18	54

Owning a Car

IF YOU DON'T ALREADY DRIVE YOU ARE probably looking forward to the day when you will. Your parents and teachers have probably already told you that driving is not just for fun, and a car is most certainly not a toy. Driving is a big responsibility. Unless you are a very careful driver, you might injure yourself and others.

A car is one of the most expensive items you will ever purchase. After the initial purchase of a car, there are also fixed ownership costs and variable operating costs. Even though you take good care of it, a car gradually loses its resale value.

The young people in this chapter explore some of the differences between buying a new car and buying a used car and between owning and leasing a car. They learn about estimating and budgeting the different operating costs and what protection different kinds of insurance offer. Maria suffers "sticker shock" when she first sees the prices at a new car dealership. She learns, however, that dealers are willing to negotiate.

With his father, Alex learns that it can be wise to buy a used car. Freda looks closely at the mileage her car gets as she does the driving for the T-shirt business. She and Hari discover that leasing a car can make sense, particularly for a business. Finally, Trevor learns about automobile insurance. In particular, he researches the different kinds of auto insurance available and how the rates are determined for a driver and a car.

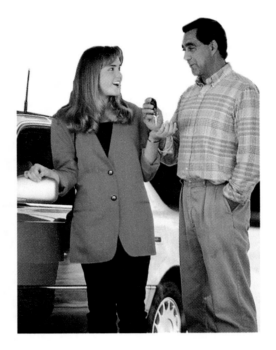

When Maria's father told her that it is expensive to own and maintain a car, she thought that he just didn't want her driving too much. Then he told her that if she could save half of her weekly paycheck for six months, he would help her buy a car. That was a challenge. Much to her surprise, she was able to set a goal and achieve it.

Maria wanted a new car, so she and her father visited several new car dealers. At first, Maria was shocked by what she found. Each new car had a sheet of paper stuck to the inside of one of the windows. On the paper was a list of various features of that car: automatic transmission, power steering, air conditioning, AM/FM stereo with cassette deck, cruise control—and on and on. Many of the features had prices listed beside them.

The total price at the bottom of the list surprised Maria the most. She wondered how ordinary people could buy new cars. Maria's father explained that the sticker price was not what the dealer actually expected to receive for the car. Then Maria was really confused. How was she to know what price to pay for a car?

Maria wondered whether there were other things that she should know before spending her hard-earned money for a car.

OBJECTIVES: *In this lesson, we will help Maria to:*

- *Find out how to shop for a new car.*
- *Calculate the percent of markup.*
- *Calculate monthly car payments.*
- *Compare terms for financing a car.*

SHOP BEFORE GOING TO A CAR DEALER

Maria has been reading consumer magazines such as *Car and Driver, Motor Trend, Consumer Reports,* and *Changing Times.* She was able to find many of these in bookstores, in newsstands, and in her local and school library. From these sources she has learned a lot about the cost and quality of different cars.

Maria still does not understand why there is such a difference between the **sticker price** that appears on a car window and what the consumer finally pays. But she found that books such as *Edmund's New Car Prices* and the *Kelley Blue Book New Car Price Manual* give useful information about what dealers pay for cars and options. The sticker price is the suggested retail price of the car. The range between the dealer's cost and the sticker price is where Maria can negotiate.

Markup is the amount that a dealer adds to his cost to arrive at the sticker price. Maria noticed that some cars and some options have a higher markup than others. She also knows that the purchase price is not everything. She will try to choose a reputable dealer with a good service department, one that will be convenient for her when service is required.

DEALING WITH DEALERS

Maria has shopped around and is prepared to bargain. She knows that she must be able to walk away from a car and go to another dealer if the price does not suit her. "Falling in love" with a car will often lead to paying too much. Maria has learned three essentials for bargaining:

1. Use competitive conditions to your advantage.
2. Know the dealer's costs.
3. Allow the dealer a reasonable profit.

What is a reasonable profit? You and the dealer might have different ideas about that. There are, of course, many other factors involved. But most of them come down to the amount of the original markup and the laws of supply and demand. In other words, if many people want a particular model car, then the dealer will not be motivated to reduce the price very much. Following are some of the tactics that car dealers might use in their efforts to attract buyers and sell cars.

Bait and Switch   Maria saw an advertisement in the newspaper that seemed too good to be true. It was. When Maria and her father visited the dealer, they found that the advertised car was not available. But the salesperson tried to interest them in another, more expensive car. This was Maria's first encounter with the bait-and-switch technique.

Financing   Maria's father explained that when the time comes for **financing,** that is, borrowing the money needed for a car, they will obtain a loan from their own bank or credit union rather than from a company recommended by the auto dealer. This is because sometimes the dealer, in cooperation with the finance company, tries to add special fees or insurance charges to the purchase price. The terms of financing include the amount to be borrowed, the purchaser's down payment, the interest rate, and the number of months of the loan. Maria will consider all of these and weigh them in light of her own financial resources. The **finance charge** is the amount that the credit will cost.

Trade-In   If Maria has a car to **trade in,** that is, to sell to the dealer when she buys a new car, then her negotiations will be even more complicated, and she will have to be even more careful. Sometimes, to get a buyer's interest, a dealer will talk about paying a high price for a trade-in and later lower the amount. In this case the dealer is using a highball strategy. In a similar way a dealer might quote you an unrealistically low price for a new car and then, when you return after shopping around, explain that, on review, the original quote must be raised. This approach is called the lowball. In most cases it is better not to mention that there is a trade-in until the price of the new car has been settled. If the price offer for your old car seems too low, you can always sell it yourself.

ALGEBRA *REVIEW*

Solve each equation or, if needed, select a variable, write an equation, and then solve.

1. $0.2 \cdot \$5647 = c$

2. $7.5\% \cdot \$7800 = c$

3. Find 2.3% of $4568.

4. Five is what percent of 15?

5. Ten is what percent of 25?

6. What percent of $7560 is $529.20?

7. Add $4500 to 9% of $4500.

8. Find 1.12% of $9800.

9. $585.12 is 12% of what number?

10. $500 is 10% of what number?

Ask Yourself

1. What should you do before shopping for a car?

2. What are some of the tactics a car dealer might use to interest a potential buyer?

3. What does financing mean?

SHARPEN YOUR SKILLS

SKILL 1

EXAMPLE 1 The sticker price for a Starfire two-door convertible is $14,255. Using her car books, Maria finds the dealer's cost is $12,450.

QUESTION What is the percent of the dealer's markup?

SOLUTION
The markup m is the difference between the sticker price s and the dealer's cost d.

$$m = s - d$$
$$m = 14{,}255 - 12{,}450$$
$$m = 1805$$

The markup is $1805.

To find p the percent of markup, Maria writes an equation using the following question to help her.

What percent of 12,450 is 1805?
$$p(12{,}450) = 1805$$
$$p = \frac{1805}{12{,}450} \qquad \text{Divide both sides by 12,450.}$$
$$p = 0.145 \qquad \text{To the nearest thousandth}$$

The percent of markup is 14.5%.

EXAMPLE 2 Maria compares the prices of the following options listed on the sticker with dealer's costs she finds in her car books.

Option	Dealer's Cost	Sticker Price
Air conditioning	$356	$420
Roof rack	116	138
Automatic transmission	645	774
AM/FM radio, tape deck	440	506

QUESTION How can Maria use a spreadsheet to show the markup and percent of markup for the options?

SOLUTION
Maria uses the formula +C3−B3 for cell D3 and +D3/B3 for cell E3. To find the totals she uses the sum function of the spreadsheet. Her spreadsheet looks like the following. Notice that she formatted column E for percent.

	A	B	C	D	E
1		Dealer	Sticker		Percent of
2	Option	Cost	Price	Markup	Markup
3	Air conditioning	356	420	64	18%
4	Roof rack	116	138	22	19%
5	Automatic transmission	645	774	129	20%
6	AM/FM radio, tape deck	440	506	66	15%
7	Total	1557	1838	281	18%

Recall the monthly payment formula for repaying a loan from Lesson 5–1.

$$M = \frac{Pr(1 + r)^n}{(1 + r)^n - 1}$$
where M = monthly payment
P = amount of loan
r = monthly interest rate
n = number of payment periods

EXAMPLE 3 Maria's friend Bryan plans to buy a new Moonbeam car with a list price of $19,490 and finance it through his credit union. He can make a down payment of 10% or 20% and then pay off the loan at a 6% annual interest rate over a period of 3 years or 5 years.

QUESTION What will be the monthly payments and the total cost of the car under the different payment plans?

SOLUTION

Bryan has four ways to pay:

10% down for 3 years
10% down for 5 years
20% down for 3 years
20% down for 5 years

He decides to use a spreadsheet to compare costs. The loan amount is the original cost minus the down payment. Use the monthly payment formula to calculate the payments for each plan. Since the APR is 6%, the monthly interest rate is $0.06 \div 12 = 0.005$. To find the total payments, remember to multiply the unrounded amount by the number of payments. The total cost includes the down payment. The finance charge is the difference between the total cost and the original cost.

	A	B	C	D	E	F	G	H	I
1	Original	Percent	Down	Loan		Monthly	Total	Total	Finance
2	Cost	Down	Payment	Amount	Years	Payment	Payment	Cost	Charge
3	19490	10%	1949	17541	3	533.63	19210.72	21159.72	1669.72
4	19490	10%	1949	17541	5	339.12	20347.00	22296.00	2806.00
5	19490	20%	3898	15592	3	474.34	17076.20	20974.20	1484.20
6	19490	20%	3898	15592	5	301.44	18086.22	21984.22	2494.22

From the spreadsheet Bryan concludes that if he can afford to make higher monthly payments he will save money.

TRY YOUR SKILLS

A car costing the dealer $12,800 has a sticker price of $15,232.

1. Find the markup.

2. Find the percent of markup.

A car costing the dealer $11,250 is marked up 18.5%.

3. Find the markup.

4. Find the sticker price.

A car is to be sold for $13,200. The annual interest rate is 8.4% with a 20% down payment for 3 years.

5. Find the monthly interest rate.

6. Find the monthly payment.

7. Find the total cost of the car.

8. Find the finance charge.

EXERCISE YOUR SKILLS

The following is a portion of the Moonbeam price list, which also appears in the Reference Section of the book.

Moonbeam	Dealer Cost	Sticker Price
Model A, 4-door wagon, 8 cylinder	$12,725	$14,722
Model B, 4-door sedan, 6 cylinder	14,062	16,295
Model C, 4-door sedan, 4 cylinder	11,053	12,755
Options		
Conventional spare tire	$ 62	$ 73
Air conditioning	695	817
Electronic climate control	850	1000
Antilock brake system	838	985
Rear window defroster	136	160
AM/FM radio and cassette player	132	155
Rear-facing third seat	132	155
Cruise control	178	210
Stripe, painted	51	61

KEY TERMS

finance charge
financing
markup
sticker price
trade-in

Find the percent of markup for each of the following. Round to the nearest tenth of a percent.

1. Model A
2. Model B
3. Model C
4. Air conditioning
5. Electronic climate control
6. Antilock brake system
7. AM/FM radio and cassette player
8. Rear facing third seat

Find the total dealer cost and total sticker price for each of the following.

9. Model B with air conditioning, rear window defroster, AM/FM radio and cassette player, cruise control

10. Model A with electronic climate control, anti-lock brakes, rear window defroster, rear-facing third seat

11. Model C with air conditioning, rear window defroster, AM/FM radio and cassette player, stripe

12. Find the percent of markup for each of the total suggested retail prices in Exercises 9, 10, and 11.

13. Find the monthly payment for Model B at the sticker price with air conditioning, rear window defroster, AM/FM radio and cassette player, and cruise control. Financing terms are 6% annual interest rate, 20% down, 3 years to pay.

14. Using the sticker price, find the monthly payment for the car and options described in Exercise 10. Terms are 8.4% annual interest rate, 15% down, 4 years to pay.

15. Using the sticker price, find the total amount paid for the car and options described in Exercise 11. Terms are 6% annual interest rate, 20% down, 5 years to pay.

Exercises 16–22 are based on the purchase of Model A at the sticker price with air conditioning, anti-lock brake system, rear window defroster, AM/FM radio and cassette player, rear-facing third seat, and cruise control.

16. Find the total dealer cost.

17. Find the total sticker price.

18. Find the percent of markup. Round to the nearest tenth of a percent.

19. Find the monthly payment at 6% APR for 6 years with 20% down.

20. Find the monthly payment at 6% APR for 4 years with 10% down.

21. In Exercises 19 and 20, which of the terms of financing has the lower monthly payment?

22. In Exercises 19 and 20, which of the terms of financing has the lower total cost?

23. Compare the terms of financing, and discuss reasons why a person might wish to select one or the other.

24. The cash price for certain used cars is shown in the following table. The annual interest rate is 9%. The payment period is 4 years. The down payment can be 10% or 20%. Use a spreadsheet program or calculator to complete the table.

Original Cost	Percent Down	Down Payment	Loan Amount	Years	Monthly Payment	Total Payment	Total Cost	Finance Charge
$7500	10%			4				
7500	20%			4				
8000	10%			4				
8000	20%			4				

25. For the used car in Exercise 24 with an original cost of $7,500, how much will be saved by making a 20% down payment?

26. Suppose the dealer agrees to finance the $8000 in Exercise 24 at a 5% annual interest rate if you make a down payment of 25%. How much will you save over putting 20% down at the 9% rate?

1. Find the value of $2500 after it has been invested at a 7% annual return and compounded annually for 5 years.

2. How long will it take for an investment to double if it increases in value by 10% a year?

3. If 1400 out of 5000 people interviewed pay their credit card bills in full each month, what percent of the sample does not pay their bills in full each month?

4. Use the formula for the Future Value of a Periodic Investment to find the value of an IRA into which annual contributions of $2000 have been made for 20 years. Assume that the investment has had an annual compounded return of 8% a year over the entire period.

5. If Marci can afford a monthly payment of $360, how much can she afford to borrow at a yearly interest rate of 9.5% for 5 years?

6. The cost function for a small business is given by the equation $c = 2.5n + 125$, where n is the number of items produced. The revenue function for the business is given by the equation $r = 6n$. Use a graphing calculator to find the intersection of the graphs of the functions, that is, the break-even point of the business.

7. Use a graphing calculator to find the linear regression equation that is the line of best fit for the following table values.

Number of Years After Start	0	0.5	1.0	1.5	2.0	2.5	3.0	3.5
Value of Mutual Fund (in $)	9	9.6	9.8	10.3	10.9	11.4	12.0	12.5

8. Use IRS Form 1040A to find the refund or payment due for the following married couple. Social Security numbers are shown in parentheses.

    (012–46–3774) (012–49–1132) Dependents:
    Alex and Nina Black   Hal, age 16   (422–55–9031)
    9500 Banyon Drive   Gladys, age 17 (422–55–9032)
    Orlando, Florida 32819   Withheld: from Alex, $3200
    Filing a joint return   from Nina, $950
    Wages: Alex, $57,500   Interest income: $4,100
              Nina, $28,000   Dividend income: $2,455

9. You are a real-estate broker who has just sold a house for $325,000. You earn 5% on the first $125,000 and 6% on any amount over $125,000. Determine your total commission.

10. Your checkbook shows a balance of $285.35. Your bank deducts a service charge of $2.50 each month but does not pay interest and does not charge for each check. The only check that you wrote during the month was for $185.44 and was mailed the day before the bank's monthly statement arrived. What will the bank show as the closing balance?

Alex thought it would be fun to help his father pick out a used van for the family. His father took him to quite a few used-car lots, where they looked at a variety of vans before they finally made the deal for the van they bought. At first, Alex did not understand why his father inspected the vans so closely and insisted on driving them, sometimes even onto the freeway and back.

Alex's father told him about the first used car that he had bought. It was only two years old and showed 14,000 miles on the odometer. Alex's father took a short test drive and didn't even notice that the odometer was not working. In fact, he didn't notice a number of things about the car. It turned out that the car was a repainted taxi with 90,000 miles on it. During the first year the car needed a new radiator, an alternator, a fuel pump, and a rebuilt transmission. These repairs were very costly.

Alex now understands why the van that his father finally purchased was thoroughly checked by their mechanic before his father bought it. He also knows that there is a lot more to buying a used car than first meets the eye.

OBJECTIVES: *In this lesson, we will help Alex to:*

- *Learn how to select a used car.*
- *Estimate the value of a used car.*
- *Understand equity and depreciation.*

EQUITY AND DEPRECIATION

When you take out a loan to buy a car, the car dealer is being paid mostly by the bank. The bank really owns the car until you finish paying for it. If for some reason you cannot continue your car payments, you will have to sell the car to pay the remainder of loan, or the bank will repossess the car—that is, take it from you. As you pay off the loan, you increase your equity in the car. **Equity** is your part of the financial worth of the car. If your car could be resold for $8000 but you still owe $3000 on a loan, then your equity in the car is $5000.

But what is a car worth at any time? You know that a car loses value as it gets older. It can be sold for less and less, until it is worth nothing. This loss in the value of the car is called **depreciation.** For a new car this loss begins immediately after you purchase the car. For a used car, depreciation is more gradual.

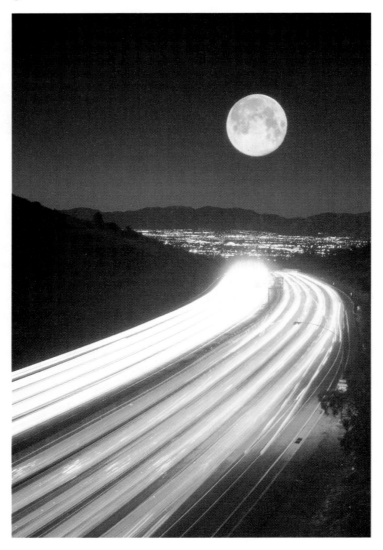

Depreciation varies from one car to another. Some cars have a reputation for lasting and therefore hold their value better than others. Also, inflation means that the price of everything goes up a little bit each year. Inflation slightly reduces the effects of depreciation. Straight-line depreciation means that a car, or whatever is being discussed, depreciates an equal amount each year. For example, if a car costs $20,000 and depreciates over a straight line for 10 years, it loses 10% of its original cost or $2000 each year.

But cars, especially new cars, do not depreciate in a straight line. They tend to lose more of their resale value in the first years and less in the later years. A car might depreciate by 25–30% in the first year and 20–25% in the second year. To Alex, this does not seem "right." An "almost new car" should be worth almost as much as a new car. In fact, people now watch their finances more closely and are often interested in buying almost new cars. Consequently, newly purchased cars do not depreciate as rapidly as brand-new cars.

WHERE TO BUY

Alex has learned that while it is fairly easy to compare the prices of new cars at different dealers, it is difficult to compare the prices of used cars. This is because you can rarely find two used cars of the same year and model with the same number of miles and the same amount of wear and tear. Fortunately, there are lists that give the book value for used cars. **Book value** is a standard estimate of the resale value for an average used car of a particular year and model. Current used car prices are found in *Edmund's Used Car Prices* and in the *Official Used Car Guide* published monthly by the National Automobile Dealer's Association. The latter publication is also know as the *Blue Book*. The book value is sometimes used for tax purposes. In practice, you can expect to pay more or less than the book value depending on the condition of the car.

Dealers   New car dealers generally keep the best used cars traded in, so you can expect to find good, but fairly expensive, used cars at new car dealers. There are many different kinds of salespeople at a used-car dealership. Some will help you find the best car to suit your needs; others just want to make a sale as quickly as possible. Until you get some experience, it is important to go slowly when looking at cars. Talk to your friends and relatives about different dealers. Ask a knowledgeable friend to go with you when you first visit car dealers. The Federal Trade Commission (FTC) has written consumer-protection regulations concerning the selling of used cars. The FTC encourages the buyer to have all verbal claims put in writing and to have a mechanic inspect the car before it is purchased.

Private Sales   Cars for sale are listed in the newspaper. If a buyer knows what he or she wants, what questions to ask, and how to look at a car, then a private sale can be a good deal. It is always important to read the car's **title** carefully to be sure that there is no mistake about the car or owner. FTC regulations do not apply to used cars if they are purchased from an individual.

Car Rental Company Sales and Auctions   Sometimes good used cars are available when rental agencies or other companies sell their cars. Rental cars are generally kept in excellent condition, but a buyer should always think twice about buying a car that has been driven by many people. Cars bought at auction or from a car-leasing agency require extra investigation because maintenance information may not be available.

Inspecting a Used Car   It should be clear that staring at an engine, kicking the tires, or driving a car around the block will not tell you what you need to know about the condition of a used car. A smart buyer conducts three types of inspections: on the lot, on the road, and in the garage.

On the lot, you should ask questions about the history of the car including replacement of parts, accidents, and the number of previous owners. On the road, you should pay attention to how the car feels and sounds. Trust your experience, and ask more questions if anything does not seem right. Finally, pay a mechanic to examine the car. You will do this only with a car that you are almost ready to buy.

In addition, the name of the previous owner must be given to you with a proof of odometer reading. It is also a good idea to call the National Highway Traffic Safety Administration (1-800-424-9393) to see if there are any safety recalls on the car.

ALGEBRA *REVIEW*

Set up an appropriate coordinate system and plot the points.

1. (1, 1000), (2, 600), (3, 400)

2. (1, 10,000), (2, 8000), (3, 6000)

3. (1, 10.5), (2, 21), (3, 42)

Find the value of *t*. Round to the nearest hundredth.

4. $t = 1.05^{10}$

5. $t = (1000)(1.006)^{24}$

6. $t = (5000)(1.005)^{36}$

Evaluate each for *x* = 3.

7. $3.5x - 1$

8. $1.05x + 2.007$

9. $1.05x + 2.25$

10. $1.005x$

Ask Yourself

1. What is equity?

2. What is depreciation?

3. Which depreciates more rapidly in the early years, a new car or a used car?

4. What is book value?

5. Why do new car dealers tend to charge more for their used cars than other dealers?

SHARPEN YOUR SKILLS

SKILL 1

There are several different ways to calculate depreciation. The method depends on the rate at which a car, or something else of value, tends to lose its value.

EXAMPLE 1 A new car sells for $20,000 and depreciates in a straight line until, after ten years, it is worth nothing.

QUESTION How would you construct a table and draw a graph to show the straight-line depreciation?

SOLUTION
The amount of depreciation is the same for each year. Depreciation begins as soon as the car is purchased and continues at the rate of 10% per year. At the end of the first year the car has depreciated $2000.

$$20,000 \cdot 0.10 = 2000$$

You can use a spreadsheet to calculate the depreciation and graph the results. Choose the number of years for x values and the resale values for y values. The graph shows the resale value as a function of time.

	A	B	C	D
1		Percent of	Amount of	Resale
2	Year	Depreciation	Depreciation	Value
3	0			20000
4	1	10%	2000	18000 ←+D3−C4
5	2	10%	2000	16000
6	3	10%	2000	14000
7	4	10%	2000	12000
8	5	10%	2000	10000
9	6	10%	2000	8000
10	7	10%	2000	6000
11	8	10%	2000	4000
12	9	10%	2000	2000
13	10	10%	2000	0

The graph of the resale value of the car is a straight line that slopes downward and to the right. This slope accounts for the name "straight-line" depreciation. The graph appears at the top of the next page.

Resale Value: Straight-Line Depreciation

EXAMPLE 2 Records for one standard model of car indicate that during 10 years it depreciates by the following percent of the original cost:

25, 20, 12, 10, 8, 7, 6, 5, 4, 3.

QUESTION How can these percents be used to form a table and a graph to represent the depreciation?

SOLUTION
We start with the original price of $20,000 and subtract the depreciation amount for each year. Use a spreadsheet as in Example 1. Again, the graph shows the resale value of the car as a function of time. In the formula for cell C4, be sure to use an *absolute* address for cell D3 so the address stays the same when the formula is copied to another cell.

	A	B	C	D
1		Percent of	Amount of	Resale
2	Year	Depreciation	Depreciation	Value
3	0	$+\$D\$3*B4$		20000
4	1	25%	5000 $+D3-C4$	15000
5	2	20%	4000	11000
6	3	12%	2400	8600
7	4	10%	2000	6600
8	5	8%	1600	5000
9	6	7%	1400	3600
10	7	6%	1200	2400
11	8	5%	1000	1400
12	9	4%	800	600
13	10	3%	600	0

Resale Value: Rapid Depreciation

(vertical axis) Resale Value (Thousands of Dollars)

(horizontal axis) Years after Purchase

SKILL 2

Comparing depreciation with loan repayment can help you understand what your car is really worth and how to use your money.

EXAMPLE 3 Bryan decides to buy the Moonbeam car discussed in Lesson 10–1, Example 3, for $19,490. He makes a down payment of 20% and obtains a 3-year loan at an APR of 6%.

QUESTION How does Bryan's equity in the car vary during the life of the car?

SOLUTION

In Lesson 10–1, Example 3, Bryan found that the monthly payment is $474.34. He uses a spreadsheet to make an amortization schedule as in Lesson 5–4, Example 2. Bryan includes an additional column to calculate the depreciated value of the car. He assumes straight-line depreciation at 10% per year, that is $19,490(0.10 \div 12)$, or $162.42 per month.

He also includes a column to calculate the equity, which is the part of the car he owns. The equity is the difference between the depreciated value and the amount you owe on the car; that is, the unpaid balance of the loan. Notice that the final payment is adjusted for a zero unpaid balance. Although the loan is repaid in 36 months, Bryan extends the spreadsheet to 60 months so he can find out his equity after 5 years.

Bryan's spreadsheet is shown on pages 470 and 471. He used formulas as shown.

	A	B	C	D	E	F	G
1	Payment	Payment	Interest	Note	Unpaid	Depreciated	
2	Number	Amount	Due	Reduction	Balance	Value	Equity
3	0				15592.00	19490.00	3898.00
4	1	474.34	77.96	396.38	15195.62	19327.58	4131.96
5	2	474.34	75.98	398.36	14797.26	19165.16	4367.90
6	3	474.34	73.99	400.35	14396.91	19002.74	4605.83
7	4	474.34	71.98	402.36	13994.55	18840.32	4845.77
8	5	474.34	69.97	404.37	13590.18	18677.90	5087.72
9	6	474.34	67.95	406.39	13183.79	18515.48	5331.69
10	7	474.34	65.92	408.42	12775.37	18353.06	5577.69
11	8	474.34	63.88	410.46	12364.91	18190.64	5825.73
12	9	474.34	61.82	412.52	11952.39	18028.22	6075.83
13	10	474.34	59.76	414.58	11537.81	17865.80	6327.99
14	11	474.34	57.69	416.65	11121.16	17703.38	6582.22
15	12	474.34	55.61	418.73	10702.43	17540.96	6838.53
16	13	474.34	53.51	420.83	10281.60	17378.54	7096.94
17	14	474.34	51.41	422.93	9858.67	17216.12	7357.45
18	15	474.34	49.29	425.05	9433.62	17053.70	7620.08
19	16	474.34	47.17	427.17	9006.45	16891.28	7884.83
20	17	474.34	45.03	429.31	8577.14	16728.86	8151.72
21	18	474.34	42.89	431.45	8145.69	16566.44	8420.75
22	19	474.34	40.73	433.61	7712.08	16404.02	8691.94
23	20	474.34	38.56	435.78	7276.30	16241.60	8965.30
24	21	474.34	36.38	437.96	6838.34	16079.18	9240.84
25	22	474.34	34.19	440.15	6398.19	15916.76	9518.57
26	23	474.34	31.99	442.35	5955.84	15754.34	9798.50
27	24	474.34	29.78	444.56	5511.28	15591.92	10080.64
28	25	474.34	27.56	446.78	5064.50	15429.50	10365.00
29	26	474.34	25.32	449.02	4615.48	15267.08	10651.60
30	27	474.34	23.08	451.26	4164.22	15104.66	10940.44
31	28	474.34	20.82	453.52	3710.70	14942.24	11231.54
32	29	474.34	18.55	455.79	3254.91	14779.82	11524.91
33	30	474.34	16.27	458.07	2796.84	14617.40	11820.56
34	31	474.34	13.98	460.36	2336.48	14454.98	12118.50
35	32	474.34	11.68	462.66	1873.82	14292.56	12418.74
36	33	474.34	9.37	464.97	1408.85	14130.14	12721.29

Callouts (formulas):
- +F3−E3
- @ROUND(E3*0.005, 2)
- +B4−C4
- +E3−D4
- +F3−162.42

	A	B	C	D	E	F	G
1	Payment	Payment	Interest	Note	Unpaid	Depreciated	
2	Number	Amount	Due	Reduction	Balance	Value	Equity
37	34	474.34	7.04	467.30	941.55	13967.72	13026.17
38	35	474.34	4.71	469.63	471.92	13805.30	13333.38
39	36	474.28	2.36	471.92	0.00	13642.88	13642.88
40	37					13480.46	13480.46
41	38					13318.04	13318.04
42	39					13155.62	13155.62
43	40					12993.20	12993.20
44	41					12830.78	12830.78
45	42					12668.36	12668.36
46	43					12505.94	12505.94
47	44					12343.52	12343.52
48	45					12181.10	12181.10
49	46					12018.68	12018.68
50	47					11856.26	11856.26
51	48					11693.84	11693.84
52	49					11531.42	11531.42
53	50					11369.00	11369.00
54	51					11206.58	11206.58
55	52					11044.16	11044.16
56	53					10881.74	10881.74
57	54					10719.32	10719.32
58	55					10556.90	10556.90
59	56					10394.48	10394.48
60	57					10232.06	10232.06
61	58					10069.64	10069.64
62	59					9907.22	9907.22
63	60					9744.80	9744.80

Notice what happens in row 39 of the spreadsheet. Bryan's last payment amount was $474.28, which he inserts in cell B39. Now his unpaid balance is zero. Also, the depreciated value and the equity are equal.

Since his loan is paid off, he only needs to calculate the depreciated value and the equity in row 40. He adjusts row 40 to calculate only the depreciated value and the equity, and copies row 40 onto rows 41–63.

Bryan graphs the unpaid balance, the depreciated value, and the equity on the same graph. He notices that the graphs intersect and observes that after 36 months, the depreciated value is equal to his equity in the car. His equity in the car increases and then decreases when the depreciated value is less than the amount of money he has spent for the car. Although the graphs of the unpaid balance and the equity appear to be a straight lines, they are not. These graphs are slightly curved.

TRY YOUR SKILLS

A new car is purchased for $16,000, with a 20% down payment, and is financed for 3 years at a rate of 7.2% per year.

1. Find the amount of the monthly payment.
2. Find the amount of interest paid in the first payment.
3. Find the amount of principal paid in the first payment.
4. Find the amount of interest and principal paid in the second payment.
5. If the car depreciates 30% the first year and 20% the second year, what is its resale value at the beginning of the third year?
6. Since the car depreciates 30% during the first year, what is the percent of depreciation during the first month?
7. Find the owner's equity in the car at the end of 1 year.

EXERCISE YOUR SKILLS

1. Why do some buyers prefer "young" used cars to new cars?

2. How can a used-car buyer reduce the chances of making a mistake when purchasing a used car?

Alex's uncle purchased a new car for $24,000. Terms of financing were 15% down, 6% annual interest, and 36 months to pay. Annual depreciation on the car as a percent of the purchase price is likely to follow the pattern 30%, 20%, 15%, 10%, 8%, 7%, 5%, 3%, 2%.

3. Find the amount of the monthly payment.

4. Find the amount of interest paid in the first payment.

5. Find the amount of principal paid in the first payment.

6. What is the estimated resale value (depreciated value) of the car at the start of the fourth year after purchase?

7. Draw a graph showing the unpaid balance as a function of time. Example 3 will remind you of how this is done. (A straight line will approximate this graph.)

8. On the same axes as the graph for Exercise 7, graph the resale value of the car based on depreciation.

9. Draw a graph showing equity as a function of time.

10. At approximately what point in time does Alex's uncle have the least amount of equity in the car?

11. At approximately what point in time does Alex's uncle have the greatest amount of equity in the car?

Alex's father purchases a used van for $12,000. He pays $4000 down and finances the car for 3 years at an annual rate of 9%. He estimates that, because it is used, the van will depreciate in a straight line over 6 years.

12. Find the monthly payment for the van.

13. Find the amount of interest paid in the first payment.

14. Find the amount of principal paid in the first payment.

15. Graph the outstanding balance of the loan as a function of time.

16. On the same axes, graph the resale value as determined by depreciation.

17. What is the approximate equity in the car after 3 years?

Bob and Dena are both planning to purchase used cars for $8000. Financing can be selected from the following terms: Pay 10% down and finance the remainder at a 6.5% annual interest rate or pay 20% down and finance the remainder at a 5.5% annual interest rate; the time of the loan can be 3, 4, or 5 years. Bob has $3000 in the bank and a good job and would like to purchase a new car in 3 years. Dena has $500 saved and can borrow a few hundred dollars from her parents. She has a part-time job and hopes to keep the car for as long as possible.

18. Which financing terms would be best for Bob and why?

19. Which financing terms would be best for Dena and why?

20. According to the terms you selected, who pays the greater total amount for the car?

21. If the car depreciates in a straight line over 5 years, what amount is Bob likely to get when he sells it?

MIXED REVIEW

1. Jason works from 10:00 A.M. until 12:30 P.M., and then from 1:00 P.M. to 5:00 P.M. He earns $9.60 an hour. What is his daily pay?

2. Emily has $568.21 in her checking account. What is the balance in the account after she pays bills totaling $178.96?

3. Find the interest earned on $790.30 for one month at a monthly interest rate of 0.45%.

4. Use the Multiples-of-Salary Chart in the Reference Section to find the amount of life insurance a 25-year-old who has gross earnings of $23,500 would need in order to provide 60% income replacement.

5. If you buy 680 shares of stock at a price of $21\frac{1}{4}$ per share, what is the total cost of the stock? The commission rate is 1.5%.

What profit does an insurance company expect to make for each 1-year term insurance policy described in Exercises 6 and 7? Assume that the direct and indirect expenses for each policy are $20. Refer to the following table. Write and solve an equation.

EXPECTED DEATHS PER 100,000 ALIVE AT SPECIFIED AGE		
Age	Expected Deaths Within 1 Year	Expected to be Alive in 1 Year
45	315	99,685
46	341	99,659
47	371	99,629
48	405	99,595
49	443	99,557

6. Face value: $100,000; Age of insured: 47; annual premium: $500

7. Face value: $180,000; Age of insured: 45; annual premium: $860

10-3 OWNING AND OPERATING COSTS

When Freda became a salesperson in the business started by her friends Evelyn and Greg, she began to use her car for business. She distributed posters and flyers and also delivered the finished T-shirts and sweatshirts.

Freda soon found that Hari, who kept the company's financial records, wanted an accurate report on the car expenses. So Freda began keeping all of her receipts for the purchase of gas and for car maintenance. Hari said that the business will at least pay for her

gas. Freda also has monthly payments on her car loan and insurance premiums to pay. She asked Hari whether the company could pay part of these bills. Hari said he would speak to an accountant to find out more about personal and business expenses.

After learning more about business expenses for a car, Hari and Freda looked into car leasing. They found that in some circumstances it makes sense financially for a business to lease, rather than own, a company car.

OBJECTIVES: In this lesson, we will help Freda to:

- *Calculate car mileage per gallon of gasoline.*
- *Compute operating costs of a car and total costs per mile.*
- *Compute the costs of leasing a car.*

MILEAGE

Hari tells Freda that she should keep all receipts for the purchase of gas and also keep track of miles driven for business. Each time she uses the car for business, she is to write down the beginning and ending odometer readings. The **odometer** indicates the number of miles a car has been driven. Freda also is to calculate **miles per gallon,** the average number of miles her car travels on each gallon of gas.

MAINTENANCE

Since Freda's car has only 12,000 miles on it, expenses for tune-ups, parts, and repairs should be low. Her car is still under **warranty,** which means that the manufacturer pays for many repairs. The warranty expires after a certain number of miles or years.

To maintain her car in good condition, Freda has the oil changed every 3000 miles. She has a tune-up every 10,000 miles and regularly checks the air pressure in the tires and the levels of engine coolant.

REPAIRS

Freda has talked with mechanics and has learned that she will eventually need a new battery, new tires, and brake linings. She will also need to replace filters and hoses. She may need replacements for or repairs on a variety of other parts including the fuel pump, alternator, and muffler. Freda plans to set aside an amount of money each month so that she will be able to pay her bills for repairs as they come up.

LEASING A CAR

Because car expenses are often high and because cars depreciate, car leasing has become popular. **Leasing** is an arrangement by which a car is rented for a monthly fee under a contract that extends for several years. The person leasing the car pays for gas; the company that leases the car pays some maintenance and repair expenses.

ASK YOURSELF

1. What is an odometer?

2. What is the difference between maintenance and repairs?

3. What is meant by "miles per gallon?"

4. What does leasing mean?

ALGEBRA *REVIEW*

1. Find the average of 4, 7, and 9.

2. Jack walked 15 miles in 5 hours. What was his average speed in miles per hour?

3. Pat ran 8 miles in 70 minutes. Find her average rate in minutes per mile.

4. Ms. Gonzales drove 140 miles in 3.5 hours. Find her average speed in miles per hour.

5. Mr. Oliver used 12 gallons of gas to drive 156 miles. What was his average miles per gallon?

Find the solution of each system of equations.

6. $x + y = 10$
   $y = 5$

7. $y = 2x + 1$
   $y = 15$

8. $y = 2x - 3$
   $x = 2.5$

SHARPEN YOUR SKILLS

SKILL 1

The number of miles that you can drive for each gallon of gas depends on the kind of driving you do. Steady driving at a moderate speed gives the best miles per gallon. Idling or moving at very low speeds in city traffic and driving at high speeds use more gas and lower the miles per gallon rate.

Miles per gallon of gas is like miles per hour except that you divide total miles by the number of gallons of gas used rather than by the hours driven.

Average Miles per Gallon

$a = m \div g$ where a = average miles per gallon
m = total miles driven
g = total gallons purchased

Freda learned from a friend the way to determine the average miles per gallon of gas.

1. Record the mileage when the gas tank is full.

2. Fill the tank the next time you purchase gas; then record the mileage, gallons purchased, and amount spent.

3. Follow this procedure several times so that you have enough miles and gallons for a valid average.

EXAMPLE 1 Freda filled her gas tank four times, recording the following information. She did not record the gallons or dollar amount the first time because she did not have a previous odometer reading and therefore could not calculate the mileage she had traveled.

Odometer Reading	Gallons of Gas	Amount
11,400	——	——
11,604	9.7	$12.03
11,842	11.2	13.89
12,027	8.6	10.67

QUESTION For the given information, what is the average miles per gallon?

SOLUTION
Freda can find the answer by using the formula $a = m \div g$.

Total miles driven: $m = 12{,}027 - 11{,}400 = 627$
Total gallons purchased: $g = 9.7 + 11.2 + 8.6 = 29.5$
Average miles per gallon: $a = 627 \div 29.5 = 21.3$

Freda averages 21.3 miles per gallon of gas.

EXAMPLE 2 Freda has purchased gas three times to drive 627 miles in Example 1.

QUESTION What is Freda's cost per mile for gas?

SOLUTION

The cost per mile, like gallons per mile, is an average. It is found by dividing total cost by the number of miles. Since we have used a for average miles per gallon, we will use b for the cost per mile.

Average Cost per Mile

$b = c \div m$ where b = average cost per mile
c = total cost
m = total miles driven

According to Freda's records,

Total cost:	$12.03 + \$13.89 + \$10.67 = \$36.59
Total miles driven:	627
Average cost per mile:	$b = 36.59 \div 627 = 0.058$

The average cost per mile for gas is $0.058 or 5.8 cents. This does not seem like a lot of money, but, as we shall see, the total cost per mile of owning and operating a car is much higher.

SKILL 2

Freda is driving about 1000 miles per month. Using this fact and the information from Example 2, she calculates that her yearly expense for gas will be $12 \cdot 1000 \cdot 0.058 = \$696$.

In addition to the purchase of gas, Freda has to pay the following.

Regular maintenance: tune-up, oil change, filters, wheel alignment. This will cost about $125 per year.

Repairs: Some can be anticipated, such as new tires, but most cannot be known in advance. Freda can be reasonably sure of some repair expenses, and she knows that these will increase with time. Freda feels that $240 per year will be a safe estimate for now.

Monthly loan payment: $304.72

Insurance premium: $428.60, four times per year

Taxes and fees: $94 per year

EXAMPLE 3 Using the information given, Freda decides to set up a budget.

QUESTION How should Freda set up a budget for her car expenses?

SOLUTION

All expenses must be annualized, that is, the total amount per year must be calculated for each expense. Gasoline, maintenance, repairs, and taxes/fees are already stated as annual. The loan payment is monthly. Insurance is paid four times a year. We multiply to annualize these expenses.

Loan payment: 12(304.72) = 3656.64
Insurance: 4(428.60) = 1714.40

Annual expenses are shown in the following table.

Annual Expenses (Dollars)					
Gasoline	**Maintenance**	**Repairs**	**Loan Payment**	**Insurance**	**Taxes/Fees**
696	125	240	3656.64	1714.40	94

After looking at the yearly budget, Freda decides that she needs a monthly budget to know how much money to put aside for monthly bills each month and to get a better idea of what the car is costing her. She divides each annual amount by 12 to find the monthly amounts.

Monthly Expenses (Dollars)					
Gasoline	**Maintenance**	**Repairs**	**Loan Payment**	**Insurance**	**Taxes/Fees**
58	10.42	20	304.72	142.87	7.83

Freda was surprised when she saw what her car was costing her each month. She hadn't realized how much it was because she pays the bills at so many different times. She used her spreadsheet to make a pie graph of the costs.

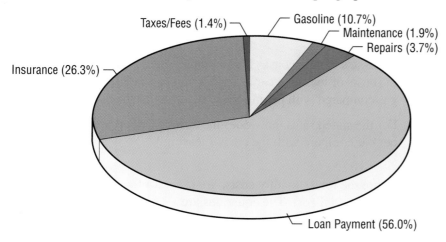

EXAMPLE 4   Freda totaled her monthly expenses and found that they came to $543.84.

QUESTION   What is Freda's average cost per mile to own and operate her car?

SOLUTION
Freda drives an average of 1000 miles per month, so the average cost per mile can be found by dividing her total costs by 1000.

$$543.84 \div 1000 = 0.544$$

Freda can't believe that it costs her more than 50 cents for every mile that she drives. She shows her figures to Hari. He has been investigating car costs and tells her that the business should pay some of her operating expenses but not the loan payment. He says that she is buying the car, so it will belong to her, not to the business.

Freda then calculates her car expenses, leaving out the loan payment.

Expenses: $58 + 10.42 + 20 + 142.87 + 7.83 = 239.12$
Average per mile: $239.12 \div 1000 = 0.239$

Freda's operating expenses are $0.239 or 23.9 cents per mile.

Hari and the others agree that it is fair to pay Freda 24 cents a mile for the business use of her car. They later discover that this is about the amount that most employees are allowed to charge when using their cars for business.

SKILL 3

Freda's cousin, Vanessa, drives for a delivery company. At first the company pays her 25 cents a mile. Then, as business increases, the company considers leasing a car. It can lease a car for $300 a month. The leasing agency pays for all expenses (except gas) and repairs. Leasing would save wear and tear on Vanessa's car, but would it save money for the company?

EXAMPLE 5   Vanessa's company asks her to prepare figures and graphs showing the costs of leasing a car at $300 a month and $0.058 per mile for gas compared with paying $0.25 per mile for the use of her car.

QUESTION   For the stated rates, how does the cost of paying Vanessa to use her car compare with the cost of leasing?

SOLUTION
First we compare the total monthly costs. We use x for the number of miles driven and y for the total cost. The equations are

$y = 0.25x$ Cost of using Vanessa's car

$y = 300 + 0.058x$ Cost of leasing plus 0.058 per mile

$0.25x = 300 + 0.058x$ Substitute 0.25x for y in the second equation.
$0.192x = 300$ Combine x terms.
$x = 1562.5$

The costs are equal at 1562.5 miles per month. Since some rounding is involved in these calculations, you can say that if Vanessa drives more than 1560 miles per month, the company will save by leasing a car. Graphing will give a better idea of the comparison.

Using a graphing calculator, you can graph these two equations as shown.

Range:

Xmin: 0   Ymin: 0
Xmax: 2000   Ymax: 500
Xscl: 100   Yscl: 50

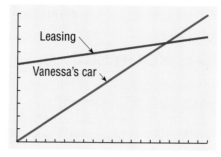

EXAMPLE 6 Vanessa's company would like to compare the cost per mile of leasing with paying $0.25 for using Vanessa's car.

QUESTION How do the costs per mile compare?

SOLUTION

In this case the cost of using Vanessa's car is fixed at 25 cents a mile. To find the cost per mile of leasing c, you must divide the monthly leasing cost by the number of miles x, and then add the cost per mile for gas.

The cost per mile equations are

$c = 0.25$ Cost per mile of using Vanessa's car

$c = \dfrac{300}{x} + 0.058$ Cost per mile of leasing a car

Solve these equations for x.

$0.25 = \dfrac{300}{x} + 0.058$ Substitute 0.25 for c in the second equation.

$0.192 = \dfrac{300}{x}$ Subtract 0.058 from both sides.

$0.192x = 300$

$x = 1562.5$

Graph the cost per mile equations on a graphing calculator using the following range values. Use X and Y for the variables x and c in the equations.

Xmin: 0   Ymin: 0
Xmax: 2000   Ymax: 1
Xscl: 100   Yscl: 0.1

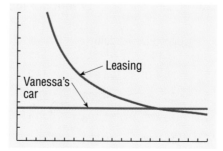

The tracing feature on the calculator can be used to find the cost per mile for different total monthly amounts. For example, the cost of leasing at 1000 miles per month is 35.8 cents per mile, and the cost at 1600 miles per month is 24.6 cents per mile.

TRY YOUR SKILLS

Mark drives 537 miles on 23.7 gallons of gas. He pays $30.57 for the gas. He calculates his yearly car expenses, including loan payments of $2560, to be $5470. He can lease a car for $300 per month.

1. Find Mark's average miles per gallon.

2. Find his average cost per gallon.

3. Find his average monthly expense.

4. Find Mark's average cost per mile if he drives 12,000 miles per year.

5. Find the cost of leasing for a month and driving 1350 miles while paying an average of 5.7 cents per mile for gas.

6. Write equations for the cost of paying Mark a per mile amount for operating expenses (not counting the loan) and for the leasing cost.

EXERCISE YOUR SKILLS

1. Car companies advertise the miles per gallon that can be expected from a car. Why should this be considered when buying a car?

2. Why is it important to take care of minor repairs promptly?

3. Why should a car owner estimate repair expenses when preparing an annual budget?

4. What are some of the differences between owning and leasing a car?

Use the information in the table to find the distance traveled, the miles per gallon, and the cost per mile for the trip.

	Odometer First	Odometer Last	Gallons of Gas	Cost of Gas	Distance	Miles per Gallon	Cost per Mile
5.	28,431	28,848	18.4	$25.15			
6.	38,715	39,326	15.7	18.95			
7.	11,477	11,628	10.1	14.00			
8.	18,388	19,374	40.6	63.74			
9.	15,428	15,639	10.9	14.38			

Isabel drives 15,000 miles a year and has the following real and estimated car costs:

Maintenance: $150 per year
Repairs: $300 per year
Loan payments: $401.30 per month
Insurance: $603.27 twice per year
Taxes: $114 per year
Gasoline: $0.06 per mile

Find each of the following for Isabel.

10. Her annual expense for gasoline

11. Her monthly expense for insurance

12. The total of her annual car expenses

13. Her average cost per mile to own and operate her car

14. Her annual cost to operate the car not including the monthly loan payment

15. Create a pie graph to show her annual expenses for her car.

Mark works for a company that will make a decision about whether to pay him 20 cents per mile to drive his car for the business or to lease a car for $200 per month and 6 cents per mile.

16. Write equations for the total monthly cost under each arrangement.

17. Solve the equations simultaneously to find the number of miles per month for which the costs will be equal.

18. Graph the two equations to show the relationship.

19. Which is the better arrangement for 1000 miles per month?

20. What is the difference in costs for a 1000-mile month?

21. Which is the better arrangement for 2000 miles per month?

22. What is the difference in costs for a 2000-mile month?

MIXED REVIEW

1. Jacob borrowed $10,000 and repaid it at the rate of $200 per month for 5 years. Find the total interest paid over the 5 years.

Phyllis bought 1200 shares of stock at $37\frac{3}{8}$ per share. In Exercises 2–4, ignore the commission costs.

2. Find the cost of the shares.

3. After 2 years, Phyllis sold the shares at $45\frac{5}{8}$ per share. What was the total sale price?

4. What was the capital gain or loss on the shares?

5. Use IRS Form 1040EZ to find the amount of the refund or payment due to the nearest dollar for a single taxpayer who had $1,600 withheld on his $13,900 income. The taxpayer has no taxable interest and cannot be claimed as a dependent by someone else.

Trevor, who is 17, is allowed to drive the family car. Sometimes, he drives his 11-year-old sister Tracey and her friends to a weekend event. Last Saturday, Trevor drove Tracey and her friends, Roy and Leah, to a movie. As he drove, another car came speeding past, rapidly changing lanes and weaving back and forth among the other cars. Just as the car passed, Trevor was distracted for an instant by the younger children in the back seat, who were wrestling over a piece of candy.

As Trevor looked up, the car that was speeding collided with another car. Both cars seemed to be heavily damaged as they skidded off the road. Trevor had to change lanes quickly and slow down to avoid another car that had moved into his lane because of the accident. His sister and her friends had their seat belts on, but they were shaken up.

As a result of seeing the accident, Trevor asked his parents to tell him about insurance. He wanted to know what it is, why it is needed, and why costs are higher for young people. His parents explained that by collecting money from thousands of car owners, the insurance company can pay for car repairs and medical bills after accidents.

Compared with the many thousands of people who purchase insurance, only a few actually have accidents. But no one knows which ones will have the accidents, so everyone should purchase insurance. Furthermore, in most states, the law requires that you have auto insurance.

OBJECTIVES: *In this lesson, we will help Trevor to:*

⬤ *Examine several types of insurance that are available for car owners.*

⬤ *Compare rates of auto insurance for different drivers on the basis of the age, sex, training, and marital status of the driver and several other factors.*

⬤ *Understand how insurance companies use accident statistics to estimate their costs.*

AUTOMOBILE INSURANCE

Most states now require that car owners purchase insurance. The reason for this is that any driver might have an accident. Depending on the circumstances, the driver might then be responsible for the financial consequences of the accident. The legal term for this responsibility is **liability.** Even if not legally liable, a driver would still want to have his or her car repaired or replaced. There are a number of different kinds of automobile insurance dealing with liability.

Bodily Injury Liability   This coverage is protection against claims or lawsuits that are brought against you by pedestrians, riders in your car, and people in other cars whose injuries were caused by an accident for which you are responsible. The amount of protection can vary from $10,000 to $1,000,000 or more. Some people choose high amounts because in serious accidents the medical bills can be enormous.

Property Damage Liability   Coverage for property damage liability is paid for the damage done to other people's property, such as a car, a building, or a fence. Property damage insurance might range from $5,000 to $50,000. It also protects you when you are driving someone else's car.

Medical Payments   Medical payment insurance covers the person driving the car and others riding in it, whether or not that driver is at fault. In Trevor's case this coverage protects him and the other members of his family when they ride in their own car or in another person's car.

Collision   Coverage for **collision** pays for damage to Trevor's family's car if it is in an accident. Some people decline this coverage if their cars are not worth much. For example, if a car is worth less than $1000, many people would not want to pay several hundred dollars a year for collision insurance.

Comprehensive Physical Damage   Coverage for comprehensive physical damage pays for damages that result from a fire, falling object, theft, flood, earthquake, and the like. The accident that Trevor witnessed was not caused by any of these events.

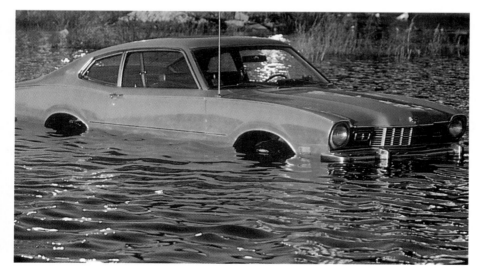

Uninsured and Underinsured Motorist Protection Coverage for uninsured and underinsured motorist protection pays for personal expenses due to an accident if it can be shown that the driver of the other car was at fault and if the other driver has little or no bodily injury liability coverage.

THE COST OF INSURANCE

Naturally, it costs more to insure a car that is worth $40,000 than one that is worth $8000. In addition to the value of cars, insurance companies keep records of all kinds of driving-related accidents. They have found that accidents are more likely to occur in some places than in others. These companies also keep track of driver characteristics and their relationship to accidents. Insurance companies charge higher rates for people and places that they find are more often involved in accidents.

In the Reference Section of the text you will find a table entitled Driver Rating Factors. The **driver rating factor** is based on characteristics such as age, sex, whether the person is married, the kind of driving done, and whether the driver has been through a driver training program. The base premium is multiplied by the appropriate factor to find the premium for a specific individual.

Ask Yourself

1. What is liability?

2. What is bodily injury liability insurance?

3. What is property damage liability insurance?

4. What does comprehensive physical damage insurance cover?

ALGEBRA *REVIEW*

Given the equation $p = rb$

1. Find p when $r = 1.8$ and $b = 525$.

2. Find p when $r = 2.25$ and $b = 980$.

3. Find p when $r = 4.4$ and $b = 1200$.

4. Find r when $p = 2015$ and $b = 3.1$.

5. In general, what happens to the value of p if r is unchanged and b decreases?

Given the equation $p = \dfrac{e}{n} + h$

6. Find p if $e = 568,500$, $n = 10,000$, and $h = 90$.

7. Find e if $p = 52.80$, $n = 10,000$, and $h = 30$.

8. Find n if $p = 825.86$, $e = 72,586,000$, and $h = 100$.

9. Find h if $p = 591$, $e = 526,000$, and $n = 65$.

10. In general what happens to p if e and h are unchanged and n increases?

Trevor is now looking at actual insurance costs. Through a friend who works for an insurance company, he has obtained the following table of basic rates.

Car Class Rating	Collision Deductible			Comprehensive Deductible		
	$100	$250	$500	$50	$250	$500
1–10	$ 500	$ 430	$ 370	$140	$130	$110
11–20	820	750	690	350	310	280
21–30	1140	1040	940	450	410	370
31–40	1280	1190	1100	670	610	560

AUTOMOBILE INSURANCE, SIX-MONTH BASIC RATE SCHEDULE

Trevor noticed that several factors affect basic rates. One is the car class rating. This rating reflects the resale value of the car. Another factor is the amount of the **deductible,** the first expenses that are not covered by insurance. The deductible is a way of reducing insurance rates. In the case of bills resulting from an accident or injury, the insured person pays the deductible amount before the insurance company begins to pay. The higher the deductible, the more you pay yourself and the lower your **premium,** that is, the amount you pay for your insurance. For example, if repairs after an accident cost $1000 and your deductible is $250, then you pay $250, and the insurance company pays $750.

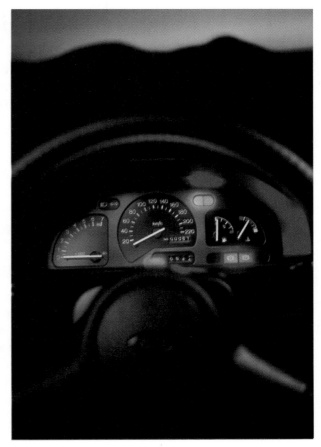

The insurance premium is dependent on the driver rating factor r and the basic rate b for the car. The six-month basic rate is found in the table at the top of this page or in the Reference Section. The driver rating factor is found in a Driver Rating Factor table in the Reference Section. Note there are three tables, one for each of these categories: married youths, unmarried youths, and all adults.

Annual Insurance Premium Formula

$p = 2rb$ where p = the annual premium paid to the company
r = the driver rating factor
b = the six-month basic rate for the car

EXAMPLE 1   Trevor is 17 years old. He has had driver training. He is single. He is not the owner or principal operator of the car that he drives. He does errands in the car. He does not use the car to drive to work.

QUESTION   According to the Driver Rating Factor tables, what is Trevor's rating factor?

SOLUTION
First find the table, Part II, Unmarried Youths. Trevor is 17, so you use the top half of the table and the section marked M. Trevor is not the owner or usual driver, but he has had driver education and does not drive to work. Thus, working your way across the table, you find that his rating is 2.45.

———— **SKILL 2** ————————————————

EXAMPLE 2   Trevor is driving his family's car, which has a car class rating of 8. He plans to purchase collision insurance with a deductible of $500.

QUESTION   How can you use the premium equation to find Trevor's annual premium?

SOLUTION
The basic rate for Trevor is found in the Automobile Insurance table at the top of the previous page. Look in the collision half of the table (left side) in the column under $500 and the row 1–10. Thus the basic rate is $430.

Using the equation

$$p = 2rb$$

gives

$$p = 2(2.45)(430) = 2107$$

The annual premium is $2107.

———— **SKILL 3** ————————————————

EXAMPLE 3   One insurance company is considering a new policy for young people aged 14 and 15. The policy will insure them for injuries sustained as passengers riding in a car. Some parents are interested in this because their children in this age range are often passengers in cars driven by new, 16- or 17-year old drivers who lack adequate insurance.

The insurance company must decide on the price of a premium. Their records show that for every 10,000 14- and 15-year-old passengers, there are 425 accidents resulting in personal injuries. The medical costs of these injuries average $3800 per person. The company must also receive $35 per policy to cover overhead and profit.

QUESTION What should be the premium for the new policy?

SOLUTION

The premium can be found by considering the total estimated expense for the population, then dividing the amount by the total number of policies paying premiums. Then the overhead must be added.

Algebraically, the cost of the premium can be expressed as

$$p = \frac{e}{n} + o$$ 　　where p = premium
　　　　　　　　e = expense paid out
　　　　　　　　n = number paying premiums
　　　　　　　　o = overhead per premium

$$p = \frac{425(3800)}{10,000} + 35$$
$$= 196.50$$

The company's figures suggest a premium of $196.50.

TRY YOUR SKILLS

1. Use the Driver Rating Factor tables to find the rating for a married female, 20 years old, who owns her own car, has not had driver training, and drives the car to work.

2. Use the Automobile Insurance table to find the annual premium for comprehensive coverage, with a $250 deductible, on a car with a class rating of 12.

3. Write an algebraic equation for finding the annual insurance premium based on the six-month basic rate for the car and the driver rating factor. Explain each part of the equation.

Insurance records show that of 10,000 drivers in a certain class, 1378 were in accidents with an average cost to the insurance company of $6089.

4. What is the average cost per person insured?

5. If the company overhead is $95 per policy, what should the company charge as a premium?

EXERCISE YOUR SKILLS

1. Why do laws require drivers to be financially responsible for damage to other people or their property?

2. Why is it important for drivers to have adequate medical payment coverage?

3. What is the difference between a premium and a deductible?

4. What is the advantage in having a deductible in insurance coverage?

Use the Driver Rating Factor tables in the Reference Section to find the driver rating factor for each of the following individuals.

5. A 19-year-old female without driver training who drives to work, and is not the owner or usual driver of the car.

6. A 19-year-old unmarried male without driver training who drives to work and is not the owner or usual driver.

7. A 17-year-old female with driver training who drives to work and is the owner.

8. A 22-year-old female with or without driver training who drives to work and is the owner.

9. The usual driver of the car, 33 years of age, who drives to work.

10. The usual driver of the car, 65 years of age, who does not drive to work.

Use the Automobile Insurance, Six-Month Basic Rate Schedule to find the total six-month premiums in Exercises 11–15.

	Driver Rating Factor	Car Class Rating	Collision Deductible	Comprehensive Deductible	Six-Month Premium
11.	2.25	15	$500	$ 50	
12.	1.00	32	250	250	
13.	2.40	9	500	500	
14.	0.80	23	100	500	
15.	1.10	10	500	250	

16. An insurance company's records show that for every 10,000 drivers in a certain category there are 1067 accidents with an average cost to the company of $12,400 each. The company charges an overhead amount of $125 per premium. On the basis of these figures, what premium should be charged for a policy in this category?

17. The AVR insurance company's records show that for every 100,000 drivers in a certain category there are 2311 accidents with an average cost to the company of $18,200. For this category the company charges an overhead of $190 per premium. What should be the premium to cover costs?

18. What are some of the differences between the ways in which an insurance company and an individual look at an insurance policy?

19. What are some of the factors that lead an insurance company to raise or lower the premiums every year?

MIXED REVIEW

1. You are a real-estate agent and you sold a house for $120,000 earning a commission of 3% on the first $100,000 and 6% on the amount above $100,000. What was your commission?

A loan of $6000 is repaid one year later in a single payment of $6600.

2. What was the interest paid on the loan?

3. What was the rate of interest on the loan?

4. A person buys a $100,000 whole life insurance policy at age 30. Use the table entitled Accumulated Cash Value of $100,000 Whole Life Policy in the Reference Section to find the cash value of the policy when the person retires 20 years later.

Find the cost of these shares of stock. The cost includes a commission of 1.5%.

5. 50 shares at 35\frac{1}{2}$ a share

6. 1,000 shares at 9\frac{7}{8}$ a share

Use the Comparison Table for Term and Whole Life Premiums on page 326, the table showing Accumulated Cash Value of $100,000 Whole Life Policy on page 345, and the formula for the Future Value of a Periodic Investment to answer Exercises 7 and 8.

7. Find how much tax-deferred money a 25-year-old person could expect to accumulate between the ages of 25 and 30 by buying a $100,000 term policy instead of a more expensive whole life policy. The money saved will be invested into an IRA that should earn 8% a year.

8. For the person in Exercise 5, compare the cash value of a whole life insurance policy after 5 years with the value of the IRA at the end of 5 years. Which seems to be the more profitable investment?

The Tierney family consists of Mr. and Mrs. Tierney and their three dependent children. Last year, the family had salary income of $47,500, interest income of $1457, and dividend income of $2869. Their IRA contribution was $1750. The allowable deductions were real-estate taxes of $1690, local income taxes of $2100, and mortgage interest on their home of $5980.

9. Determine whether the family is better off itemizing their deductions or using the standard deduction. What is the deduction that the family should take?

10. Use IRS Form 1040 and Schedule A to find the Tierneys' income tax for the previous year.

1. When financing a car, what factors cause monthly payments to be higher and what factors cause them to be lower?

2. What is meant by straight-line depreciation?

3. If you purchase a new car, what will probably be the most costly factor in operating the car?

4. How would you describe the depreciation of most cars?

5. What are some of the causes contributing to the variation in insurance rates?

The following is a portion of the price list for a new car.

Item	Dealer's Cost	Sticker Price
4-door sedan	$13,460	$16,152
Air conditioning	420	483
AM/FM radio, tape deck	750	877.50
Full-size spare tire	120	160

6. Find the markup for the car without options.

7. Find the percent of markup of the car without options.

8. Find the markup and percent of markup for each of the options.

9. Find the markup and percent of markup for the car with all options.

A car with a selling price of $16,800 can be financed for

(a) 20% down and 6% annual interest rate with three years to pay

(b) 10% down and 7.2% annual interest rate with four years to pay

10. Find the monthly interest rate for an annual rate of 6%.

11. Find the monthly payment under terms (a).

12. Find the monthly payment under terms (b).

13. Find the amount of interest and principal in the first month's payment under terms (a).

14. Find the amount of interest and principal in the first month's payment under terms (b).

15. Find the annual depreciation of the car for 10 years with straight-line depreciation.

16. Draw a graph showing the resale value of the car with the following annual depreciation percents, starting with the purchase of the car: 25, 20, 15, 12, 10, 8, 5, 3, 2.

17. With the depreciation given in Exercise 16, what will be the resale value of the car at the beginning of the fifth year?

The odometer of a car read 10,356 at the start of a trip and 12,621 at the end of the trip. During the trip, 92.1 gallons of gas were purchased at a total cost of $114.20.

18. Find the average miles per gallon for the trip.

19. Find the average cost per mile for the trip.

Betty drives an average of 900 miles a month. Her average cost per mile for gas is 5.5 cents. In addition to the gas expense, she estimates $140 per year for maintenance and $200 per year for repairs. Her monthly loan payment is $186.74, her insurance premium is $715.50 every six months, and she pays $60 a year in taxes.

20. Find the estimated amounts that Betty spends on gas per month and per year.

21. How much does Betty pay per year in insurance premiums? per month?

22. Set up a yearly budget showing Betty's car expenses.

23. Make a pie graph of Betty's yearly car expenses.

24. Find Betty's per mile expense for owning and operating her car.

25. On the same coordinate system, graph Betty's per mile cost of operating her car (not including the loan expense) and the cost of leasing a car for $200 a month and paying 5.5 cents a mile for gas.

26. When would it be more cost efficient for Betty to lease?

27. Use the Driver Rating Factor table to find the rating of a 21-year-old unmarried male who owns his own car, has not had driver training, and uses the car for work.

28. Use the Driver Rating Factor table to find the rating of a 17-year-old unmarried female who does not own her own car, has had driver training, and does not use the car for work.

29. Find the annual insurance premium for a person with a driver rating factor of 2.10 and a six-month base rate of $675.

30. Suppose the driver in Exercise 29 has a six-month base rate of $710. Find her annual insurance premium.

31. Use the Driver Rating Factor table to find the rating of a 25-year-old female who drives to work.

32. Find the annual insurance premium for the woman in Exercise 31 if the six-month base rate is $525.

A car with a base price of $12,200 has a sticker price of $14,030.

1. Find the markup.
2. Find the percent of markup.

Kayla borrows $12,000 at a monthly rate of 0.5% for 36 months.

3. Find the amount of the monthly payment.
4. Find the amount of interest paid in the first month.

Sam records the following odometer readings: start, 34,560; end, 36,723. While traveling this distance, he uses 82.6 gallons of gas and spends $106.65.

5. How many miles did he travel?
6. Find his average number of miles per gallon of gas.
7. Find his average cost per mile of driving.

The following table shows Sandra's monthly car budget in dollars. She drives an average of 1000 miles per month. She has also found that she can lease a car for $240 a month and pay 6 cents per mile for gas.

Gasoline	Maintenance	Repairs	Loan Payment	Insurance	Taxes
$52	$15	$20	$245.30	$150.85	$11

8. Make a pie graph of Sandra's annual costs.
9. How much does Sandra pay in a year on her car loan?
10. Find her average cost per mile of owning and operating the car.
11. Find her average cost per mile of operating the car without the purchasing cost.
12. Write an equation for the monthly cost of operating the car.
13. Write an equation for the monthly cost of leasing a car.
14. Find the solutions common to the equations in Exercises 12 and 13.
15. What does the common solution mean to Sandra?
16. Find the annual insurance premium for a situation in which a driver with a rating of 2.55 purchases insurance with a six-month base rate of $627.80.

CUMULATIVE REVIEW

1. Your family's take-home pay is $3950 a month. How much can you afford to spend for credit card payments each month?

2. Use the Rule of 72 to determine how long it will take your $2300 savings account to double in value if it is growing at a rate of 4.5%.

3. Boris intends to buy a computer system for $1300. At one store he can finance the purchase in monthly installments at 12% over 2 years. At another store he can finance the purchase at 9% over 2 years. How much money will he save by financing the purchase at 9%?

4. Complete the following chart that shows the interest cost for two months on a credit card when you pay 10% of the amount due to the nearest dollar each month and your card carries an APR of 9%.

Month	Balance	Interest	Amount Owed	Payment
1	$1000.00			
2	907.50			

5. Find the total interest paid in Exercise 4.

6. Use the Comparison Table for Term and Whole Life Premiums in the Reference Section to find the amount paid in premiums in 30 years for a whole life insurance policy of $200,000 purchased at the age of 25.

Use the Tax Rate Schedule in the Reference Section to find the tax owed to the nearest dollar by each of the following taxpayers.

7. a single person with a taxable income of $105,950

8. a married couple, filing jointly, with a taxable income of $137,846

9. Use IRS Form 1040EZ and the Tax Table in the Reference Section to find the tax paid by a single person with a salary of $15,950 and interest income of $380. The person cannot be claimed as a dependent by anyone else.

10. A car with a base price of $6200 has a sticker price of $7500. Find the amount and percent of markup.

11. A used car purchased for $6000 is expected to depreciate in a straight line over five years. Find its resale value after two years.

12. A car uses 43.5 gallons of gasoline to cover 1155 miles. Find the average number of miles traveled on one gallon of gas.

13. Which equation or equations could represent a linear regression equation for the line of best fit?

    a. $y = x^a + b$　　　　　b. $-a = bx - y$

    c. $y = b^x + a$　　　　　d. $y = a + bx$

14. Use the Tax Tables in the Reference Section to find the tax owed by a head of household with a taxable income of $20,300.

PROJECT 10–1: Investigating Percent of Markup

Work with a group of four or five of your classmates to complete this activity.

1. Use *Edmund's New Car Prices* or the *Kelley Blue Book New Car Price Manual* as a reference.
2. Each student in the group selects ten automobiles, representing two or three manufacturers. Choose a wide range of models and prices.
3. Each student chooses a list of options for his or her automobiles.
4. Calculate the percent of markup for the automobiles and for each of the options.
5. Prepare a table showing the markup for different models and calculate averages for the different manufacturers.
6. Prepare a similar table for the options by manufacturer.

PROJECT 10–2: Percent of Markup for Used Cars

Work with a group of four or five classmates.

1. Have each member of a group select 20 cars advertised in the newspaper or on a used-car lot.
2. Consult a reference book such as *Edmund's* or the *Blue Book* for used-car prices to use in calculating the trade-in value of the cars.

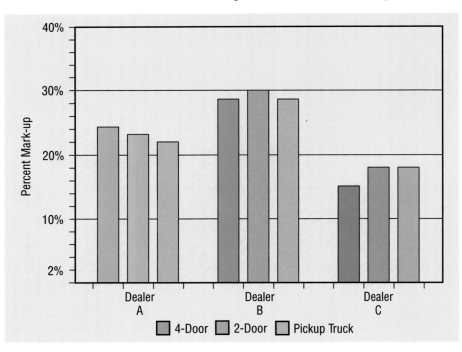

3. Use the prices listed in the newspaper or on the cars in the lot to calculate the percent of markup over the book value the dealer is using.
4. Prepare a table of the trade-in values and the percent of markup for the cars you researched.
5. Draw bar graphs, similar to the one shown, indicating the percent of markup by dealer and car model.

Some equations and their graphs belong to families. The equations in the same family all have some common characteristics. Recall that the slope-intercept form of the equation of a line is $y = mx + b$, where m is the slope and b is the y-intercept. To determine slope from a graph of a line, remember that slope $= \dfrac{\text{change in } y}{\text{change in } x}$. The y-intercept is the point where the graph intersects the y-axis.

Find the equation of each graph then note the family characteristic.

Example

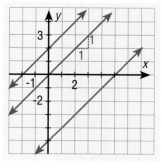

From the graph, the slope of each line is 1, so $m = 1$ in each case. The y-intercepts are 2, 0, and -5. Therefore the equations are

$$y = x + 2$$
$$y = x$$
$$y = x - 5$$

The family characteristic is that they all have the same slope.

1.

2.

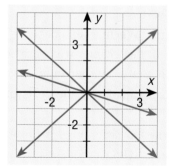

Graph the equations and note the family characteristics.

Example $y = x^2 - 3x$
    $y = -x^2 + 3x$

The second is the reflection of the first in the x-axis. Note that

$$-(x^2 - 3x) = -x^2 + 3x$$

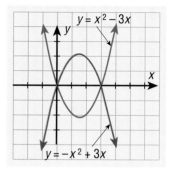

3. $y = x^2$
    $y = 3x^2$
    $y = 0.5x^2$

4. $y = x^2$
    $y = x^2 + 2$
    $y = x^2 - 3$

Travel Plans

AMERICANS LIKE TO TRAVEL, ESPECIALLY AT VACATION time. We visit relatives, beaches, mountains, and cities. We play and watch all sorts of games and sports. We visit places that were important in our nation's history and marvel at the natural beauty of different parts of the country. There really is a lot to see and do. For some, getting away from home and the everyday routine is in itself reason to travel.

In any case, travel is a big business. The travel and tourism industry is happy to cater to our needs and desires. Places to stay, amusements, souvenirs, good things to eat – these and much more call out to us from brochures, from roadside signs, and in all kinds of advertising.

But travel is not free. Whether reasonable or expensive, it always costs money. In this chapter we consider some travel costs. We work with Betty and her family, reading maps, figuring distances, and calculating the costs of automobile travel.

We consider Sylvia and her mother, who travel not only for fun but also because Sylvia's mother is a travel agent and wants first-hand experience about different places. Sylvia helps her mother by using spreadsheets to budget travel costs and a graphing calculator to compare the expenses and earnings from tours.

Reading airline schedules, with the various abbreviations, takes some skill. With Ramón, a college student, we compare the costs of travel by air and by a car.

11–1 Making Travel Plans

11–2 Travel Costs: Different Perspectives

11–3 Flying Saves Time and Sometimes Money

MAKING TRAVEL PLANS

Betty likes to travel. She and her family take vacation trips almost every summer. One of her favorite trips was to the West Coast. Betty remembers seeing giant sequoia trees that are the oldest living things on earth, more than 5000 years old. Betty's father bought her a small redwood box and a model truck with tiny redwood logs on the back.

On other trips, Betty saw the Liberty Bell in Philadelphia and stood at the top of the Empire State Building in New York. In Boston she really liked the science museum. She also remembers how her little brother left his teddy bear in a taxicab.

Betty helps her family plan their trips. They usually drive between 300 and 500 miles a day for two days. Then they stay for a few days to relax, see the sights, and visit. They usually take a different way home to see other sights.

Betty's mother teaches history at the high school and frequently knows interesting stories about the people who have lived in the area they are visiting. Betty's father makes up riddles and games for them to play as they drive along. She expects to enjoy the trip this year more than ever.

OBJECTIVES: *In this lesson, we will help Betty to:*

Use map-reading skills to plan a trip within the United States.
Use a mileage chart to determine distances.
Determine the cost of driving different distances.

WHICH WAY TO GO

After deciding where you are going, the next big travel decision is selecting the route, that is, which roads you will take. There are almost always several ways to drive to any given place. Each route has advantages and disadvantages.

If you want to get to your destination in the least amount of time, then you will probably take the big **interstate highways.** These do not have stop lights and with two or three lanes in each direction, traffic generally keeps moving. But sometimes it is more interesting, at least for a time, to take what are called **secondary roads.** These are well-marked federal and state highways that pass through towns and let you see more of what an area is really like. But traveling a distance on secondary roads can make a trip take a very long time.

Betty's family has found that it helps to have an interstate atlas along on any trip. This is a booklet with detailed road maps of every state. The **mileage chart** printed in the atlas gives the distances between many cities. Other mileage distances either are marked on the map or can be estimated on the basis of the map distance and the scale showing miles as distances.

When Betty and her family plan a trip, they have a lot to consider. Betty now helps with some of the driving. They usually switch drivers after one or two hours. When they stop to change drivers, they stretch and sometimes have a snack.

The family has learned to have a destination but also to be flexible when traveling. They sometimes find the most interesting places unexpectedly and stay longer than they had planned. On other occasions they find that a city or attraction is not what they had hoped, and they move along sooner than planned. There are also difficulties beyond their control. Once they had to wait for four hours while the car was repaired. Another time they traveled back roads for two hours to get around rush hour traffic. They have learned to turn problems into adventures and disappointments into opportunities.

Ask Yourself

1. How do you decide which routes to take when traveling?

2. Why would people take secondary roads instead of interstate highways?

3. What are two unexpected events that might change the timing of a trip?

ALGEBRA *REVIEW*

If $d \div a = g$ and $g \cdot c = t$, find each of the following.

1. g when $d = 500$ and $a = 25$

2. d when $g = 22.5$ and $a = 19.6$

3. a when $d = 300$ and $g = 28$

4. t when $g = 52$ and $c = 1.35$

5. c when $t = 127$ and $g = 105$

6. g when $t = 10.56$ and $c = 1.28$

For four numbers a, b, c, and d,

7. Write an expression for the average of a and b.

8. Write an expression for the average of a, b, c, and d.

9. If the average of a, b, and c is d, what is the average of a, b, c, and d?

10. If the average of a and b is 10, the average of c and d is 12, find the average of a, b, c, and d.

SHARPEN YOUR SKILLS

SKILL 1

EXAMPLE 1 Betty and her family have decided to drive from their home in Indianapolis, Indiana, to visit their cousins in Milledgeville, Georgia.

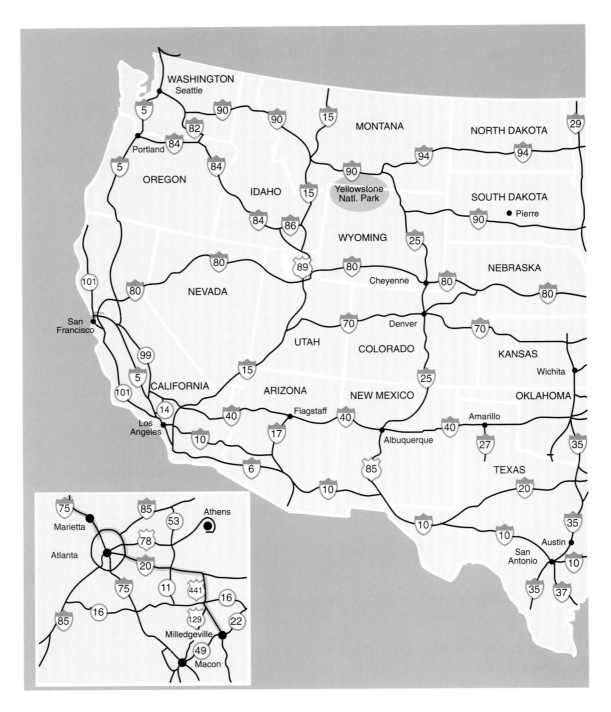

CHAPTER 11 TRAVEL PLANS

QUESTION   What highways can they take to drive from Indianapolis to Milledgeville?

SOLUTION

The map below shows the interstate highways in the United States. The route that they will take is highlighted: first, south on 65 through Louisville to Nashville; then southeast on 24 to Chattanooga and then 75 to Atlanta. The trip from Atlanta to Milledgeville is highlighted on the inset map of the area around Atlanta. They first take Interstate 20 east and then 441, a secondary road, south.

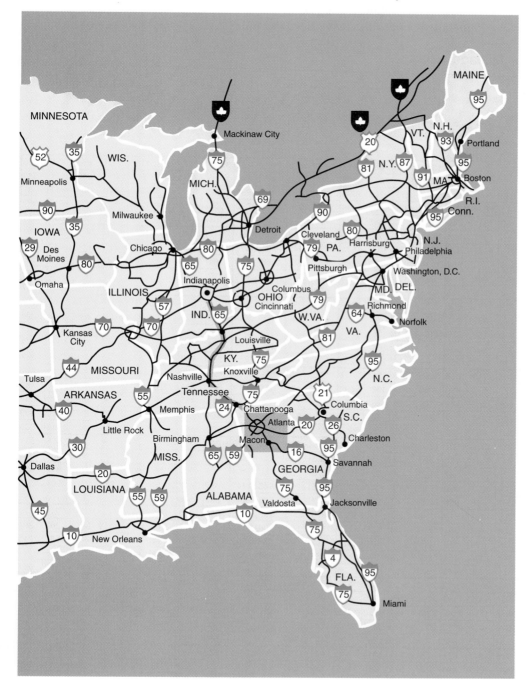

EXAMPLE 2 The mileage chart shows the distances between many major cities.

QUESTION How far is it from Indianapolis, Indiana, to Atlanta, Georgia?

SOLUTION

To answer the question, we use the mileage chart. Atlanta is the third city down the left-hand column. We read across the Atlanta row to the Indianapolis column. The cell where this row and column intersect shows the mileage between the two cities: 493 miles.

United States Mileage Chart

	Atlanta, GA	Boston, MA	Cheyenne, WY	Chicago, IL	Cincinnati, OH	Cleveland, OH	Dallas, TX	Denver, CO	Des Moines, IA	Detroit, MI	Indianapolis, IN	Kansas City, MO	Louisville, KY	Memphis, TN	Milwaukee, WI	Minneapolis, MN	New Orleans, LA	Omaha, NE	Philadelphia PA	Pittsburgh, PA
Albuquerque, NM	1381	2172	517	1281	1372	1560	638	417	977	1525	1266	782	1301	1010	1319	1190	1134	858	1899	1619
Amarillo, TX	1097	1897	511	1043	1096	1285	358	423	742	1269	991	547	1019	726	1084	975	850	643	1624	1344
Atlanta, GA		1037	1442	674	440	672	795	1398	870	699	493	798	382	371	761	1068	479	986	741	687
Austin, TX	919	1911	994	1110	1083	1327	193	906	877	1315	1037	682	982	615	1184	1129	517	837	1615	1367
Birmingham, AL	150	1165	1347	642	465	709	645	1286	787	724	475	697	364	246	728	1006	342	898	869	741
Boston, MA	1037		1907	963	840	628	1748	1949	1280	695	906	1391	941	1296	1050	1368	1507	1412	296	561
Charleston, SC	289	929	1722	877	603	730	1072	1678	1150	842	696	1078	591	660	964	1282	720	1266	633	666
Cheyenne, WY	1442	1907		954	1174	1279	869	100	627	1211	1068	650	1161	1101	987	788	1361	495	1678	1390
Chicago, IL	674	963	954		287	335	917	996	327	266	181	499	292	530	87	405	912	459	738	452
Cleveland, OH	672	628	1279	335	244		1159	1321	652	170	294	779	101	468	374	692	786	693	567	287
Columbus, OH	533	735	1235	308	108	139	1028	1229	618	192	171	656	209	576	395	713	894	750	462	182
Dallas, TX	795	1748	869	917	920	1159		781	684	1143	865	489	819	452	991	936	496	644	1452	1204
Denver, CO	1398	1949	100	996	1164	1321	781		669	1253	1058	600	1120	1040	1029	841	1273	537	1691	1411
Des Moines, IA	870	1280	627	327	571	652	684	669		584	465	195	566	599	361	252	978	132	1051	763
Detroit, MI	699	695	1211	266	259	170	1143	1253	584		278	743	360	713	353	671	1045	716	573	287
Flagstaff, AZ	1704	2495	757	1604	1695	1883	961	657	1300	1848	1589	1105	1624	1333	1642	1481	1457	1171	2222	1942
Harrisburg, PA	700	373	1579	639	468	314	1383	1592	952	474	534	1019	569	931	726	1044	1142	1084	102	189
Indianapolis, IN	493	906	1068	181	106	294	865	1058	465	278		485	111	435	268	586	796	587	633	353
Jackson, MS	391	1406	1257	742	655	899	404	1169	809	914	646	644	554	212	824	1036	178	845	1110	939
Kansas City, MO	798	1391	650	499	591	779	489	600	195	743	485		520	451	537	447	806	201	1118	838
Knoxville, TN	193	911	1372	527	253	485	837	1328	800	512	346	728	241	385	614	932	596	916	615	511
Louisville, KY	382	941	1161	292	101	345	819	1120	566	360	111	520		367	379	697	685	687	668	388
Mackinaw City, MI	935	916	1291	387	495	439	1261	1341	673	284	460	864	562	880	368	508	1247	805	842	556
Miami, FL	655	1504	2097	1329	1095	1264	1300	2037	1525	1352	1148	1448	1037	997	1416	1723	856	1641	1208	1200
Minneapolis, MN	1068	1368	788	405	692	740	936	841	252	671	586	447	697	826	332		1214	357	1143	857
New Orleans, LA	479	1507	1361	912	786	1030	496	1273	978	1045	796	806	685	390	994	1214		1007	1211	1070
Norfolk, VA	540	558	1764	831	604	508	1329	1758	1141	666	700	1162	642	877	918	1236	1019	1273	263	384
Pierre, SD	1361	1726	434	763	1050	1098	943	518	492	1029	944	592	1055	1043	690	394	1394	391	1501	1215
Pittsburgh, PA	687	561	1390	452	287	129	1204	1411	763	287	353	838	388	752	539	857	1070	895	288	
Portland, ME	1139	106	1986	1042	942	707	1850	2028	1359	775	1001	1486	1043	1398	1129	1447	1609	1491	398	663
Portland, OR	2601	3046	1159	2083	2333	2418	2009	1238	1786	2349	2227	1809	2320	2259	2010	1678	2505	1654	2821	2535
San Antonio, TX	983	1988	1027	1187	1160	1404	270	939	954	1392	1114	759	1059	692	1261	1206	550	914	1692	1444
San Francisco, CA	2496	3095	1188	2142	2362	2467	1753	1235	1815	2399	2256	1835	2349	2125	2175	1940	2249	1683	2866	2578
Seattle, WA	2618	2976	1228	2013	2300	2348	2078	1307	1749	2279	2194	1839	2305	2290	1940	1608	2574	1638	2751	2465
Tulsa, OK	772	1537	765	683	736	925	257	681	443	909	631	248	659	401	757	695	647	387	1264	984
Washington, DC	608	429	1611	671	481	346	1319	1616	984	506	558	1043	582	867	758	1075	1078	1116	133	221
Wichita, KS	903	1587	583	696	787	975	365	509	392	940	681	197	710	532	734	644	815	298	1314	1034

EXAMPLE 3 Betty estimates that the driving distance from Atlanta to Milledgeville is 115 miles. The family car gets an average of 20 miles per gallon of gas. They expect to pay about $1.35 a gallon for gas while on the trip.

QUESTION How much should they budget for gas between Indianapolis and Milledgeville?

SOLUTION

The total distance d is the distance from Indianapolis to Atlanta plus the distance from Atlanta to Milledgeville

$493 + 115 = 608$ miles

To find the number of gallons of gas to be used, divide the total distance by the average miles per gallon.

$g = \dfrac{d}{a}$ where d = total distance in miles
a = average miles per gallon
g = number of gallons needed

$g = \dfrac{608}{20} = 30.4$ gallons

Next, we find the total cost.

$t = gc$ where g = number of gallons
c = average cost per gallon
t = total cost for gas

$t = 30.4(1.35) = 41.04$

They should budget about $41 for gas.

1. Use the map on pages 502–503 to find the quickest and most direct route to drive from New Orleans, Louisiana, to Jacksonville, Florida.

2. Use the map on pages 502–503 to find the fastest and most direct route to drive from Columbia, South Carolina, to Birmingham, Alabama.

3. Use the mileage chart on page 504 to find the distance from Milwaukee, Wisconsin, to Portland, Oregon.

4. Use the mileage chart on page 504 to find the distance from Chicago, Illinois, to Cleveland, Ohio.

5. If you drive 387 miles and get an average of 21.5 miles per gallon of gas, how many gallons of gas do you use?

6. If you used 24 gallons of gas and got an average of 18.5 miles per gallon of gas, how many miles did you drive?

7. If you use 29.6 gallons of gas and pay an average of $1.39 a gallon, what is the total cost of the gas?

8. If you purchase 13.8 gallons of gas at a total cost of $15.85, what is the average price per gallon?

9. Write an equation for finding gallons of gas when you know total miles and miles per gallon. Explain what each variable in the equation stands for.

EXERCISE YOUR SKILLS

1. What are three factors that might affect your travel plans?
2. What are some advantages of traveling by car?
3. Why should you use a map to plan before you begin a car trip?

Use the maps provided in the lesson to find a direct route for each trip.

4. Norfolk, Virginia, to Valdosta, Georgia
5. Macon, Georgia, to Boston, Massachusetts
6. Miami, Florida, to Savannah, Georgia
7. Dallas, Texas, to Nashville, Tennessee
8. Use maps to find a direct route from where you live to Macon, Georgia.

Use the mileage chart in the lesson or the one in the Reference Section of the book to find the following distances.

9. Tulsa, Oklahoma, to Denver, Colorado
10. Albuquerque, New Mexico, to Cheyenne, Wyoming
11. Mackinaw City, Michigan, to Boston, Massachusetts
12. Flagstaff, Arizona, to Cleveland, Ohio
13. Minneapolis, Minnesota, to Tulsa, Oklahoma
14. Pierre, South Dakota, to Dallas, Texas

Find the cost of driving the following distances with the miles per gallon and the cost of gasoline as shown. Round each number of gallons to the nearest tenth and each cost to the nearest cent.

	Miles	Miles per Gallon	Number of Gallons	Cost per Gallon	Total Cost
15.	540	20.5		$1.29	
16.	2110	17.3		1.19	
17.	750	26.4		1.24	
18.	1235	22.6		1.15	
19.	1827	19.7		1.35	
20.	926	27.5		1.27	

A family from Philadelphia, Pennsylvania, plans a 9-day vacation. Their destination is Nashville, Tennessee, but they plan to stay overnight three times on the way and twice on their return trip. Work in groups of two or three to do the following exercises. Use information from travel magazines and talk with experienced travelers, as needed, to help you answer the following.

21. Plan the trip, indicating the miles to be covered each day, where they would stay each night, and what things they might do and see along the way.

22. Plan the trip back using a different route but also indicating distances, places of interest, and where they would stay overnight.

23. Assuming that they drive an average of about 80 miles a day while in the Nashville area, make an estimate of their total miles for the trip.

24. Estimate miles per gallon and cost per gallon, and on the basis of these estimates, determine an approximate cost for gasoline for the trip.

MIXED REVIEW

1. With $778.22 in your checking account you write a check for $105.89 to Grocery Mart for your weekly groceries. Show how you would enter this check in your check register.

Your credit account balance from the first of the month through the 12th was $1298. You made a payment of $325 on the 13th. You made no further charges or payments for the remaining 17 days of the month. The monthly finance charge is 1.5%.

2. What was your average daily balance?

3. What was the finance charge for the month?

The following table shows the Consumer Price Index for selected years between 1984 and 1992. The prices given in Exercises 4–6 are for 1984. Find the approximate cost that you might expect to pay for each item in 1992.

CONSUMER PRICE INDEX (1984–1992)					
Year	All Items	Food	Shelter	Apparel	Medical Care
1984	100	100	100	100	100
1985	107.6	105.6	107.7	105	113.5
1986	109.6	109	115.8	105.9	122
1987	113.6	113.5	121.3	110.6	130.1
1988	118.3	118.2	127	115.4	138.6
1989	124	125.1	132.8	118.6	149.3
1990	130.7	132.4	140	124.1	162.8
1991	136.2	135.8	146.3	128.8	177
1992	140.7	138.3	152.2	131	189.4

4. a $100,000 house   5. $250 in clothes

6. $2500 in medical care

Use the Tax Table in the Reference Section to find the tax owed by each of the following taxpayers.

7. a single person with taxable income of $35,900

8. a married couple, filing jointly, with taxable income of $48,548

A perspective is a point of view. A city looks quite different from the window of a bus than from an airplane. Similarly, the cost of a restaurant meal appears different, depending on whether you are the person having dinner or the one who owns the restaurant. In this lesson we will consider, from different perspectives, some of the costs of vacationing.

Ever since her parents were divorced, Sylvia has helped her mother with the cooking and cleaning at home. Consequently she and her mother really enjoy a vacation when they can eat out and let someone else wash the dishes. They also like camping out. There is a great feeling of freedom and relaxation that comes from being out of doors all day long.

Sylvia's mother has another reason for traveling. She is a travel agent and needs first-hand information about some of the accommodations and attractions that she recommends to her clients. She and Sylvia watch their costs very closely, both for themselves and also so that they can tell clients of the travel agency how to budget for a trip.

OBJECTIVES: *In this lesson, we will help Sylvia and her mother to:*

- *Compare and calculate the costs of food and lodging.*
- *Prepare a vacation budget.*
- *Compute costs for tour vacations.*

MAKING CHOICES ON A BUDGET

Sylvia and her mother have discovered that no one who travels likes expensive surprises. For example, many people do not mind paying a little extra for a good meal or accommodations with special features. But they want to know in advance what to expect. For that reason, Sylvia helps her mother prepare a detailed budget before a trip, and they also make reservations to be sure that they will not be disappointed. A **reservation** is an establishment's assurance that a motel room or a table in a restaurant will be held for you at a particular date or time. Sometimes you must use your credit card to guarantee a reservation. Even campgrounds accept reservations, and with so many people now using these facilities it is important to reserve a spot.

When Sylvia and her mother go camping or stay in a cabin with kitchen facilities, they purchase groceries and cook some meals for themselves. However, they prefer restaurants and have become skilled in predicting the prices of meals. They set up budgets for people with different tastes and means so that Sylvia's mother can make recommendations. They are always collecting brochures about restaurants and points of interest.

Everyone who travels says, "What shall we do today?" Sylvia has learned that the costs of entertainment vary greatly. Walking in the woods is free; the fee to enter a zoo or museum is usually reasonable; a special evening tour can be costly. Going to the beach is much cheaper than skiing. Gifts and souvenirs are always fun to buy, but once again, the costs can add up. Sylvia's mother says that they must stick to their budget and be careful to avoid impulse buying.

Ask Yourself

1. How can you plan your lodging expenses when you travel?

2. How can you save money on food costs when you travel?

3. What are several less expensive forms of entertainment that you enjoy?

ALGEBRA *REVIEW*

For numbers x, y, and z,

1. If $x = 10.50$ and $y = 11.90$, find the average of x and y.

2. If $x = 13.20$ and the average of x and y is 15.00, find y.

3. If $x = 8.76$, $y = 9.49$, and $z = 11.75$, find the average of x, y, and z.

4. If $x = 45.60$, $y = 58.20$, and the average of x, y, and z is 51.42, find z.

5. If $a = b + 1$ and $b = c + 1$, find the average of a, b, and c.

Graph each system of equations and find the coordinates of the intersection of the graphs.

6. $y = 10$ and $y = x + 2$

7. $y = 30 + 3x$ and $y = 5x$

8. $y = 1000 + 40x$ and $y = 75x$

SHARPEN YOUR SKILLS

SKILL 1

Sylvia has found the following lodging information in a travel guide.

LOTUS INN
LOTUS, IOWA

I-123 at 33rd Place, S.W.
Exit 3   555-1234
Location: 3322 South Street,
Southwest Lotus 50010
Downtown: 3.5 mi
Ball park: 1 mi
Zoo: 2 mi.
Airport: 7 mi
Restaurants & shopping nearby.

Rates: Tax 8%

1 person single	2 people double bed
$37.00	$43.00

2 people 2 dble bds	4 people 2 dble bds
$48	$58

PLEASURE LODGE
MONMOUTH, IOWA

Near International Airport
Indoor Pool   HBO
Health Club   Sauna
Location: 1001 Sandwich Road
Exit Stream Road,
Follow Airport signs.
Complimentary: airport limo,
light breakfast, cable TV.

Rates: Tax 8%. XP $8
Teens free. RB $8

1 person	Standard: $65 Luxury: $90
2 people	Standard: $75 Luxury: $105

All rooms have two double beds.

(XP means extra person)
(RB means roll-away bed)

EXAMPLE 1 Sylvia and her mother want to plan their lodging costs.

QUESTION What would be the cost for Sylvia and her mother to stay at each of these motels for 2 nights? They would take the least expensive room with two double beds.

SOLUTION
At Lotus Inn, two beds for two people cost $48 per night.

$2(48) = 96$ Rate for two nights
$0.08(96) + 96 = 103.68$ The tax is 8%.

The cost at Lotus Inn is $103.68.

At Pleasure Lodge teens can stay free of charge and the least expensive room costs $65.

$2(65) = 130$ Rate for two nights
$1.08(130) = 140.40$ Multiply by 1.08 for room and tax
                     (100% + 8% = 108%).

The cost at Pleasure Lodge is $140.40.
Sylvia and her mother would stay at Lotus Inn.

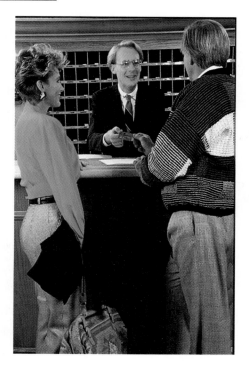

EXAMPLE 2 Sylvia and her mother want to use the actual cost of their meals while traveling as a reasonable basis for a budget for clients.

QUESTION How can they arrive at a reasonable estimate of the cost for meals?

SOLUTION
The best method is to record a number of prices for meals and then find the average. The following are the amounts spent by Sylvia and her mother. Sylvia used her spreadsheet skills to write formulas that would give totals and averages.

	A	B	C	D	E	F	G
1		Monday	Tuesday	Wednesday	Thursday	Total	Average
2	Breakfast	7.50	12.75	6.28	11.32	37.85	9.46
3	Lunch	12.85	14.50	6.70	9.43	43.48	10.87
4	Dinner	52.45	32.60	44.76	18.40	148.21	37.05
5	Total	72.80	59.85	57.74	39.15	229.54	57.39

+F2/4

@SUM(B2..B4)

On the basis of this record of expenses, Sylvia's mother will recommend to her clients that, for a similar trip, they plan to budget $10 per day for breakfast, $11 per day for lunch, and $37 per day for dinner, or $58 total per day.

Notice that $229.54 is the total of all the daily totals and also the total of all the meal totals. These two totals are equal because they are simply different ways of adding the same numbers. If you were preparing this with a calculator rather than a spreadsheet, this total would provide a check on your calculations.

SKILL 2

EXAMPLE 3 Her mother was so impressed with Sylvia's computer skills that she asked Sylvia to prepare a table in which to record the expenses for their trip.

QUESTION How can a spreadsheet be used to record travel expenses?

SOLUTION
Sylvia set up a budget as follows. Her mother suggests that her clients prepare a budget like this as a target, using their own numbers, and then take a blank form on the trip and fill in the actual amounts.

	A	B	C	D	E	F	G
1		Day 1	Day 2	Day 3	Day 4	Total	Average
2	Food	72.80	59.85	57.74	39.15	229.54	57.39
3	Lodging	56.70	56.70	28.65	31.50	173.55	43.39
4	Car	19.20	0.00	15.60	10.00	44.80	11.20
5	Entertainment	32.00	29.00	8.50	14.75	84.25	21.06
6	Gifts	0.00	13.00	21.68	8.90	43.58	10.90
7	Total	180.70	158.55	132.17	104.30	575.72	143.93

Sylvia's mother sponsors tours for people who like someone else to do the planning. Some of these trips include costs for travel, accommodations, meals, entrance to some attractions, and guided tours. Some costs are fixed, such as chartering a plane. Other costs, such as meal expenses, vary with the number of people on the tour. You can use break-even analysis to determine how many people are needed to make a profit.

EXAMPLE 4 Sylvia's mother has found that for a certain tour, the fixed cost is $10,000, and the cost per person is $150. She charges $720 per person for the tour.

QUESTION How can Sylvia help her mother determine the number of people needed to break even?

SOLUTION

Graphing equations for income (revenue) and expenses (cost) will show profit and loss.

The equation for income is

$y = 720x$ where x = number of people
y = total revenue

The equation for expenses is

$y = 10,000 + 150x$ where $10,000$ = fixed expense
x = number of people
y = total expenses

Graph the equations with the following range values.

Xmin: -2	Ymin: -2500
Xmax: 30	Ymax: 20000
Xscl: 2	Yscl: 2500

Using the trace function shows a break-even point at about 17.5.

Solve the system of two equations using substitution.

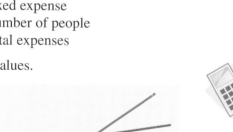

X = 17.536842   Y = 12626.526

$$y = 720x$$
$$y = 10,000 + 150x$$
$$720x = 10,000 + 150x$$
$$570x = 10,000$$
$$x = 17.5 \qquad \text{To the nearest tenth}$$

To break even, Sylvia's mother will need 18 people on the tour.

TRY YOUR SKILLS

The Downtown Motor Inn charges the following: Single, $47; double, $57; tax, 6.5%.

1. Find the cost of a single for 1 night, with tax.

2. Find the cost of a double for 3 nights, with tax.

3. The table in Example 2 shows amounts spent for restaurant meals by Sylvia and her mother. Find the average amount per person for each meal.

4. On the basis of the information in Example 4, what will be Sylvia's mother's profit if 20 people purchase the tour?

EXERCISE YOUR SKILLS

The following summarizes the costs at two different hotels.

Milledgeville Lodge
Rates: single, $48
          2 adults, $56
          Teens, free
Tax: 8% on room only
Free in-room movies
Game machines

Georgia Inn
Rates: single, $39
          2 adults, $48
          Teens, $10
Tax: 8% on room only
Video rental: $7.00
Miniature golf, $3.00

Find each of the following costs.

1. Two adults at the Milledgville Lodge for 3 nights

2. Single at the Georgia Inn for 2 nights, with 1 video rental

3. Two adults and 2 teens at the Georgia Inn for 4 nights, with 2 rounds of miniature golf per person and 2 video rentals

Use a calculator to find the totals for each day, the totals for each meal, and the average cost for each meal; each cost shown is for a family of four.

		Monday	Tuesday	Wednesday	Thursday	Total	Average
4.	Breakfast	$17.50	$22.75	$19.54	$23.32		
5.	Lunch	28.75	24.50	15.27	17.43		
6.	Dinner	78.45	32.60	58.36	98.63		
7.	Total						

8. How can you double check your totals?

9. Using these averages and the rates listed above, prepare a food and lodging budget for a family of four (2 adults, 2 teens) to stay at the Milledgeville Lodge for 3 days.

A travel agent prepares a tour package, charging $560 per person. The tour has fixed costs of $7500 and a per person cost of $180.

10. Write an equation for the total revenue of the travel agency.

11. Write an equation for the total expenses of the travel agency.

12. Set up a range and use a graphing calculator to graph the equations.

13. Use the graphs to find the break-even point for the tour.

14. Find the break-even point by solving the system of equations.

15. What will be the profit or loss if 15 people take the tour?

16. What will be the profit or loss if 28 people take the tour?

John and Harriet Gomez operate a restaurant. Their average costs per day are $500 fixed, $10 per person having dinner. The average amount spent by a person having dinner is $20.

17. How many dinner patrons are needed for the Gomezes to break even?

18. What is the financial result of having only 35 dinner patrons?

19. How does the Gomezes' thinking about the price of dinner differ from the thinking of one of their patrons coming for dinner?

20. How might lowering the average price of dinner lead the Gomezes to earn less money from their restaurant?

21. How might lowering the average price of dinner lead the Gomezes to earn more money from their restaurant?

22. Without actually lowering the prices, how might the Gomezes determine whether this would be a good strategy?

A tour company arranges a bus tour. The advertised price is $445 per person. The tour has fixed costs of $9100 and a per person cost of $225.

23. Write an equation for the amount paid in to the tour company.

24. Write an equation for the amount paid out by the tour company.

25. The number of people able to take the tour is limited by the fact that the bus has a maximum capacity of 45 passengers. Will the tour company be able to make a profit?

26. Suppose 50 people sign up for this tour. Accommodating the extra people will raise the fixed costs to $12,350 and lower the cost per person to $210. The price of the tour must remain at $445 per person. Should the tour company accept reservations from 50 people, or maintain a limit of 45 people? Explain your answer.

1. Marsha's job at the library pays her $12.50 plus $1\frac{1}{2}$ times her regular wage rate for each hour over 35 hours. How much did she earn in a week during which she worked 42 hours?

2. Keisha earns $9.50 per hour. She works $37\frac{1}{2}$ hours per week for 49 weeks and received 3 weeks of paid vacation. Her other benefits cost her employer $4000 per year. What is the annual cost to the company of her salary and benefits?

Liam's bank pays interest of 0.25% per month on his checking account. It imposes a monthly service charge of $6 for balances under $200, $5 for balances between $200 and $499, and no service charge for balances over $499. For balances under $500 there is a charge of $0.02 for each check. Find the new balance in a checking account with the given balance and number of checks written.

3. $501; 22 checks

4. $400; 15 checks

5. Determine which of the following loans has a greater total cost.

    a. $12,000 at 14% for 5 years

    b. $15,000 at 11% for 4 years

The annual depreciation on a new $20,000 car is 25% immediately upon purchase. At the end of each year that follows, the percent of depreciation from the original price changes as follows: 20%, 12%, 10%, 8%, 7%, 6%, 5%, 4%, 3%. Find the value of the car at the times indicated in Exercises 6–8.

6. Immediately after purchase

7. In 1 year

8. In 5 years

Use the Rule of 72 to find how many years it will take for $3000 to grow to $24,000 at the given rates of interest.

9. 5.25%

10. 7.5%

11. 9%

José bought 1500 shares of stock at $43\frac{1}{8}$ per share. In Exercises 12–14 ignore the commission costs.

12. Find the cost of the shares.

13. After three years, José sold the shares at $57\frac{3}{8}$ per share. What was the total sale price?

14. What was the capital gain or loss on the shares?

A loan of $10,000 is repaid one year later in a single payment of $10,900.

15. What was the interest paid on the loan?

16. What was the rate of interest on the loan?

FLYING SAVES TIME AND SOMETIMES MONEY

Ramón, like Sylvia and Betty, has taken family vacations by car. He has always enjoyed visiting new places, especially those that played a part in history, such as the Alamo, and those that have great natural beauty, such as Niagara Falls. He loves art and frequently sketches what he sees.

Now Ramón goes to college and sometimes flies to school. For this reason he has become familiar with airline timetables and air fares.

He lives in Chicago and goes to school in New Orleans. He occasionally visits friends in other parts of the country.

It takes some practice to read airline schedules. Comparing times and connections can be tricky. In addition, there is quite a range in air fares. Ramón, like many people, depends on a travel agent to make arrangements. But there are always several options, and planning ahead can often save time and money.

OBJECTIVES: *In this lesson, we will help Ramón to:*

- *Understand what is involved in air travel.*
- *Make travel plans based on airline schedules.*
- *Compare the costs of traveling by air and by car.*

MAKING AIR TRAVEL PLANS

Ramón likes to fly because it saves time. When school is out, he wants to get home. When he is home on vacation, he likes to stay until the last possible minute. Nevertheless, he has learned to plan ahead. He realizes that a lot of students go back to school at the same time, so it is important to buy a ticket early. This saves money and helps him get the most convenient flight.

As Ramón became more involved in travel matters, checking flight times and rates became a hobby. He was interested not just in his own trips, but also in how the airlines operate. He has learned to consider time zones. Airlines always list departure and arrival times according to local time. This is helpful because most people want to know the local time when they will arrive.

People also want to know the length of a trip. The airline schedule doesn't tell you this. You have to figure it out yourself, making sure to include the change in time from the place of departure to that of arrival. A **time zone** is a geographic area that has the same time. Dallas and Chicago are in the same time zone. Dallas and Los Angeles are in different time zones.

Ramón knows that several major airlines fly many times a day to major cities. Smaller cities are often connected by smaller regional airlines. As he started checking schedules, Ramón was surprised to learn how many ways there are to get from one place to another.

The **advance purchase** of a ticket, that is, buying a ticket 14–60 days ahead can sometimes save money. On a round trip, staying over Saturday night can also save money. One-way fares are usually priced higher than round-trip fares. Some tickets that are purchased under discount arrangements are **nonrefundable,** that is, you cannot get your money back or use the tickets on another flight. Tickets for which you pay full fare can be used on flights other than the one specified on the ticket. Sometimes an airline will have special rates to attract customers.

Ask Yourself

1. What is the obvious advantage of air travel?

2. What is a disadvantage of air travel?

3. Why is it important to check schedules and costs before taking a trip?

ALGEBRA *REVIEW*

The following are distances along a line:

> **A to B, 300 miles**
> **B to C, 400 miles**
> **C to D, 1000 miles**

1. What is the least possible distance from *A* to *C?*

2. What is the greatest possible distance from *A* to *C?*

3. Find the distance from *B* to *D,* passing through *C.*

4. Find the least possible distance from *D* to *B.*

5. Draw a diagram showing the least possible distance from *A* to *D.*

6. What is the least distance from *A* to *D?*

7. If the distance from *A* to *C* is 700 miles and the distance from *A* to *D* is 1700 miles, draw a diagram showing the arrangement of *A, B, C,* and *D.*

SKILL 1

The airline schedule shown is part of one that Ramón obtained in his effort to learn about airlines. A longer schedule appears in the Reference Section.

WINDWARD AIRLINES SCHEDULE

Leave	Arrive	Stops/Via	Rmks
From Chicago, IL			
To Muscle Shoals, AL			491 mi
1:00p0	5:10p	ATL	L
4:55p0	9:00p	ATL	D X6
To Myrtle Beach, SC			743 mi
8:10a0	1:18p	ATL	B
9:54a0	3:15p	ATL	S
Eff. May 15			
1:00p0	6:59p	ATL	L
6:44p0	11:45p	ATL	D X6
To Naples, FL			1136 mi
8:15a0	1:48p	MCO	B
2:10p0	7:25p	MCO	S
To Nashville, TN			401 mi
5:30a0	8:45a	CVG	S X67
6:15a0	9:24a	CVG	S X67
7:40aM	10:55a	CVG	X67
9:57a0	12:59p	CVG	X67
11:35aM	2:50p	CVG	X67
3:20pM	6:55p	CVG	X6
3:50p0	6:55p	CVG	
5:35p0	8:43p	CVG	X6
5:40pM	8:43p	CVG	X67
Eff. May 1			
To Nassau, Bah			1301 mi
8:10a0	1:55p	ATL	
To New Orleans, LA			831 mi
6:15a0	10:14a	CVG	B/S X67
8:10a0	12:11p	ATL	B
9:25aM	1:35p	CVG	L6
Eff. May 6			
9:57a0	1:35p	CVG	L

Leave	Arrive	Stops/Via	Rmks
10:18a0	2:45p	DFW	X67
11:35a0	3:55p	ATL	L
1:00p0	5:20p	ATL	L
3:14p0	7:20p	ATL	S6
3:20pM	7:45p	CVG	DX6
3:50p0	7:45p	CVG	D
6:44p0	10:35p	ATL	D
8:45p0	12:59a	DFW	S
To New York, NY/			
Newark, NJ			734 mi
L-LaGuardia; J-Kennedy;			
E-Newark			
7:40aM	12:25pL	CVG	X67
11:35aM	4:40pL	CVG	X67
11:35aM	4:55pE	CVG	X67
3:20pM	7:45pE	CVG	D X6
5:35p0	10:20pL	CVG	
5:40pM	10:20pL	CVG	X67
Eff. May 1			
To Norfolk/Virginia Beach/			
Williamsburg, VA			707 mi
6:30a0	12:29p	ATL	B X7
9:54a0	3:10p	ATL	S
11:35aM	3:55p	CVG	X67
1:00p0	6:25p	ATL	L
5:35p0	9:40p	CVG	
5:40pM	9:40p	CVG	X67
To Oakland, CA			1843 mi
Also see San Francisco, San Jose			
8:00a0	12:05p	SLC	B
11:45a0	3:45p	SLC	
3:10p0	7:10p	SLC	D
6:15p0	9:55p	SLC	

Leave	Arrive	Stops/Via	Rmks
To Oklahoma City, OK			695 mi
8:30a0	12:35p	DFW	B
10:18a0	2:15p	DFW	X67
12:10p0	4:15p	DFW	X7
3:50p0	9:25p	DFW	S
8:45p0	12:35a	DFW	S
To Ontario, CA			1707 mi
Also see Los Angeles, Burbank,			
Long Beach and Orange County			
8:00a0	11:55a	SLC	B
8:30a0	1:10p	DFW	
11:45a0	3:25p	SLC	
12:10p0	4:20p	DFW	X7
3:50p0	8:10p	DFW	
6:15p0	9:45p	1	
To Orange County, CA			1732 mi
Also see Los Angeles, Burbank,			
Long Beach and Ontario			
6:15a0	12:05p	CVG	X67
8:00a0	12:05p	SLC	B
8:30a0	1:15p	DFW	
9:25aM	3:40p	CVG	6
Eff. May 6			
11:45a0	3:40p	SLC	
3:10p0	7:35p	SLC	S
3:50p0	8:10p	DFW	
To Orlando, FL			995 mi
5:30a0	10:00a	CVG	B X67
8:15a0	11:42a		B
9:25aM	2:35p	CVG	L6
Eff. May 6			
9:45a0	2:54p	ATL	S
9:57a0	2:35p	CVG	L

Ramón has learned that each commercial airport in the world is assigned a three-letter code. Usually, the code makes it easy to recognize the name of the city or airport that it represents. However, sometimes the reference is not so obvious. Following are codes from the schedule and the cities that they represent. In the schedule, the codes refer to places at which the flight stops on its way to the final destination. A flight that does not make any stops is called a **non-stop flight. A connection** is a stop at which you must change planes in the course of your trip. If your flight is a **direct flight,** the plane will have one or more intermediate stops but you will not change planes.

ATL	Atlanta, Georgia	CVG	Cincinnati, Ohio
DFW	Dallas/Ft. Worth, Texas	MCO	Orlando, Florida
SLC	Salt Lake City, Utah		

Other symbols in the schedule and their meanings are as follows.

B, L, S, D   Breakfast, Lunch, Snack, Dinner

X6 and X7   The flight travels every day except 6 (Saturday) and 7 (Sunday).

EXAMPLE 1 Ramón is making travel arrangements to go back to school. He wants information about flights from Chicago, Illinois, to New Orleans, Louisiana. Chicago has two airports: O'Hare (O) and Midway (M).

QUESTIONS 1. Which flight will get him to New Orleans the earliest in the day? At what time and from which Chicago airport does the flight leave? Where does it stop along the way?

2. Find the same information for the latest flight he can take.

3. About how long is the time for the trip?

SOLUTIONS

By examining the schedule we obtain the information.

1. The flight with the earliest arrival time leaves O'Hare at 6:15 A.M. and arrives in New Orleans at 10:14 A.M. It passes through Cincinnati. This flight does not travel on Saturday or Sunday.

2. The latest flight leaves O'Hare at 8:45 P.M. and arrives in New Orleans at 12:59 A.M. the following morning. It stops at the Dallas/Ft. Worth airport.

3. Calculating several of the times shows that the average length of the trip is about 4 hours. Since Chicago and New Orleans are both on Central Time, it is not necessary to adjust the time as calculated from the schedule.

You cannot always calculate the length of a flight directly from the departure and arrival times. Use the time zone map as needed to determine the duration or length of flights.

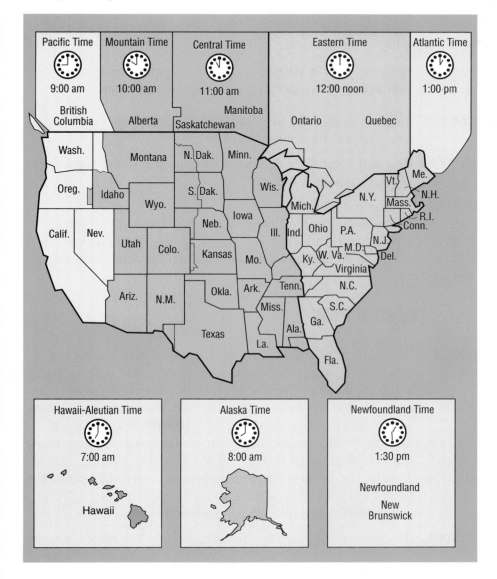

EXAMPLE 2 Ramón wants to find out how to adjust the travel time for flights traveling from west to east.

QUESTION How long is the first flight listed from Chicago to Norfolk/Virginia Beach?

SOLUTION

To determine the effect of time change on the length of flight, it is necessary first to find out the time zones of the various cities. Use the time zone map to do this.

Chicago is in the Central Time Zone. Norfolk is in the Eastern Time Zone. The clocks in Norfolk are an hour later than those in Chicago. To make the arrival time correspond to Chicago time, subtract 1 hour from the length of the trip according to the schedule.

Using Chicago time, you can see that the trip lasts from 6:30 A.M. until 11:29 A.M. Subtracting 1 hour gives a trip of 4 hours and 59 minutes.

EXAMPLE 3 Ramón now wants to work with travel time for flights traveling west from one time zone to another.

QUESTION How long is the flight that leaves Chicago at 3:10 P.M. and arrives in San Francisco at 7:10 P.M.?

SOLUTION

The clocks in San Francisco are 2 hours earlier than those in Chicago. When it is 7:10 in San Francisco, it is 9:10 in Chicago. Therefore you must add 2 hours to the flight time based on the schedule. The total actual flight time is 6 hours.

───── **SKILL 3** ─────────────────────

EXAMPLE 4 Several times, Ramón has driven to college to save money and also to bring things, such as his stereo system, that were too bulky to take on the plane. Now he has a chance to drive someone's car from New Orleans to Chicago at the start of the summer vacation. He wonders how much it would save. He can get a special rate flight for $300. Each day that he arrives later will keep him from a summer job at which he will earn $65 a day. His friend's car gets 20 miles per gallon of gas. Ramón expects to pay $1.25 per gallon for gas. He plans to drive for 2 days. Overnight lodging will cost $40 and he is budgeting $25 per day for food.

QUESTION How much will he save by driving?

SOLUTION

From the mileage chart in Lesson 11–1, Ramón finds the distance from New Orleans to Chicago is 912 miles. Driving 912 miles will take $912 \div 20 = 45.6$ gallons of gas.

Cost of gas: $1.25(45.6) = \$57$
Lodging for one night: $40
Meals for two days: $2(25) = \$50$
Total cost of driving: $57 + 40 + 50 = \$147$

Since the cost of flying is $300, it seems that Ramón will save 300 − 147 = $153 by driving. However, he will lose two days pay which is 2(65) = $130. His actual cost of driving is 147 + 130 = $277. So he will save only 300 − 277 = $23. Ramón decides saving $23 is not worth a tiring trip by car.

TRY YOUR SKILLS

Ramón plans to take a flight to visit a friend in Orlando, Florida. His friend can pick him up at about noon. Ramón's ticket costs $625. Use the Windward Airlines Schedule in the Reference Section.

1. At what time does the most convenient flight leave Chicago?
2. From what airport does it leave?
3. Is it a direct flight?
4. How long is the flight?
5. Does it seem likely that Ramón could save money by driving? Why or why not?

EXERCISE YOUR SKILLS

Ramón's sister is planning a trip from Chicago to Williamsburg, Virginia. She wants to arrive between 2:00 and 4:00 in the afternoon. Use the Windward Airlines Schedule in the Reference Section.

1. Which departure times could she take?
2. From which airport does the earlier flight leave?
3. From which airport does the later flight leave?

KEY TERMS

advance purchase
connection
direct flight
nonrefundable
non-stop flight
time zone

4. How long is each flight?

5. Why might she choose the first flight?

6. Why might she choose the second flight?

7. If you want to compare the cost of air travel with car travel, what are some of the costs that must be included in each?

8. Why do you think airlines often charge people less if they pay for a particular flight weeks or months in advance?

9. What is one disadvantage of buying an airline ticket months in advance?

Use the Windward Airlines Schedule in the Reference Section to do the following.

10. Find the shortest flight from Chicago to San Francisco.

11. Find the earliest flight from Chicago to San Diego that goes through Dallas/Ft. Worth.

12. Find the latest flight from Chicago to Portland, Oregon.

13. Find the shortest flight from Chicago to Ontario, California.

14. Estimate the average flight time from Chicago to Phoenix.

15. Estimate the average flight time from Chicago to San Diego.

16. What is the air mileage for a trip from Chicago to Palm Springs, California?

Name the time zone in which each city is located.

17. Nashville, Tennessee

18. Newark, New Jersey

19. El Paso, Texas

20. Oakland, California

21. Denver, Colorado

22. Boston, Massachusetts

Ramón's friend Michael is planning a trip from Chicago to Portland, Oregon. Use the information in the Reference Section as needed. If he drives, he expects that it will take 4 days. He will have to spend $70 a day for meals and lodging. His car gets 20 miles to the gallon of gas, and he expects to pay an average of $1.30 per gallon. He can get an airline ticket for $780.

23. What is the air mileage for the trip?

24. What is the driving mileage for the trip?

25. What will be the approximate cost of the trip by car?

26. Does he save by traveling by car or flying? How much?

MIXED REVIEW

Your credit card carries a monthly finance charge of 1.25%. You pay 10% of the amount owed to the nearest dollar each month. Your beginning balance is $1676. Complete the following table for 3 months. Assume that you make no further purchases during that time.

	Month	Balance	Interest	Amount Owed	Payment
1.	1	$1676			
2.	2				
3.	3				

Use the Tax Table in the Reference Section to determine the tax owed by each of the following taxpayers.

4. A married person, filing separately, with $45,750 in taxable income

5. A head of household with $72,000 in taxable income

Use the information in the table below to find the distance traveled, the miles per gallon, and the cost per mile of the trip.

	Odometer First	Odometer Last	Gallons of Gas	Cost of Gas	Distance (miles)	Miles per Gallon	Cost per Mile
6.	26,580	27,000	21	$24.50			
7.	250	866	19.5	27.56			
8.	86,290	87,326	51.6	49.75			

9. Cheryl prepares flower arrangements for Niven's Flower Store. She receives $7.50 for each budget arrangement, $12.25 for each regular arrangement, and $18.50 for each deluxe arrangement. How much did she earn in a week in which she assembled 30 budget arrangements, 15 regular arrangements, and 12 deluxe arrangements?

10. Myrna sells annual memberships in an exercise club. Her boss has offered her a choice between Plan 1, no salary but a straight $40 for each membership sold, and Plan 2, a weekly base salary of $200 plus $30 for each membership sold. How many memberships would Myrna have to sell to make the same amount of money under either plan?

11. Joseph's annual salary is $68,250. How much does Joseph pay for FICA taxes each year?

12. Heidi's annual salary is $137,500. How much does Heidi pay for FICA taxes each year?

13. Oshi's annual salary is $52,500. How much does Oshi pay for FICA taxes each year?

1. What is an advantage in traveling by interstate highways?

2. What is an advantage in traveling by secondary roads?

3. Why should plans for travel by car be somewhat flexible?

4. What is an advantage of air travel over travel by car?

5. What is an advantage in traveling by car?

6. In addition to times of departure and arrival, what must you check when calculating the length of time for a flight?

Use the U.S. interstate map in the Reference Section to identify which routes you would take to travel from Dallas, Texas, to the following cities.

7. New Orleans, Louisiana

8. Richmond, Virginia

9. Boston, Massachusetts

10. Little Rock, Arkansas

11. Chicago, Illinois

12. Kansas City, Missouri

Use the mileage chart in the Reference Section to find the distances by road between the following.

13. Austin, Texas, and Detroit, Michigan

14. Cincinnati, Ohio, and Flagstaff, Arizona

15. Kansas City, Missouri, and Miami, Florida

16. Louisville, Kentucky, and Philadelphia, Pennsylvania

17. Find the cost of traveling the distances for Exercises 13–16 if you get 23.5 miles per gallon and gas costs an average of $1.45 per gallon.

18. A family of five wants to share a room at the Honolulu Royale. The rates are $86.00 for two people, $10 for each additional person, and $4.00 for a rollaway bed; the tax is 9.6%. Assuming that they need one rollaway bed, how much will it cost the family to stay for 5 nights?

Use the following for Exercises 19–21.

ATLANTA PLAZA Rates: 1 pers, $65–83; 2 people, dbl, $73–91
Tax 8%, Addit. Pers $8. Children free. Rollaway $6

---

19. What is the least that it would cost two adults and one child to stay for 4 nights, assuming that each has his or her own bed?

20. What is the least that it would cost one person to stay for 2 nights?

21. What is the most that it would cost two people to stay 2 nights?

Fnd the total amount spent each day, the total for each meal, and the average cost of a meal.

		Monday	Tuesday	Wednesday	Thursday	Total	Average
22.	Breakfast	$15.20	$22.75	$12.28	$23.78		
23.	Lunch	25.47	30.80	11.45	17.56		
24.	Dinner	89.63	71.29	67.40	38.59		
25.	Total						

Harvey's Tours arranges guided tours to different parts of the country. The 4-day bus tour of New England has a fixed cost of $4200 and a per person cost of $160. The charge for the tour is $450 per person.

26. Write an equation for the expenses of the tour.

27. Write an equation for the revenue received by Harvey's Tours.

28. Graph the equations.

29. Find the approximate break-even point.

30. Find the break-even point algebraically.

31. How much profit will Harvey's make if 20 people go on the tour?

Find the time zone for each city.

32. Washington, D.C.

33. Eugene, Oregon

34. Butte, Montana

35. Tulsa, Oklahoma

Use the Windward Airlines Schedule in the Reference Section to do the following.

36. Estimate the average length of a trip from Chicago to San Antonio.

37. Find the departure time of the earliest flight from Chicago to Oklahoma City, Oklahoma.

38. Find the arrival time of the latest flight from Chicago to Roanoke, Virginia.

39. Find the length of the first flight from Chicago to Raleigh/Durham, North Carolina.

Consuela is planning a trip from Chicago to San Antonio, Texas. Use the information in the Reference Section as needed. If she drives, she expects that it will take 3 days. She will have to spend $80 a day for meals and hotels. Her car gets 22 miles to the gallon of gas, and she expects to pay an average of $1.35 per gallon. She can get an airline ticket for $660.

40. What is the air mileage for the trip?

41. What is the driving mileage for the trip?

42. What will be the approximate cost of the trip by car?

43. About how much money does she save traveling by car?

1. What are two possible causes of delays in traveling by car?

2. Why should travelers estimate costs before starting on a car trip?

3. Use the interstate map to plan the most direct route from Indianapolis, Indiana, to Yellowstone National Park in Wyoming.

Complete the table to find the number of gallons and total cost.

	Miles	Miles per Gallon	Number of Gallons	Cost per Gallon	Total Cost
4.	1768	22.6		$1.47	
5.	2591	25.6		1.29	

Use the following descriptions to answer Exercises 6 and 7.

HUMPHREY LODGE
Rates: Double $56
Addit. Pers: $8
Tax: 9%

McCARTHY INN
Rates: Double $86
Children free
Tax: 8%

6. Find the cost of three adults sharing a room at the Humphrey Lodge for 2 nights.

7. Find the cost of two adults and two children for 3 nights at the McCarthy Inn.

Alex and Alicia run nature tours to different parts of the country. Three days in the Everglades, leaving from Miami, cost them a flat rate of $1500 and $125 per person. The charge per person is $225.

8. Write an equation for the cost of the tour to Alex and Alicia.

9. Write an equation for the income for the tour.

10. Graph the equations.

11. Use the graph to find the break-even point for the tour.

12. Find the break-even point algebraically.

13. What is the amount of profit or loss if ten people go on the tour?

Use the Windward Airlines schedule in the Reference Section to do the following.

14. Find the length of time of the first flight from Chicago to Myrtle Beach.

15. Find the length of time of the last flight from Chicago to Norfolk, Virginia.

CUMULATIVE REVIEW

Assume that 40 hours = 1 workweek, 52 weeks = 1 year, and 12 months = 1 year. Find the hourly, weekly, monthly, and yearly salaries for the following pay rates.

1. $17,680 per year   **2.** $2000 per month   **3.** $5.50 an hour

4. Your checkbook shows a balance of $178.64. You have outstanding checks that total $53.21. The bank statement shows a balance of $229.85. Find the amount of a bank fee that has not yet been recorded in your checkbook.

You have a savings account that contains $6000. Your bank pays interest at an annual rate of 5.5% compounded as given. Find the value of the account at the end of 3 years.

5. annually   **6.** quarterly

Jackie and Joshua have gone into business selling doughnuts. They charge 75¢ for each doughnut. Their fixed costs are $400. Each doughnut costs $0.36.

7. How many doughnuts must they sell before they begin to make a profit?

8. How much profit will they have after selling 1500 doughnuts?

9. The previous balance of a credit card statement is $930. The monthly interest rate is 1.5% and you pay $125 plus all of the interest that has accumulated for that month. What is the new balance?

10. Use the Credit Scoring Table on page 293 of Lesson 6-6 to find the probability that a person with a score of 115 will repay a loan.

An insurance company expects 99.804% of all 36-year-old people to live at least one more year. The direct and indirect expenses for issuing one policy are $25.

11. Find the break-even premium for a 1-year $150,000 term insurance policy on a 36-year-old person.

12. What profit does the company expect to make on a $200,000 1-year term policy for a 36-year-old person who is charged with a premium of $550?

13. How many bonds can you buy with $20,000 given that the price of the bond is quoted at $101\frac{3}{4}$, that is, at $1017.50 for each bond? Ignore the effect of commissions.

14. Find the loss incurred in buying 100 shares of stock at $63\frac{3}{8}$ and then selling them 1 year later at $57\frac{5}{8}$. The commission rate on each transaction is 2%.

15. Use the Tax Tables in the Reference Section to find the tax to be paid by a single person who has $29,760 of taxable income.

A car is purchased by a dealer for $12,600 and sold for $16,000.

16. Find the markup.   **17.** Find the percent of markup.

18. If a used car is purchased for $8000 and depreciates in a straight line for 8 years, what is its value after 3 years?

PROJECT 11–1: Planning a Camping Trip

Work with a group of classmates to plan a 2-week camping trip for a family of four.

1. Decide where you want to go and what sights you want to see.
2. Plan a driving route to include campsites along the way.
3. Estimate the cost of gasoline, lodging, and food for the trip, including the cost of camping at the sites you choose.
4. Find the approximate cost of renting a bus with a driver for a camping trip similar to the one you planned above.
5. Estimate the fixed and variable costs.
6. Set a fair per-person price for the trip.
7. Write cost and revenue equations.
8. Graph the cost and revenue equations.
9. Find the break-even point for the trip.

PROJECT 11–2: Planning an Airline Trip

Work with a group of classmates to find bargain air fares.

1. Have each person choose five cities to visit.
2. Get prices from various airlines and travel agents.
3. Find the savings achieved by purchasing in advance and by staying particular lengths of time.
4. Have individuals present their findings and discuss different arrangements with the group.
5. Prepare a list of tips to travelers about how to save money on air fare.

The algebraic expressions x, $x^2 + x$, and $2x^2 - 3x + 1$ are called *polynomials*. Polynomials can be multiplied by using rules such as the distributive property.

Multiply.

Example $(x - 1)(x - 2)$

$$
\begin{aligned}
(x - 1)(x - 2) &= (x - 1)x - (x - 1)2 && \text{Distributive property}\\
&= x^2 - x - 2x + 2 && \text{Distributive property}\\
&= x^2 - 3x + 2 && \text{Collect like terms.}
\end{aligned}
$$

1. $(x - 1)(x + 2)$ **2.** $(x + 3)(x - 2)$ **3.** $(x + 3)(x + 3)$

4. $(x - 2)(x + 2)$ **5.** $(x + 1)(x - 5)$ **6.** $(x - 3)(x - 2)$

Notice in your answers to Exercises 1–6 that the third term is equal to the product of the two constant terms, and the coefficient of the x term is equal to the sum of the two constant terms. You can use this pattern to factor some polynomials, that is, to express them as the product of two binomials.

Factor.

Example $x^2 - x - 2$

$x^2 - x - 2 = (x + 1)(x - 2)$ The constant terms required must have a product of -2 and a sum of -1. Two numbers with these properties are 1 and -2.

7. $x^2 + x - 2$ **8.** $x^2 + 5x + 6$ **9.** $x^2 - 4x + 3$

10. $x^2 + 2x + 1$ **11.** $x^2 + x - 20$ **12.** $x^2 - 1$

The method of factoring can be used to find the points at which the graph of a quadratic function crosses the x-axis. By setting the value of the function equal to 0 and factoring you can find the x values for which the value of the function (y value), is 0.

Find the x values that make the function equal 0.

Example: $f(x) = x^2 - 6x$

$$
\begin{aligned}
x^2 - 6x &= 0 && \text{Set the function equal to 0.}\\
x(x - 6) &= 0 && \text{Factor.}\\
x = 0 \quad &\text{or} \quad x - 6 = 0\\
x = 0 \quad &\text{or} \quad x = 6
\end{aligned}
$$

13. $y = x^2 + 5x + 4$ **14.** $f(x) = x^2 + 2x - 3$

15. $f(x) = x^2 - x - 6$ **16.** $y = x^2 - 9$

HOME
OWNERSHIP

The American Dream

THE SELF–CONTAINED SINGLE–FAMILY home with its nuclear family unit—Mom, Dad, two children, and a dog—is portrayed as every middle-class American's fondest wish. The fact that not every family even wants children, a dog, or a house of its own is overlooked when the message is broadcast. Of the families that do wish to pursue this "American dream," many will find home ownership to be expensive and laden with responsibilities. Even if they can afford to buy a house, there is always something in it that needs maintenance, repair, or updating.

Because many families still want to own a house no matter what the cost, they must turn to banks and other lending institutions for financing. At a time when a single-family, three-bedroom house in a moderate neighborhood carries a price tag of over $100,000 in many cities, most buyers must obtain a mortgage to be able to afford to buy a house.

The teenagers in this chapter will examine the process of buying a house and some of the many responsibilities associated with home ownership.

Vernon's family decided to buy a house when they moved from Chicago to Dallas. They considered the advantages of buying as compared with renting and the differences between new and older houses.

When Olivia's family is thinking about moving, she becomes aware of all the costs associated with buying a house, such as the fees and points. She also learns about making a down payment and the total financed price.

Rosa researches the different kinds of mortgages that are available and how a family can determine how much it can afford to borrow.

Hari discovers why homeowner's insurance is extremely important, especially when the temperature falls and the water pipes freeze.

When Vernon's family moved to Dallas from Chicago, where they had lived in a three-bedroom apartment, they decided to consider buying a house. They were moving because Vernon's mother was transferred to Dallas by her company. Vernon's father, who is an accountant, was confident that he could find another accounting job in their new city. It seemed unlikely that the company would move Vernon's mother again soon, so they thought that it was an excellent time to buy a house.

Vernon was very excited about having his own room in a new house. He had shared a bedroom with his younger brother Kev since their sister needed a room of her own. The brothers never liked sharing the bedroom. Kev always complained about Vernon's choice of music, so Vernon could not play his CD player while Kev was studying. Vernon finally bought a pair of headphones to solve that problem. The neighbors were grateful, too!

Vernon also liked the idea that they might have space for a real recreation room. It would allow him to have more than one friend at a time over to his house. His mother and father were pleased that they could decide what kinds of carpeting, drapes, and appliances they would have. They also liked controlling the temperature themselves. In rented apartments it had always seemed too hot or too cold.

The family knew that buying a house would certainly cost more money than renting an apartment. A lot of responsibility comes with home ownership, too. Vernon and Kev would have to help care for the yard. They might all have to help make repairs and improvements.

To get a whole closet to himself and to have a place where he could close the door and get away from his brother, Vernon was willing to do a lot of painting and lawn mowing. He decided to find out all that he could about home ownership in order to help his parents make the buying decision.

Consider some of the advantages and disadvantages of owning a house.
Understand the advantages and disadvantages of buying a new house.
Recognize the relationship between the selling price of a house and its size, location, age, and amenities.

CHOOSING A PLACE TO LIVE

Vernon's family was ready to buy a house. Like many, they chose to buy a house partly for noneconomic reasons—such as the freedom to have pets, the space to plant a garden, and the ability to make home improvements. Vernon and Kev were eager to have bedrooms of their own, and this was now possible.

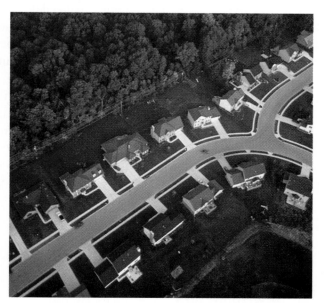

Home ownership usually costs more than renting, but it does allow you to build up equity. **Equity** is the difference between what the house (or other property) is worth and what is owed on the mortgage. The monthly payments decrease the debt and increase the amount of equity the buyer has in the property. When their house is paid for in full, Vernon's family will have 100% equity in it.

Home ownership has been a good investment in recent years. Many people have been able to experience a large **capital gain,** or increase in the value of their property, when selling their houses. Vernon's parents do not intend to sell this house until their three children are through college and perhaps have homes of their own. By then, their equity in the house will be substantial.

ADVANTAGES OF OWNING A HOME

Home ownership often provides a feeling of security and independence.
Home ownership provides a means of forced savings by building up equity.
Home ownership has proven to be a good investment over the years because of increasing property values.
Interest paid on a mortgage is tax deductible.

DISADVANTAGES OF OWNING A HOME

The money invested in the house is not readily available for other purchases.
An owner assumes responsibility for financing and maintaining a house; renters leave those responsibilities to the owner.
Home ownership makes moving more difficult because the old house must be sold.
Home ownership may require larger monthly payments (including utilities, property taxes, and insurance) than renting requires.

Vernon's parents felt that they were ready for home ownership because they met the following conditions.

- They could make the monthly payments out of their current income.
- They planned to live in the same place for three years or longer.
- They were prepared to take on the responsibility for the financing, repair, and maintenance of a house.

Owning a house often provides individuals and family members with a sense of pride, enjoyment, and satisfaction. Vernon was not sure he would actually enjoy mowing the lawn, but he would derive a lot of satisfaction from being able to have his own room!

When they examined the housing market in their new city, Vernon and his family realized that the next major decision they would have to make would be whether to buy a new house or an older one. A new house often costs more than an older one, but the repairs are usually less expensive. However, the buyer may have to make a substantial investment in landscaping, window treatments, carpeting, and major appliances.

Although older houses are usually less expensive than new ones, they frequently need repairs such as a new roof, a new furnace, new plumbing, or electrical work. Sometimes the buyer will want to add a room or modify existing ones. Many of the improvements or changes that the buyer makes may not appeal to the next owner. Therefore these improvements may not increase the cash value of the house.

ALGEBRA *REVIEW*

Solve for y if $x = 10.18$. Round your answers to the nearest hundredth.

1. $y = 177.23 + 0.76x$

2. $y = 5.787 - 0.0065x$

3. $y = \dfrac{109.75}{x}$ 4. $y = \dfrac{0.098}{x}$

5. $x = 15.666 + 0.123y$

6. $x = 14.75y - 12.7$

7. $x = \dfrac{1.65y}{43.47}$ 8. $x = \dfrac{177.711}{y}$

9. $x = \dfrac{3.81y}{542.6}$ 10. $x = \dfrac{2301.46}{y}$

Ask Yourself

1. What are some noneconomic reasons people may choose to buy a house?

2. What is equity?

3. What are some of the advantages of buying a new house instead of an older one? What are the disadvantages?

4. What conditions should you meet before buying a house?

SHARPEN YOUR SKILLS

SKILL 1

The price of a house is related to many factors, including size, age, location, and amenities. You can use *regression analysis* to determine a correlation between the price and one of these factors. Remember that a graphing calculator can automatically calculate *a*, the *y* intercept, and *b*, the slope of a line that best fits a set of data. The calculator also determines a value for *r*, the *correlation coefficient*, which is a measure of how good the fit is. If *r* is close to $+1$ or -1, the fit is very good.

EXAMPLE 1 The most frequently used measure of the size of a house is its floor area. Vernon collected the information shown about nine houses that are for sale in his new city.

Price (in thousands of dollars)	Area (in square feet)
$348	4350
91	2800
260	3600
150	3100
164	3100
185.9	3400
102.4	2850
375	4375
136.3	3000

QUESTION What is a linear model of the form $y = a + bx$ that will allow Vernon to predict the number of square feet in a house in his city if he knows the selling price? Is there a good correlation between price and area?

SOLUTION

Use the linear regression function on your graphing calculator to find the line that best fits the data. Begin by entering the ordered pairs (x, y) of data, where *x* represents the price and *y* represents the area in each house. Your calculator will display the following information:

$a = 2235.432486$
$b = 5.768568698$
$r = 0.9916637943$

The linear equation that best relates the price and the number of square feet is

$y = 2235.432486 + 5.768568698x$

The graph of the equation is shown in a scatter plot of the data.

Xmin: -25 Ymin: -500
Xmax: 400   Ymax: 5000
Xscl: 50   Yscl: 1000

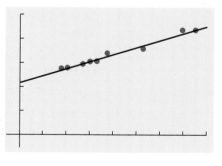

Since the value of *r* is close to 1, there is an excellent linear correlation between the price and the number of square feet in the house. This means that as the price increases, the area increases.

You can use a regression equation to determine missing information. You can use regression analysis to make a prediction about a house, such as its square footage, if you know its price.

EXAMPLE 2 Vernon found an advertisement for a house in his new city that appealed to him. Although the price of the house is listed at $290,000, the number of square feet is misprinted in the ad.

QUESTION What should the area of this house be?

SOLUTION
Use the regression equation that you found in Example 1. Since the original data had only four digits, round a and b to four significant digits. Remember to use 290 for $290,000 since the original data gave the prices of houses in thousands.

$y = 2235 + 5.769x$ Substitute 290 for x.
$y = 2235 + 5.769(290)$ Solve for y.
$y = 3908.01$

Vernon estimates that there are approximately 3900 ft^2 in this house.

SKILL 3

You can use a regression equation to evaluate information such as whether a house is appropriately priced.

EXAMPLE 3 Vernon's family looked at a house with 2700 ft^2 that was priced at $150,000.

QUESTION Is this house fairly priced?

SOLUTION
Use the regression equation that you used in Example 2.

$y = 2235 + 5.769x$ Substitute 2700 for y.
$2700 = 2235 + 5.769x$ Solve for x.
$2700 - 2235 = 5.769x$ Subtract 2235 from both sides.
$$\frac{2700 - 2235}{5.769} = x$$ Divide both sides by 5.769.
$80.603 = x$

Remember the price of houses in the original data was given in thousands so the result 80.603 gives a price of $80,603. At $150,000, this house appears to be significantly overpriced. A price of $80,603 seems more appropriate.

Since the price of a house is related to many factors, you can determine many different correlations between various factors.

EXAMPLE 4 Priscilla has collected the information shown about the average prices of houses in towns outside a major business financial center to which people commute each day.

QUESTIONS What is a linear model of the form $y = a + bx$ that will allow Priscilla to predict the distance from the financial center to a house in her county if she knows the selling price of the house? Is there a good correlation between the price and the distance from the financial center?

Average Price (in thousands of dollars)	Distance from Financial Center (in miles)
$400	24
375	27
360	29
325	33
295	36
280	42
200	50
125	60

SOLUTION

Use the linear regression function on your graphing calculator to find the linear equation that best relates the price (in thousands) and the location.

Your calculator will display the following values for a, b, and r.

$a = 76.62623558$
$b = -0.1322075783$
$r = -0.995827639$

The linear equation that best relates the price and the distance from the financial center to four significant digits is

$y = 76.63 - 0.1322x$

A graph of the linear regression equation is shown on a scatter plot of the data.

Xmin: −25   Ymin: −5
Xmax: 500   Ymax: 75
Xscl: 50   Yscl: 10

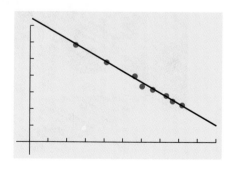

Notice that the slope of the line is a negative number. Therefore the line falls downward to the right. From the graph, you can see that the distance y decreases as the price x increases.

Since the value of r is −0.995827639, there is a very good *negative* linear correlation between price and the distance from the financial center. This means that as the price increases, the distance decreases.

Carla noticed that the following amenities are used as selling points in advertisements for homes: hardwood floors, walk-in closets, professional landscaping, fireplace, island kitchen, central air conditioning, skylight, underground sprinkler system, security system, patio/deck, pool, and spa/jacuzzi. She collected the information shown about houses selling in her town. Give answers to four significant digits.

Price (in thousands of dollars)	Number of Amenities
$200	10
102	3
111	4
88	1
120	4
145	7
125	5
162	8
175	9
130	6
173	8
210	11

1. What is a linear model of the form $y = a + bx$ that will allow Carla to predict the number of amenities in a house in her town if she knows the selling price of the house?

2. Is there a good correlation between the price and the number of amenities?

3. Graph the linear regression equation on a scatter plot of the datas.

4. Estimate the number of amenities in a house in Carla's town that is selling for $190,000.

5. Do you think that a $120,000 house with five amenities is fairly priced?

EXERCISE YOUR SKILLS

1. What are some of the disadvantages of owning a house?

2. In your opinion, do the advantages of owning a house outweigh the disadvantages or vice versa? Explain.

3. Why does owning a house usually cost more than renting an apartment?

Juanita collected the information shown about houses selling in her town. Give answers to four significant digits.

Price (in thousands of dollars)	Age (in years)
$108	47
200	10
95	50
150	28
125	44
215	4
240	0*
255	0*
225	2
140	35

*Brand new house

4. What is a linear model of the form $y = a + bx$ that will allow Juanita to predict the age of a house in her town if she knows the selling price of the house?

5. Is there a good correlation between the price and the age of a house?

6. Graph the linear regression equation on a scatter plot of the data.

7. Estimate the age of a house in Juanita's town that is selling for $185,000.

8. Do you think that a $160,000 house that is 40 years old is fairly priced?

Daniel collected the information shown about houses selling in his town. Give answers to four significant digits.

Price (in thousands of dollars)	Number of Bathrooms
$188	1
200	1
290	2.5
210	1.5
310	2.5
325	3
380	4
250	2
350	3.5

9. What is a linear model of the form $y = a + bx$ that will allow Daniel to predict the number of bathrooms in a house in his town if he knows the selling price of the house?

10. Is there a good correlation between the price and the number of bathrooms in a house?

11. Graph the linear regression equation on a scatter plot of the data.

12. Estimate the number of bathrooms in a house in Daniel's town that is selling for $225,000. Round your answer to the nearest half bathroom.

13. Do you think that a $300,000 house with 1.5 bathrooms is fairly priced?

A realtor conducted a study to determine whether the price of houses in his city is related to traffic patterns on the street on which the house is located. He used the traffic scoring system shown. He then made the chart shown to organize the data. Give answers to four significant digits.

1.	Quiet cul-de-sac
2.	Very little traffic
3.	Moderate traffic
4.	Heavy traffic

Price (in thousands of dollars)	Traffic Patterns
$200	3
150	4
300	1
350	1
275	2
250	2
225	3
175	4

14. What is a linear model of the form $y = a + bx$ that will allow the realtor to predict the traffic patterns on the street on which a house is located if he knows the selling price of the house?

15. Is there a good correlation between the price and the traffic patterns on the street on which a house is located?

16. Graph the linear regression equation on a scatter plot of the data.

17. Estimate the traffic pattern score on the street on which a house selling for $325,000 is located.

18. Do you think that a $280,000 house with a traffic pattern score of 4 is fairly priced?

MIXED REVIEW

1. With $987.44 in your checking account, you write a check for $122.87. What is the new balance?

2. Use the Rule of 72 to determine how long it will take Elliott's $2400 savings account to double in value if it is growing at a rate of 3.0%?

3. Determine the effective rate of interest if the APR is 16.5%.

4. Marcia has a balance of $1232 on her VISA card, which carries an APR of 15%. If she makes no further purchases with this card, how many months will it take her to pay off the balance? She can afford monthly payments of $160.

5. Last month, the Reynolds family spent $250 on the car payment, $315 on credit card payments, $400 on utilities, $620 on rent, $810 on food, and $2256 on everything else. To the nearest tenth of a percent, what percent of their take-home pay did the family spend on credit payments?

6. Use the United States Mileage Chart in the Reference Section to find the distance from Columbus, Ohio, to Atlanta, Georgia.

7. The Uptown Motor Inn charges $42 for a single room and $52 for a double room. The tax is 5%. Find the cost paid by Mr. and Mrs. Engel for a stay of four nights.

12-2 BUYING A HOUSE: HOW TO DETERMINE THE COSTS

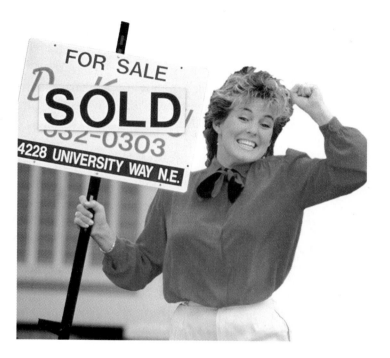

Olivia's parents would like to buy another house. If the family lived a little closer to the school, they would allow Olivia to drive her brother Orson and herself to school. Olivia's mother has been working for several years and recently received a substantial increase in her salary, so this may be a good time to make a move. They have often talked about wanting more space, especially since Olivia and Orson became teenagers. The children are both pleased about the prospect of having groups of friends over on the weekends.

Olivia has noticed that adults become very secretive when they talk about how much things cost—except, of course, when they talk about how little things used to cost! Olivia will ask Vernon whether he knows about the kinds of costs they will encounter if they do decide to buy another house. From what Vernon has already said, Olivia knows that her parents might be in for quite a surprise. If they think the cost of a bicycle has gone up in 35 years, just wait until they see what has happened to housing prices!

OBJECTIVES: *In this lesson, we will help Olivia to:*

- *Recognize the major costs to consider when purchasing a house, including the price, the mortgage loan, points, down payment, closing costs, and the cost of new furnishings.*
- *Compute the monthly payments and total financed price of a house.*

THE COST OF OWNING YOUR HOME

Since 1980 the prices of houses have consistently outpaced consumer incomes. Since Olivia has studied the effects of inflation and the changes in prices as measured by the Consumer Price Index, she understands why close to two thirds of the families buying houses have two incomes. Six out of ten first-time buyers are couples in which both partners hold jobs.

Because of the high cost of new houses, most first-time buyers must make compromises such as the following:

- Buy a condominium as a first home.
- Commute longer distances to buy less expensive housing.
- Live in smaller and older houses that may need repairs.
- Buy a manufactured house, generally called a mobile home or prefabricated home.
- Accept a more moderate life-style to pay for shelter.

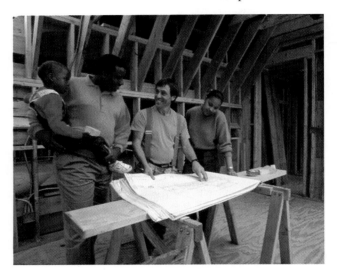

How Much to Spend Olivia's family banker told her parents that there is a rule of thumb that a family should look for housing priced within $2\frac{1}{2}$ times their annual income. Since Olivia's mother and father are both working, they can spend a large sum of money on a house. However, they certainly cannot pay that much in cash.

The Mortgage The family will take a **mortgage,** which is a loan on which both principal (the amount borrowed) and interest are paid back over a fixed number of years, usually 15 to 30 years. The amount of the mortgage is based on the **appraised value** of the house. A mortgage contract is signed by the lender and the borrower. The amount of the monthly mortgage payment is based on the amount of the loan, the interest rate, and the amortization period of the loan. The lender may include property taxes and the cost of insurance in the mortgage payment. Over the course of the loan, the borrower may end up making total payments that equal between 2 and 4 times the purchase price of the house. However, the alternative of saving until they can pay cash for a house would prevent many families from ever being able to purchase a house.

Another important factor that they must consider when borrowing for a real-estate purchase are the points that the lender may charge. **Points** are a one-time charge that is paid at the time of the mortgage closing. One *point* is equal to 1% of the value of the mortgage. Points provide additional revenue to the bank, and some banks require a larger number of points to be paid in exchange for a lower interest rate over the term of the loan.

The Down Payment When Olivia's parents buy the house, they will make a **down payment** (pay part of the purchase price). Their down payment could range from 0 to 30% or more of the purchase price. The higher their original down payment, the lower their monthly mortgage expense will be. When they sell their other house, they can use the equity that they had in it as the down payment on the new house.

The Closing Costs The family will also have to have enough cash available for various *closing costs* that occur at the time the sale is completed. **Closing costs** are the charges and fees associated with the transfer of ownership of a home to a new buyer. These costs are incurred by the buyer for the work involved in preparing and processing all of the documents needed to complete the purchase of the home. A *buyer's costs* may include the following:

- Title search
- Mortgage preparation
- Appraisal fee
- Recording fees
- Credit report
- Legal fees

The *seller's costs* include the realtor's commission, the preparation of the deed, and any real estate taxes that are due. These costs can range from several hundred to several thousand dollars. The realtor who is handling a sale can estimate the closing costs when final arrangements for the sale are being made.

Additional Costs Olivia's family knows that their expenses do not end at the closing. When they move into the house, certain other expenses will be incurred. The costs of curtains, draperies, floor coverings, and landscaping are in this category. Also, the family may be required to submit deposits to the utility companies that supply electricity, water, and telephone service. If they are not changing utility companies, they may be able to apply the deposits made when they bought their first house. They do know that they will want new curtains, bedspreads, draperies, and several new carpets for the new house. They may even want a few new pieces of furniture. They would like to get a new bedroom suite for Olivia and replace the couch that the dog chewed. Therefore they must also plan to have the resources available to meet these costs.

Ask Yourself

1. What are points?

2. What is the name given to costs incurred by the buyer for work involved in preparing and processing all the documents for a house purchase?

3. Name three compromises that first-time buyers might make because the prices of new houses are so high.

ALGEBRA *REVIEW*

The following table shows average home prices in thousands of dollars in two counties at different times.

	1978	1983	1988	1993
Bell	43	49	61	52
Mann	25	34	47	40

1. When was the highest average price reached in Bell County?

2. What year shows the greatest difference in prices for the two counties?

3. Draw a double bar graph to represent the information given in the table.

SHARPEN YOUR SKILLS

To find the monthly mortgage payment, use the monthly payment formula that you used in Chapter 5.

Monthly Payment Formula

$$M = \frac{Pr(1 + r)^n}{(1 + r)^n - 1}$$

where P = amount of loan
r = monthly interest rate
n = number of payment periods
M = monthly payment

The **total financed price** is the total purchase price plus the financing costs and the down payment. The total financed price T is found by multiplying the monthly payment M by the number of months n over the course of the loan and then adding the down payment d.

$$T = nM + d$$

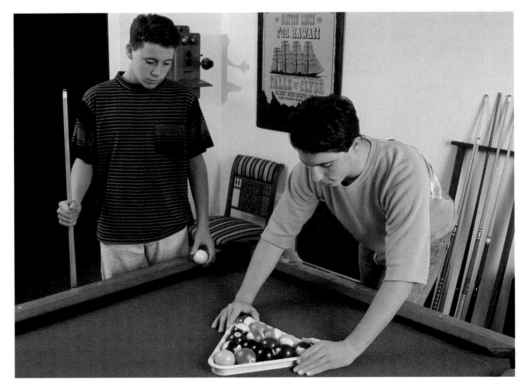

Recall that the calculation of the monthly payment is given to many decimal places in your calculator. Use that answer to calculate the total financed price by using the answer key $\boxed{\text{ANS}}$ on your calculator. Remember that the last monthly payment is adjusted to the actual total.

Olivia has noticed that house listings give the purchase price of the house. However, the listings do not tell her how much the monthly mortgage payments will be. The house listing shown is very appealing to her.

**GIANT TEEN ROOM
ON 3RD FLOOR!**
Teens can entertain their guests in their own suite. Parents will enjoy the built-in sauna, spa, pool, and large paneled billiard room. A steal at $765,000.

EXAMPLE 1 Olivia knows that her parents plan to make a down payment of 20% on the house that they purchase. They want to amortize their loan over 30 years, so they will apply for a 30-year mortgage. The prevailing interest rate on a home loan is 12%.

QUESTIONS What are the monthly payment and the total financed price of the house for $765,000 under these conditions? What is the ratio of the total financed price to the purchase price?

SOLUTIONS
Determine the amount that they will borrow.

$$0.20(765,000) = 153,000 \qquad \text{They make a down payment of 20\%.}$$

Therefore they must borrow $765,000 - 153,000 = 612,000$.

Use the monthly payment formula.

$$M = \frac{612,000(0.01)(1 + 0.01)^{360}}{(1 + 0.01)^{360} - 1} \qquad P = 612,000, \, r = (0.12 \div 12) = 0.01,$$
$$n = 30 \cdot 12 = 360$$
$$M = 6295.11 \qquad \text{To the nearest cent}$$

The monthly payment is $6295.11. Olivia is certain that her parents will not share her enthusiasm for a house with such a large monthly payment!

In the course of 30 years the total financed price will be

$$T = nM + d$$
$$T = 360\,\boxed{\text{ANS}} + 153,000$$
$$T = 2,419,239.27$$

The total financed price is $2,419,239.27.

The ratio of the total financed price to the purchase price is

$$\frac{2,419,239.27}{765,000} = 3.2 \qquad \text{To the nearest tenth}$$

The total financed price of $2,419,239.27 is more than three times the original cost of the house.

As you learned in Chapter 5, the total payment on a loan for a given amount of money over a specified amortization period is always lower when the interest rate is lower. Olivia has heard that even a *small* decrease in the interest rate on a home loan can create a *large* decrease in the total financed price.

EXAMPLE 2 The mortgage plans that are available to Olivia's family require that they make a down payment of 20% of the purchase price. The house that they want to buy costs $250,000. They have found rates for 30-year fixed mortgages that range from 9% at her mom's credit union to 12% at a local bank.

QUESTIONS How does the total financed price on a $250,000 home with a 20% down payment and a 30-year mortgage differ as interest rates vary from 9% to 12% in increments of 0.5%? In each case, what is the savings over a 12% loan?

SOLUTION

Determine the 20% down payment and the loan amount.

Down payment: $0.2(250,000) = 50,000$
Loan amount: $250,000 - 50,000 = 200,000$

Use a spreadsheet program to find the total financed price, to the nearest dollar, and the savings over a 12% loan as the interest rate varies from 9% to 12% in increments of 0.5%. For cell B5 the formula is:

$$200000*(A5/12)*(1+(A5/12))^{360}/((1+(A5/12))^{360}-1)$$

Format column A for three fixed places. Format column B for two fixed places and format columns C and D for zero fixed places. In the formula for cell D5, be sure to use an address for cell C11 that stays the same when the formula is copied, called an *absolute* address.

	A	B	C	D
1		$200,000 mortgage with $50,000 down payment		
2		30-year	30-year	
3	Interest	Monthly	Total Financed	Savings
4	Rate	Payment	Price	Over 12%
5	0.090	1609.25	629328	161273
6	0.095	1681.71	655415	135186
7	0.100	1755.14	681852	108749
8	0.105	1829.48	708612	81989
9	0.110	1904.65	735673	54928
10	0.115	1980.58	763010	27591
11	0.120	2057.23	790601	0

@ROUND(B5*360+50000,0)

+A5+0.005

+C11−C5

Olivia used the graph function of her computer spreadsheet program to graph the relationship between the interest rate and the total financed price.

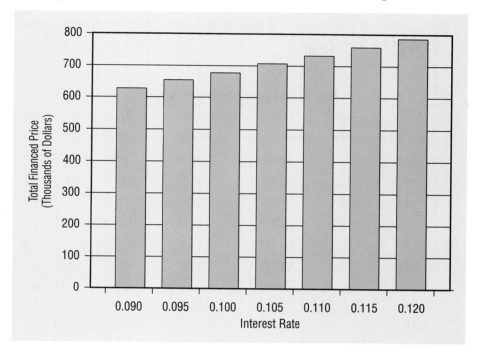

Over the course of 30 years, they will pay $161,273 more with a loan at 12% than at 9%. The savings with a 9% loan over a 12% loan for $200,000 over 30 years is almost as large as the amount that is borrowed!

Note that Olivia's parents will pay back significantly more than they borrowed over the course of 30 years. At 9% they will pay back 629,328 ÷ 250,000 = 2.5 times the cost of the house. At 12% they would pay back 790,601 ÷ 250,000 = 3.2 times the cost of the home.

─────── **SKILL 3** ───────────────────

Olivia's parents learned that the total financed price is much smaller on 15-year mortgages, but the monthly payments are larger. For example, the monthly payment on a $200,000 loan at 9% is $1609.25 for 30 years and $2028.53 for 15 years. The 15-year mortgage costs $419.28 more each month.

EXAMPLE 3   Although her family cannot afford the larger monthly payment, Olivia would like to know what the savings over the course of the loan would be if they could afford the monthly payments on a 15-year mortgage.

QUESTION   What is the total financed price of a $250,000 home with a loan of $200,000 and a $50,000 down payment for 15 years as interest rates vary from 9% to 12% in increments of 0.5%? In each case, what is the savings over a 30-year loan?

SOLUTION

Olivia used the spreadsheet program from Example 2 to find the total financed price, to the nearest dollar, and the savings over a 30-year mortgage as the interest rate varies from 9% to 12% in increments of 0.5%. She copied and adjusted the formulas as necessary.

	A	B	C	D	E	F
1	$200,000 mortgage with $50,000 down payment					
2		15-year	15-year	30-year	30-year	Savings
3	Interest	Monthly	Total Financed	Monthly	Total Financed	Over
4	Rate	Payment	Price	Payment	Price	30-year
5	0.090	2028.53	415136	1609.25	629328	214192
6	0.095	2088.45	425921	1681.71	655415	229494
7	0.100	2149.21	436858	1755.14	681852	244994
8	0.105	2210.80	447944	1829.48	708612	260668
9	0.110	2273.19	459175	1904.65	735673	276498
10	0.115	2336.38	470548	1980.58	763010	292462
11	0.120	2400.34	482061	2057.23	790601	308540

Formula pointing to B5: `200000*(A5/12)*(1+(A5/12))^180/((1+(A5/12))^180−1)`

Formula pointing to C5: `@ROUND(B5*180+50000,0)`

Formula pointing to F5: `+E5−C5`

She used the graph function of her computer spreadsheet program to graph the relationship between the interest rate and the total financed price for the 15-year and 30-year mortgages.

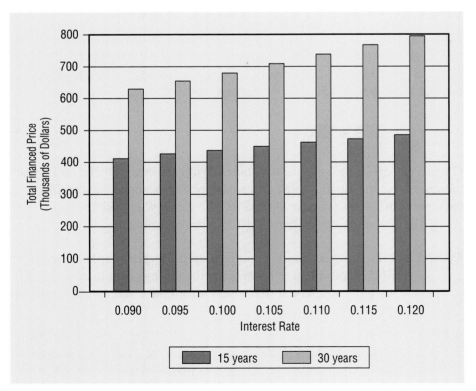

For example, the monthly payment on a $200,000 mortgage at 9% for 15 years would cost Olivia's family over $400 more each month than the same loan for 30 years. However, they would save $214,192 over the course of the loan.

SKILL 4

As you learned in Chapter 5, the higher the down payment on a loan, the lower the total payment.

EXAMPLE 4 Olivia's family planned $50,000 (20% of the purchase price of $250,000) for the down payment on their home. However, Olivia would like to know what the savings over the course of the loan would be if they made a down payment of $75,000, which is 30% of $250,000, and borrowed only $175,000.

QUESTION What is the total financed price of a mortgage for $175,000 for 30 years with a $75,000 down payment as the interest rate varies between 9% and 12% in increments of 0.5%? In each case, determine the savings over a $200,000 mortgage for 30 years with a $50,000 down payment.

SOLUTION
Olivia used a spreadsheet program to find the total financed priced with a mortgage of $175,000 for 30 years with a $75,000 down payment, to the nearest whole dollar, and the savings over a mortgage of $200,000 for 30 years with a $50,000 down payment as the interest rate varies from 9% to 12% in increments of 0.5%.

Interest Rate	Total Financed Price 30% Down	Total Financed Price 20% Down	Savings Over 20% Down
0.090	$581912	$629328	$47416
0.095	604738	655415	50677
0.100	627870	681852	53982
0.105	651286	708612	57326
0.110	674964	735673	60709
0.115	698884	763010	64126
0.120	723026	790601	67575

Then she used the graph function of her spreadsheet program to graph the relationship between the interest rate and the total financed price for down payments of 20% and 30%.

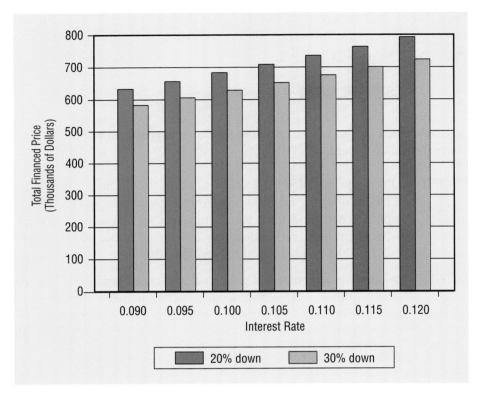

At 30% down the family would make a down payment of $25,000 more than at 20% down. However, at a rate of 9%, for example, the total financed price of the house would be

$629,328 − $581,912 = $47,416

less than the total financed price with 20% down.

In general, you will obtain the lowest total financed price on a house if you select a mortgage with the *lowest* interest rate, the *shortest* amortization period, and the *largest* down payment.

SKILL 5

Some banks charge *points,* which are a one-time charge that is paid at the time of the mortgage closing. As mentioned earlier in this lesson, one point is equal to 1% of the value of the mortgage. When points are charged, the points p must be added to the other costs to find the *total financed price T.*

$$T = nM + d + p$$

EXAMPLE 5 A local bank is offering a choice of the following 30-year fixed mortgage rates:

> A: 10% with 0 points
> B: 9.75% with 1 point
> C: 9.5% with $1\frac{1}{2}$ points

QUESTION If you would like to buy a house that costs $125,000 with a loan of $100,000 and a down payment of $25,000, what is your lowest-priced option?

SOLUTION
Use the monthly payment formula to find each monthly payment. Then find the total financed price. Remember to use the monthly payment price in your calculator, not the rounded amount.

For plan A:

$$M_A = \frac{100{,}000(0.10 \div 12)(1 + 0.10 \div 12)^{360}}{(1 + 0.10 \div 12)^{360} - 1}$$
$P = 100000; \; n = 30 \cdot 12 = 360;$
$r = 0.10 \div 12$

$= 877.57$ To the nearest cent

$T_A = 360(M_A) + 25{,}000 + 0$ $T = nM + d + p$
$= 340{,}926$ To the nearest dollar

For plan B:

$$M_B = \frac{100{,}000(0.0975 \div 12)(1 + 0.0975 \div 12)^{360}}{(1 + 0.0975 \div 12)^{360} - 1}$$

$= 859.15$ To the nearest cent

$T_B = 360(M_B) + 25{,}000 + 0.01(100{,}000)$
$= 335{,}296$ To the nearest dollar

For plan C:

$$M_C = \frac{100{,}000(0.095 \div 12)(1 + 0.095 \div 12)^{360}}{(1 + 0.095 \div 12)^{360} - 1}$$

$= 840.85$ To the nearest cent

$T_C = 360(M_C) + 25{,}000 + 0.015(100{,}000)$
$= 329{,}208$ To the nearest dollar

The lowest-priced option is plan C for $329,208. Therefore if you have the money to pay $1\frac{1}{2}$ points up front, you can save money over the course of the loan.

It is important to explore and evaluate all of your options before selecting a mortgage plan.

1. Determine the total financed price on a $180,000 house if you finance it at 15% for 30 years with a 25% down payment.

2. Calculate the ratio of the total financed price to the purchase price in Exercise 1.

Use a spreadsheet program to determine how the total financed price on a $240,000 loan for 15 years with a down payment of $60,000 differs as interest rates vary from 6% to 8% in increments of 0.5%. In each case, what is the savings over an 8% loan?

	Interest Rate	Total Financed Price	Savings Over 8%
3.	6%	$424,546	
4.	6.5%		
5.	7%		
6.	7.5%		
7.	8%		

8. If you would like to buy a house that costs $160,000 with a loan of $125,000 and a down payment of $35,000, which of the following is your lowest-priced option for a 15-year fixed mortgage?

    A: 12% with 0 points
    B: 11.5% with 1 point
    C: 11% with 2 points

EXERCISE YOUR SKILLS

1. What are the major costs to consider in buying a house?

2. How does the amount of the down payment on a house affect the amount that is borrowed in the form of a mortgage?

3. The total financed price for a house is approximately how many times the purchase price?

4. Determine the monthly payment and total financed price on a $188,000 house if you finance it at 9.5% for 30 years with a down payment of 25%.

5. Determine the ratio of the total financed price to the purchase price of the house in Exercise 4.

KEY TERMS

appraised value
closing costs
down payment
mortgage
points
total financed price

Use a spreadsheet program to determine the total financed price of a $200,000 house with each combination of down payment, interest rate, and amortization period.

	Percent Down	Down Payment	Loan Amount	Interest Rate	Number of Years	Total Financed Price
6.	20%	$40,000	$160,000	10%	15	
7.	20%			10%	30	
8.	20%			11%	15	
9.	20%			11%	30	
10.	30%			10%	15	
11.	30%			10%	30	
12.	30%			11%	15	
13.	30%			11%	30	

14. In Exercises 6–13, which combination of percent down, interest rate, and amortization period yields the lowest total financed price? Which yields the highest total financed price?

Use a spreadsheet program to determine the total financed price of a $120,000 home with each combination of down payment, interest rate, and amortization period.

	Percent Down	Down Payment	Loan Amount	Interest Rate	Number of Years	Total Financed Price
15.	25%			8%	30	
16.	20%			8%	30	
17.	25%			9%	30	
18.	20%			9%	30	
19.	25%			9%	15	
20.	20%			9%	15	
21.	25%			8%	15	
22.	20%			8%	15	

23. In Exercises 15–22, which combination of percent down, interest rate, and amortization period yields the lowest total financed price? Which yields the highest total financed price?

24. Use the graph function of your computer spreadsheet program to graph the relationship between interest rate and total financed price on a $150,000 house financed for 15 years with a 25% down payment as interest rates vary between 6% and 10% in increments of 1%.

25. If you would like to buy a house that costs $225,000 with a loan of $175,000 and a down payment of $50,000, which of the following is your lowest-priced option for a 15-year fixed mortgage?

    A: 8% with 0 points
    B: 7.5% with 1 point
    C: 7% with 2 points

26. If you would like to buy a house that costs $185,000 with a loan of $150,000 and a down payment of $35,000, which of the following is your lowest-priced option for a 30-year fixed mortgage?

    A: 9% with 1 point
    B: 8.5% with 2 points
    C: 8% with 3 points

MIXED REVIEW

1. A bank charges $0.03 for each of the first 25 checks written each month and $0.055 for each check above 25. What is the monthly charge for 33 checks?

2. Determine the monthly payment that must be made to reduce a $2222.44 MasterCard balance to zero in 3 years. The card carries an APR of 15%.

3. Your VISA card carries an APR of 21%. Your balance is currently $1200.50 and you pay 10% of the amount owed to the nearest dollar each month. Complete the following table for the next three months. Assume that you make no further purchases during that time.

Month	Balance	Interest	Amount Owed	Payment
1	$1200.50			
2				
3				

4. Use the Comparison Table for Term and Whole Life Premiums in the Reference Section to find the yearly premium for $350,000 of whole life insurance purchased at the age of 40.

5. How many whole shares of stock in DWB company can you purchase for $50,000 at the current price of $65\frac{1}{2}$ per share? Ignore the cost of commissions.

A car that cost the dealer $13,900 has a sticker price of $16,150.

6. Find the mark-up.   **7.** Find the percent of mark-up.

8. An airplane departed from Newark Airport in New Jersey at 1:00 p.m. local time and arrived in Austin, Texas, at 3:15 p.m. local time. How long was the flight?

12-3 THE MORTGAGE: HOW MUCH CAN YOU BORROW?

Rosa had never given the price of buying a home much thought until Vernon and Olivia started discussing it. Her family had bought their house when she was much younger, and she and her sister Consuela had always shared a room. It is a large room for a bedroom, with two closets. Rosa and Consuela are good friends as well as sisters, so they have been able to accommodate each other's belongings and attitudes about the room fairly well.

Consuela left for college last fall, so Rosa now has a lot more space and privacy. Consuela left a lot of her posters and other "kid's stuff" at home when she left. She will probably be coming home in the summer to find a job before she goes back to college, but she is talking about spending next year in France if she can arrange the appropriate work-study program.

Rosa herself will be leaving for college next year. Her parents are talking about selling this house and buying one that is more convenient to downtown and their father's job. Although Rosa has not been particularly interested in mortgages, she wants Olivia and Vernon to tell her what they have learned. Rosa's father is a veteran, and the house in which they now live was bought with a VA loan. Rosa has never cared about the family mortgage before, but that attitude may change since she will be relying on her parents to help support her when she goes to college. If the family buys a different house, their whole financial plan may undergo some significant changes.

OBJECTIVES: In this lesson, we will help Rosa to:

- *Learn about the types of mortgages that are available to home buyers, including the fixed-rate mortgage, the adjustable-rate mortgage, the balloon mortgage, the Federal Housing Administration loan, and the Veterans Administration loan.*
- *Compute the price of a home loan for which a family can qualify, on the basis of their monthly income.*
- *Understand the importance of knowing whether a mortgage may be prepaid or assumed.*

FINANCING THE PURCHASE OF A HOME

When Rosa's family bought their first house, interest rates and house prices were quite high. As they consider their second house purchase, they will notice that in recent years, interest rates have varied substantially. Over the past several years, interest rates on home loans have ranged from as high as 16% to as low as 6%. Financing for a home purchase is as important as the purchase price or the construction costs.

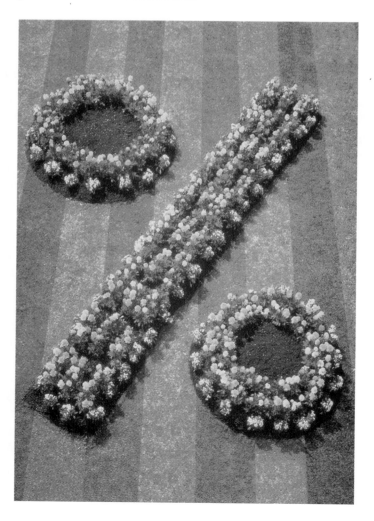

Rosa's family will consider fixed-rate mortgages and adjustable-rate mortgages. **Fixed-rate mortgages (FRMs)** are amortized with a constant monthly payment over a fixed number of years (usually 15, 20, 25, or 30). If the family selects a fixed-rate mortgage, their monthly mortgage payment will remain the same despite rising (or falling) interest rates. Most conventional fixed-rate mortgages require that the buyer make a down payment of at least 20% of the appraised value of the house and borrow no more than 80%.

At one time, a fixed-rate mortgage was the only choice available. As interest rates began to rise and vary widely, mortgage lenders developed an alternative method of providing mortgages. This alternative is called a variable or adjustable-rate mortgage. An **adjustable-rate mortgage (ARM)** is a loan with an interest rate that can be adjusted up or down by the lending institution over the life of the loan. In most cases the starting interest rate on an ARM is lower than the rates offered on fixed-rate mortgages. Most adjustable mortgages have a **cap** on upward and downward rate adjustments, which specifies the largest and smallest rates that the lending institution can charge the borrower at any time during the course of the loan. Usually, the adjustments are yearly and are based on the **prime interest rate.** This rate varies from bank to bank, but often the rate set by the largest banks is used.

Another type of mortgage is the **balloon mortgage,** in which after a certain time the entire remaining balance becomes due. When a borrower expects to sell the house after a short period of time, a balloon mortgage may be desirable because the interest rate for the years before the balloon payment becomes due may be very low.

GOVERNMENT-INSURED REAL ESTATE LOANS

Some fixed-rate mortgages are backed by insurance provided by the federal government.

Federal Housing Administration Loans   In **Federal Housing Administration loans,** the Federal Housing Administration (FHA) of the Department of Housing and Urban Development provides federal insurance on certain loans that are obtained through an approved lending institution such as a bank. A buyer who can make only a small down payment can sometimes obtain an FHA loan. With FHA loans, the borrower can borrow 95–100% of the sales price.

Veterans Administration Loans   Rosa's father qualifies for a **Veterans Administration loan** because he is a veteran. People who served in the armed forces during World War II, the Korean War, the Vietnam War, and certain cold war periods receive special loan privileges from the Veterans Administration (VA). If a veteran fails to repay the loan, the Veterans Administration has the right to deduct the amount not repaid from any pension or other compensation received.

An individual with a VA loan must also pay a small monthly charge for insurance in addition to the regular monthly payment. A down payment is recommended, but the VA loan law permits the Veterans Administration to guarantee loans up to a specified maximum with no down payment if the lender is willing to make the loan for the full amount of the purchase price. Rosa's father can qualify for the VA loan only if his monthly payments do not exceed a specified amount of his take-home pay. This standard applies to FHA loans as well.

QUALIFYING FOR A MORTGAGE

Since housing costs have gone up so much since they bought their first house, Rosa's parents will have to determine how much they can borrow. They know that under current mortgage lending standards, buyers will qualify for a mortgage as long as the monthly cost of the loan does not exceed 28% of their gross monthly income. The bank that lends them the money for their new house will want to be sure that they can afford to make the payments each month. Bankers have found that 28% of gross income is a reasonable limit to place on the size of the mortgage payment.

PREPAYING A MORTGAGE AND ASSUMING A MORTGAGE

As with installment loans, it is important to know whether your mortgage has a **prepayment clause** that allows you to make additional payments to principal or prepay the balance of the mortgage. Many banks allow you to make additional payments to principal at any time with no penalty. If you can make prepayments without a penalty, you will save a lot of interest if you can pay an extra amount each month. It is possible to pay off a 30-year mortgage in only 20 years by making one extra payment per year.

Sometimes mortgages have an **assumability clause** that enables a buyer to assume your mortgage when you sell your house. This may make it easier for you to sell your house.

ALGEBRA *REVIEW*

Solve for x.

1. $0.3x = 1200$

2. $0.4x = 2100$

3. $0.25x = 356$

4. $0.82x = 3500$

5. $x = 21.6\%$ of $2500

6. $x = 0.29\%$ of $1030

7. $x = 0.018(123{,}456.78)$

8. $y = \dfrac{4x(1 + q)}{2t^4 r}$

9. $z = \dfrac{5ab^{36}}{x(c + 2)^{36}}$

10. $\sqrt{t} = \dfrac{5xy^{42}z}{36}$

Ask Yourself

1. How do fixed-rate and adjustable-rate mortgages differ?

2. What is the main advantage of an adjustable-rate mortgage?

3. Who is eligible for a VA loan?

4. What percent of gross income is the limit that banks place on the size of a borrower's mortgage payment?

SHARPEN YOUR SKILLS

SKILL 1

As you learned in Chapter 5, the monthly payment formula can be solved for M, giving the amount formula. This formula is used to determine the amount P that can be borrowed if the monthly payment M, the monthly interest rate r, and the number of months n of the loan are all known.

Amount Formula

$$P = \frac{M[(1 + r)^n - 1]}{r(1 + r)^n}$$

where P = amount of loan
r = monthly interest rate
n = number of payment periods
M = monthly payment

EXAMPLE 1 Rosa's parents have a gross monthly income of $3600. Their bank will lend them the money to buy a house if their monthly mortgage payment is no more than 28% of their monthly income.

QUESTIONS How large can their monthly mortgage payment be? How much money can they borrow at 9% for 30 years?

SOLUTION
The family's monthly mortgage payment can be no more than $0.28(3600) = 1008$.

$$P = \frac{M[(1 + r)^n - 1]}{r(1 + r)^n}$$ Use the amount formula.

$$P = \frac{1008[(1 + 0.0075)^{360} - 1]}{0.0075(1 + 0.0075)^{360}}$$ $M = 1008$, $r = 0.09 \div 12 = 0.0075$, $n = 30 \cdot 12 = 360$

$$P = 125{,}276.12$$

They qualify for a loan of $125,276.12 at 9% for 30 years.

Their loan represents 80% of the purchase price x.

$$0.80x = 125{,}276.12$$
$$x = 125{,}276.12 \div 0.80$$
$$x = 156{,}595.15$$

They can purchase a house priced no more than $156,595.15.

Even if you qualify, on the basis of your monthly income, to buy a house for a specific amount of money, you must still have the cash for a down payment.

EXAMPLE 2 On the basis of their monthly income, Rosa's family will qualify to purchase a house priced at $156,595.15. Rosa's parents will have $28,000 in cash after they sell their current house.

QUESTIONS Will they have enough money to make a 20% down payment on a house selling for $156,595.15? If not, what is the most expensive house that they can consider?

SOLUTION

$$0.20(156,595.15) = 31,319.03 \qquad \text{Find 20\% of the purchase price.}$$

Since they have only $28,000 for a down payment, they *cannot* afford to buy a house priced at $156,595.15 because the down payment must be $31,319.03.

Use the amount of money that they have for a down payment to determine the price x of a house that they can afford.

$$0.20x = 28,000 \qquad \text{20\% of } x \text{ is \$28,000.}$$
$$x = 28,000 \div 0.20$$
$$x = 140,000$$

They *can* afford a house priced at $140,000.

The borrower must meet all of the requirements set out by the bank or financial institution from which mortgage money will be borrowed. Meeting the 28% of gross monthly income requirement is not sufficient if you do not have a down payment of 20% of the purchase price. Similarly, it is not sufficient for you to have 20% of the purchase price. You must also meet the 28% of gross monthly income requirements.

TRY YOUR SKILLS

1. To qualify for a loan at the Scelfo family's local bank, a borrower's monthly mortgage payment must be no more than 28% of the family's gross monthly income. The bank is currently offering mortgages for 30 years at 15%. If the Scelfo family has a monthly income of $5200, for how large a loan will they qualify?

2. If their loan in Exercise 1 represents 80% of the purchase price of a house, what is the most expensive house that the family can consider?

3. If their bank requires a 20% down payment and the family has $30,000 in cash, will they be able to purchase the house in Exercise 2?

EXERCISE YOUR SKILLS

1. Why is the interest rate on a mortgage an important factor in a home purchase?

2. In times of inflation, would you rather have an adjustable-rate mortgage or a fixed-rate mortgage? Explain.

3. Who provides the insurance for loans made under the FHA and VA programs?

4. Why do lending institutions set a limit on the amount of money that a family can borrow?

To qualify for a loan at the local bank, a borrower's monthly mortgage payment must be no more than 28% of his or her family's gross monthly income. The bank is currently offering mortgages for 30 years at 9%. On the basis of monthly income, for how large a loan will each family qualify? If the loan represents 80% of the purchase price of a house, what is the most expensive house that each family can consider?

	Monthly Income	Mortgage Payment	Loan Amount	House Price
5.	$2500			
6.	5900			
7.	3500			
8.	1200			
9.	2650			
10.	2899			
11.	5245			
12.	4350			
13.	3495			
14.	1600			
15.	5777			

16. If the bank requires that a borrower make a down payment of 20% of the purchase price of his or her house and the family in Exercise 5 has $50,000 to use as a down payment, will they qualify for the loan? If not, what is the most expensive house that they can consider?

17. If the bank requires that a borrower make a down payment of 20% of the purchase price of his or her house and the family in Exercise 6 has $32,000 to use as a down payment, will they qualify for the loan? If not, what is the most expensive house that they can consider?

18. If the bank requires that a borrower make a down payment of 20% of the purchase price of his or her house and the family in Exercise 7 has $8000 to use as a down payment, will they qualify for the loan? If not, what is the most expensive house that they can consider?

19. If the bank requires that a borrower make a down payment of 20% of the purchase price of his or her house and the family in Exercise 8 has $20,000 to use as a down payment, will they qualify for the loan? If not, what is the most expensive house that they can consider?

20. If the bank requires that a borrower make a down payment of 20% of the purchase price of his or her house and the family in Exercise 9 has $24,000 to use as a down payment, will they qualify for the loan? If not, what is the most expensive house that they can consider?

21. If you want to make a down payment of 30% of the purchase price of your house and you have $50,000 in cash, can you purchase a house priced at $165,000?

1. Carin saves $85 per week. She wants to buy a computer that costs $2750. For how many weeks must she save before she can buy the computer?

2. How much will $4859 be worth at the end of 3 years, given that the annual interest rate is 6%, compounded semiannually?

3. Use the formula for the Future Value of a Periodic Investment to find how much an IRA will be worth in 35 years if the annual contribution is $1,500 and if the compounded annual return is 10.5%.

The table below shows the recent 5-year record of a mutual fund.

Date	Time After Start (in Years)	Price of Mutual Fund (in dollars)
Jul 1, 1990	0	$30.00
Jan 1, 1991	0.5	31.40
Jul 1, 1991	1.0	35.80
Jan 1, 1992	1.5	36.20
Jul 1, 1992	2.0	34.10
Jan 1, 1993	2.5	38.40
Jul 1, 1993	3.0	41.40
Jan 1, 1994	3.5	40.25
Jul 1, 1994	4.0	45.10
Jan 1, 1995	4.5	46.03
Jul 1, 1995	5.0	47.30

4. Use a graphing calculator to find the linear regression equation that is the line of best fit for the data.

5. Use the equation that you found in Exercise 4 to predict a possible value for the mutual fund in January of 1997.

Use IRS Form 1040EZ and the tax table in the Reference Section to find the tax paid by the following single persons with the given salaries and interest income. The person cannot be claimed as a dependent by anyone else.

6. A salary of $12,000 and no interest income

7. A salary of $8,800 and interest income of $213

8. Use the odometer readings and gasoline purchases shown below to determine the distance traveled, the number of miles obtained per gallon of gas, and the cost per mile for a weekend trip of the Baxter family.

   1st odometer reading: 15,401.8
   2nd odometer reading: 15,853.9
   16 gallons of gas costing $20.75

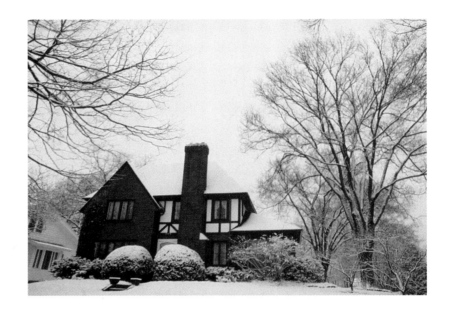

H ari first realized the importance of homeowner's insurance on a cold New Year's Day three years ago. His family got up to find that all the water pipes had frozen. Since it was their custom to spend the day with Hari's grandparents, who live in a nearby town, they decided to leave early and stop for coffee on the way.

Hari's father does not think very clearly without his morning coffee. He opened the cold water tap in the bathtub to help drain the frozen pipes but forgot to open the drain as well. When Hari opened the front door that evening when the family arrived home, he could hear the water splashing over the side of the bathtub. The bathroom floor was a small lake! Water had run down the hall, soaking the carpets in two bedrooms, and was slowly seeping into the dining room.

While Hari was eager to help clean up the mess, his mother suggested that they first call the insurance company to find out what they should do. The insurance agent told her to take photographs of all the damaged areas before moving or cleaning anything. She was also advised to make a list of all damaged items and any parts of the house that needed repair because of the flood. Hari's mother and the agent then agreed on a time at which an insurance adjuster could come to assist the family in filing an insurance claim.

Hari's father pointed out that even though they have a $250 deductible clause in their insurance policy, their settlement from the insurance company would still pay for a new carpet for Hari's room. Hari was thrilled to help pick it out. He chose a color that went well with his computer desk chair and a type that would not build up static electricity when he walked on it.

- *Understand why homeowner's insurance is important.*
- *Examine the different categories of coverage in a typical insurance policy.*
- *Learn about two important factors that affect how much an insurance company pays on claims—the deductible and the replacement cost.*
- *Recognize the value of keeping good records to document a claim.*
- *Examine the costs of home maintenance and repairs.*
- *Calculate some of the tax advantages of home ownership.*

THE ROLE OF HOMEOWNER'S INSURANCE

A house is the single largest purchase that most families make. Unfortunately, as Hari discovered, unforeseen events may result in damage to this valuable asset and its contents. Incidents such as theft or an injury on their property can also occur. An injured person could hold Hari's family liable and require them to pay medical expenses.

Homeowner's insurance generally covers the following potential disasters: fire and lightning, theft, windstorm, hail, explosion, smoke, cave-ins, frozen pipes, falling objects, and water damage. Sometimes there is an extra charge for certain types of coverage. For example, it may be difficult or very costly to obtain flood insurance in areas that are prone to floods.

Homeowner's insurance protects Hari's family against the financial hardship created by these potentially costly incidents. Requirements for homeowner's insurance vary by state. In the state in which Hari lives, the cost of homeowner's insurance is included in the monthly mortgage payment that his family makes to their bank. The bank then pays the insurance company.

PURCHASING DECISIONS

You may buy insurance from one of three sources: from an independent insurance agent who sells insurance for many companies, from an agent who sells only one company's insurance, or from the insurance company itself. It is often advantageous to find a local representative who can help file claims, clarify questionable points, and accelerate the payment of claims.

Hari's family made choices about the coverage that they wanted when they purchased their homeowner's insurance policy. **Coverage** refers to the items eligible for financial compensation and the dollar limits on that compensation. Decisions that the family had made on coverage affected their premium. The **premium** is the amount that they pay to the insurance company for the coverage that they selected. Some of the options from which they had to choose when they selected their insurance are the following.

Dwelling (Coverage A)   Dwelling coverage includes the dwelling described in the policy, any structures connected to it, outdoor equipment, building equipment used to maintain the property, and permanently attached fixtures. Rented or borrowed property is not included.

Other Structures (Coverage B)   Structures that are detached from the dwelling are insured for 10% of the amount of the insurance on the dwelling.

Personal Property (Coverage C)   Personal property coverage includes all items owned, regardless of where they are located. If they are not located on the premises, they are covered up to a total of 10% of the dwelling coverage. Since many items are excluded, it is important to read this section of a homeowner's policy carefully.

Loss of Use (Coverage D)   If the water had damaged Hari's house so much that his family had been unable to live in it, they would have been covered for temporary living expenses up to the limit specified in their policy.

Personal Liability (Coverage E)   If a visitor had slipped and fallen on the flooded bathroom floor, Hari's family would have been held liable, or responsible, for any injury that the person suffered as a result. Their insurance company would have been required to defend them in court if the injured person had decided to bring a lawsuit against them.

Medical Payments to Others (Coverage F)   If an injured visitor had been admitted to the hospital, Coverage F would have helped pay for medical expenses, even if Hari's family was not held personally liable.

REIMBURSEMENT FACTORS

The amount of the reimbursement that Hari's family will receive for any claim will be determined by the original cost of the damaged items, the deductible amount, and the kind of insurance they have.

The amount that Hari's family pays—in their case, $250—for repairing the eligible damage before the insurance policy begins to reimburse them is the **deductible.** The deductible applies to each accident or loss. If Hari's father makes the same mistake one week later, they will pay the deductible again. In general, the higher the deductible, the lower the insurance premium.

Buildings and other permanent structures (Coverage A and B) automatically have **replacement cost** coverage. If the flooded bathroom floor and part of the walls were completely damaged, the insurance would cover the actual amount that it costs to repair or replace these items. Personal property, however, is not automatically covered at replacement cost. It is covered for the purchase price at the time it was purchased. For example, if the damaged carpeting in Hari's room cost $675 when it was installed four years ago but costs $790 to replace at the time of the flood, a standard policy would have reimbursed the family for $675. However, Hari's family paid an additional premium to ensure that their personal property was covered at replacement cost.

KEEPING RECORDS

When the insurance adjuster arrived to inspect the damage, Hari quickly learned how important it was to keep detailed information about what the family owned. They had kept the receipts that showed where they had purchased the carpeting and exactly what kind it was. This was fortunate because they had purchased high-quality carpeting.

OTHER FINANCIAL RESPONSIBILITIES

There are many costs associated with maintaining a home that are not covered by homeowner's insurance. Many of these costs can be anticipated, so you can put money aside to cover them. Hari's family has periodic service check-ups of the following.

- Heating and air-conditioning systems
- All parts of the house for termite inspection and extermination of other pests if necessary
- Gutters, drains, and roof
- Plumbing and electrical systems

Appliances such as the refrigerator, range, freezer, clothes dryer, washing machine, television, hot water heater, dishwasher, and garbage disposal will not last forever and will need to be replaced at some time. The family must also maintain their lawn and the outside of their home.

Property taxes and mortgage interest can be used to reduce your taxable income. These items can be deducted on your federal tax return, saving you a considerable amount of money, particularly in the early years when your mortgage interest is high.

Ask Yourself

1. What is the advantage of buying replacement cost insurance?

2. What is included in personal property coverage?

3. Which kinds of coverage protect the family against liability for injury to a nonfamily member?

SKILL 1

EXAMPLE 1 Hari's family pays an additional annual premium of $35 on their homeowner's insurance for replacement cost coverage on their personal property. Their deductible is $150. They filed two insurance claims this year for personal property damage. One was for $200, the other for $800. They estimated that they would have been reimbursed 10% less for their claims if they had not had the replacement cost coverage.

QUESTION Was the extra premium that they paid for replacement cost coverage worth it?

SOLUTION

	With Extra Coverage	Without Extra Coverage	
Claim 1	$200	$180	200 − 0.1(200) = 180
Less deductible	− 150	− 150	
Reimbursement	$ 50	$ 30	
Claim 2	$800	$720	800 − 0.1(800) = 720
Less deductible	− 150	− 150	
Reimbursement	$650	$570	

Total personal property damage claims
With replacement cost coverage: 50 + 650 = $700
Without replacement cost coverage: 30 + 570 = $600
Advantage of extra coverage: 700 − 600 = $100

They paid $35 to receive an extra $100 in reimbursements. The additional coverage was worth it because they had a net savings of 100 − 35 = $65.

SKILL 2

In Chapter 7 you learned to find the future value of money that you are investing periodically. The formula is as follows.

Future Value of a Periodic Investment Formula

$$A = \frac{p[(1 + r)^n - 1]}{r}$$

where A = future value of periodic investment
p = the investment made at the end of each period
r = periodic interest rate
n = number of periods

This formula can be solved for p, which will allow you to calculate the periodic investment (or payment) if you know the future value, the periodic interest rate, and number of periods for which the money will be earning interest.

$$A = \frac{p[(1 + r)^n - 1]}{r}$$

$$Ar = p[(1 + r)^n - 1] \qquad \text{Multiply both sides by } r.$$

$$\frac{Ar}{(1 + r)^n - 1} = p \qquad \text{Divide both sides by } (1 + r)^n - 1.$$

Periodic Investment Formula

$$p = \frac{Ar}{(1 + r)^n - 1}$$

where p = periodic investment
A = future value of periodic investment
r = periodic interest rate
n = number of periods

EXAMPLE 2 Hari's parents would like to set up a savings account from which they can draw money to replace major appliances or make home improvements when it becomes necessary. For purposes of setting up this savings account, they are assuming that they will have to replace their appliances in ten years, although their appliances will probably last longer. They know that appliance costs will probably increase in ten years, so they estimated the cost of each appliance at that time as follows:

Appliance/Home Improvement	Estimated Cost in 10 Years
Refrigerator	$ 1500
Freezer	800
Dryer	450
Washer	600
Television	1250
Disposal	250
Hot water heater	380
Dishwasher	850
Air conditioner	3000
Carpets	6500
Total:	$15,580

QUESTION How much should Hari's parents save each month to have $15,580 in their account at the end of ten years if their account carries an APR of 6% and interest is paid monthly?

SOLUTION

$$p = \frac{Ar}{(1 + r)^n - 1}$$ Use the periodic investment formula.

$$p = \frac{15{,}580(0.005)}{(1 + 0.005)^{120} - 1}$$ Substitute $A = 15{,}580$, $r = 0.06 \div 12 = 0.005$, and $n = 10 \bullet 12 = 120$.

$$p = 95.07$$ To the nearest cent

Hari's parents will have to save $95.07 each month to have $15,580 in their savings account at the end of ten years.

SKILL 3

EXAMPLE 3 Hari's parents want to know how much they will save in federal taxes this year by deducting their property taxes and the interest charges on their mortgage on their return. Their property taxes are 1.25% of the assessed value of their home, which is $105,000. They will pay $6200 in mortgage interest this year.

QUESTION If they are in a 28% tax bracket, how much can they save in federal income tax this year?

SOLUTION

Property tax cost: $0.0125(105{,}000) = 1312.50$
Save 28% of deduction: $0.28(1312.50 + 6200.00) = 2103.50$

They may save as much as $2103.50 on their federal income tax by deducting their property taxes and mortgage interest on their tax return. If they have no other large deductions, such as medical expenses, these tax savings will be less.

TRY YOUR SKILLS

1. Justin's family has homeowner's insurance for their house. They paid $100 this year for replacement cost coverage on their personal property. Their deductible is $200. They filed an insurance claim for $1500 this year on a stereo system that was stolen and for which they paid $980 some years ago. Was the extra premium that they paid for replacement cost coverage worth it?

2. If you estimate that it will cost $8000 to replace all the carpeting in your house in 5 years and you wish to save for that expense now, how much should you place in your savings account each month if it has an APR of 3%?

3. What would you have to save each month in Exercise 2 if the APR on your account were 4% instead of 3%?

Calculate the maximum tax savings for the following deductions. Assume that each family is in a 28% tax bracket and that their property tax is the given percent of the assessed value of their home.

	Assessed Value	Property Tax	Amount of Property Tax	Mortgage Interest	Total Deductions	Tax Savings
4.	$150,000	1.8%		$11,205		
5.	225,000	0.9%		15,876		
6.	85,000	1.2%		6,788		

EXERCISE YOUR SKILLS

1. Why is it more difficult to obtain flood insurance in areas that are prone to flooding?

2. Why is it important to keep receipts from household appliances and other furnishings that you purchase?

3. Should you clean up the mess from a disaster before or after the insurance adjuster comes to your home?

4. What costs are not covered by homeowner's insurance?

KEY TERMS

coverage
deductible
homeowner's
   insurance
premium
replacement cost

Li-Ming's family has homeowner's insurance for their house. They paid $150 this year for replacement cost coverage on their personal property. Their deductible is $250. They filed an insurance claim for $1275 this year on a piece of furniture that was damaged by a flood and for which they paid $790 some years ago.

5. What amount did they recover?

6. What amount would they have recovered if they had not had replacement cost coverage?

7. Was the extra premium that they paid for replacement cost coverage worth it?

Phillipa's family has homeowner's insurance for their house. They paid $60 this year for replacement cost coverage on their personal property. Their deductible is $150. They filed an insurance claim for $400 this year on an awning that was blown down by a hurricane and for which they paid $225 some years ago.

8. What amount did they recover?

9. What amount would they have recovered if they had not had replacement cost coverage?

10. Was the extra premium that they paid for replacement cost coverage worth it?

Complete the following chart to find the amount that you must save each month to replace each appliance or make each home improvement in 10 years. Assume that your account carries an APR of 5%.

	Appliance/ Home Improvement	Estimated Cost in 10 Years	Monthly Savings Required
11.	Refrigerator	$1850	
12.	Water heater	375	
13.	Garage	3500	
14.	Carpeting	8200	
15.	Dishwasher	850	
16.	Kitchen cabinets	8700	
17.	Bedroom furniture	1695	
18.	Garbage disposal	280	
19.	Replacement roof	3600	
20.	Outdoor deck	520	
21.	Large-screen television	3350	
22.	Microwave oven	420	

Calculate the maximum tax savings for the following deductions. Assume that each family is in a 28% tax bracket and that its property tax is the given percent of the assessed value of its house.

	Assessed Value	Property Tax	Amount of Property Tax	Mortgage Interest	Total Deductions	Tax Savings
23.	$140,000	2.40%		$10,625.00		
24.	120,000	1.58%		9,850.75		
25.	205,000	1.25%		16,143.75		
26.	83,500	1.05%		6,575.63		
27.	102,000	0.85%		8,032.50		
28.	165,600	0.70%		13,041.00		

MIXED REVIEW

1. Use the Rule of 72 to determine how long it will take Tiffany's $5500 savings account to double. It is growing at a rate of 5.25% per year.

2. The fixed costs of James's business are $275 per week. The variable costs are $5.20 per item. James produces 90 items in 1 week. What is his total cost for the week?

3. A family's take-home pay is $4400.00 per month. How much can it afford to spend on credit payments each month, excluding a home mortgage?

4. Use the Driver Rating Factors Table in the Reference Section to find the driver rating factor for the usual driver of the car who is 50 years of age and drives to work.

Use the United States Mileage Chart in the Reference Section to find the distance between the following pairs of cities.

5. Austin, Texas, and Philadelphia, Pennsylvania

6. Pierre, South Dakota, and New Orleans, Louisiana

The charges at two motels are as follows.

Cozy Inn	**Stay-A-While Lodge**
Rates: single, $45	Rates: single, $35
2 adults, $60	2 adults, $45
teens, free	teens, $10 each
Room tax: 6%	Room tax: 6%
Free in-room movies	Game machines
Swimming pool	Free continental breakfast

Find the motel costs for each individual or family.

7. a family of two adults and one teenager at Stay-A-While Lodge for 2 nights

8. a single person at Cozy Inn from Friday afternoon until Monday morning

1. What are the advantages and disadvantages of owning a house?

2. How do fixed-rate and adjustable-rate mortgages differ?

3. What percent of gross income is the limit that banks place upon the size of a borrower's mortgage payment?

Karen collected the information shown about houses selling in her town. Give answers to four significant digits.

4. What is a linear model of the form $y = a + bx$ that will allow Karen to predict the number of bedrooms in a house in her town if she knows the selling price of the house?

5. Graph the linear regression equation on a scatter plot of the data.

6. Is there a good correlation between the price and the number of bedrooms in a house?

7. Estimate the number of bedrooms in a house in Karen's town that is selling for $260,000. Round your answer to the nearest whole bedroom.

Price (in thousands of dollars)	Number of Bedrooms
$150	2
195	3
68	1
210	3
310	5
270	4
135	2
190	2

Use a spreadsheet program to determine how the total financed price on a $210,000 mortgage for 30 years with a down payment of $50,000 differs as interest rates vary from 9% to 11% in increments of 0.5%. In each case, what is the savings over an 11% loan?

	Interest Rate	Total Financed Price	Savings Over 11%
8.	9.0		
9.	9.5		
10.	10.0		
11.	10.5		
12.	11.0		

To qualify for a loan at the local bank, a borrower must have a monthly mortgage payment that is no more than 28% of his or her family's gross monthly income. The bank is currently offering mortgages for 30 years at 8%. On the basis of monthly income, for how large a loan will each family qualify? If the loan represents 80% of the purchase price of a house, what is the most expensive house that each family can consider?

	Monthly Income	Mortgage Payment	Loan Amount	House Price
13.	$3200			
14.	2700			
15.	3300			
16.	1750			
17.	6200			
18.	4350			
19.	2375			
20.	3000			
21.	2244			
22.	2890			

23. If the bank requires that a borrower make a down payment of 20% of the purchase price of a house and the family in Exercise 13 has $25,000 to use as a down payment, will they qualify for the loan? If not, what is the most expensive house that they can consider?

24. Avram's family has homeowner's insurance for their house. They paid $125 this year for replacement cost coverage on their personal property. Their deductible is $250. They filed an insurance claim for $1600 this year on some sterling silver flatware that was stolen and for which they paid $1020 some years ago. Was the extra premium that they paid for replacement cost coverage worth it?

25. If you estimate that it will cost $7200 to replace all the carpeting in your house in 5 years and you wish to save for that expense now, how much should you place in your savings account each month if it has an APR of 4.5%?

Calculate the maximum tax savings for the following deductions. Assume that each family is in the 28% tax bracket and that their property tax is the given percent of the assessed value of their home.

	Assessed Value	Property Tax	Amount of Property Tax	Mortgage Interest	Total Deductions	Tax Savings
26.	$165,000	1.78%		$11,375		
27.	245,000	0.89%		16,988		
28.	75,000	1.12%		5,899		

CHAPTER 12 TEST

Ben collected the following information about houses selling in his town. Give answers to four significant digits.

1. What is a linear model of the form $y = a + bx$ that will allow Ben to predict the number of acres of land on which a house stands in his town if he knows the selling price of the house?

2. Is there a good correlation between the price and the number of acres?

3. Graph the linear regression equation on a scatter plot of the data.

4. Estimate the number of acres of land to the nearest tenth for a house selling for $220,000.

Price (in thousands of dollars)	Acres of Land
$145	0.4
190	0.6
65	0.3
205	0.7
315	1.5
275	1.3
140	0.5
185	0.8

Find the total financed price of a $270,000 house with each of the following combinations of down payment, interest rate, and amortization period.

	Percent Down	Down Payment	Loan Amount	Interest Rate	Number of Years	Total Financed Price
5.	20%	$54,000	$216,000	10%	15	
6.	20%			10%	30	
7.	30%			11%	15	
8.	20%			11%	30	

To qualify for a loan at the local bank, a borrower must have a monthly mortgage payment that is no more than 28% of the family's gross monthly income. The bank is currently offering mortgages for 15 years at 7.5%. On the basis of monthly income, for how large a loan will each family qualify? If the loan represents 80% of the purchase price of a house, what is the most expensive house that each family can consider?

	Monthly Income	Mortgage Payment	Loan Amount	House Price
9.	$4300			
10.	1950			
11.	6500			

12. If you estimate that it will cost $3800 to stain the exterior of your house in 3 years and you wish to save for that expense now, how much should you place in your savings account each month if it has an APR of 4.0%?

13. Calculate the federal income tax savings for a family in a 28% tax bracket with a house assessed at $162,000 with property tax of 2.1% and mortgage interest of $10,500.

CUMULATIVE REVIEW

1. Determine your monthly car payment if you borrow $8,500 at 9.5% annual interest for 5 years.

2. Determine the plan that would provide the lowest total financed price for a $18,900 car. Assume that the loan in Plan 1 is from a bank or other lending institution and that the loans in Plans 2 and 3 are from the car manufacturer's financing subsidiary.

	Plan 1	Plan 2	Plan 3
Rebate	$1500	$1000	0
Rate	10.25%	7.9%	4.5%
Time	36 months	36 months	36 months
Loan Amount	$17,400	$17,900	$18,900
Monthly Payment			
Total Financed Price			

Your credit card carries a monthly finance charge of 1.75%. You pay 10% of the amount owed to the nearest dollar each month. Your beginning balance is $1,240. Complete a table with the following column heads for 3 months.

	Month	Balance	Interest	Amount Owed	Payment
3.	1	$1240			
4.	2				
5.	3				

6. Your credit card account balance from 2/1–2/13 was $550. Then you made a payment of $100 on 2/14. You made no further charges or payments for the remainder of the month. Determine your closing balance for February if your card carries a monthly finance charge of 1.25% and it is not a leap year.

7. Use the Multiples of Salary Chart in the Reference Section to find the amount of life insurance a 45-year-old person who has gross earnings of $40,000 would need in order to provide 60% income replacement.

8. If you buy 750 shares of stock at a price of $75\frac{1}{4}$, what will the total cost of the stock be? The commission rate is 1%.

9. Determine the distance traveled, the mileage per gallon of gas, and the cost per mile based on the following information.

    1st odometer reading: 35,461.9
    2nd odometer reading: 35,953.5
    17.4 gallons of gas at a cost of $19.87

10. Use the cost of gasoline per mile found in Exercise 9 to find the total operating expenses and the cost per mile for operating a car for 11,750 miles with $250.00 in maintenance costs and $274.50 in repair costs.

PROJECT 12–1: **Shopping for a House**

1. Find newspaper advertisements for 20 houses that are for sale in your town.
2. For each house, find the number of square feet, the number of bedrooms, and the number of bathrooms.
3. Make a poster presenting the information you gather.
4. Run a regression analysis to find a linear model of the form $y = a + bx$ that will allow you to predict the number of square feet in a house in your town if you know its selling price.
5. Run a regression analysis to find a linear model of the form $y = a + bx$ that will allow you to predict the number of bedrooms in a house in your town if you know its selling price.
6. Run a regression analysis to find a linear model of the form $y = a + bx$ that will allow you to predict the number of bathrooms in a house in your town if you know its selling price.
7. Use your results from Exercise 4 to predict how many square feet are in a house in your town that is selling for $150,000.
8. Use your results from Exercise 5 to predict how many bedrooms are in a house in your town that is selling for $175,000.
9. Use your results from Exercise 6 to predict how many bathrooms are in a house in your town that is selling for $120,000.
10. Select a house that has fewer square feet than your model would predict based on its price. Determine whether the house has other advantages or amenities that warrant the higher price.
11. Select a house that has more bedrooms than your model would predict based on its price. Determine whether the house has any major drawbacks.

Recall that the *principal square root* of a number is either positive or zero.

$$\sqrt{9} = 3 \qquad \sqrt{9} \neq -3 \qquad \sqrt{5-5} = 0$$

A *radical equation* contains a variable under a radical sign. To solve a radical equation involving one or more square roots, isolate one square root on one side of the equation. Then square each side of the equation. The resulting equation may have solutions that are not solutions of the original equation. Therefore, it is important to check all solutions in the original equation.

Solve.

Examples

a.
$$2\sqrt{x} + 1 = 3\sqrt{x} - 4$$
$$1 + 4 = 3\sqrt{x} - 2\sqrt{x}$$
$$5 = \sqrt{x}$$
$$25 = x$$

Check $2\sqrt{25} + 1 \ ? \ 3\sqrt{25} - 4$
$$2 \cdot 5 + 1 \ ? \ 3 \cdot 5 - 4$$
$$10 + 1 \ ? \ 15 - 4$$
$$11 = 11$$
25 is a solution

b.
$$\sqrt{x} = x - 2$$
$$x = x^2 - 4x + 4 \quad \text{Square both sides.}$$
$$0 = x^2 - 5x + 4$$
$$0 = (x - 4)(x - 1) \quad \text{Factor.}$$
$$0 = x - 4 \quad \text{or} \quad 0 = x - 1$$
$$4 = x \quad \text{or} \quad 1 = x$$

Check $\sqrt{4} \ ? \ 4 - 2 \qquad \sqrt{1} \ ? \ 1 - 2$
$$2 = 2 \qquad\qquad 1 \neq -1$$
4 is a solution.   1 is not a solution.

1. $\sqrt{x} - 3 = 2$

2. $3\sqrt{x} - 2 = \sqrt{x} + 6$

3. $\sqrt{x + 1} = x - 5$

4. $\sqrt{x - 1} = 0$

5. $\sqrt{4 + x} = x - 2$

6. $\sqrt{x^2} - 5 = 4$

A *rational equation* contains one or more rational expressions. The denominators in a rational equation may be eliminated by multiplying both sides of the equation by each denominator. Check each solution in the original equation.

Solve.

Example $\dfrac{14}{x} = x + 5$

$$14 = x^2 + 5x \qquad\qquad \text{Multiply both sides by } x.$$
$$0 = x^2 + 5x - 14 \qquad \text{Subtract 14 from both sides.}$$
$$0 = (x + 7)(x - 2) \qquad \text{Factor.}$$
$$0 = x + 7 \quad \text{or} \quad 0 = x - 2$$
$$-7 = x \quad\quad \text{or} \quad 2 = x \qquad \text{Check in the original equation.}$$

7. $\dfrac{6}{x} = 3$

8. $\dfrac{x}{3} = \dfrac{x - 4}{2}$

9. $\dfrac{1}{x} = \dfrac{x}{4}$

10. $\dfrac{x}{2} + \dfrac{5}{x} = \dfrac{7}{2}$

11. $\dfrac{x}{3} = \dfrac{x + 3}{5}$

12. $\dfrac{3}{x + 2} = \dfrac{x + 2}{3}$

CHAPTER 13

Apartments and Other Housing Alternatives

MANY YOUNG PEOPLE FIND THE IDEA of moving away from their parents and establishing a residence of their own to be very appealing. Whether they are going to college, into the work force, or a combination thereof, young people gain a feeling of independence from living on their own. With the expression of that feeling of independence comes responsibility. If they do not go to the grocery store, no food will appear in the refrigerator!

Most young people share an apartment, rented house, or other living quarters with others. Many aspects of sharing living space with another person become immediately apparent. Roommates learn very quickly whether their living habits differ. They must decide who makes sure that the rent and the telephone bill are paid on time, who buys the groceries, who prepares the meals, and how much noise, quiet, clutter, fastidiousness, dirty laundry, television, extra company, immodesty, late-night snacking, and so on each person can tolerate. How does each roommate adjust so that the group can live harmoniously?

After renting for a while, many young people want to purchase a home. A condominium or a mobile home may fit their needs and match their ability to pay.

In this chapter our high school students will learn about the factors involved in renting an apartment or a house. They will study the purchase of smaller-scale housing for people who want the advantages of home ownership without paying the price for a full-sized house.

13–1 **The Hunt for an Apartment: How Much Can You Afford in Rent?**

13–2 **Selecting an Apartment: What Should You Do Before Signing a Lease?**

13–3 **Choosing a Place to Live: Should You Rent or Buy?**

THE HUNT FOR AN APARTMENT: HOW MUCH CAN YOU AFFORD IN RENT?

Evelyn and Freda enjoy being in business together. Their painted T-shirts and sweatshirts are selling very well. They hope that their sales will continue over the summer before they both enroll at the community college in the fall. They are planning to add beach towels with hand-painted beach scenes to their product list for the summer to offset the loss of sweatshirt sales during June, July, and August.

Evelyn is planning to study business management at the community college. Freda will pursue advertising and marketing. They have found that they are quite compatible and have the same attitudes about many things. They have even discussed taking an apartment together when they attend college. The drive from home to the community college takes 45 minutes. Although they can live at home with their parents during college, the commute will be inconvenient. Evelyn is also anxious to be out on her own.

There are several living arrangements from which Evelyn and Freda can choose if they do move away from home. The community college has some very nice dormitories on campus. However, they are not sure that a dormitory is the best choice for them because they will need space in which to continue their business.

There are also a number of reasonably priced apartments near campus. Most of the apartments are furnished, several have access to a swimming pool, and some are conveniently located near shopping areas. As summer approaches, Evelyn and Freda will discover other factors that they must consider before they present their plans to their parents.

OBJECTIVES: *In this lesson, we will help Evelyn and Freda to:*

- *Consider the advantages of rental housing.*
- *Recognize what is involved in sharing living space with a roommate.*
- *Recognize different types of rental housing.*
- *Consider the cost of renting as a percent of income.*

LOOKING FOR AN APARTMENT

Evelyn and Freda have decided that they want a two-bedroom apartment near campus. It would be ideal if the apartment were well kept, sunny, and had room for their business. They are finding a great variety in number of bedrooms, features, and rents in the newspaper advertisements. They see ads for expensive small apartments in ideal locations and affordable large apartments that are located out of walking distance from campus. They wonder whether apartments are available that are between these two extremes. They notice the abbreviations that appear in these ads. They recognize "bdrms" as bedrooms and "LR" and "K" as living room and kitchen, respectively. They now know that "hdwd flrs," "Ht/wtr," and "Fully crptd" stand for hardwood floors, hot water, and fully carpeted, respectively.

They realize that they will have to base their decision on a number of factors including the **rent,** the location, the number and size of the rooms, the condition of the apartment, the availability of the apartment, whether it is furnished, and whether they will have to pay for **utilities** (heat, water, electricity) separately.

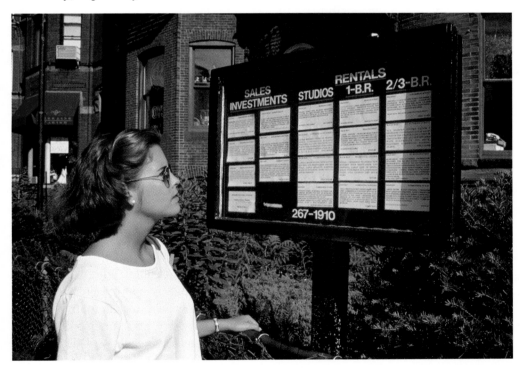

FACTORS TO CONSIDER IN DECIDING TO RENT

Renting an apartment has many advantages over buying a house. They include the following:

* Knowing exactly how much you must spend each month
* Not being tied down to one location
* Not being responsible for maintenance of the grounds
* Getting to know the community well before making a commitment to stay there
* Spending less for housing than buyers do

Evelyn and Freda know that they want to share an apartment. They are already good friends, and sharing will cut their costs in half. However, they are aware that they should discuss the practical arrangements ahead of time. Evelyn's older cousin, Ed, had problems when he and his best friend moved in together. They did not anticipate the kinds of decisions they would have to make to live together amicably. For instance, they had very different opinions about when they could have visitors. Ed's roommate had so many parties that Ed found it very difficult to get all his studying done. They finally became so angry with each other that they are no longer on speaking terms! In order not to endanger their friendship, Evelyn and Freda will discuss some of these potential problems ahead of time.

The issues that they will consider include the following.

* *Living habits*   How will they set hours for listening to music, watching TV, and other activities that might interrupt the other person? How will they divide the responsibilities for washing dishes, making beds, cleaning the bathroom, vacuuming, and so on?
* *Independence or companionship*   Do they want to spend time together or lead relatively separate lives? Do they want to eat meals together? Do they want to do their laundry at the same time?
* *Sharing expenses*   How will they divide the rent, utility bills, food costs, new furniture, household supplies, and the security deposit?
* *Overnight guests*   Who may stay, and how often?
* *Ending the arrangement*   Can either one move out at any time? How would the remaining roommate replace the one who leaves?

HOUSING ALTERNATIVES

Before making their decision to share an apartment, Freda and Evelyn considered several other possibilities:

Living with Parents   Many college students live at home with their parents. This can be a good alternative when the college is nearby and the family enjoys being together. For Evelyn and Freda, however, this alternative was not practical. It would mean a long commute every day, and they would have to own and maintain cars to make the 45-minute drive each way. This would definitely cut into their studying time, the time that they hope to devote to their business, and their sense of belonging to the college community.

Single Rooms   Another option that they considered was renting single rooms in a home or hotel. They would want rooms with kitchen and laundry facilities. Evelyn and Freda think that they need more space than single rooms would provide. However, a single room can be a good alternative, especially for someone who does not want to invest in furniture and household items or who is not planning to live in one place for long.

Dormitory   The dormitories on campus contain sleeping rooms, usually for two people, with separate desks and closet space for each. Some have eating facilities, and some do not. Evelyn and Freda thought about this alternative because it is generally less expensive than private housing. Their reason for eliminating this option was the same as their reason for eliminating single rooms—they would simply have less space than they would like.

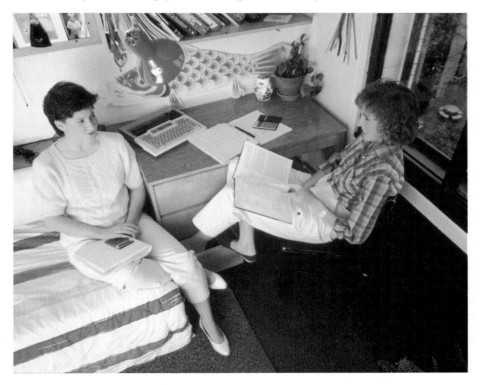

Apartments   An apartment building or complex is a multiple-unit dwelling that may contain as few as two living units or as many as hundreds. Apartments can be found in a high-rise building, house, converted warehouse, over the garage of a single-family house, or in the coach house of a large, old estate. Most apartments are rented unfurnished, but some are available with furniture. They seldom have any private yard space, but other conveniences such as a parking space and coin-operated washing machines are often provided.

HOW TO CHOOSE AN APARTMENT

Having decided that what they really want is an apartment, Freda and Evelyn now must go about finding one. For them, the most important factor in choosing an apartment is location—they want to live near campus. Next, they know that they want two bedrooms, a common living room and kitchen, and a dining room. They really need only one bathroom. They also need a place to do their laundry and a safe place to keep their bikes.

Other people looking for apartments will have different needs. Elderly people or people with limited mobility may prefer a ground-floor apartment. Commuters may want to be close to train and bus lines. A couple with children may want to be near schools and shopping. A single person may choose a large building or one with a common pool or tennis courts so that he or she can meet new people. Apartments are available for people with special needs. For example, apartments with ramps, low counters, and extra-wide doorways are available for the wheelchair-bound.

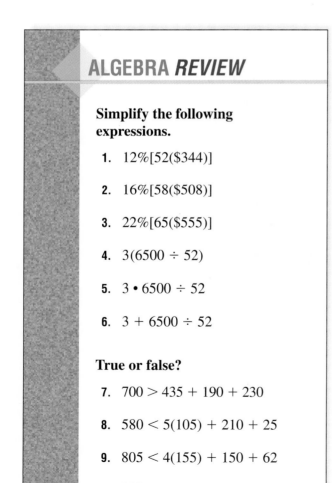

ALGEBRA *REVIEW*

Simplify the following expressions.

1. $12\%[52(\$344)]$

2. $16\%[58(\$508)]$

3. $22\%[65(\$555)]$

4. $3(6500 \div 52)$

5. $3 \cdot 6500 \div 52$

6. $3 + 6500 \div 52$

True or false?

7. $700 > 435 + 190 + 230$

8. $580 < 5(105) + 210 + 25$

9. $805 < 4(155) + 150 + 62$

10. $980 > 7(110) + 126 + 22$

Ask Yourself

1. What are four common types of housing arrangements from which college students may choose?

2. What are two economic advantages of renting instead of buying?

3. What are two noneconomic advantages of renting instead of buying?

4. Why should you discuss potential problems with a prospective roommate before deciding whether or not to share an apartment?

5. What issues should you discuss with a prospective roommate before you decide whether or not to share an apartment?

SHARPEN YOUR SKILLS

SKILL 1

One guideline that financial advisers provide for renters is that they should spend no more than *one week's gross income* on their monthly rent.

EXAMPLE 1 Evelyn and Freda estimate that the yearly income from their business will be $18,000 next year. Each will also receive $2500 per year from her parents for living expenses at college.

QUESTION How much can they afford to pay in rent each month?

SOLUTION

Annual income: $18,000 + 2(2500) = 23,000$

Weekly income: $23,000 \div 52 = 442.31$

Therefore they should spend no more than $442.31 on their monthly rent.

EXAMPLE 2 Evelyn found an advertisement for an apartment with a monthly rental charge of $550.

QUESTION Can they afford this apartment? If not, how much yearly income would they need to do so?

SOLUTION

They cannot afford this apartment because their weekly income is only $442.31. They would need a gross income for one week of $550 or a yearly income of $52(550) = $28,600$ to afford this apartment.

SKILL 2

There are costs associated with living in an apartment in addition to the rent. If you use a **broker** to help you find an apartment, you may have to pay the broker a fee. When you rent an apartment, you probably will have to pay your landlord a **security deposit,** which will be returned to you when you move out if you have done no damage to the apartment. You must also consider the costs of utilities, which might or might not be included in your rent.

EXAMPLE 3 Miranda used a broker to find an apartment that rents for $525 per month. The broker charges a fee of 10% of one year's rent. When she signs the lease, she must pay a security deposit equal to two month's rent in addition to the first month's rent. If she moves out at the end of one year and does no damage to the apartment, her security deposit will be returned. However, she will be unable to use that money during the year.

QUESTIONS What is Miranda's total cash outlay for the year? How much will she have to pay before she moves into the apartment?

SOLUTIONS

Total rent for the year:	$12(525) = 6300$
Broker's fee:	$0.1(6300) = 630$
Security deposit:	$2(525) = 1050$
Total cash outlay for the year:	$6300 + 630 + 1050 = 7980$

She must pay the broker's fee of $630, the security deposit of $1050, and the first month's rent of $525 before she moves into the apartment. Therefore she must pay $630 + 1050 + 525 = \$2205$ in advance.

EXAMPLE 4   Jorge pays $425 in rent and has a high electric bill of approximately $200 each month because, owing to the hot, sticky climate in which he lives, he uses the air conditioner all the time. His monthly water and gas bills are $50 and $30, respectively. He has the opportunity to move to another apartment in which utilities (electricity, water, and gas) are included in the monthly rent of $600.

QUESTION   Will Jorge save money by moving to this new apartment even if he spends $100 on relocation expenses?

SOLUTION
Current monthly expenses: $425 + 200 + 50 + 30 = 705$

The new apartment will cost him $705 - 600 = \$105$ less each month. His moving expenses will be recovered in the first month. Therefore he will save money by moving.

TRY YOUR SKILLS

Determine how much monthly rent you can afford for each gross annual income.

1. $37,500   **2.** $52,500   **3.** $22,750

Determine the gross annual income that you must earn to be able to afford each monthly rent.

4. $610   **5.** $295   **6.** $805

7. Room and board (food) cost $6000 in a dormitory on campus for the nine months of the school year. Jacob would like to rent a furnished apartment near campus with his old friend Leo. He found one that rents for $400 per month, including electricity, water, and gas. However, he must sign a lease for 12 months. He estimates that grocery expenses for two people will be about $350 per month. Should he rent the apartment? Why or why not?

EXERCISE YOUR SKILLS

1. What are some of the advantages of sharing an apartment with another person?

2. What are some of the disadvantages of sharing an apartment with another person?

3. If you had a choice, would you prefer to buy a house or rent an equally comfortable apartment? Why?

Determine how much monthly rent you can afford for each gross annual income.

4. $66,500	5. $22,000	6. $35,000	7. $29,600
8. $19,750	9. $45,095	10. $33,900	11. $18,000

Determine the gross annual income that you must earn to be able to afford each monthly rent.

12. $595	13. $700	14. $375	15. $775
16. $525	17. $975	18. $300	19. $645

20. Thelma used a broker to find an apartment that rents for $675 per month. The broker charges a fee of 10% of one year's rent. When she signs the lease, she must pay a security deposit equal to one month's rent as well as the first and last month's rent. What is her total cash outlay for the year? How much will she have to pay before she moves into the apartment?

21. Rosalyn used a broker to find an apartment that rents for $505 per month. The broker charges a fee of 15% of one year's rent. When she signs the lease, she must pay a security deposit equal to two months' rent as well as the first and last month's rent. What is her total cash outlay for the year? How much will she have to pay before she moves into the apartment?

22. Howard pays $535 in rent and has monthly electric, water, and gas bills of $175, $69, and $75, respectively. He has the opportunity to move to another apartment in which utilities are included in the monthly rental fee of $900. His relocation bills will be approximately $75. Will he save money by moving to this apartment?

23. Caleb pays $760 in rent and has monthly electric, water, and gas bills of $150, $38, and $65, respectively. He has the opportunity to move to another apartment in which utilities are included in the monthly rental fee of $875. His relocation bills will be approximately $150. Because of the distant location of the new apartment, the monthly transportation cost increases by $30. Will he save money by moving to this apartment?

24. Room and board (food) cost $6500 in a dormitory on campus for the nine months of the school year. Martha would like to rent a furnished apartment near campus with her friend McKee. She found one that rents for $425 per month, including electricity, water, and gas. Martha expects to pay about $350 per month for their food bills. However, she must sign a lease for 12 months. Should she rent the apartment? Why or why not?

25. Room and board (food) costs $5000 in a dormitory on campus for the nine months of the school year. Damien would like to rent a furnished apartment near campus. He found one that rents for $290 per month, including electricity, water, and gas. Groceries will cost about $200 per month. However, he must sign a lease for 12 months. Should he rent the apartment? Why or why not?

MIXED REVIEW

1. If you can afford a monthly payment of $285, how much money can you borrow at 8.75% for 5 years?

A car that cost a dealer $13,750 is marked up 21%.

2. Find the mark-up.

3. Find the sticker price.

4. Use the Driver Rating Factors Table in the Reference Section to find the driver rating factor for a 20-year-old unmarried female without driver training who drives to work and is not the owner or usual driver.

Use the Windward Airlines Schedule in the Reference Section to find each of the following:

5. The earliest flight from Chicago to Phoenix, Arizona.

6. The flight from Chicago to Phoenix that takes the least amount of time.

7. A family's monthly income is $4,575. How large can their monthly mortgage payment be?

8. If the prevailing annual interest rate is 10% and the family of Exercise 7 would like a 15-year mortgage, how much can the family borrow?

9. The monthly take-home pay of a family is $3300. Credit-card payments are $540. Determine, to the nearest tenth of a percent, how much the family is spending on its credit payments. Is the family in danger of credit overload?

10. Students who are selling painted T-shirts have 25 hours of labor costs at $6.00 per hour; advertising, energy, and transportation costs of $50.00; and variable costs of $3.00 for each T-shirt produced. The price of each T-shirt is $7.00. Write the cost function and the revenue function. Determine the break-even point algebraically.

SELECTING AN APARTMENT: WHAT SHOULD YOU DO BEFORE SIGNING A LEASE?

Ana is working as a hostess at the seafood restaurant at which Sylvia and her father sometimes dine. The restaurant is near the mall, where Ursula works as a clerk at Joyful Toys. Sometimes the three friends get together after work and do some shopping at the mall or go to the movies.

Ana and Ursula have been thinking about sharing an apartment after high school graduation. Ana will be going on vacation for a month this summer, but the restaurant manager told her that she can have her job back when she returns. Ursula enjoys her job and plans to continue working at the toy store after graduation. She may apply for an assistant manager's position at the end of the year.

Ana and Ursula have been good friends since they were 10 years old and spent two weeks at Girl Scout camp together. They discovered that they could prepare a pretty good stew over an open campfire and they both liked their marshmallows lightly toasted, not burned. Perhaps they would be compatible roommates. Sylvia has promised to show them all the best places to buy towels, dishes, blankets, and sheets for their new apartment.

OBJECTIVES: In this lesson, we will help Ana and Ursula to:

- *Investigate potential sources of information on rental housing.*
- *Learn how to inspect an apartment.*
- *Read and comprehend a lease agreement.*
- *Recognize the relationship between the monthly rental charge and the number of amenities, size, location, number of bedrooms, and so on.*

HELPERS IN THE SEARCH

Ana does not have to move out of her family's house, but she wants to give independence a try. If she shares an apartment with Ursula for a while, she may discover that being an independent adult carries more responsibility than she is prepared to accept. But she will not know that until she tries. Her parents will support her efforts at independence and help her to cope with the adult decisions that she will be making. They have suggested several sources of information that may be useful to her in her search for an apartment.

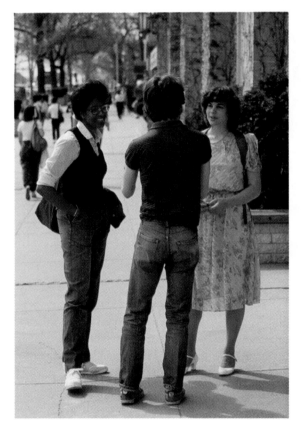

Friends   It is often helpful to ask friends whether they know of any available apartments to rent. If they are currently renting, they can also fill you in on the procedures to follow and any problems that they have had.

Newspapers   Reading the advertisements for apartments in the real-estate section of the newspaper will give Ana and Ursula an idea of the rents being charged for assorted types and sizes of apartments in various locations.

Real-Estate Agencies   Owners of apartments or apartment buildings often list available rental units with a local real-estate broker. **Real-estate agents** are licensed by the state to help people find an apartment. They may charge the renter a fee based on the amount of rent paid for the apartment they have shown.

INSPECTING THE APARTMENT

When Ana and Ursula find an apartment in which they are interested, they should inspect it carefully and thoroughly. They should look into the closets, peer under the sink, tour the basement, flush the toilet, and test the shower. They should ask themselves whether the cabinet space is sufficient and the refrigerator is large enough. Other questions that they should ask themselves include: Will the size and shape of the rooms fit their needs? Are the locks and security systems adequate? Are storage facilities, laundry facilities, and parking areas available?

If they are seriously considering a particular apartment, they should ask other tenants in the building questions such as the following.

- Is routine maintenance adequate and prompt?
- Are complaints responded to quickly, and is the owner accessible?
- Is the building noisy?
- Have rent increases been excessive?
- Does the building have a high turnover of tenants?

Turnover refers to people moving in and out in rapid succession. In general, it is less desirable to live in a building with a high turnover rate than in one with a more stable population.

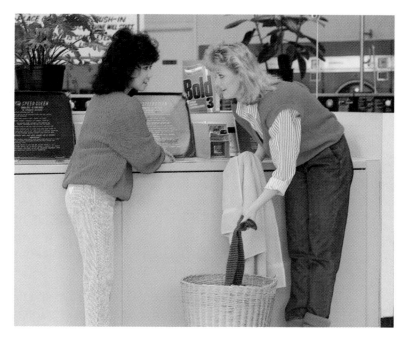

READING THE LEASE

When Ana finds an apartment, she probably will be required to sign a **lease,** which is an agreement between the landlord and her in which she agrees to pay for the use of the apartment. The landlord agrees to allow her to occupy the apartment as long as she complies with the provisions of the lease. The lease imposes financial obligations on her as a tenant; therefore she must be sure that she understands all of the clauses in the lease before she signs it. Some of the important clauses of a lease include the following.

Term of Lease   This clause sets the length of time your lease will be in force.

Rent   You agree to pay a specified amount on a certain day in each month during the term of the lease. You are legally obligated to pay the rent for the entire term of the lease, even if you want to move out before the expiration of the term of the lease.

Security Deposit   The security deposit, usually equal to one month's rent, is held by the landlord (the owner of the apartment), to pay for any damages to the apartment caused by the tenant. The security deposit is collected when you sign the lease. It will be returned to you when the lease ends and you leave the apartment in the same condition in which you found it. Therefore, before renting an apartment, you should make a list of existing damages and have the landlord sign it. This will assist you in settling claims for any damage found in the apartment when you leave and ask for your security deposit refund.

Condition of Premises   You agree that the premises are acceptable "as is." Sometimes, the landlord may agree to paint the apartment or to fix existing damage before you move in.

Repair   You agree not to damage the apartment, which is the landlord's property.

Evaluate y for $x = 15.82$. Round answers to the nearest hundredth.

1. $y = 37 - 0.17x$

2. $y = \dfrac{225.72}{x}$

3. $x = 49.975 + 0.246y$

4. $x = \dfrac{4.78y}{98.98}$

Determine a linear equation that fits the points.

5.

x	2	4	6
y	7	11	15

6.

x	-2	-1	0
y	-5	-4	-3

7. The cost of renting an apartment in a building is $100 plus $150 per room. Write an equation for the cost.

The yearly cost of home heating oil in one neighborhood was $120 plus $1.00 per square foot.

8. Find the cost of heating a 1200 ft^2 home.

9. Find the cost of heating an 850 ft^2 home.

10. Write an equation for the yearly cost of heating oil.

Use; Sublet You agree that only you and your immediate family will live with you. If you are sharing an apartment with a roommate, you both must sign the lease. Some leases may prohibit pets from living with you. In the event that you want to leave the apartment before the expiration of the lease term, you may want to arrange for someone else to move in and assume your obligations to pay rent. This is called a sublet arrangement. Most leases prohibit you from subletting your apartment without the landlord's consent.

Plurals If you have a roommate, both of you share responsibility for obeying the lease. If your roommate fails to pay his or her portion of the monthly rent, you are obligated to pay the entire monthly rent due under the lease. Therefore you should choose your roommates carefully.

Access; Quiet Enjoyment You give the landlord permission to enter your apartment. As long as you obey the terms of the lease, the landlord agrees to allow you to enjoy your apartment and not to disturb your occupancy of the apartment.

Compliance You agree to obey the laws regulating pets, sanitation, noise, and the rights of other tenants in the apartment building.

Default You agree that if you fail to pay your rent or to otherwise comply with the terms of the lease, the landlord has the right to terminate the lease and to commence eviction proceedings. In addition, the landlord may have the right to retain your personal property in your apartment until the rent that is owed is paid in full.

Ana realizes that the terms of a lease are quite serious. She will select her apartment carefully, knowing that if she signs a lease for one year, she must be willing to make that year's commitment as an adult.

Ask Yourself

1. What are three sources of information that you can consult if you are looking for an apartment?

2. What should you inspect in any apartment before renting it?

3. What is a lease? Why does a person who is renting an apartment have to sign one?

SHARPEN YOUR SKILLS

SKILL 1

Ana noticed that the following amenities are used as selling points in advertisements for apartments: extra storage, attached parking, fireplace, exercise facility, spa, pool, sauna, playground, tennis courts, clubhouse, cable TV or satellite hookup, washer/dryer, central air conditioning, and patio.

EXAMPLE 1 Ana collected the information shown about some apartments that are for rent in her town.

Monthly Rent	Number of Amenities
$ 600	5
1100	10
475	3
980	8
1050	9
450	3
675	6
350	1
570	6
700	7

QUESTIONS What is a linear model of the form $y = a + bx$ that will allow Ana to predict the number of amenities in an apartment in her town if she knows the monthly rental charge? Is there a good correlation between the monthly rental charge and the number of amenities?

SOLUTIONS

Use the linear regression function on your graphing calculator to find the line that best fits the data. The calculator also determines the value for r, the correlation coefficient, which is a measure of how well the line fits the data. If r is close to $+1$ or -1, the fit is very good. If r is close to 0, there is virtually no fit. Begin by entering the ordered pairs (x, y) of data, where x represents the rental charge and y represents the number of amenities in each apartment. Your calculator will display the following information.

$a = -1.414050978$
$b = 0.0103799295$
$r = 0.9556044872$

The linear equation that best relates the rental charge x and the number of amenities y is

$y = -1.414050978 + 0.0103799295x$

Since the value of r is close to $+1$, there is a very good correlation between monthly rental charge and number of amenities.

EXAMPLE 2 Ursula found an advertisement for an apartment that looks very appealing to her. The monthly rental charge for this apartment is listed at $625.

QUESTION Estimate the number of amenities in this apartment using the regression equation that you derived in Example 1.

SOLUTION

Substitute $x = 625$ into the equation from Example 1 and solve for y.

$$y = -1.414050978 + 0.0103799295x$$
$$y = -1.414050978 + 0.0103799295(625)$$
$$y = 5.07 \quad \text{To the nearest hundredth}$$

There should be approximately five amenities in this apartment.

EXAMPLE 4 Ana looked at an apartment with four amenities and a monthly rental charge of $800.

QUESTION Is this apartment fairly priced?

SOLUTION

Substitute $y = 4$ into the equation from Example 1 and solve for x.

$$y = -1.414050978 + 0.0103799295x$$
$$4 = -1.414050978 + 0.0103799295x$$
$$4 + 1.414050978 = 0.0103799295x$$
$$\frac{4 + 1.414050978}{0.0103799295} = x$$
$$522 = x \quad \text{To the nearest dollar}$$

This apartment appears to be significantly overpriced. A monthly rental charge of $522 seems more appropriate for this apartment.

EXAMPLE 4 Ana and Ursula are aware that there is a large variation in the number of apartments in each apartment building. They collected the information shown about some apartments that are for rent in their town.

QUESTION Is there a linear relationship between the monthly rental charge and the number of apartments in the building?

Monthly Rent	Number of Apartments in Building
$800	12
750	180
500	160
550	8
375	210
400	2
400	40
675	50

SOLUTION

Use the linear regression function on your graphing calculator to find the line that best fits the data. Begin by entering the ordered pairs (x, y) of data, where x represents the rental charge and y represents the number of amenities in each apartment. Your calculator will display the following information:

$$a = 110.9617225$$
$$b = -0.0507177033$$
$$r = -0.0988009201$$

Since the value of r is -0.10 to the nearest tenth, there is almost no linear relationship between monthly rental charge and number of apartments in the building.

TRY YOUR SKILLS

For Evelyn and Freda the most important factor in choosing an apartment is location. They want to be in walking distance from the campus because neither will have a car at college. Freda collected the information shown about eight two-bedroom rental apartments near school.

Monthly Rent	Distance From School (in miles)
$400	0.4
525	0.2
295	1.1
350	0.6
600	0.1
575	0.1
410	0.3
375	0.5

1. What is a linear model of the form $y = a + bx$ that will allow Freda to predict the distance from school of a rental apartment if she knows the monthly rental charge?

2. Is there a good correlation between the monthly rental charge and the distance from campus?

3. Estimate the distance from campus, to the nearest tenth of a mile, of an apartment that has a monthly rental charge of $435.

4. Do you think that an apartment with a monthly rental charge of $590 that is located 0.7 mile from campus is fairly priced?

EXERCISE YOUR SKILLS

1. If you are about to rent an apartment, why should you read the lease very carefully before you sign it?

2. In general, why would it be less desirable to live in an apartment building with a high rate of turnover than in one with a more stable population?

3. If you are considering renting an apartment in a certain building, why should you talk to the other tenants of the building before you make your final decision?

4. If you rent an apartment, why should you make a list of everything that is wrong with it before you sign the lease? Why should you ask the landlord to sign that list?

Rufus collected the information shown about apartments that are for rent in his town.

Monthly Rent	Number of Bedrooms
$ 900	3
1000	3
600	2
295	0*
450	1
1500	4
310	0*
425	1
575	2

*Studio or efficiency

5. What is a linear model of the form $y = a + bx$ that will allow Rufus to predict the number of bedrooms in an apartment in his town if he knows the monthly rental charge for the apartment?

6. Is there a good correlation between the monthly rental charge and the number of bedrooms?

7. Estimate the number of bedrooms in an apartment in Rufus' town that has a monthly rental charge of $500. Express your answer to the nearest whole number of bedrooms.

8. Do you think that an apartment with one bedroom and a monthly rental charge of $675 is fairly priced?

Rakesh collected the information shown about apartments that are for rent in his town.

Monthly Rent	Area (in square feet)
$495	675
550	780
395	500
650	900
440	610
900	1030
975	1200
760	995

9. What is a linear model of the form $y = a + bx$ that will allow Rakesh to predict the area in an apartment in his town if he knows the monthly rental charge for the apartment?

10. Is there a good correlation between the monthly rental charge for the apartment and the area?

11. Estimate the area in square feet in an apartment in Rakesh's town that has a monthly rental charge of $500.

12. Do you think that an apartment with 920 ft² and a monthly rental charge of $650 is fairly priced?

Gila collected the information shown about apartments that are for rent in her town.

Monthly Rent	Distance from Hardware Store (in miles)	Number of Bathrooms	Number of Closets
$1200	0.1	2.5	5
1100	4.1	2	4
400	0.2	1	1
375	1.0	1	1
400	1.9	1.5	2
700	3.1	2	3
1200	1.2	3	6
1575	1.8	3	6

13. Is there a good correlation between the monthly rental charge and the distance from the hardware store?

14. Is there a good correlation between the monthly rental charge and the number of bathrooms?

15. Is there a good correlation between the monthly rental charge and the number of closets?

MIXED REVIEW

1. Jeremy is a real-estate agent who has sold a house for $179,500. He earns 4.5% commission on the first $100,000 and 6.5% commission on any amount over $100,000. How much did he earn from the sale of this house?

2. Gladys has a balance of $1,455 on her MasterCard, which carries an APR of 17.5%. If she makes no further purchases with this card, how many months will it take her to pay off the balance? She can afford monthly payments of $170.

3. Use the Multiples-of-Salary Chart in the Reference Section to find the amount of life insurance that a 45-year-old man who has gross earnings of $65,000 would need in order to provide an income replacement of 60%.

4. Find the amount of the income tax refund or payment due to the nearest dollar for a family of two parents and three dependent children with wages of $44,000, $300 interest income, and dividends of $270. The family had medical expenses of $4,200, mortgage interest of $3,967 that they paid on their home, real estate taxes of $2100, state income tax payments of $1,680, and charitable contributions of $900. The husband made a tax-deductible contribution of $950 into his IRA. The wife did not work outside the home. The amount of withholding for the year was $5,985, and the filing status was married filing jointly.

Mr. and Mrs. Anderson are purchasing a car for $13,000. Find the monthly payment under the terms indicated in Exercises 5 and 6.

5. 8.5% annual interest rate, 20% down, 4 years to pay

6. 11% annual interest rate, 15% down, 5 years to pay

In Exercises 7 and 8, find the total financed price T for a home that costs $80,000 with a down payment of $20,000. Use the formula $T = nM + d + p$, where n is the total number of monthly payments, M is the monthly payment, d is the down payment, and p is the cost of points paid on the loan.

7. Bank A provides a 7% 30-year mortgage with $1\frac{1}{2}$ points.

8. Bank B provides a $7\frac{1}{2}$% 30-year mortgage with no points.

9. If your gross annual income is $43,100, how much monthly rent can you afford?

10. The following table shows the correspondence between the prices of several homes and the floor area of those homes. Use a graphing calculator to find a linear model of the form $y = a + bx$ that best represents the data.

Floor Area (in square feet)	1510	1800	2230	3100	4500
Price (in thousands of dollars)	90	105	113	129	146

CHOOSING A PLACE TO LIVE: SHOULD YOU BUY OR RENT?

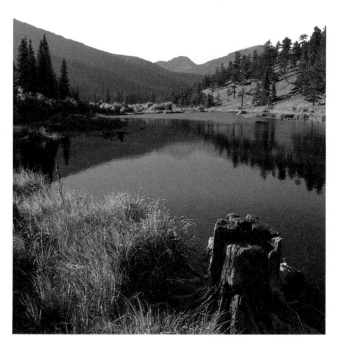

Yvette and Yvonne will be attending a large university in the Colorado mountains. Yvette wants to study architecture, and Yvonne will probably major in art history and secondary education. Since they are twins and are very close friends as well, they want to attend the same college.

The twins are now looking for a place to live in Colorado. Their parents are thinking about buying a condominium for the twins to live in rather than renting an apartment. There does not seem to be much to choose from in rental apartments in the resort and university town in the Colorado mountains.

If they buy a condominium, the twins' parents will make payments to build up their equity. If the twins rent an apartment, their parents will make rental payments every month but will have no equity invested after their daughters finish college. After their daughters attend the university, the parents can sell the condo. Alternatively, they can retain ownership of the condominium and use it as a vacation home in the summer and rent it to other university students during the school year.

Yvonne likes the idea of living in a condominium. She would like to learn to ski, but she may not be able to convince Yvette to learn too. Yvette is a serious student and plans to spend a lot of her time studying in the library.

OBJECTIVES: *In this lesson, we will help Yvette and Yvonne to:*

⬤ *Investigate the financial aspects of condominium ownership.*
⬤ *Compare the costs of renting or buying a particular home.*

HOUSING ALTERNATIVES

Rather than pay rent, Yvette's and Yvonne's parents prefer to buy their daughters' dwelling at school for the following reasons:

- The interest on their mortgage and their property taxes are tax deductible.*
- The money that they spend on mortgage payments provides them with equity in the property.

  *If the dwelling is not the family's primary residence, the tax advantages may be diminished.

A condominium may be an apartment in a multiple-unit building, a townhouse in a development, or a semidetached living unit with common areas. A **condominium** (also called a *condo*) is a hybrid of individual ownership and common living. As in a house, when you own a condominium, you own the individual unit in which you live. A condo, however, also offers the advantages of shared living. All the condo owners in a particular building or development share the common costs of maintaining the building or development and its common areas. Condominium ownership involves (1) individual ownership of one unit and (2) joint ownership with the other unit owners of the facilities and common areas (yard, pool, patio, hallways, and so on). All of the unit owners are members of the **owners' association,** which is responsible for management of the building, grounds, and common areas. The purchase of a condominium means that the buyer automatically shares in the administration of the building and property. The condominium owner pays a monthly fee to cover the costs of operating and maintaining all of the common areas.

ADVANTAGES AND DISADVANTAGES OF CONDOMINIUMS

Advantages	Disadvantages
• You build equity rather than just pay rent.	• You cannot always control major maintenance expenses, such as roof repair.
• You share costs of maintaining the building and common areas.	• The owners' association may need to approve changes that affect common areas, such as painting the exterior of your unit.
• You are not responsible for outside maintenance, such as lawn care and snow removal.	
• The condominium may include use of a swimming pool, party room, or other recreational facility.	• Resale of a condominium may be slower than resale of a single-family home.

Another housing alternative is a *mobile home*. A **mobile home** is a portable structure built on a chassis, designed to be used without a permanent foundation. It is usually parked in a permanent or long-term location. Unlike a recreational vehicle, it is designed as a year-round dwelling. In recent years mobile homes have grown in popularity among families with lower incomes. The mobile home owner must also have land on which to park the mobile home. Generally, the mobile home owner will rent space in a mobile home park and pay a monthly rental fee. The monthly rental fee for the land is likely to be less than the rental for an apartment. This is especially true in recreational and retirement areas or in other places where temporary housing is in demand. Ownership of a mobile home, therefore, combines features of ownership (the mobile home) and rental (the land).

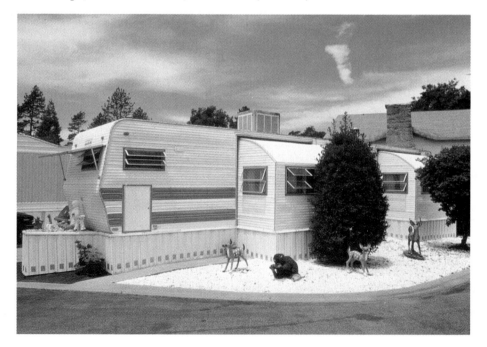

ADVANTAGES AND DISADVANTAGES OF MOBILE HOMES

Advantages
- The cost is usually much lower than for other forms of housing.
- A mobile home can be moved to a new location.
- The costs to maintain a mobile home are low.
- The small, compact size of a mobile home makes it easy to keep clean and tidy.
- Without much yard space, there is little outdoor maintenance needed.

Disadvantages
- You cannot park mobile homes wherever you please—there are strict zoning regulations governing where you can put them.
- Living space in a mobile home is limited.
- There is usually little yard space around a mobile home.
- Mobile homes are not well-protected in storms because of lightweight construction.

ALGEBRA *REVIEW*

Evaluate.

1. 105% of 1500

2. 105% of (105% of 1500)

3. 98% of (98% of 975)

4. $1450(1.08)^2$

5. $1450(1.08)^4$

6. $1450(1.08)^6$

Evaluate for x or y.

7. $x = \dfrac{98{,}000(0.01)(1 + 0.01)^{180}}{(1 + 0.01)^{180} - 1}$

8. $y = \dfrac{66{,}000(0.015)(1 + 0.015)^{360}}{(1 + 0.015)^{360} - 1}$

Use the values for x and y from Exercises 7 and 8 to find q and t.

9. $q = 180x + 24{,}500.00$

10. $t = 360y + 16{,}500.00$

Yvonne's and Yvette's family have decided to buy a condominium for the twins. The students will not have to take care of maintaining a yard and will have the use of the common facilities. Yvette is looking forward to picking out some appropriate furniture for it. Yvonne is looking forward to the ski season.

Ask Yourself

1. What is a condominium?

2. What advantages are there to buying a condominium rather than renting an apartment?

3. Why do many people with lower incomes choose to buy a mobile home?

SHARPEN YOUR SKILLS

SKILL 1

EXAMPLE 1   Although Yvonne's and Yvette's parents have decided to buy a condominium, they are still interested in comparing the monthly cost of buying a condo with the monthly rental charge on an apartment. They have located an apartment that rents for $480 per month and a condo with a purchase price of $80,000.

QUESTIONS   What is their monthly mortgage payment on the condo if they make a 20% down payment and borrow the remainder at 9% for 30 years? Which alternative would require a greater cash expenditure per month?

SOLUTIONS

If they make a down payment of 20%, they will borrow 80% of the purchase price.

$0.8(80,000) = 64,000$

$$M = \frac{Pr(1 + r)^n}{(1 + r)^n - 1}$$ Use the monthly payment formula.

$$M = \frac{64,000(0.0075)(1 + 0.0075)^{360}}{(1 + 0.0075)^{360} - 1}$$ $P = 64,000$; $r = (0.09 \div 12) = 0.0075$; $n = 30 \cdot 12 = 360$

$M = 514.96$ To the nearest hundredth

The monthly mortgage payment on the condo is $514.96, which is $514.96 − $480 = $34.96 more than the monthly rental charge on the apartment.

SKILL 2

EXAMPLE 2   Yvonne and Yvette will be at the university for 4 years. They can obtain only a one-year lease on the apartment in Example 1. When they renew the lease, the landlord will raise the rent by 5% each year.

QUESTION   Are the monthly mortgage payments on the condo larger than all the monthly rent payments over the course of 4 years?

SOLUTION

The mortgage payments on the condominium are fixed at $514.96 per month.

$4 \cdot 12 \cdot 514.96 = 24,718.08$

The mortgage payments for 4 years are $24,718.08.

The monthly rent on the apartment increases 5% each year.

Year 1: 480.00
Year 2: 480.00(1.05) = 504.00
Year 3: 480.00(1.05)^2 = 529.20
Year 4: 480.00(1.05)^3 = 555.66

Rental payments for 4 years are calculated as follows:

$$12(480.00 + 504.00 + 529.20 + 555.66) = 24,826.32$$

The rental payments for 4 years are $24,826.32.

The rental payments are $24,826.32 - 24,718.08 = \$108.24$ more than the mortgage payments for 4 years.

Yvonne wonders why any college student would rent an apartment if the monthly mortgage payments over the course of 4 years are less. Yvette points out that the sum of the monthly mortgage payments is not the only factor to be considered when making a decision on whether to buy or rent. To buy a condominium, you must have enough money up front for a down payment and closing costs. In addition, your monthly costs are increased by the fee paid to the condominium association and property taxes paid to the town. Finally, moving from an apartment is much simpler than selling a condo when the student is ready to leave the area.

--- **SKILL 3** ---

EXAMPLE 3 Sari found a condominium in which she wants to live. The owner will rent it to her for $1000 per month or sell it to her for $145,000.

QUESTION If she obtains a 15-year mortgage at 9% with a 20% down payment, will the total financed price or the total rental payments be greater over the 15 years? Assume that she can have a series of 3-year leases and that the rent increases by 10% at the end of each 3-year period.

SOLUTION
If she makes a down payment of 20%, she will borrow 80% of the purchase price.

$$0.8(145,000) = 116,000$$

$$M = \frac{Pr(1 + r)^n}{(1 + r)^n - 1}$$ Use the monthly payment formula.

$$M = \frac{116,000(0.0075)(1 + 0.0075)^{180}}{(1 + 0.0075)^{180} - 1}$$ $P = 116,000$; $r = (0.09 \div 12) = 0.0075$; $n = 15 \cdot 12 = 180$

$$M = 1176.55$$ To the nearest hundredth

Sari's monthly mortgage payment is $1176.55. In the course of 15 years she will pay $\boxed{\text{ANS}}$ (180) = $211,778.86 in mortgage payments. Remember to use the calculation of the monthly mortgage payment to many decimal places. The down payment is 0.2(145,000) = $29,000.

Total financed price of condo: 211,778.86 + 29,000 = $240,778.86

The rent will increase by 10% at the end of each 3-year period.

Years 1–3:	1000
Years 4–6:	$1000(1.1) = 1100.00$
Years 7–9:	$1000(1.1)^2 = 1210.00$
Years 10–12:	$1000(1.1)^3 = 1331.00$
Years 13–15:	$1000(1.1)^4 = 1464.10$

In each 3-year period she will make 36 payments. For 15 years the calculations are as follows:

36(1000.00 + 1100.00 + 1210.00 + 1331.00 + 1464.10) = 219,783.60

The rental payments for 15 years would be $219,738.60.

The total financed price of the condo is 240,778.86 − 219,783.60 = $20,995.26 more than the rental payments for 15 years.

Although it appears that Sari will pay less if she rents, rather than buys, the condo, renting might not be the best choice for her. There are tax advantages to buying that will effectively increase her income. After 15 years she will own the condominium and will no longer have monthly mortgage payments, whereas if she rents, she will have to continue to pay rent. The value of the condo is likely to increase over 15 years, so she may be able not only to recover the total financed price, but also to make a profit. If she rents an apartment for 15 years, she will never be able to recover any of the rent payments she makes.

___ **SKILL 4** ___

EXAMPLE 4 Sari would like to estimate the price for which she will be able to sell her house at the end of 15 years.

QUESTION If the value of her condo increases by 5% each year, for what price will she be able to sell it at the end of 15 years?

SOLUTION
At the end of 15 years she will be able to sell her condo for

$145,000(1.05)^{15} = \$301,444.59$

Since the total financed price of her condo is $240,778.86, she will make a profit of

301,444.59 − 240,778.86 = $60,665.73

TRY YOUR SKILLS

Rico found a condominium in which he wants to live. The owner will rent it to him for $675 per month or sell it to him for $125,000.

1. Assume that he can have a series of 3-year leases and that the rent increases by 8% at the end of each 3-year period. What is the total of all rental payments for 30 years?

2. If he obtains a 30-year mortgage at 12% with a 20% down payment, what is the total financed price?

3. Will the total financed price or the total rental payments be greater over the 30 years?

4. Determine the value of Rico's condo at the end of 30 years if it increases in value by 4% each year.

EXERCISE YOUR SKILLS

1. Why is it necessary to have an owners' association for condominium complexes?

2. Condominium owners' associations often have some strict rules that the individual home owners are obligated to obey. For example, the association may demand that each condo have the same exterior color. Why would associations have such rules?

3. If you had to choose between renting or buying a particular condo, which option would you choose? Why?

Compute the monthly payment for each condo if you can obtain a 30-year mortgage at 9% with 20% down for each. Compare your results with the rental charge for each condo, and state which alternative would require a greater cash expenditure per month.

4. Purchase price of $90,000; monthly rental charge of $750

5. Purchase price of $44,000; monthly rental charge of $220

6. Purchase price of $24,000; monthly rental charge of $230

7. Purchase price of $85,000; monthly rental charge of $435

8. Purchase price of $120,000; monthly rental charge of $1000

9. Purchase price of $140,000; monthly rental charge of $750

Assume that you can have a series of 3-year leases and that the rent increases by 9% at the end of each 3-year period.

10. In Exercise 4, will the total financed price or the total rental payments be greater over the 30 years?

11. In Exercise 5, will the total financed price or the total rental payments be greater over the 30 years?

12. In Exercise 6, will the total financed price or the total rental payments be greater over the 30 years?

13. In Exercise 7, will the total financed price or the total rental payments be greater over the 30 years?

14. In Exercise 8, will the total financed price or the total rental payments be greater over the 30 years?

15. In Exercise 9, will the total financed price or the total rental payments be greater over the 30 years?

16. How much will the condominium in Exercise 4 be worth at the end of 30 years if it increases in value by 4% each year?

17. How much will the condominium in Exercise 5 be worth at the end of 30 years if it increases in value by 4% each year?

18. How much will the condominium in Exercise 6 be worth at the end of 30 years if it increases in value by 5% each year?

19. How much will the condominium in Exercise 7 be worth at the end of 30 years if it increases in value by 5% each year?

20. How much will the condominium in Exercise 8 be worth at the end of 30 years if it increases in value by 6% each year?

21. How much will the condominium in Exercise 9 be worth at the end of 30 years if it increases in value by 6% each year?

MIXED REVIEW

1. Use the Rule of 72 to determine how long it will take Gregory's $3200 savings account to double in value. It is growing at an annual rate of 4.5%.

2. You have a loan of $22,500.00 at an annual rate of 8% for 4 years for which you have been making monthly payments of $549.29. How much money will you save if you prepay it at the end of 37 months?

3. Determine the monthly payment that must be made to reduce a $1888.88 VISA balance to zero in 28 months. The card carries an APR of 13.9%.

4. How many shares of stock in SRB company can you purchase for $25,000 given the current price of the stock is $52 $\frac{1}{2}$ per share? Ignore commission costs.

Calculate the tax savings for the following deductions. Assume that each family is in the 28% tax bracket and that their property tax is the given percentage of the assessed value of their home.

	Assessed Value	Property Tax	Amount of Property Tax	Mortgage Interest	Total Deductions	Tax Savings
5.	$100,000	2.0%		$ 8215		
6.	68,000	1.2%		4273		
7.	209,000	1.9%		14562		

1. What are the advantages and disadvantages of sharing an apartment with another person?

2. What should you inspect in any apartment before renting it?

3. What are the advantages and disadvantages of owning, rather than renting, a condominium?

Determine how much rent you can afford for each gross annual income.

4. $62,900	5. $42,500	6. $63,100	7. $80,000
8. $120,000	9. $38,500	10. $54,000	11. $95,750

Determine the gross annual income that you must earn to be able to afford each monthly rent.

12. $785	13. $1320	14. $1095	15. $840
16. $560	17. $490	18. $870	19. $390

20. Jillian used a broker to find an apartment that rents for $635 per month. The broker charges a fee of 10% of one year's rent. When she signs the lease, she must pay a security deposit equal to two months' rent as well as the first and last month's rent. What is her total cash outlay for the year? How much will she have to pay before she moves into the apartment?

21. Arnie pays $725 in rent and has monthly electric, water, and gas bills of $150, $75, and $90, respectively. He has the opportunity to move to another apartment in which utilities are included in the monthly rental fee of $990. His relocation bills will be approximately $75. Will he save money by moving to this apartment?

Room and board (food) cost $5800 in a dormitory on campus for the nine months of the school year.

22. Myrna would like to rent a furnished apartment near campus with her friend Lois. She found one that rents for $410 per month, including electricity, water, and gas. However, she must sign a lease for 12 months. She estimates that grocery expenses for two people will be about $300 per month. Should Myrna rent the apartment? Why or why not?

23. Erik is considering a furnished apartment exactly like the one that Myrna and Lois are considering. However, the rent is $460 per month and he will have no roommate. Should he rent the apartment? Why or why not?

David collected the following information about apartments that are for rent in his town.

Monthly Rent	Area (in square feet)
515	660
560	750
375	520
680	920
460	630
950	1095
995	1180
750	975

24. What is a linear model of the form $y = a + bx$ that will allow David to predict the area in square feet in an apartment in his town if he knows the monthly rental charge for the apartment?

25. Is there a good correlation between the monthly rental charge and the area?

26. Estimate the area in square feet in an apartment in David's town that has a monthly rental charge of $600. Express your answer to the nearest square foot.

27. Do you think that an apartment with 780 ft^2 and a monthly rental charge of $575 is fairly priced?

Carmen and Bicki are going to be roommates at college. They have found an apartment that rents for $600 per month and a condominium with a purchase price of $85,000.

28. What is their monthly mortgage payment on the condo if they make a 20% down payment and borrow the remainder at 9% for 30 years? Which alternative would require a greater cash expenditure per month?

29. Assume that Carmen and Bicki can have a series of 3-year leases and that the rent increases by 10% at the end of each 3-year period. Will the total financed price or the total rental payments be greater over the 30 years?

30. How much will Carmen and Bicki's condominium be worth at the end of 30 years if it increases in value by 6% each year?

31. Carmen and Bicki find another bank that will allow them to make a 25% down payment on the condominium and borrow the remainder at 8% for 15 years. What would be their monthly expenditure under this option?

32. Compare the total financed price in Question 28 to that in Question 31. Which is greater?

Determine how much rent you can afford for each gross annual income.

1. $79,200   2. $34,500   3. $55,600   4. $69,600

Determine the gross annual income that you must earn to be able to afford each monthly rent.

5. $695   6. $670   7. $365   8. $875

9. Patricia used a broker to find an apartment that rents for $525 per month. The broker charges a fee of 16% of 1 year's rent. When she signs the lease, she must pay a security deposit equal to 1 month's rent, as well as the first and last month's rent. What is her total cash outlay for the year? How much will she have to pay before she moves into the apartment?

Leonard collected the following information about houses selling in his town.

Monthly Rent	Number of Amenities
$ 625	6
1140	10
500	3
990	8
1095	9
480	3
325	1
750	8

10. What is a linear model of the form $y = a + bx$ that will allow Leonard to predict the number of amenities in an apartment?

11. Is there a good correlation between the price and the number of amenities?

12. Estimate the number of amenities in an apartment with a monthly rental charge of $600. Round your answer to the nearest whole amenity.

Jared found a condominium in which he wants to live. The owner will rent it to him for $850 per month or sell it to him for $140,000.

13. If he obtains a 15-year mortgage at 9% with a 20% down payment, will the total financed price or the total rental payments be greater over the 15 years? (Assume that he can have a series of 3-year leases and that the rent increases by 8% at the end of each 3-year period.)

14. Assuming the mortgage and lease conditions in Question 13, which alternative (renting or making monthly mortgage payments) would require a greater cash expenditure per month during the first 3-year lease?

15. Which alternative would require a greater cash expenditure per month during the fifth 3-year lease?

16. How much will Jared's condominium be worth at the end of 15 years if it increases in value by 4% each year?

CUMULATIVE REVIEW

1. You work 35 hours per week for 49 weeks. You get 3 weeks of paid vacation and 25% of your base salary in fringe benefits. You earn $4,850 per month. Determine your yearly earnings, including benefits.

2. The fixed costs for your business are $180 per week. The variable costs are $4.90 per item. What is the cost if you produce 95 items in 1 week?

3. Using the information in Exercise 2, determine your profit or loss for 1 week if you sell all the items for $8.75 each.

4. Determine the plan that would provide the lowest total financed price for a $20,595 car. Assume that the loan in Plan 1 is from a bank or other lending institution and that the loans in Plans 2 and 3 are from the car manufacturer's financing subsidiary.

	Plan 1	Plan 2	Plan 3
Rebate	$2000	$1200	0
Rate	11.25%	8.9%	3.5%
Time	36 months	36 months	36 months
Loan Amount	$18,595	$19,395	$20,595
Monthly Payment			
Total Financed Price			

5. Determine which plan yields the highest profit to the company in Exercise 4.

6. Use the Multiples of Salary Chart in the Reference Section to find the amount of life insurance needed by a 35-year-old man who has gross earnings of $30,000 in order to provide 75% income replacement.

7. If you buy 850 shares of stock at a price of $23\frac{1}{4}$, what will the total cost of the stock be? Ignore commission costs.

8. A family that is in the 31% tax bracket pays $5,300 in mortgage interest and a property tax that is 1.42% of the assessed valuation of their home, which is $230,000. Find their possible income-tax savings.

9. Determine how much monthly rent you can afford if your gross annual income is $55,500.

Compute the monthly payment for each condo on which you can obtain a 30-year mortgage at 9% with 20% down. Compare your results with the rental charge for each condo and state which alternative would require a greater cash expenditure per month.

10. Purchase price of $125,800; monthly rental charge of $800

11. Purchase price of $66,000; monthly rental charge of $600

12. If your condominium is currently worth $75,000 and it increases in value by 4% each year, how much will it be worth at the end of 20 years?

PROJECT 13–1: Utilities

Select one utility that is used in your home. Make telephone calls and visits to the utility's local office to answer the following questions.

1. How often does the utility company send bills to its customers?
2. How does the company measure its customers' usage? Does it have a meter in each customer's home?
3. In what units is each commodity measured? What is the cost per unit?
4. What state or local taxes are added to each bill?
5. Collect your family's bills for this utility for at least 3 months and find the average monthly usage and cost.

PROJECT 13–2: Choosing an Apartment

1. Find newspaper advertisements for 20 apartments that are for rent in your town.
2. For each apartment, find the number of square feet and the number of bedrooms.
3. Make a poster presenting the information that you gather.
4. Run a regression analysis to find a linear model of the form $y = a + bx$ that will allow you to predict the number of square feet in an apartment in your town if you know its monthly rental charge.
5. Is there a strong correlation between the number of square feet and the monthly rental charge?
6. Run a regression analysis to find a linear model of the form $y = a + bx$ that will allow you to predict the number of bedrooms in an apartment in your town if you know its monthly rental charge.
7. Is there a strong correlation between the number of bedrooms and the monthly rental charge?
8. Use your results from Exercise 4 to predict how many square feet are in an apartment in your town that rents for $600 per month.
9. Use your results from Exercise 6 to predict how many bedrooms are in an apartment in your town that rents for $450 per month.

Like distance, the absolute value of a number is always nonnegative. The absolute value of 5 can be thought of as the distance of 5 from 0 on the number line. There are two points that give this distance: 5 and -5. Therefore,

$$|5| = 5 \quad \text{and} \quad |-5| = 5$$

Solve each equation for x.

Example
$$|x + 1| - 3 = 5$$
$$|x + 1| = 8 \qquad \qquad \text{Add 3 to each side.}$$
$$x + 1 = 8 \quad \text{or} \quad x + 1 = -8 \qquad \text{Separate into two parts.}$$
$$x = 7 \quad \text{or} \qquad \quad x = -9 \qquad \text{Check in original equation.}$$

1. $|x| = 10$
2. $|x + 2| = 5$
3. $|x - 1| + 2 = 8$

4. $|x - 4| = 1$
5. $|x| + 1 = 5$
6. $|x| - 2 = 2.5$

An absolute value inequality has two conditions that must be satisfied.

Graph the solution of each inequality on a number line.

Examples   **a.** $|x + 1| < 5$ **b.** $|x + 1| \geq 5$

The equation $|x + 1| = 5$ means $x + 1 = 5$ or $x + 1 = -5$.
Thus $x = 4$ and $x = -6$ are boundary points for each inequality.

a. For $|x + 1| < 5$, points inside these boundaries satisfy the inequality.

```
    ◄─┼─┼─○─┼─┼─┼─┼─┼─┼─┼─┼─○─┼─┼─┼─►
     -8 -7 -6 -5 -4 -3 -2 -1  0  1  2  3  4  5  6  7
```

b. For $|x + 1| \geq 5$, points on and outside these boundaries satisfy the inequality.

```
    ◄─┼─┼─●─┼─┼─┼─┼─┼─┼─┼─┼─●─┼─┼─┼─►
     -8 -7 -6 -5 -4 -3 -2 -1  0  1  2  3  4  5  6  7
```

7. $|x| < 5$
8. $|x - 1| < 7$
9. $|x| > 2$
10. $|x + 1| \geq 3$

Graph each equation on a coordinate system.

Example $y = |x|$

$$y = x \quad \text{for} \quad x \geq 0$$
$$y = -x \quad \text{for} \quad x < 0$$

Thus we have the graph shown.

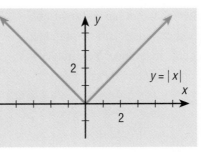

11. $y = -|x|$
12. $y = |x - 1|$

13. $x = |y|$
14. $y = 2|x|$

Budgeting

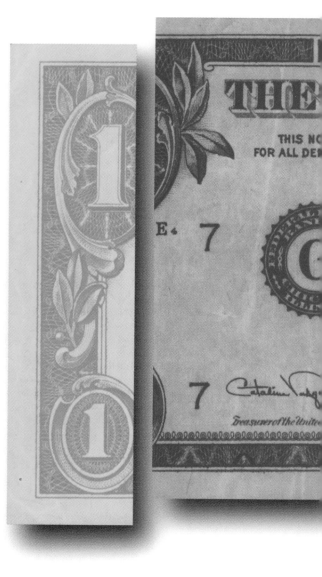

IN THIS FINAL CHAPTER, WE WILL HELP the young people from the previous chapters use what they have learned in the construction of a budget to control their expenses. Especially expenses that recur, such as housing, food, and transportation.

Each individual and family has its own budgetary problems and its own solutions to those problems. Nevertheless, the solutions have common characteristics. The primary characteristic is that a budget is a means of taking control of your money rather than allowing it to take control of you.

This control is easiest to exert over monthly bills. However, you will need to make provisions for large or unexpected expenses. Earlier chapters have shown how some unexpected expenses can be managed by buying adequate insurance. You may also need to create and maintain a cash reserve.

Beyond paying your monthly bills, you will want to spend some of your money on personal needs and wants. An intelligently devised budget can help you to do this without damaging your economic health and security. It will even show you ways in which you can put aside some money for savings.

A budget should respect the needs and wishes of all family members. The adults of the family bear the largest responsibility, since they have more experience and knowledge than the younger family members.

This chapter will concentrate on some aspects of consumer information and money management not covered in earlier chapters. It will also provide specific examples of sample budgets that you may be able to use as models for preparing your own budget.

BUDGETING HOUSING EXPENSES

eople are always asking, "Where does the money go?" You have probably heard someone say something like, "Last week I had $50. Now it's gone, and I can't believe I spent it all." As you grow older, that $50 can become $100 or $500. To use money successfully, you must keep track of how you spend it.

Sylvia's parents learned the hard way. Her mother and father had a lot of disagreements about money before their divorce. Now Sylvia and her mother watch their expenses closely. Last year, Sylvia bought a dress on the layaway plan.

Sylvia is not sure how she will meet her expenses in college. She hopes that she will not have to drop out after a year or two as some of her friends did. Her father will help her, but she does not yet know just what all of her expenses will be. Further, she does not know how much she might need to earn herself to cover the expenses that her mother and father cannot pay.

Maria has had some experience in planning her finances by making car and insurance payments. As a result, she thinks that it must be much more complicated for her parents, who have to pay for a house, car, food, clothing, medical care, and even unexpected expenses. Maria and Sylvia both realize that among all the expenses that an individual or family has to budget carefully for, housing expenses are very important.

OBJECTIVES: *In this lesson, we will help Sylvia and her mother:*

- *Decide why a budget is needed.*
- *Study what goes into a budget.*
- *Learn how to calculate property taxes.*
- *Control monthly housing expenses.*

BUDGET GUIDELINES

A **budget** is a plan of action that shows how you will plan the distribution of your money to meet expenses and achieve financial goals. Sometimes Sylvia wishes she had never heard of budgets. First you have to have a plan. Then you have to write down all the amounts you spend and what they were for. Then you have to compare your expenditures with your plan. Then you have to figure out what went wrong. (Something always does seem to go wrong.) Sylvia's mother insists that without planning, they both would be worse off. In fact, Sylvia and her mother have worked out the following guidelines to help them budget their money.

- Have clear long-term and short-term goals.
- Recognize your standard of living and how it affects goals.
- See how goals change depending on your stage in life.
- Identify the alternatives when faced with decisions and choices.

MAKING THE GUIDELINES WORK

Establishing Goals　　One of Sylvia's long-range goals is to graduate from college. She knows that she will have a much better chance of finding work that is satisfying and that pays well if she completes college.

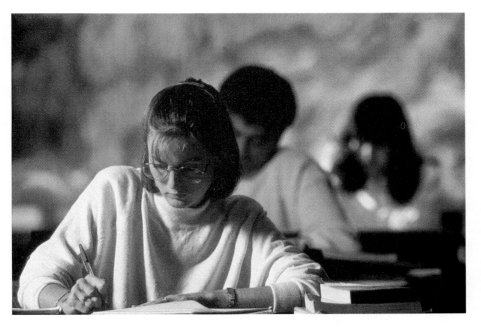

Sylvia's mother is helping her understand that, to meet her long-term goals, she must be willing to work toward some short-term goals, such as the preparation of their monthly and yearly budgets. These budgets will allow them to save some of the money that Sylvia will need for college.

Standard of Living　　In discussions with their parents, both Sylvia and Maria have heard the expressions "standard of living," and "quality of life." A **standard of living** is the economic level at which an individual or family lives. People with a high standard of living can buy more of the things that most of us want.

Quality of life is another matter. It refers to the happiness and satisfaction that we find in our lives. This doesn't depend as much on money as on whether we are doing things that enrich our lives. For most of us, both standard of living and quality of life depend, in part, on whether we have adequate control over our income and spending.

Stage in Life   Your stage in life plays a large part in your budgeting plans. Single people just starting out may spend much of their disposable income on entertainment and career clothes. A young family may have less disposable income because they need to spend more on insurance and medical care and to save for their children's education. A retired person living on a fixed income may not be able to afford luxury items.

Your stage in life can play a big role in determining your housing needs as well. A single person may live in a small apartment or share expenses with roommates in order to live in a city. Married couples with children will probably be more concerned about the school system available to them than couples without children. Elderly people may be more concerned about the ease of access in their homes, climate, public transportation, and the medical services available in their community.

Preferences   Different people have different styles of living. Some do not want the responsibilities that come with home ownership. Of those who buy, some like old houses, some like new. Yvette and Yvonne want a condo with easy maintenance, a pool, and a health club. Sylvia's mother wants a small house in a clean, quiet neighborhood.

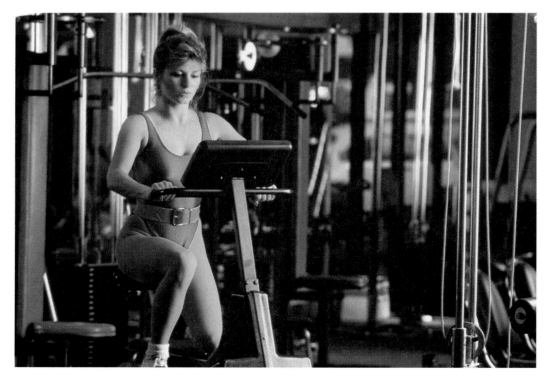

HOUSING EXPENSES

The more that Sylvia and her mother study budget guidelines, the more they notice how much those guidelines force them to concentrate on one particular aspect of their lives: their choice of housing. They already know that they should try not to spend more than 25 to 30 percent of their income on rent or mortgage payments. But what about other housing expenses? For example, if they decide to buy rather than rent, how much can they expect to pay in real-estate taxes? Whether they own or rent, they want to insure their belongings. If they own, they will have to have home insurance as well.

And what about **utilities**? Most home owners must pay for energy to regulate the temperature of their homes. This may mean heat in winter and air conditioning in summer. Other utility costs are for electricity, telephone, water and sewer, and regular monthly services. Some people might include cable TV in this category.

A home owner should also budget for home maintenance. This would include plumbing, electrical repairs, and perhaps landscaping. Sylvia and her mother know that maintenance expenses are not always predictable and may be higher if they buy an older home with appliances that will need to be replaced in a year or two. They will need to save to make sure they have money in the event of an expensive maintenance problem.

Of course, even utility expenses are not exactly the same every month. Extended use of the television, air conditioners, and other appliances can add considerably to your electric bill, and we all know how telephone costs vary according to use.

Ask Yourself

1. What is a budget?

2. Why is a budget useful?

3. What is the difference between standard of living and quality of life?

4. Why are housing expenses such an important factor in the preparation of a budget?

ALGEBRA *REVIEW*

Write each as an equation.

1. 6 is 5% of x

2. p is 10% of 80

3. 2.5 is x percent of 500

4. Find p in Exercise 2.

5. Find x in Exercise 3.

Write an equation and solve.

6. 50 is what percent of 75?

7. 2.8 is 5% of what number?

8. What is 2.6% of 40?

Write a linear equation in two variables to express the relationship.

9.

cost (c)	$8.00	$14.00
tax (t)	$0.28	$ 0.49

10.

Selling price (e)	$90	$110
Sale price (a)	$76.50	$ 93.50

SKILL 1

The federal government with all its services is financed largely through taxes on the incomes of individuals. Many states tax incomes and also raise money through sales taxes. Local governments, that is, cities and towns, raise large amounts of money through taxes on property.

There are two concepts involved in property taxes: assessed value and mill rate. The **assessed value** is the amount that the local government assigns to a house or other property for tax purposes. Generally, this amount is less than the owner would receive if the house were sold. It is important that the assessed value of similar properties be consistent throughout the tax area.

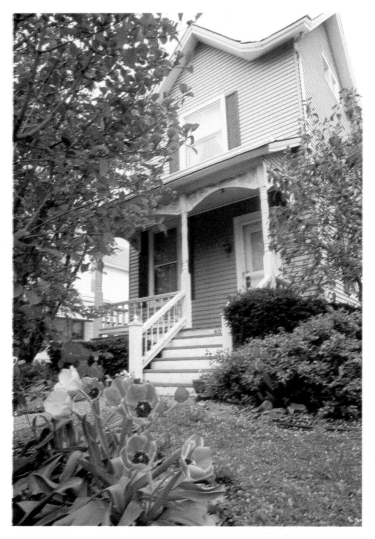

Houses may be reassessed after major improvements, such as a new room or the addition of a deck.

The **mill rate** is the amount of tax per thousand dollars of assessed value. In this respect it is somewhat like percent except that percent is in hundredths while mill rate is in thousandths.

EXAMPLE 1 Sylvia and her mother are considering buying a house that is assessed at $74,000. The tax rate in the town is 18.3 mills.

QUESTION What property taxes will Sylvia's mother have to pay?

SOLUTION
For every thousand dollars, Sylvia's mother will have to pay $18.30 in taxes.

$$T = am$$ where T = property tax
a = assessed value in thousands
m = mill rate

$$T = 74(18.3)$$
$$= 1354.20$$

The property tax is $1354.20 per year.

The expenses you have for operating your home cannot be avoided, but they can be controlled. Heating bills, for example, can be budgeted in such a way that the oil company sends you a monthly bill rather than a bill each time a delivery is made. This flattens out the payments and helps you avoid surprises. A careful look at telephone bills will tell whether you are making more long-distance calls than you need. Many times you can send a message by mail and save money. Finally, if you compare electric bills from month to month, you will see whether you have saved money by conserving the use of electricity.

How can a home owner begin to keep track of utility and other operating expenses? The best way to start is to prepare a table like the one shown.

EXPENSES	
Bill Type	**Payment Amount**
Energy for heat or air conditioning	
Electricity for household appliances	
Water	
Telephone	
Home maintenance	

In the table above, home maintenance is not a utility expense. It refers to expenses such as repairs and removal of leaves in the fall and snow in the winter. Like a utility expense, home maintenance is a regular expense and should be included in a budget.

EXAMPLE 2 Sylvia's mother remembers keeping a record of daily expenses on their recent trip and using the results to prepare a travel budget for her clients at her travel agency.

QUESTION How can Sylvia and her mother use a table of monthly housing expenses to prepare a budget for those expenses?

SOLUTION

They use a spreadsheet program to track their expenses for 4 months and then find the average of those expenses.

	A	B	C	D	E	F	G
1		Feb.	March	April	May	Total	Average
2	Heat/Air cond.	132	87	68	56	343	85.75
3	Electricity	50	48	49	52	199	49.75
4	Water	17	15	14	18	64	16.00
5	Telephone	38	45	62	19	164	41.00
6	Home Maintenance	62	17	40	0	119	29.75
7	Total	299	212	233	145	889	222.25

@SUM(B2..E2) → F2
+F2/4 → G2
@SUM(B2..B6) → B7

On the basis of this 4-month record, Sylvia and her mother decide to budget $230 per month for utilities and home maintenance. They will pay particular attention to the energy budget, since the trial period did not include all of the winter months.

Using the ideas discussed in Chapter 12, they also plan to budget no more than about 28% of their family income, or about $600 a month, to cover the mortgage payments on the house they hope to buy. If they decide to rent instead of buy, then they will look for a house or apartment that rents for $600 per month. Then their entire housing budget, including the monthly mortgage payment or rent, will be $230 + $600, or $830.

TRY YOUR SKILLS

Write each decimal number as a mill rate.

1. 0.0183   2. 0.0200   3. 0.0016   4. 0.0102

5. The mill rate in Smithtown is 23.4. What will be the yearly property tax on a house in Smithtown that has an assessed value of $65,000?

6. Mr. and Mrs. Gant pay an annual property tax of $1827.50. Their house has an assessed value of $85,000. Find the mill rate.

7. Construct a monthly housing budget for a family that spends the following amounts in a 4-month period: energy, $120, $108, $130, $126; electricity, $31, $28, $37, $40; water, $12, $13.50, $15.50, $21; telephone, $80, $36, $47, $51; home maintenance, $43, $72, $18, $10.

8. Suppose that the family of Exercise 7 has an annual income of $32,000. They wish to not spend more than 25% of their income on monthly rent. Find the total monthly housing budget of the family when the monthly rent is included.

EXERCISE YOUR SKILLS

KEY TERMS

assessed value
budget
mill rate
quality of life
standard of living
utilities

1. Why do you think that the assessed value of a house is usually less than the amount that could actually be received from the sale of the house?

2. What factors might cause the mill rate of a community to rise or fall?

3. Do you think that 4 months is a satisfactory period of time to obtain a useful estimate of your monthly housing expenses? For which expenses would it be satisfactory and for which expenses would it not be satisfactory?

4. What will be the annual property tax on a house that is assessed at $78,000 with a mill rate of 25.7?

5. What will be the quarterly payment of a property tax on a house with an assessed value of $41,000 and a mill rate of 13.6?

What is the mill rate of a house assessed at the following values with the given annual property tax?

6. $53,000; $678.40

7. $160,000; $1328

8. $112,000; $2430.40

9. $99,550; $1582.85

Construct a monthly budget for the housing expenses of each of the following families based on their actual expenditures over the indicated time periods.

10.

The McGrew Family	Sept.	Oct.	Nov.	Dec.	Total	Average
Heat/Air cond.	$71	$94	$99	$118		
Electricity	34	55	40	48		
Water	24	16	18	18		
Telephone	34	32	59	74		
Home Maintenance	9	28	43	40		
Total						

11.

The Lopez Family	June	July	Aug.	Sept.	Total	Average
Heat/Air cond.	$16	$58	$76	$ 70		
Electricity	59	98	64	71		
Water	19	21	30	28		
Telephone	40	22	35	33		
Home Maintenance	0	31	87	104		
Total						

12. Construct a monthly budget for a family that has the following pattern of expenditures: energy, $1400 per year; electricity, $36 per month; water $70 every 3 months; telephone, $28 per month; home maintenance, $600 per year.

13. Suppose that the income of the McGrew family in Exercise 10 is $1600 per month and that they wish to budget no more than 25% of their income for monthly rent payments. Find the family's total monthly housing budget including monthly rent payments.

14. The Lopez family in Exercise 11 has an income of $40,000 a year and will not qualify for a mortgage if their mortgage payments are greater than 28% of their monthly income. Find the family's total monthly housing budget including mortgage payments.

MIXED REVIEW

1. Alicia has her own business making figurines in her basement and selling them from her home. She sells the figurines at $35 each. Last year her weekly production was 10, not including two weeks that she took off for vacation. The cost of materials was $3.75 per figurine and 95% of the figurines were sold. What were her earnings for the year?

2. The island of Manhattan was bought from the Manhattan Indians by Peter Minuit, reportedly for the equivalent of $24. If instead of being used to buy Manhattan, the $24 had been invested at 6% compounded annually for 370 years (until the year 1996), do you think that the 1996 value of the investment would be worth more or less than the current value of the island of Manhattan? Explain your answer.

3. A stock's closing price was $52\frac{7}{8}$, $\frac{3}{4}$ of a point lower than yesterday's closing price. What was yesterday's closing price?

4. A car that was bought as a used car for $9000 depreciated over a 10-year period at the rate of 10% of the original price per year. What was the value of the car at the end of the seventh year?

Brenda drives 10,000 miles a year. Last year her expenses for maintenance and repairs were $380 and her insurance was $1392. There are no loan payments on the car. Her cost of gasoline was $0.05 per mile. Find each of the following.

5. Her total annual operating expenses

6. Her average cost per mile to operate the car

7. Use the US Mileage Chart in the Reference Section to find approximately how far Tulsa, Oklahoma, is from Boston, Massachusetts.

8. A motel charges $40 a night for a single individual and $50 for a couple. Mr. and Mrs. Price stayed at the motel. The charge for a video rental was $5.00 and the room tax was 6%. The couple rented two movies. Find the cost for them to stay at the motel for 3 nights.

14-2 FOOD BUDGETS AND UNIT PRICES

When Olivia's parents were making plans to sell their house, they reviewed all of their finances. Since Olivia was interested, they decided that she was old enough to learn about household budgets. Olivia was surprised to discover how many decisions must be made when buying food and how much food costs. This doesn't mean that good food is necessarily expensive. Sometimes just the opposite is true. Some very nutritious foods are reasonably priced. Some prepared foods that contain a lot of fat or preservatives may be expensive.

Olivia not only studied the family's food budget, but also did some of the shopping, comparing prices and determining unit prices. She even began to feel like a parent when her little brother wanted to buy a lot of expensive snacks. She told him that the snacks were not good for him nor for their food budget.

Olivia's friend Jonathan helps keep the financial records for a local retail food store called Mrs. Harney's Pie Shop. The pies are not dessert pies but meat pies; they contain beef and chicken. Jonathan showed Olivia one reason why Mrs. Harney's unit costs are higher when she produces the pies in smaller quantities.

OBJECTIVES: *In this lesson, we will help Olivia and Jonathan to:*

Use proportions to determine unit costs based on retail prices.
Develop a food budget for a family.
Determine a business person's unit costs that incorporate the effect of fixed costs.

WHAT IS THERE TO EAT?

There was a time when American families ate at least dinner together. Now many families are like Olivia's. Very often, some are working or busy at mealtimes, and the family seems to be going in all directions. So families don't have as many meals together as they once did.

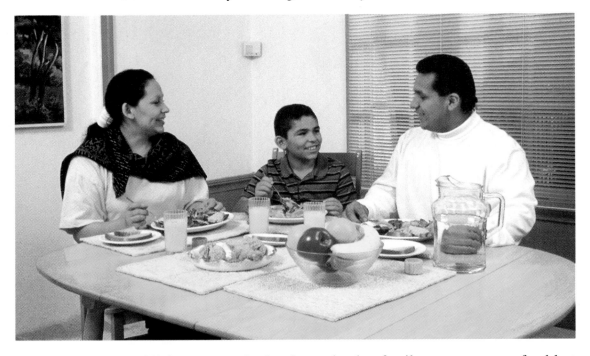

Olivia was surprised to learn that her family spent more on food last year than on any other single item in their budget. They frequently eat out, and Olivia's father regularly brings home prepared meals. That may change with the purchase of a new home. The family will have to plan the budget more carefully.

To save on food, Olivia's family is trying to follow these guidelines:

- Make a weekly meal plan that includes nutritional values, personal preferences, preparation time, and cost.
- Watch for sales on products that are needed.
- Make a list and stick to it. Avoid impulse buying.
- Compare prices among brands and stores.
- Buy in quantity when possible.
- Use coupons.

As Olivia continued shopping, she became more interested in unit prices. The **unit price** is the price per unit of measurement, for example, $1.25 per pound or $0.23 per ounce. Unit prices are usually marked, sometimes on the item, often on the store shelf beneath the product. However, this is not the case in all stores or for all products.

To know what you are buying, it is necessary to know something about measurements. As you know, there are two widely used systems of measurement. The system that is most common in the United States uses inches, feet, and miles for distance; ounces and pounds for weight; and pints, quarts, and gallons for liquid volume. The **metric system** uses centimeters, meters, and kilometers for distance; grams and kilograms for mass; and liters for liquid measure. The table below shows some of the conversions between the two systems.

Common Conversion Factors
1 centimeter (cm) = 0.39 inches (in.)
1 meter (m) = 39.4 inches
1 kilometer (km) = 0.62 miles (mi)
1 gram (g) = 0.035 ounces (oz)
1 kilogram (kg) = 2.2 pounds (lb)
1 liter (L) = 1.06 quarts (qt)

Ask Yourself

1. What are three things to consider in preparing a food budget?

2. What is an example of a nutritious food that is not expensive?

3. What is an example of an expensive food that is not nutritious?

4. What does unit price mean?

5. About how many grams are in an ounce?

6. In many communities supermarkets are required to display unit prices, which is an additional cost for the supermarket. If they were allowed to do so, do you think that many supermarkets would discontinue unit pricing?

ALGEBRA *REVIEW*

Solve the proportion $\dfrac{a}{b} = \dfrac{x}{y}$ **for**

1. x 2. b 3. a

Solve the proportion $\dfrac{2}{x} = \dfrac{y}{7}$ **for**

4. x 5. y

6. For the proportion in Exercises 4 and 5, substitute values for x and make a table for x and y.

7. Graph the points found in Exercise 6.

8. Use a graphing calculator to graph the proportion in Exercises 4 and 5.

9. What is the difference between your graph and that of the graphing calculator?

The Pythagorean theorem states that in a right triangle, $c^2 = a^2 + b^2$, where c is the length of the hypotenuse and a and b are the lengths of the legs.

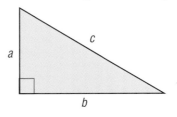

10. Find c if $a = 3$ and $b = 4$.

11. Find b if $a = 5$ and $c = 13$.

12. Find c if $a = 40$ and $b = 9$.

13. Find a if $c = 25$ and $b = 7$.

SKILL 1

Olivia went to the supermarket with her shopping list. She was surprised to find that some unit prices were not listed. Fortunately, she had her calculator.

EXAMPLE 1 Olivia found the following three sizes of Whole Grain Flakes cereal.

    a. 12 oz for $1.98

    b. 18 oz for $2.69

    c. 24 oz for $3.39

QUESTION What is the unit price for each package?

SOLUTION
The unit price in this case is the price per ounce. You can think of this as a proportion problem. Remember that a proportion is an equality of two ratios. Use this method to find the unit price of each of the packages.

a. $\dfrac{1.98}{12} = \dfrac{u}{1}$ The price of 12 ounces ($1.98) divided by 12 equals the price of 1 ounce (u) divided by 1.

$u = 0.165$

The unit price is $0.165 per ounce.

b. $\dfrac{2.69}{18} = \dfrac{u}{1}$

$u = 0.149$

The unit price is $0.149 per ounce.

c. $\dfrac{3.39}{24} = \dfrac{u}{1}$

$u = 0.141$

The unit price is $0.141 per ounce.

The 24-ounce size has the lowest unit price. In terms of price, it is the best buy.

SKILL 2

Olivia has found a number of imported items that are measured in the metric system. In fact, she has developed an interest in comparing measurements. Now her friends ask her to do conversions.

EXAMPLE 2 A friend of Olivia's visiting from Argentina bought 12 pounds of potatoes and wondered how many kilograms that is.

QUESTION How would you convert 12 pounds to kilograms?

SOLUTION

To convert 12 pounds to kilograms, use the pound/kilogram conversion on one side of a proportion.

$$\frac{2.2 \text{ lb}}{1 \text{ kg}} = \frac{12 \text{ lb}}{t \text{ kg}}$$

$$2.2t = 1 \cdot 12, \text{ or } 12 \qquad \text{If } \frac{a}{b} = \frac{c}{d}, \text{ then } ad = bc.$$

$$t = 12 \div 2.2$$

$$t = 5.5 \text{ kg}$$

12 pounds is equal to 5.5 kilograms

SKILL 3

In Chapter 4, you learned how a business calculates its variable costs by multiplying the number of items produced by its unit cost. For simplicity, it was assumed that the unit cost was constant. However, when a business needs to construct an accurate budget, it needs to account for the fact that the unit cost changes with the change in production volume.

Olivia's friend Jonathan is keeping the financial records at Mrs. Harney's Pie Shop. He taught Olivia something about unit costs from the seller's side. It is not only packaging that helps to determine unit cost but also quantity.

EXAMPLE 3 Including equipment, rent, salaries, and utilities, Mrs. Harney's Pie Shop has fixed expenses of $1250 a week. When fixed expenses are ignored, the unit cost is $2.30 per pie.

QUESTION Including fixed and unit costs, what is the unit cost for the pies in quantities of 1000, 2000, 3000, and 4000?

SOLUTION

The total cost is found by using the equation

$$c = f + un \qquad \text{where } c \text{ is the total cost}$$

f is the fixed cost

u is the unit cost per pie, not including fixed expenses

n is the number of pies produced

Substituting the given quantities gives the following values for the total cost.

$$c = 1250 + 2.30(1000) = \quad 3{,}550$$
$$c = 1250 + 2.30(2000) = \quad 5{,}850$$
$$c = 1250 + 2.30(3000) = \quad 8{,}150$$
$$c = 1250 + 2.30(4000) = \quad 10{,}450$$

The unit cost can be found by dividing each total cost by the given quantity.

$3{,}550 \div 1000 = \$3.55$ per pie for 1000 pies

$5{,}850 \div 2000 = \$2.925$ per pie for 2000 pies

$8{,}150 \div 3000 = \$2.717$ per pie for 3000 pies

$10{,}450 \div 4000 = \$2.613$ per pie for 4000 pies

You can see how the unit cost decreases as the number of pies increases.

TRY YOUR SKILLS

1. Find the unit price of 8 ounces of cheese for $3.65.
2. Find the unit price of 0.75 liters of skim milk for $0.60.
3. Find the better buy for ketchup: 14 oz for $1.65 or 20 oz for $2.15.
4. Convert 54 kilograms to pounds.
5. Convert 17 ounces to grams.

EXERCISE YOUR SKILLS

1. What might lead a family to overspend its food budget?
2. What are three ways to economize when shopping?
3. Why are some foods more expensive than others?

The following are the prices and sizes of products that you might find in a grocery store. For each one, use a proportion to calculate the unit price and decide which is the best buy in terms of cost.

4.	Oil	8 oz	10 oz	15 oz
		$1.10	$1.29	$1.89
5.	Orange juice	6 oz	12 oz	18 oz
		$0.79	$1.29	$1.79
6.	Tuna	6 oz	13 oz	16 oz
		$0.59	$1.56	$1.76
7.	Raisins	12 oz	14 1-oz pkgs	16 oz
		$1.98	$3.04	$3.20
8.	Honey	4 oz	8 oz	1 lb
		$0.80	$0.96	$1.33

Convert each of the following.

9. 6.4 kilograms to pounds
10. 2.4 ounces to grams
11. 4 quarts to liters
12. 1.5 pints to liters

Mrs. Harney has expanded her business to include lasagna and manicotti take-out. Fixed costs are $940 per week for each of the dishes (including rent, salaries, utilities, and equipment). When fixed costs are ignored, the unit costs are $1.75 for each lasagna dinner and $1.45 for each manicotti dinner.

13. Allowing for the fixed costs, find the unit cost for lasagna for the quantities 500, 1000, 1500, and 2000.
14. Allowing for the fixed costs, find the unit cost for manicotti for the same quantities.

MIXED REVIEW

1. A company plans to hire a new production manager at a starting salary of $53,000 a year. Of the company's 260 operating days, the new manager will have 10 vacation days and 8 holidays. How much of the manager's salary is the company paying for non-working days?

2. Find the amount of money Caitlyn can borrow at an annual rate of 9.5% for 4 years. She can afford a monthly payment of $280.

3. Use the formula $t = 11,743.50 + 0.31(I - 51,900)$ to find the tax owed by a single taxpayer with a taxable income of $132,000.

Terry collected the following information about the amenities offered by several homes that are available for purchase. These included items such as fireplaces, security systems, and jacuzzis.

Number of amenities	7	7	5	3	2	1	1	1
Price, in thousands	$210	$234	$170	$160	$110	$100	$125	$96

4. Use a graphing calculator to represent these data by a linear model of the form $y = a + bx$.

5. Use the linear model that you found in Exercise 4 to decide whether you think that $200,000 is a fair price for a house with six amenities.

6. Determine the total financed price on a $110,000 house that you plan to finance at 7.2% for 30 years with a 30% down payment.

Randolph Dillon has set a budget for himself of $620 a month but spent $648 during the month.

7. By how much was he over budget?

8. By what percent was he over budget?

9. Randolph works from 3:00 P.M. until 6:00 P.M. 5 days per week. How many hours per week does he work?

10. How do you write $97.83 in words on a check?

11. Angela earns $125 in gross income each week. She is single and claims no exemptions. How much will her employer withhold for income tax?

Mr. and Mrs. Schwartz collected the following information about two-bedroom apartments facing south on different floors of a high-rise apartment building.

Floor	1	5	10	15	20	25	30
Monthly Rent	$380	$430	$440	$410	$470	$500	$510

12. What is a linear model of the form $y = a + bx$ that will allow the Schwartzes to predict the rent for any apartment in the same line?

13. Suppose that two apartments become available, one on the 17th floor for $430 and the other on the 26th floor for $505. Which apartment seems more attractively priced?

A budget is like a road map. Both can help you get where you want to go, but only if you use them and refer to them from time to time. Is this the road we planned to take? Will it take us where we want to go? Is this the amount we planned to spend? Are we staying within our budget? In spite of our efforts to live within our budgets, we sometimes find that "there is too much month left at the end of the money."

A budget is a plan of action that shows how you will spend and receive money over a specific time period. To budget well, you must first believe that budgeting is important. The young people and families in our stories have shown us and told us many of the benefits of budgeting. Here are some of their reminders.

- Budgets summarize your finances.
- Budgets encourage a sensible use of income.
- Budgets require an examination of goals, values, and priorities.
- Budgets help you get the maximum value from expenditures.
- Budgets help you live within your income.
- Budgets help you adjust to unforeseen expenses.
- Budgets help you prepare for income taxes.

In this lesson you will see what an overall family budget might look like and how a family might use it to review expenses and manage money effectively.

OBJECTIVES: In this lesson, we will:

● *Consider the reasons why a budget is useful.*
● *Review the categories of a typical family budget.*
● *Learn to make adjustments in a budget.*

A **cash shortage** means that you don't have enough money to cover some of your expenses. Cash shortages can occur in any family. Often they are due to large, irregular expenses such as automobile repairs. Sometimes a cash shortage occurs because a person loses his or her job. You can try to plan for irregular expenses by setting money aside in a special savings account.

The table shows typical budgets for three different levels of after-tax income.

	Level A	Level B	Level C
Total family budget	$14,808	$19,045	$27,181
Food	3,639	4,473	5,961
Housing	4,773	6,321	9,271
Clothing	891	1,183	2,036
Transportation	3,715	4,907	6,655
Health care	770	887	1,173
Entertainment	747	1,008	1,689
Personal care	153	196	299
Miscellaneous	120	$70	97

Some families will need to include other items in their budget, such as life insurance on the principal wage earner.

The pie chart shows the results of a government survey on American spending decisions in a recent year.

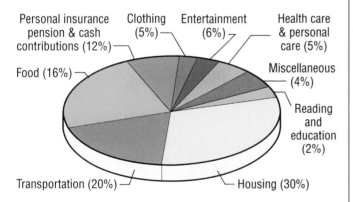

Personal insurance pension & cash contributions (12%) Clothing (5%) Entertainment (6%) Health care & personal care (5%)

Food (16%) Miscellaneous (4%)

Reading and education (2%)

Transportation (20%) Housing (30%)

Ask Yourself

1. What are three reasons for preparing a budget?

2. What are two expenditures that might cause a cash shortage?

3. What are the three largest items in most family budgets?

SKILL 1

Many people like to know not just how much they are spending in different budget categories, but also what the percentages are for the different categories. If one budget area such as food or housing takes too large a percentage of your money, this could be a reason why you have trouble with other categories.

EXAMPLE 1 Use the Level A budget on page 637.

QUESTION How would you find the percentages for the categories listed in the budget?

SOLUTION

The base for computing the percentages is the total budget, $14,808. The food amount is $3639.

What percent of $14,808 is $3639?

$$x \cdot 14{,}808 = 3639 \qquad \text{Let } x \text{ represent the percent.}$$

$$x = \frac{3639}{14{,}808}$$

$$x = 0.246$$

Food is about 25% of the total family budget. The rest of the percentages rounded to the nearest percent are given in the following table.

BUDGET CATEGORIES AND PERCENTAGES		
	Budget	**Percent**
Total family budget	$14,808	100%
Food	3,639	25%
Housing	4,773	32%
Clothing	891	6%
Transportation	3,715	25%
Health care	770	5%
Entertainment	747	5%
Personal Care	153	1%
Miscellaneous	120	1%

It is important and useful to know what percent each category of expense represents compared to the entire budget. However, the real test of your budget is how well you stick to it. Are you spending more than you should or less than you are allowed?

EXAMPLE 2 It is most important to keep track of your expenses and then, at the end of each month, determine whether you are over or under budget.

QUESTION Determine the amount by which the given actual expense is over or under the budgeted amount for the Level A budget on page 637.

SOLUTION
The third column shows the actual amount spent minus the amount budgeted. A negative amount means that the item is under budget. A positive amount means that the budgeted amount has been exceeded.

	Budget	**Actual**	**Under (−)/Over (+)**
Total family budget	$14,808	$15,111	+$303
Food	3,639	4,217	+ 578
Housing	4,773	4,792	+ 19
Clothing	891	1,003	+ 112
Transportation	3,715	3,419	− 296
Health Care	770	625	− 145
Entertainment	747	805	+ 58
Personal care	153	160	+ 7
Miscellaneous	120	90	− 30

TRY YOUR SKILLS

A recently married couple, the Hollidays, have an annual take-home pay of $22,500. Use the given budget percentages to calculate the amount budgeted for the year in each category. Round the amounts to the nearest dollar.

1. Food, 22%
2. Housing, 28%
3. Clothing, 5%
4. Transportation, 20%
5. Health care, 4%
6. Entertainment, 7%
7. Personal care, 2%
8. Miscellaneous, 1%
9. Savings, 11%

10. The Hollidays expect to start having children next year. Therefore, they have decided to budget $320 a year for the premiums of a 5-year renewable term $150,000 insurance policy on Mr. Holliday's life. Can the family budget for life insurance in this category for savings?

11. Mr. Holliday, whose employer does not offer a retirement plan, would like to contribute $2000 each year to an IRA (Individual Retirement Arrangement). Can he afford to do this and also pay for the insurance premiums in the category for savings?

EXERCISE YOUR SKILLS

1. About what percent of a budget might transportation take?
2. What percent of a budget would seem too high for housing?
3. Why is it helpful to find percentages as well as dollar amounts for budget categories?
4. How do you calculate the amount that you are over or under budget?

For each category in the Level B budget table, calculate the percent of the total to the nearest tenth of a percent.

		Budget	Percent of Total
5.	Total family budget	$19,045	
6.	Food	4,473	
7.	Housing	6,321	
8.	Clothing	1,183	
9.	Transportation	4,907	
10.	Health care	887	
11.	Entertainment	1,008	
12.	Personal care	196	
13.	Miscellaneous	70	

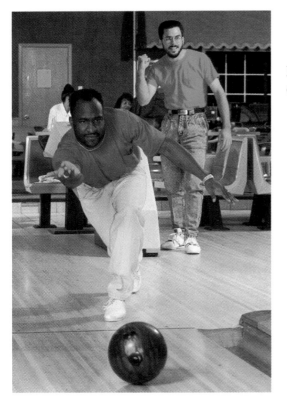

14. For each category in the Level C budget on page 637, calculate the percent of the total to the nearest tenth of a percent.

For each category in the Level C budget, calculate the amount over or under budget.

		Budget	Actual	Under(−)/Over(+)
	Total family budget	$27,181	$27,895	
15.	Food	5,961	5,088	
16.	Housing	9,271	9,271	
17.	Clothing	2,036	2,874	
18.	Transportation	6,655	7,120	
19.	Health care	1,173	920	
20.	Entertainment	1,689	2,145	
21.	Personal care	299	327	
22.	Miscellaneous	97	150	

23. List some of the differences in the ways in which a parent and a teenager think about a family budget. Prepare to discuss these with your classmates in small groups.

MIXED REVIEW

1. How does the phrase "for deposit only" protect a person who is endorsing a check?

2. A person saved $60 per week for 52 weeks in order to buy a one-year Certificate of Deposit (CD) that earned 6% a year. Find out how much was saved, how much interest is earned by the CD in 1 year, and the total amount of money in the CD account at the end of 1 year.

3. Arleigh wants to pay off the current balance of $3500 on his VISA card in $2\frac{1}{2}$ years. The annual interest rate is 17%. How large must the monthly payment be?

4. The annual dividend for Acme Enterprises is $1.20 per share. What additional information do you need in order to calculate the current rate of return on 500 shares of Acme Enterprises?

5. Find the amount of the income tax refund or payment due for a family of two parents and one dependent child with wages of $32,000, taxable interest of $650, and dividends of $300. The family had medical expenses of $5200, mortgage interest of $2900 that they paid on their home, state income tax payments of $1080, and charitable contributions of $90. The wife made a tax-deferred contribution of $550 into her IRA. The amount of withholding tax for the year was $2401.

6. André is using a broker to find an apartment that rents for $480 per month. The broker charges a fee of one month's rent. At the time that he signs the lease, André must pay a security deposit equal to one-and-one-half of one month's rent, as well as the rent for the first month. How much will he have to pay before he moves into the apartment?

7. How might a scatter plot of the 5-year performance of a mutual fund help you predict the fund's future performance?

A married couple has a combined take-home salary of $58,000. The couple is interested in buying a house with a 30-year mortgage but does not want to spend more than 25% of its take-home pay on the mortgage payments. The interest rate at the local bank is 7%.

8. What is the maximum amount that they should pay in monthly mortgage payments to the nearest dollar?

9. With a 25% down payment, what is the maximum price to the nearest dollar that they can afford to pay for a house?

Compute the monthly payment for each condo, assuming that you can obtain a 30-year mortgage at 9% with 20% down for each condo. Compare your results with the rental charge for each condo and state which alternative would require a greater cash expenditure each month.

10. Purchase price of $60,000; monthly rental charge of $500

11. Purchase price of $100,000; monthly rental charge of $800

1. What will be the annual property tax on a house that is assessed at $170,000 with a mill rate of 19.7?

2. What will be the quarterly payment of a property tax on a house with an assessed value of $96,000 and a mill rate of 27.6?

What is the mill rate of a house assessed at the following values?

3. $87,000; $974.40

4. $201,000; $4482.30

5. $407,000; $7692.30

6. $62,450; $612.01

7. Construct a monthly budget for the expenses incurred by the Matsuda family to operate their home based on actual expenditures over the time periods shown.

	Dec.	Jan.	Feb.	March	Total	Average
Heat	$115	$123	$134	$109		
Electricity	47	42	56	39		
Water	19	25	26	19		
Telephone	56	82	38	50		
Home maintenance	0	74	23	120		
Total						

8. Construct a monthly operating budget for a family that has the following pattern of expenditures: energy, $2600 per year; electricity, $48 per month; water $68 every 3 months; telephone, $49 per month; home maintenance, $725 per year.

For each item listed, calculate the unit price and decide which is the best buy in terms of cost.

	A	B	C
9.	16 oz	12 oz	8 oz
	$1.29	$1.19	$0.89
10.	$14\frac{1}{2}$ oz	12 oz	8 oz
	$1.59	$1.38	$0.95
11.	6 oz	10 oz	12 oz
	$0.95	$1.49	$1.69
12.	16 oz	9 oz	12 oz
	$2.15	$1.15	$1.75

Convert each of the following. Round to the nearest hundredth.

13. 6 quarts to liters

14. 7.5 ounces to grams

15. 12 pounds to kilograms

16. 3 kilograms to pounds

17. Marshall Mills sells professionally made videos of local events such as graduations and weddings. For any event his fixed costs are $950 and his unit cost is $2.40 per cassette. Write an equation for his costs.

Allowing for the fixed costs in Exercise 17, find the unit cost for the following number of video tapes sold for any event.

18. 20   **19.** 75   **20.** 100   **21.** 130

For each category in the budget table, calculate the percent of the total to the nearest tenth of a percent.

		Budget	Percent
22.	Total family budget	$1655	
23.	Food	410	
24.	Housing	455	
25.	Clothing	120	
26.	Transportation	385	
27.	Health care	70	
28.	Entertainment	60	
29.	Personal care	30	
30.	Miscellaneous	125	

For each of the following categories, calculate the amount over or under budget.

		Budget	Actual	Under(−)/Over(+)
31.	Total family budget	$1655	$1570	
32.	Food	410	365	
33.	Housing	455	455	
34.	Clothing	120	150	
35.	Transportation	385	347	
36.	Health care	70	25	
37.	Entertainment	60	76	
38.	Personal care	30	52	
39.	Miscellaneous	125	100	

Shauna has an annual take-home pay of $26,664. Use the given percentages to calculate the annual amount to the nearest dollar for each category.

40. Food, 21%   **41.** Housing, 28%

42. Car, 18%   **43.** Clothing, 6%

44. Health care, 10%   **45.** Savings, 7%

46. Entertainment, 8%   **47.** Personal care, 2%

1. Find the property tax on a house with a mill rate of 19.5 and an assessed value of $45,000.

2. Find the monthly payments in property taxes on a house assessed at $95,000 with a mill rate of 24.2.

Find the best buy based on unit price.

	A	B	C
3.	16 oz	$14\frac{1}{2}$ oz	12 oz
	$1.39	$1.29	$1.19
4.	1 quart	$\frac{1}{2}$ gallon	gallon
	$1.65	$2.98	$4.20

Perform each of the following conversions.

5. 85 grams to ounces

6. 2.5 kilograms to pounds

7. Judith Whitaker sells cookies. Her fixed costs are $80, and her unit cost is $0.16 per cookie. Write an equation for her costs.

Allowing for the fixed costs in Test Item 7, find the unit cost for the following number of cookies baked.

8. 100   9. 400   10. 1000

The Meyer family has a total annual take-home pay of $24,000. Use the given budget percentages to calculate the annual amount to the nearest dollar for each category.

11. Food, 22%   12. Housing, 30%

13. Car, 28%   14. Clothing, 5%

15. Insurance and pension, 11%   16. Savings, 4%

For each category, calculate the amount over or under budget.

		Budget	Actual	Under(−)/Over(+)
17.	Total family budget	$2635	$2577	
18.	Food	650	680	
19.	Housing	790	765	
20.	Clothing	185	110	
21.	Transportation	415	380	
22.	Health care	85	412	
23.	Entertainment	225	50	
24.	Personal care	35	30	
25.	Miscellaneous	250	150	

CUMULATIVE REVIEW

1. Devore has $25,500 in a CD that pays 4.8% interest, compounded quarterly. How much is her CD worth at the end of 2 years?

Your fixed costs for your business are $200 per week. Your unit cost is $3.15 per item.

2. What is the total cost of producing 100 items?

3. Each item sells for $5.75. How many must you sell to break even?

You have $90,000 to invest. Find the whole number of shares that you can buy in each of the companies below.

4. SmartMart at $73\frac{5}{8}$

5. Ocean Science Ind. at $33\frac{3}{4}$

Use the Tax Rate Schedule in the Reference Section to find the tax owed by each taxpayer.

6. A married person, filing separately, with $106,945 in taxable income

7. A head of household with $123,780 in taxable income

8. Use the Six-Month Basic Rate Schedule in Reference Section to find the 6-month premium for an insurance policy based on the following information: driver-rating factor, 2.25; collision deductible, $100; car-class rating, 25; comprehensive deductible, $50.

9. The Li family has a monthly income of $3560. It can pay 28% of this income on mortgage payments. With a 20% down payment on a 30-year mortgage at 7.6% per year, what is the maximum amount that the family can pay for a house?

10. Use the Monthly Payment Formula to calculate the monthly mortgage payment for a condo that is selling for $79,500. Assume that the down payment is 20% and that the 30-year mortgage has an annual interest rate of 8%. Compare the rental costs of $600 per month with the monthly payments.

11. Judy's monthly take-home pay is $2900. Credit card payments are $265. Determine, to the nearest tenth of a percent, how much Judy is spending on her credit payments. Is Judy in danger of credit overload?

Find the purchase price of these shares of stock at the prices shown. Ignore the effect of commissions. Find the capital gain or capital loss at the current market value one year later by subtracting the purchase price from the current market value. Calculate the percent change.

	Company	Number of Shares	Original Purchase Price/ Share	Total Price	Current Market Price/ Share	Total Value	Gain or Loss	Percent of Change
12.	XYZ Inc	100	$59\frac{1}{2}$		76			
13.	Whiz Co	55	$46\frac{1}{4}$		$38\frac{1}{2}$			
14.	AnimFrm	2,500	$6\frac{3}{8}$		$11\frac{1}{8}$			

PROJECT 14–1: **Electricity**

Electricity is delivered to almost every home, office, and factory in the United States. Power lines are often buried underground. But the costs of electricity are not invisible. We pay in several ways for the power that we use. First, we pay in hard-earned money that goes to the power companies for generating electricity. Second, we pay in the depletion of our natural resources such as coal, oil, and natural gas that may be burned to create electric power. Finally, we pay a price in the pollution that sometimes arises as fuels are converted to electric power.

1. Make an inventory of the electric appliances used in your home. List the kitchen appliances, radios, televisions, VCRs, lighting fixtures, computers, and so on. You will be surprised at how long your list becomes.

2. Find out the name of the company from which your household purchases electricity.

3. How much does the utility company charge its customers per kilowatt-hour? Does the company raise the rates at peak times of the day or year?

4. Find out from the utility company how it creates the electricity that it sells to you. What are the by-products of generating electricity by this method? What is the effect on our supplies of natural resources?

5. Find out how the electricity reaches your home from the place where it is generated.

6. Collect three or four monthly electricity bills for your household. Add the costs and divide to find the average monthly cost. Find out whether your usage is higher at certain seasons of the year. How do fluctuations in electricity use affect your household budget?

The graphing calculator is used in many of the lessons in *Mathematics of Money with Algebra.* If you are just getting acquainted with this kind of calculator, the following specific instructions for the *TI-81* and the *Casio fx7700G* may be helpful. These directions should provide you with enough information to help you with other models of graphing calculators. If you need more information, be sure to consult your calculator's manual.

Notice that there is a label on each key and labels above each key. To use the labels above a key, press [SHIFT] or [2ND] and then the key you want, or [ALPHA] and the key you want. The labels above the keys are the same color as the corresponding shift key. In this section, labels on the keys are shown in boxes ☐ and labels above the keys are shown in brackets [ ].

fx–7700G

[AC] is used to turn on the calculator and to clear the screen of the calculator. [SHIFT] [OFF] is used to turn off the calculator. [EXE] is used to execute a calculation.

TI–81

[ON] is used to turn on the calculator. [CLEAR] is used to clear the screen of the calculator. [2ND] [OFF] is used to turn off the calculator. [ENTER] is used to enter a calculation.

Menus

Menus which give you several choices are used on most graphing calculators. Check the mode menu, which is used to set special features, before you begin to use your calculator.

fx–7700G

You make menu selections with the function keys [F1] — [F6]. The label printed above each function key is active unless menu selections appear along the bottom of the screen.

The following should be displayed when you press [MDISP].

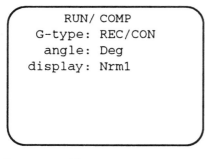

```
      RUN/ COMP
 G-type: REC/CON
   angle: Deg
 display: Nrm1
```

To correct settings use:
[MODE] [1] for run (RUN).
[MODE] [+] for computation (COMP).
[MODE] [SHIFT] [+] for rectangular coordinates (REC).
[MODE] [SHIFT] [5] for connect (CON).
[SHIFT] [DISP] [F3] [EXE] for norm 1 (Nrm 1).

TI–81

Menus appear on the screen when special keys such as [MODE] are pressed. Use the arrow keys [◀], [▶], [▲], and [▼] to highlight your selection, then press [ENTER].

The following should be displayed when you press [MODE].

```
 NORM Sci Eng
 FLOAT 0123456789
 Rad DEG
 FUNCTION Param
 CONNECTED Dot
 SEQUENCE Simul
 GRID OFF Grid on
 RECT Polar
```

To make a correction, use the arrow keys to highlight the correct choice. Then press [ENTER]. To leave mode, press [2ND] [QUIT].

Correcting Errors

If you make an error when entering a calculation, you can replay the calculation and use the insert and delete keys to make a correction. The arrow keys ◀, ▶, ▲, and ▼ are used to position the cursor.

fx–7700G

Use the left arrow key ◀ to replay the previous line. Use the delete key DEL to delete a character at the cursor location. Use the insert key SHIFT [INS] to insert a character at the cursor location.

TI–81

Use the up arrow key ▲ to replay the previous line so you can edit it. Use the delete key DEL to delete a character at the cursor location. Use the insert key INS to insert a character at the cursor location.

Evaluating Expressions

You can use a graphing calculator to evaluate mathematical expressions. The graphing calculator uses the usual rules for the order of operations, with one exception. Implied multiplication will be calculated before any multiplication, division, addition, or subtraction that is shown by an operation symbol.

Setting the Range

You must set values for the portion of the graph that is displayed in the viewing screen. The range values define a viewing rectangle. The RANGE key is used on both the *fx-7700G* and the *TI-81* to set the viewing rectangle.

```
Range
Xmin:-4.7      ←Minimum value on x-axis →
  max:4.7      ←Maximum value on x-axis→
  scl:1.       ←Scale used on x-axis    →
Ymin:-3.1      ←Minimum value on y-axis →
  max:3.1      ←Maximum value on y-axis→
  scl:1.       ←Scale used on y-axis    →
INIT
```

```
Range
Xmin=-4.8
Xmax=4.7
Xscl=1
Ymin=-3.1
Ymax=3.2
Yscl=1
Xres=1
```

Graphing Equations

To graph an equation, the equation must be in the form $y = mx + b$. If necessary, solve the equation for y before inputting it into the calculator.

fx–7700G

To enter an equation, press GRAPH.

```
Graph y=
```

Enter the equation using the key labeled X,θ,T for the variable *x.* Then press EXE to display the graph.

TI–81

To enter an equation, press Y=.

```
:Y₁ =
:Y₂ =
:Y₃ =
:Y₄ =
```

Enter the equation using the key labeled X|T for the variable *x.* Then press GRAPH to display the graph.

Trace Feature

Solutions to an equation can be found by using the trace feature to identify coordinates of points on the graph. After pressing [TRACE], you will see a blinking cursor on the screen. Use [◄] and [►] to move the trace cursor along the graph. The x- and y-coordinates of the point shown by the blinking cursor are displayed at the bottom of the screen. For the *fx7700G*, the trace label is above the [F1] key.

Graphing More than One Equation

You can graph more than one equation at a time on a graphing calculator.

fx–7700G

To graph more than one equation at the same time, press [SHIFT] [EXE] after entering the first equation. To graph $y = 3x$ and $y = x$, use:

[GRAPH] 3 [X, θ, T] [SHIFT] [EXE]
[GRAPH] [X, θ, T] [EXE]

To clear the screen before entering a new graph, press [F5] (cls) [EXE]. To switch between the graph screen and the text screen, press the [G↔T] key

TI–81

When you press [Y=] there is space for four equations to be entered, y_1, y_2, y_3, and y_4.

To clear the screen before entering a new graph, erase any previous equations. Press [Y=], then place the cursor over any character in the equation you want to erase, and press [CLEAR]. To keep the equation, but not have a graph drawn, place the cursor on the equals symbol, and press [ENTER]. To switch between the graph screen and the text screen, use [CLEAR] and [GRAPH].

Drawing a Scatter Plot

You can enter data and draw a scatter plot. First set the range.

fx–7700G

Set the mode menu as follows:
[MODE] [÷] for regression mode (REG).
[MODE] [4] for linear regression (LIN).
[MODE] [SHIFT] [1] for storing statistical data (STO).
[MODE] [SHIFT] [3] for drawing a statistical graph (DRAW).
Use [SHIFT] [,] and [F1] (DT) to enter the data. To enter (2, 3), use 2 [SHIFT] [,] 3 [F1] (DT). A dot is placed on the graph screen as each data point is entered.

TI–81

Pressing [2ND] [STAT] accesses the statistics menu. To enter data, use [2ND] [STAT] [►] [►] [1] (Edit). To leave the data screen, press [2ND] [QUIT]. To clear statistical data, use [2ND] [STAT] [►] [►] (Data) [2] (ClrStat) [ENTER]. To draw a scatter plot, use [2ND] [STAT] [►] (Draw) [2] (Scatter) [ENTER].

Graphing a Linear Regression Equation

fx–7700G

Use [MODE] [÷] [F6] (REG) to get the linear regression menu for $y = a + bx$, where r is the correlation coefficient. To calculate a, b, and r, use [F1] (A) [EXE] [F2] (B) [EXE] [F3] (R) [EXE]. To graph the line of best fit, use [GRAPH] [SHIFT] [F4] (Line) [1] [EXE].

TI–81

To calculate a, b, and r for a linear regression $y = a + bx$ where r is the correlation coefficient, use [2ND] [STAT] [2] (LinReg) [ENTER]. To graph the line of best fit, use [Y=] [VARS] [►] [►] (LR) [4] (RegEQ) [GRAPH].

SINGLE Persons—WEEKLY Payroll Period

(For Wages Paid in 19–)

If the wages are—		And the number of withholding allowances claimed is—										
At least	But less than	0	1	2	3	4	5	6	7	8	9	10
		The amount of income tax to be withheld is—										
$0	$50	$0	$0	$0	$0	$0	$0	$0	$0	$0	$0	$0
50	55	1	0	0	0	0	0	0	0	0	0	0
55	60	1	0	0	0	0	0	0	0	0	0	0
60	65	2	0	0	0	0	0	0	0	0	0	0
65	70	3	0	0	0	0	0	0	0	0	0	0
70	75	4	0	0	0	0	0	0	0	0	0	0
75	80	4	0	0	0	0	0	0	0	0	0	0
80	85	5	0	0	0	0	0	0	0	0	0	0
85	90	6	0	0	0	0	0	0	0	0	0	0
90	95	7	0	0	0	0	0	0	0	0	0	0
95	100	7	1	0	0	0	0	0	0	0	0	0
100	105	8	1	0	0	0	0	0	0	0	0	0
105	110	9	2	0	0	0	0	0	0	0	0	0
110	115	10	3	0	0	0	0	0	0	0	0	0
115	120	10	4	0	0	0	0	0	0	0	0	0
120	125	11	4	0	0	0	0	0	0	0	0	0
125	130	12	5	0	0	0	0	0	0	0	0	0
130	135	13	6	0	0	0	0	0	0	0	0	0
135	140	13	7	0	0	0	0	0	0	0	0	0
140	145	14	7	1	0	0	0	0	0	0	0	0
145	150	15	8	1	0	0	0	0	0	0	0	0
150	155	16	9	2	0	0	0	0	0	0	0	0
155	160	16	10	3	0	0	0	0	0	0	0	0
160	165	17	10	4	0	0	0	0	0	0	0	0
165	170	18	11	4	0	0	0	0	0	0	0	0
170	175	19	12	5	0	0	0	0	0	0	0	0
175	180	19	13	6	0	0	0	0	0	0	0	0
180	185	20	13	7	0	0	0	0	0	0	0	0
185	190	21	14	7	1	0	0	0	0	0	0	0
190	195	22	15	8	1	0	0	0	0	0	0	0
195	200	22	16	9	2	0	0	0	0	0	0	0
200	210	23	17	10	3	0	0	0	0	0	0	0
210	220	25	18	11	5	0	0	0	0	0	0	0
220	230	26	20	13	6	0	0	0	0	0	0	0
230	240	28	21	14	8	1	0	0	0	0	0	0
240	250	29	23	16	9	2	0	0	0	0	0	0
250	260	31	24	17	11	4	0	0	0	0	0	0
260	270	32	26	19	12	5	0	0	0	0	0	0
270	280	34	27	20	14	7	0	0	0	0	0	0
280	290	35	29	22	15	8	2	0	0	0	0	0
290	300	37	30	23	17	10	3	0	0	0	0	0
300	310	38	32	25	18	11	5	0	0	0	0	0
310	320	40	33	26	20	13	6	0	0	0	0	0
320	330	41	35	28	21	14	8	1	0	0	0	0
330	340	43	36	29	23	16	9	2	0	0	0	0
340	350	44	38	31	24	17	11	4	0	0	0	0
350	360	46	39	32	26	19	12	5	0	0	0	0
360	370	47	41	34	27	20	14	7	0	0	0	0
370	380	49	42	35	29	22	15	8	2	0	0	0
380	390	50	44	37	30	23	17	10	3	0	0	0
390	400	52	45	38	32	25	18	11	5	0	0	0
400	410	53	47	40	33	26	20	13	6	0	0	0
410	420	55	48	41	35	28	21	14	8	1	0	0
420	430	56	50	43	36	29	23	16	9	2	0	0
430	440	58	51	44	38	31	24	17	11	4	0	0
440	450	59	53	46	39	32	26	19	12	5	0	0
450	460	61	54	47	41	34	27	20	14	7	0	0
460	470	64	56	49	42	35	29	22	15	8	1	0
470	480	67	57	50	44	37	30	23	17	10	3	0
480	490	70	59	52	45	38	32	25	18	11	4	0
490	500	73	60	53	47	40	33	26	20	13	6	0
500	510	75	63	55	48	41	35	28	21	14	7	1
510	520	78	66	56	50	43	36	29	23	16	9	2
520	530	81	68	58	51	44	38	31	24	17	10	4
530	540	84	71	59	53	46	39	32	26	19	12	5
540	550	87	74	61	54	47	41	34	27	20	13	7
550	560	89	77	64	56	49	42	35	29	22	15	8
560	570	92	80	67	57	50	44	37	30	23	16	10
570	580	95	82	70	59	52	45	38	32	25	18	11
580	590	98	85	73	60	53	47	40	33	26	19	13

(Continued on next page)

SINGLE Persons—WEEKLY Payroll Period
(For Wages Paid in 19--)

If the wages are—		And the number of withholding allowances claimed is—										
At least	But less than	0	1	2	3	4	5	6	7	8	9	10
		The amount of income tax to be withheld is—										
$590	$600	$101	$88	$75	$63	$55	$48	$41	$35	$28	$21	$14
600	610	103	91	78	66	56	50	43	36	29	22	16
610	620	106	94	81	68	58	51	44	38	31	24	17
620	630	109	96	84	71	59	53	46	39	32	25	19
630	640	112	99	87	74	61	54	47	41	34	27	20
640	650	115	102	89	77	64	56	49	42	35	28	22
650	660	117	105	92	80	67	57	50	44	37	30	23
660	670	120	108	95	82	70	59	52	45	38	31	25
670	680	123	110	98	85	72	60	53	47	40	33	26
680	690	126	113	101	88	75	63	55	48	41	34	28
690	700	129	116	103	91	78	65	56	50	43	36	29
700	710	131	119	106	94	81	68	58	51	44	37	31
710	720	134	122	109	96	84	71	59	53	46	39	32
720	730	137	124	112	99	86	74	61	54	47	40	34
730	740	140	127	115	102	89	77	64	56	49	42	35
740	750	143	130	117	105	92	79	67	57	50	43	37
750	760	145	133	120	108	95	82	70	59	52	45	38
760	770	148	136	123	110	98	85	72	60	53	46	40
770	780	151	138	126	113	100	88	75	63	55	48	41
780	790	154	141	129	116	103	91	78	65	56	49	43
790	800	157	144	131	119	106	93	81	68	58	51	44
800	810	159	147	134	122	109	96	84	71	59	52	46
810	820	162	150	137	124	112	99	86	74	61	54	47
820	830	165	152	140	127	114	102	89	77	64	55	49
830	840	168	155	143	130	117	105	92	79	67	57	50
840	850	171	158	145	133	120	107	95	82	69	58	52
850	860	173	161	148	136	123	110	98	85	72	60	53
860	870	176	164	151	138	126	113	100	88	75	62	55
870	880	179	166	154	141	128	116	103	91	78	65	56
880	890	182	169	157	144	131	119	106	93	81	68	58
890	900	185	172	159	147	134	121	109	96	83	71	59
900	910	187	175	162	150	137	124	112	99	86	74	61
910	920	190	178	165	152	140	127	114	102	89	76	64
920	930	193	180	168	155	142	130	117	105	92	79	67
930	940	196	183	171	158	145	133	120	107	95	82	69
940	950	199	186	173	161	148	135	123	110	97	85	72
950	960	202	189	176	164	151	138	126	113	100	88	75
960	970	205	192	179	166	154	141	128	116	103	90	78
970	980	208	194	182	169	156	144	131	119	106	93	81
980	990	211	197	185	172	159	147	134	121	109	96	83
990	1,000	214	200	187	175	162	149	137	124	111	99	86
1,000	1,010	217	203	190	178	165	152	140	127	114	102	89
1,010	1,020	220	206	193	180	168	155	142	130	117	104	92
1,020	1,030	224	210	196	183	170	158	145	133	120	107	95
1,030	1,040	227	213	199	186	173	161	148	135	123	110	97
1,040	1,050	230	216	202	189	176	163	151	138	125	113	100
1,050	1,060	233	219	205	192	179	166	154	141	128	116	103
1,060	1,070	236	222	208	194	182	169	156	144	131	118	106
1,070	1,080	239	225	211	197	184	172	159	147	134	121	109
1,080	1,090	242	228	214	200	187	175	162	149	137	124	111
1,090	1,100	245	231	217	203	190	177	165	152	139	127	114
1,100	1,110	248	234	220	206	193	180	168	155	142	130	117
1,110	1,120	251	237	223	209	196	183	170	158	145	132	120
1,120	1,130	255	241	227	213	199	186	173	161	148	135	123
1,130	1,140	258	244	230	216	202	189	176	163	151	138	125
1,140	1,150	261	247	233	219	205	191	179	166	153	141	128
1,150	1,160	264	250	236	222	208	194	182	169	156	144	131
1,160	1,170	267	253	239	225	211	197	184	172	159	146	134
1,170	1,180	270	256	242	228	214	200	187	175	162	149	137
1,180	1,190	273	259	245	231	217	203	190	177	165	152	139
1,190	1,200	276	262	248	234	220	206	193	180	167	155	142
1,200	1,210	279	265	251	237	223	209	196	183	170	158	145
1,210	1,220	282	268	254	240	226	212	198	186	173	160	148
1,220	1,230	286	272	258	244	230	216	202	189	176	163	151
1,230	1,240	289	275	261	247	233	219	205	191	179	166	153
1,240	1,250	292	278	264	250	236	222	208	194	181	169	156

$1,250 and over   Use Table 1(a) for a **SINGLE person** on page 26. Also see the instructions on page 24.

MARRIED Persons—WEEKLY Payroll Period
(For Wages Paid in 19--)

If the wages are—		And the number of withholding allowances claimed is—										
At least	But less than	0	1	2	3	4	5	6	7	8	9	10
		The amount of income tax to be withheld is—										
$0	$125	$0	$0	$0	$0	$0	$0	$0	$0	$0	$0	$0
125	130	1	0	0	0	0	0	0	0	0	0	0
130	135	2	0	0	0	0	0	0	0	0	0	0
135	140	3	0	0	0	0	0	0	0	0	0	0
140	145	3	0	0	0	0	0	0	0	0	0	0
145	150	4	0	0	0	0	0	0	0	0	0	0
150	155	5	0	0	0	0	0	0	0	0	0	0
155	160	6	0	0	0	0	0	0	0	0	0	0
160	165	6	0	0	0	0	0	0	0	0	0	0
165	170	7	0	0	0	0	0	0	0	0	0	0
170	175	8	1	0	0	0	0	0	0	0	0	0
175	180	9	2	0	0	0	0	0	0	0	0	0
180	185	9	3	0	0	0	0	0	0	0	0	0
185	190	10	3	0	0	0	0	0	0	0	0	0
190	195	11	4	0	0	0	0	0	0	0	0	0
195	200	12	5	0	0	0	0	0	0	0	0	0
200	210	13	6	0	0	0	0	0	0	0	0	0
210	220	14	8	1	0	0	0	0	0	0	0	0
220	230	16	9	2	0	0	0	0	0	0	0	0
230	240	17	11	4	0	0	0	0	0	0	0	0
240	250	19	12	5	0	0	0	0	0	0	0	0
250	260	20	14	7	0	0	0	0	0	0	0	0
260	270	22	15	8	2	0	0	0	0	0	0	0
270	280	23	17	10	3	0	0	0	0	0	0	0
280	290	25	18	11	5	0	0	0	0	0	0	0
290	300	26	20	13	6	0	0	0	0	0	0	0
300	310	28	21	14	8	1	0	0	0	0	0	0
310	320	29	23	16	9	2	0	0	0	0	0	0
320	330	31	24	17	11	4	0	0	0	0	0	0
330	340	32	26	19	12	5	0	0	0	0	0	0
340	350	34	27	20	14	7	0	0	0	0	0	0
350	360	35	29	22	15	8	1	0	0	0	0	0
360	370	37	30	23	17	10	3	0	0	0	0	0
370	380	38	32	25	18	11	4	0	0	0	0	0
380	390	40	33	26	20	13	6	0	0	0	0	0
390	400	41	35	28	21	14	7	1	0	0	0	0
400	410	43	36	29	23	16	9	2	0	0	0	0
410	420	44	38	31	24	17	10	4	0	0	0	0
420	430	46	39	32	26	19	12	5	0	0	0	0
430	440	47	41	34	27	20	13	7	0	0	0	0
440	450	49	42	35	29	22	15	8	1	0	0	0
450	460	50	44	37	30	23	16	10	3	0	0	0
460	470	52	45	38	32	25	18	11	4	0	0	0
470	480	53	47	40	33	26	19	13	6	0	0	0
480	490	55	48	41	35	28	21	14	7	1	0	0
490	500	56	50	43	36	29	22	16	9	2	0	0
500	510	58	51	44	38	31	24	17	10	4	0	0
510	520	59	53	46	39	32	25	19	12	5	0	0
520	530	61	54	47	41	34	27	20	13	7	0	0
530	540	62	56	49	42	35	28	22	15	8	1	0
540	550	64	57	50	44	37	30	23	16	10	3	0
550	560	65	59	52	45	38	31	25	18	11	4	0
560	570	67	60	53	47	40	33	26	19	13	6	0
570	580	68	62	55	48	41	34	28	21	14	7	1
580	590	70	63	56	50	43	36	29	22	16	9	2
590	600	71	65	58	51	44	37	31	24	17	10	4
600	610	73	66	59	53	46	39	32	25	19	12	5
610	620	74	68	61	54	47	40	34	27	20	13	7
620	630	76	69	62	56	49	42	35	28	22	15	8
630	640	77	71	64	57	50	43	37	30	23	16	10
640	650	79	72	65	59	52	45	38	31	25	18	11
650	660	80	74	67	60	53	46	40	33	26	19	13
660	670	82	75	68	62	55	48	41	34	28	21	14
670	680	83	77	70	63	56	49	43	36	29	22	16
680	690	85	78	71	65	58	51	44	37	31	24	17
690	700	86	80	73	66	59	52	46	39	32	25	19
700	710	88	81	74	68	61	54	47	40	34	27	20
710	720	89	83	76	69	62	55	49	42	35	28	22
720	730	91	84	77	71	64	57	50	43	37	30	23
730	740	92	86	79	72	65	58	52	45	38	31	25

(Continued on next page)

MARRIED Persons—WEEKLY Payroll Period
(For Wages Paid in 19--)

If the wages are—		And the number of withholding allowances claimed is—										
At least	But less than	0	1	2	3	4	5	6	7	8	9	10
		The amount of income tax to be withheld is—										
$740	$750	$94	$87	$80	$74	$67	$60	$53	$46	$40	$33	$26
750	760	95	89	82	75	68	61	55	48	41	34	28
760	770	97	90	83	77	70	63	56	49	43	36	29
770	780	98	92	85	78	71	64	58	51	44	37	31
780	790	100	93	86	80	73	66	59	52	46	39	32
790	800	103	95	88	81	74	67	61	54	47	40	34
800	810	106	96	89	83	76	69	62	55	49	42	35
810	820	108	98	91	84	77	70	64	57	50	43	37
820	830	111	99	92	86	79	72	65	58	52	45	38
830	840	114	101	94	87	80	73	67	60	53	46	40
840	850	117	104	95	89	82	75	68	61	55	48	41
850	860	120	107	97	90	83	76	70	63	56	49	43
860	870	122	110	98	92	85	78	71	64	58	51	44
870	880	125	113	100	93	86	79	73	66	59	52	46
880	890	128	115	103	95	88	81	74	67	61	54	47
890	900	131	118	106	96	89	82	76	69	62	55	49
900	910	134	121	108	98	91	84	77	70	64	57	50
910	920	136	124	111	99	92	85	79	72	65	58	52
920	930	139	127	114	101	94	87	80	73	67	60	53
930	940	142	129	117	104	95	88	82	75	68	61	55
940	950	145	132	120	107	97	90	83	76	70	63	56
950	960	148	135	122	110	98	91	85	78	71	64	58
960	970	150	138	125	112	100	93	86	79	73	66	59
970	980	153	141	128	115	103	94	88	81	74	67	61
980	990	156	143	131	118	105	96	89	82	76	69	62
990	1,000	159	146	134	121	108	97	91	84	77	70	64
1,000	1,010	162	149	136	124	111	99	92	85	79	72	65
1,010	1,020	164	152	139	126	114	101	94	87	80	73	67
1,020	1,030	167	155	142	129	117	104	95	88	82	75	68
1,030	1,040	170	157	145	132	119	107	97	90	83	76	70
1,040	1,050	173	160	148	135	122	110	98	91	85	78	71
1,050	1,060	176	163	150	138	125	112	100	93	86	79	73
1,060	1,070	178	166	153	140	128	115	103	94	88	81	74
1,070	1,080	181	169	156	143	131	118	105	96	89	82	76
1,080	1,090	184	171	159	146	133	121	108	97	91	84	77
1,090	1,100	187	174	162	149	136	124	111	99	92	85	79
1,100	1,110	190	177	164	152	139	126	114	101	94	87	80
1,110	1,120	192	180	167	154	142	129	117	104	95	88	82
1,120	1,130	195	183	170	157	145	132	119	107	97	90	83
1,130	1,140	198	185	173	160	147	135	122	109	98	91	85
1,140	1,150	201	188	176	163	150	138	125	112	100	93	86
1,150	1,160	204	191	178	166	153	140	128	115	102	94	88
1,160	1,170	206	194	181	168	156	143	131	118	105	96	89
1,170	1,180	209	197	184	171	159	146	133	121	108	97	91
1,180	1,190	212	199	187	174	161	149	136	123	111	99	92
1,190	1,200	215	202	190	177	164	152	139	126	114	101	94
1,200	1,210	218	205	192	180	167	154	142	129	116	104	95
1,210	1,220	220	208	195	182	170	157	145	132	119	107	97
1,220	1,230	223	211	198	185	173	160	147	135	122	109	98
1,230	1,240	226	213	201	188	175	163	150	137	125	112	100
1,240	1,250	229	216	204	191	178	166	153	140	128	115	102
1,250	1,260	232	219	206	194	181	168	156	143	130	118	105
1,260	1,270	234	222	209	196	184	171	159	146	133	121	108
1,270	1,280	237	225	212	199	187	174	161	149	136	123	111
1,280	1,290	240	227	215	202	189	177	164	151	139	126	114
1,290	1,300	243	230	218	205	192	180	167	154	142	129	116
1,300	1,310	246	233	220	208	195	182	170	157	144	132	119
1,310	1,320	248	236	223	210	198	185	173	160	147	135	122
1,320	1,330	251	239	226	213	201	188	175	163	150	137	125
1,330	1,340	254	241	229	216	203	191	178	165	153	140	128
1,340	1,350	257	244	232	219	206	194	181	168	156	143	130
1,350	1,360	260	247	234	222	209	196	184	171	158	146	133
1,360	1,370	262	250	237	224	212	199	187	174	161	149	136
1,370	1,380	265	253	240	227	215	202	189	177	164	151	139
1,380	1,390	268	255	243	230	217	205	192	179	167	154	142
1,390	1,400	271	258	246	233	220	208	195	182	170	157	144

$1,400 and over   Use Table 1(b) for a **MARRIED person** on page 26. Also see the instructions on page 24.

SINGLE Persons—MONTHLY Payroll Period

(For Wages Paid in 19--)

If the wages are—		And the number of withholding allowances claimed is—										
At least	But less than	0	1	2	3	4	5	6	7	8	9	10
		The amount of income tax to be withheld is—										
$0	$210	$0	$0	$0	$0	$0	$0	$0	$0	$0	$0	$0
210	220	1	0	0	0	0	0	0	0	0	0	0
220	230	2	0	0	0	0	0	0	0	0	0	0
230	240	4	0	0	0	0	0	0	0	0	0	0
240	250	5	0	0	0	0	0	0	0	0	0	0
250	260	7	0	0	0	0	0	0	0	0	0	0
260	270	8	0	0	0	0	0	0	0	0	0	0
270	280	10	0	0	0	0	0	0	0	0	0	0
280	290	11	0	0	0	0	0	0	0	0	0	0
290	300	13	0	0	0	0	0	0	0	0	0	0
300	320	15	0	0	0	0	0	0	0	0	0	0
320	340	18	0	0	0	0	0	0	0	0	0	0
340	360	21	0	0	0	0	0	0	0	0	0	0
360	380	24	0	0	0	0	0	0	0	0	0	0
380	400	27	0	0	0	0	0	0	0	0	0	0
400	420	30	1	0	0	0	0	0	0	0	0	0
420	440	33	4	0	0	0	0	0	0	0	0	0
440	460	36	7	0	0	0	0	0	0	0	0	0
460	480	39	10	0	0	0	0	0	0	0	0	0
480	500	42	13	0	0	0	0	0	0	0	0	0
500	520	45	16	0	0	0	0	0	0	0	0	0
520	540	48	19	0	0	0	0	0	0	0	0	0
540	560	51	22	0	0	0	0	0	0	0	0	0
560	580	54	25	0	0	0	0	0	0	0	0	0
580	600	57	28	0	0	0	0	0	0	0	0	0
600	640	61	32	3	0	0	0	0	0	0	0	0
640	680	67	38	9	0	0	0	0	0	0	0	0
680	720	73	44	15	0	0	0	0	0	0	0	0
720	760	79	50	21	0	0	0	0	0	0	0	0
760	800	85	56	27	0	0	0	0	0	0	0	0
800	840	91	62	33	3	0	0	0	0	0	0	0
840	880	97	68	39	9	0	0	0	0	0	0	0
880	920	103	74	45	15	0	0	0	0	0	0	0
920	960	109	80	51	21	0	0	0	0	0	0	0
960	1,000	115	86	57	27	0	0	0	0	0	0	0
1,000	1,040	121	92	63	33	4	0	0	0	0	0	0
1,040	1,080	127	98	69	39	10	0	0	0	0	0	0
1,080	1,120	133	104	75	45	16	0	0	0	0	0	0
1,120	1,160	139	110	81	51	22	0	0	0	0	0	0
1,160	1,200	145	116	87	57	28	0	0	0	0	0	0
1,200	1,240	151	122	93	63	34	5	0	0	0	0	0
1,240	1,280	157	128	99	69	40	11	0	0	0	0	0
1,280	1,320	163	134	105	75	46	17	0	0	0	0	0
1,320	1,360	169	140	111	81	52	23	0	0	0	0	0
1,360	1,400	175	146	117	87	58	29	0	0	0	0	0
1,400	1,440	181	152	123	93	64	35	5	0	0	0	0
1,440	1,480	187	158	129	99	70	41	11	0	0	0	0
1,480	1,520	193	164	135	105	76	47	17	0	0	0	0
1,520	1,560	199	170	141	111	82	53	23	0	0	0	0
1,560	1,600	205	176	147	117	88	59	29	0	0	0	0
1,600	1,640	211	182	153	123	94	65	35	6	0	0	0
1,640	1,680	217	188	159	129	100	71	41	12	0	0	0
1,680	1,720	223	194	165	135	106	77	47	18	0	0	0
1,720	1,760	229	200	171	141	112	83	53	24	0	0	0
1,760	1,800	235	206	177	147	118	89	59	30	0	0	0
1,800	1,840	241	212	183	153	124	95	65	36	6	0	0
1,840	1,880	247	218	189	159	130	101	71	42	12	0	0
1,880	1,920	253	224	195	165	136	107	77	48	18	0	0
1,920	1,960	259	230	201	171	142	113	83	54	24	0	0
1,960	2,000	269	236	207	177	148	119	89	60	30	1	0
2,000	2,040	280	242	213	183	154	125	95	66	36	7	0
2,040	2,080	291	248	219	189	160	131	101	72	42	13	0
2,080	2,120	302	254	225	195	166	137	107	78	48	19	0
2,120	2,160	314	260	231	201	172	143	113	84	54	25	0
2,160	2,200	325	270	237	207	178	149	119	90	60	31	2
2,200	2,240	336	281	243	213	184	155	125	96	66	37	8
2,240	2,280	347	292	249	219	190	161	131	102	72	43	14
2,280	2,320	358	304	255	225	196	167	137	108	78	49	20
2,320	2,360	370	315	261	231	202	173	143	114	84	55	26
2,360	2,400	381	326	271	237	208	179	149	120	90	61	32

(Continued on next page)

SINGLE Persons—**MONTHLY** Payroll Period

(For Wages Paid in 19--)

If the wages are—		And the number of withholding allowances claimed is—										
At least	But less than	0	1	2	3	4	5	6	7	8	9	10
		The amount of income tax to be withheld is—										
$2,400	$2,440	$392	$337	$282	$243	$214	$185	$155	$126	$96	$67	$38
2,440	2,480	403	348	294	249	220	191	161	132	102	73	44
2,480	2,520	414	360	305	255	226	197	167	138	108	79	50
2,520	2,560	426	371	316	261	232	203	173	144	114	85	56
2,560	2,600	437	382	327	272	238	209	179	150	120	91	62
2,600	2,640	448	393	338	283	244	215	185	156	126	97	68
2,640	2,680	459	404	350	295	250	221	191	162	132	103	74
2,680	2,720	470	416	361	306	256	227	197	168	138	109	80
2,720	2,760	482	427	372	317	262	233	203	174	144	115	86
2,760	2,800	493	438	383	328	273	239	209	180	150	121	92
2,800	2,840	504	449	394	339	285	245	215	186	156	127	98
2,840	2,880	515	460	406	351	296	251	221	192	162	133	104
2,880	2,920	526	472	417	362	307	257	227	198	168	139	110
2,920	2,960	538	483	428	373	318	263	233	204	174	145	116
2,960	3,000	549	494	439	384	329	275	239	210	180	151	122
3,000	3,040	560	505	450	395	341	286	245	216	186	157	128
3,040	3,080	571	516	462	407	352	297	251	222	192	163	134
3,080	3,120	582	528	473	418	363	308	257	228	198	169	140
3,120	3,160	594	539	484	429	374	319	265	234	204	175	146
3,160	3,200	605	550	495	440	385	331	276	240	210	181	152
3,200	3,240	616	561	506	451	397	342	287	246	216	187	158
3,240	3,280	627	572	518	463	408	353	298	252	222	193	164
3,280	3,320	638	584	529	474	419	364	309	258	228	199	170
3,320	3,360	650	595	540	485	430	375	321	266	234	205	176
3,360	3,400	661	606	551	496	441	387	332	277	240	211	182
3,400	3,440	672	617	562	507	453	398	343	288	246	217	188
3,440	3,480	683	628	574	519	464	409	354	299	252	223	194
3,480	3,520	694	640	585	530	475	420	365	311	258	229	200
3,520	3,560	706	651	596	541	486	431	377	322	267	235	206
3,560	3,600	717	662	607	552	497	443	388	333	278	241	212
3,600	3,640	728	673	618	563	509	454	399	344	289	247	218
3,640	3,680	739	684	630	575	520	465	410	355	301	253	224
3,680	3,720	750	696	641	586	531	476	421	367	312	259	230
3,720	3,760	762	707	652	597	542	487	433	378	323	268	236
3,760	3,800	773	718	663	608	553	499	444	389	334	279	242
3,800	3,840	784	729	674	619	565	510	455	400	345	290	248
3,840	3,880	795	740	686	631	576	521	466	411	357	302	254
3,880	3,920	806	752	697	642	587	532	477	423	368	313	260
3,920	3,960	818	763	708	653	598	543	489	434	379	324	269
3,960	4,000	829	774	719	664	609	555	500	445	390	335	280
4,000	4,040	840	785	730	675	621	566	511	456	401	346	292
4,040	4,080	851	796	742	687	632	577	522	467	413	358	303
4,080	4,120	863	808	753	698	643	588	533	479	424	369	314
4,120	4,160	875	819	764	709	654	599	545	490	435	380	325
4,160	4,200	888	830	775	720	665	611	556	501	446	391	336
4,200	4,240	900	841	786	731	677	622	567	512	457	402	348
4,240	4,280	913	852	798	743	688	633	578	523	469	414	359
4,280	4,320	925	864	809	754	699	644	589	535	480	425	370
4,320	4,360	937	877	820	765	710	655	601	546	491	436	381
4,360	4,400	950	889	831	776	721	667	612	557	502	447	392
4,400	4,440	962	901	842	787	733	678	623	568	513	458	404
4,440	4,480	975	914	854	799	744	689	634	579	525	470	415
4,480	4,520	987	926	866	810	755	700	645	591	536	481	426
4,520	4,560	999	939	878	821	766	711	657	602	547	492	437
4,560	4,600	1,012	951	890	832	777	723	668	613	558	503	448
4,600	4,640	1,024	963	903	843	789	734	679	624	569	514	460
4,640	4,680	1,037	976	915	855	800	745	690	635	581	526	471
4,680	4,720	1,049	988	928	867	811	756	701	647	592	537	482
4,720	4,760	1,061	1,001	940	879	822	767	713	658	603	548	493
4,760	4,800	1,074	1,013	952	892	833	779	724	669	614	559	504
4,800	4,840	1,086	1,025	965	904	845	790	735	680	625	570	516
4,840	4,880	1,099	1,038	977	916	856	801	746	691	637	582	527
4,880	4,920	1,111	1,050	990	929	868	812	757	703	648	593	538
4,920	4,960	1,123	1,063	1,002	941	881	823	769	714	659	604	549
4,960	5,000	1,136	1,075	1,014	954	893	835	780	725	670	615	560

$5,000 and over   Use Table 4(a) for a **SINGLE person** on page 26. Also see the instructions on page 24.

MARRIED Persons—MONTHLY Payroll Period
(For Wages Paid in 19--)

If the wages are—		And the number of withholding allowances claimed is—										
At least	But less than	0	1	2	3	4	5	6	7	8	9	10
		The amount of income tax to be withheld is—										
$0	$520	$0	$0	$0	$0	$0	$0	$0	$0	$0	$0	$0
520	540	2	0	0	0	0	0	0	0	0	0	0
540	560	5	0	0	0	0	0	0	0	0	0	0
560	580	8	0	0	0	0	0	0	0	0	0	0
580	600	11	0	0	0	0	0	0	0	0	0	0
600	640	16	0	0	0	0	0	0	0	0	0	0
640	680	22	0	0	0	0	0	0	0	0	0	0
680	720	28	0	0	0	0	0	0	0	0	0	0
720	760	34	4	0	0	0	0	0	0	0	0	0
760	800	40	10	0	0	0	0	0	0	0	0	0
800	840	46	16	0	0	0	0	0	0	0	0	0
840	880	52	22	0	0	0	0	0	0	0	0	0
880	920	58	28	0	0	0	0	0	0	0	0	0
920	960	64	34	5	0	0	0	0	0	0	0	0
960	1,000	70	40	11	0	0	0	0	0	0	0	0
1,000	1,040	76	46	17	0	0	0	0	0	0	0	0
1,040	1,080	82	52	23	0	0	0	0	0	0	0	0
1,080	1,120	88	58	29	0	0	0	0	0	0	0	0
1,120	1,160	94	64	35	5	0	0	0	0	0	0	0
1,160	1,200	100	70	41	11	0	0	0	0	0	0	0
1,200	1,240	106	76	47	17	0	0	0	0	0	0	0
1,240	1,280	112	82	53	23	0	0	0	0	0	0	0
1,280	1,320	118	88	59	29	0	0	0	0	0	0	0
1,320	1,360	124	94	65	35	6	0	0	0	0	0	0
1,360	1,400	130	100	71	41	12	0	0	0	0	0	0
1,400	1,440	136	106	77	47	18	0	0	0	0	0	0
1,440	1,480	142	112	83	53	24	0	0	0	0	0	0
1,480	1,520	148	118	89	59	30	1	0	0	0	0	0
1,520	1,560	154	124	95	65	36	7	0	0	0	0	0
1,560	1,600	160	130	101	71	42	13	0	0	0	0	0
1,600	1,640	166	136	107	77	48	19	0	0	0	0	0
1,640	1,680	172	142	113	83	54	25	0	0	0	0	0
1,680	1,720	178	148	119	89	60	31	1	0	0	0	0
1,720	1,760	184	154	125	95	66	37	7	0	0	0	0
1,760	1,800	190	160	131	101	72	43	13	0	0	0	0
1,800	1,840	196	166	137	107	78	49	19	0	0	0	0
1,840	1,880	202	172	143	113	84	55	25	0	0	0	0
1,880	1,920	208	178	149	119	90	61	31	2	0	0	0
1,920	1,960	214	184	155	125	96	67	37	8	0	0	0
1,960	2,000	220	190	161	131	102	73	43	14	0	0	0
2,000	2,040	226	196	167	137	108	79	49	20	0	0	0
2,040	2,080	232	202	173	143	114	85	55	26	0	0	0
2,080	2,120	238	208	179	149	120	91	61	32	3	0	0
2,120	2,160	244	214	185	155	126	97	67	38	9	0	0
2,160	2,200	250	220	191	161	132	103	73	44	15	0	0
2,200	2,240	256	226	197	167	138	109	79	50	21	0	0
2,240	2,280	262	232	203	173	144	115	85	56	27	0	0
2,280	2,320	268	238	209	179	150	121	91	62	33	3	0
2,320	2,360	274	244	215	185	156	127	97	68	39	9	0
2,360	2,400	280	250	221	191	162	133	103	74	45	15	0
2,400	2,440	286	256	227	197	168	139	109	80	51	21	0
2,440	2,480	292	262	233	203	174	145	115	86	57	27	0
2,480	2,520	298	268	239	209	180	151	121	92	63	33	4
2,520	2,560	304	274	245	215	186	157	127	98	69	39	10
2,560	2,600	310	280	251	221	192	163	133	104	75	45	16
2,600	2,640	316	286	257	227	198	169	139	110	81	51	22
2,640	2,680	322	292	263	233	204	175	145	116	87	57	28
2,680	2,720	328	298	269	239	210	181	151	122	93	63	34
2,720	2,760	334	304	275	245	216	187	157	128	99	69	40
2,760	2,800	340	310	281	251	222	193	163	134	105	75	46
2,800	2,840	346	316	287	257	228	199	169	140	111	81	52
2,840	2,880	352	322	293	263	234	205	175	146	117	87	58
2,880	2,920	358	328	299	269	240	211	181	152	123	93	64
2,920	2,960	364	334	305	275	246	217	187	158	129	99	70
2,960	3,000	370	340	311	281	252	223	193	164	135	105	76
3,000	3,040	376	346	317	287	258	229	199	170	141	111	82
3,040	3,080	382	352	323	293	264	235	205	176	147	117	88
3,080	3,120	388	358	329	299	270	241	211	182	153	123	94
3,120	3,160	394	364	335	305	276	247	217	188	159	129	100
3,160	3,200	400	370	341	311	282	253	223	194	165	135	106

(Continued on next page)

MARRIED Persons—MONTHLY Payroll Period

(For Wages Paid in 19--)

If the wages are—		And the number of withholding allowances claimed is—										
At least	But less than	0	1	2	3	4	5	6	7	8	9	10
		The amount of income tax to be withheld is—										
$3,200	$3,240	$406	$376	$347	$317	$288	$259	$229	$200	$171	$141	$112
3,240	3,280	412	382	353	323	294	265	235	206	177	147	118
3,280	3,320	418	388	359	329	300	271	241	212	183	153	124
3,320	3,360	424	394	365	335	306	277	247	218	189	159	130
3,360	3,400	430	400	371	341	312	283	253	224	195	165	136
3,400	3,440	439	406	377	347	318	289	259	230	201	171	142
3,440	3,480	450	412	383	353	324	295	265	236	207	177	148
3,480	3,520	461	418	389	359	330	301	271	242	213	183	154
3,520	3,560	472	424	395	365	336	307	277	248	219	189	160
3,560	3,600	483	430	401	371	342	313	283	254	225	195	166
3,600	3,640	495	440	407	377	348	319	289	260	231	201	172
3,640	3,680	506	451	413	383	354	325	295	266	237	207	178
3,680	3,720	517	462	419	389	360	331	301	272	243	213	184
3,720	3,760	528	473	425	395	366	337	307	278	249	219	190
3,760	3,800	539	485	431	401	372	343	313	284	255	225	196
3,800	3,840	551	496	441	407	378	349	319	290	261	231	202
3,840	3,880	562	507	452	413	384	355	325	296	267	237	208
3,880	3,920	573	518	463	419	390	361	331	302	273	243	214
3,920	3,960	584	529	475	425	396	367	337	308	279	249	220
3,960	4,000	595	541	486	431	402	373	343	314	285	255	226
4,000	4,040	607	552	497	442	408	379	349	320	291	261	232
4,040	4,080	618	563	508	453	414	385	355	326	297	267	238
4,080	4,120	629	574	519	465	420	391	361	332	303	273	244
4,120	4,160	640	585	531	476	426	397	367	338	309	279	250
4,160	4,200	651	597	542	487	432	403	373	344	315	285	256
4,200	4,240	663	608	553	498	443	409	379	350	321	291	262
4,240	4,280	674	619	564	509	455	415	385	356	327	297	268
4,280	4,320	685	630	575	521	466	421	391	362	333	303	274
4,320	4,360	696	641	587	532	477	427	397	368	339	309	280
4,360	4,400	707	653	598	543	488	433	403	374	345	315	286
4,400	4,440	719	664	609	554	499	444	409	380	351	321	292
4,440	4,480	730	675	620	565	511	456	415	386	357	327	298
4,480	4,520	741	686	631	577	522	467	421	392	363	333	304
4,520	4,560	752	697	643	588	533	478	427	398	369	339	310
4,560	4,600	763	709	654	599	544	489	434	404	375	345	316
4,600	4,640	775	720	665	610	555	500	446	410	381	351	322
4,640	4,680	786	731	676	621	567	512	457	416	387	357	328
4,680	4,720	797	742	687	633	578	523	468	422	393	363	334
4,720	4,760	808	753	699	644	589	534	479	428	399	369	340
4,760	4,800	819	765	710	655	600	545	490	436	405	375	346
4,800	4,840	831	776	721	666	611	556	502	447	411	381	352
4,840	4,880	842	787	732	677	623	568	513	458	417	387	358
4,880	4,920	853	798	743	689	634	579	524	469	423	393	364
4,920	4,960	864	809	755	700	645	590	535	480	429	399	370
4,960	5,000	875	821	766	711	656	601	546	492	437	405	376
5,000	5,040	887	832	777	722	667	612	558	503	448	411	382
5,040	5,080	898	843	788	733	679	624	569	514	459	417	388
5,080	5,120	909	854	799	745	690	635	580	525	470	423	394
5,120	5,160	920	865	811	756	701	646	591	536	482	429	400
5,160	5,200	931	877	822	767	712	657	602	548	493	438	406
5,200	5,240	943	888	833	778	723	668	614	559	504	449	412
5,240	5,280	954	899	844	789	735	680	625	570	515	460	418
5,280	5,320	965	910	855	801	746	691	636	581	526	472	424
5,320	5,360	976	921	867	812	757	702	647	592	538	483	430
5,360	5,400	987	933	878	823	768	713	658	604	549	494	439
5,400	5,440	999	944	889	834	779	724	670	615	560	505	450
5,440	5,480	1,010	955	900	845	791	736	681	626	571	516	462
5,480	5,520	1,021	966	911	857	802	747	692	637	582	528	473
5,520	5,560	1,032	977	923	868	813	758	703	648	594	539	484
5,560	5,600	1,043	989	934	879	824	769	714	660	605	550	495
5,600	5,640	1,055	1,000	945	890	835	780	726	671	616	561	506
5,640	5,680	1,066	1,011	956	901	847	792	737	682	627	572	518
5,680	5,720	1,077	1,022	967	913	858	803	748	693	638	584	529
5,720	5,760	1,088	1,033	979	924	869	814	759	704	650	595	540
5,760	5,800	1,099	1,045	990	935	880	825	770	716	661	606	551

$5,800 and over   Use Table 4(b) for a **MARRIED person** on page 26. Also see the instructions on page 24.

19-- Tax Table

If line 5 (Form 1040EZ), line 22 (Form 1040A), or line 37 (Form 1040) is–		And you are–			
At least	But less than	Single	Married filing jointly *	Married filing separately	Head of a house-hold
		Your tax is–			
0	5	0	0	0	0
5	15	2	2	2	2
15	25	3	3	3	3
25	50	6	6	6	6
50	75	9	9	9	9
75	100	13	13	13	13
100	125	17	17	17	17
125	150	21	21	21	21
150	175	24	24	24	24
175	200	28	28	28	28
200	225	32	32	32	32
225	250	36	36	36	36
250	275	39	39	39	39
275	300	43	43	43	43
300	325	47	47	47	47
325	350	51	51	51	51
350	375	54	54	54	54
375	400	58	58	58	58
400	425	62	62	62	62
425	450	66	66	66	66
450	475	69	69	69	69
475	500	73	73	73	73
500	525	77	77	77	77
525	550	81	81	81	81
550	575	84	84	84	84
575	600	88	88	88	88
600	625	92	92	92	92
625	650	96	95	96	96
650	675	99	99	99	99
675	700	103	103	103	103
700	725	107	107	107	107
725	750	111	111	111	111
750	775	114	114	114	114
775	800	118	118	118	118
800	825	122	122	122	122
825	850	126	126	126	126
850	875	129	129	129	129
875	900	133	133	133	133
900	925	137	137	137	137
925	950	141	141	141	141
950	975	144	144	144	144
975	1,000	148	148	148	148

1,000

At least	But less than	Single	Married filing jointly	Married filing separately	Head of a house-hold
1,000	1,025	152	152	152	152
1,025	1,050	156	156	156	156
1,050	1,075	159	159	159	159
1,075	1,100	163	163	163	163
1,100	1,125	167	167	167	167
1,125	1,150	171	171	171	171
1,150	1,175	174	174	174	174
1,175	1,200	178	178	178	178
1,200	1,225	182	182	182	182
1,225	1,250	186	186	186	186
1,250	1,275	189	189	189	189
1,275	1,300	193	193	193	193

If line 5 (Form 1040EZ), line 22 (Form 1040A), or line 37 (Form 1040) is–		And you are–			
At least	But less than	Single	Married filing jointly	Married filing separately	Head of a house-hold
		Your tax is–			
1,300	1,325	197	197	197	197
1,325	1,350	201	201	201	201
1,350	1,375	204	204	204	204
1,375	1,400	208	208	208	208
1,400	1,425	212	212	212	212
1,425	1,450	216	216	216	216
1,450	1,475	219	219	219	219
1,475	1,500	223	223	223	223
1,500	1,525	227	227	227	227
1,525	1,550	231	231	231	231
1,550	1,575	234	234	234	234
1,575	1,600	238	238	238	238
1,600	1,625	242	242	242	242
1,625	1,650	246	246	246	246
1,650	1,675	249	249	249	249
1,675	1,700	253	253	253	253
1,700	1,725	257	257	257	257
1,725	1,750	261	261	261	261
1,750	1,775	264	264	264	264
1,775	1,800	268	268	268	268
1,800	1,825	272	272	272	272
1,825	1,850	276	276	276	276
1,850	1,875	279	279	279	279
1,875	1,900	283	283	283	283
1,900	1,925	287	287	287	287
1,925	1,950	291	291	291	291
1,950	1,975	294	294	294	294
1,975	2,000	298	298	298	298

2,000

At least	But less than	Single	Married filing jointly	Married filing separately	Head of a house-hold
2,000	2,025	302	302	302	302
2,025	2,050	306	306	306	306
2,050	2,075	309	309	309	309
2,075	2,100	313	313	313	313
2,100	2,125	317	317	317	317
2,125	2,150	321	321	321	321
2,150	2,175	324	324	324	324
2,175	2,200	328	328	328	328
2,200	2,225	332	332	332	332
2,225	2,250	336	336	336	336
2,250	2,275	339	339	339	339
2,275	2,300	343	343	343	343
2,300	2,325	347	347	347	347
2,325	2,350	351	351	351	351
2,350	2,375	354	354	354	354
2,375	2,400	358	358	358	358
2,400	2,425	362	362	362	362
2,425	2,450	366	366	366	366
2,450	2,475	369	369	369	369
2,475	2,500	373	373	373	373
2,500	2,525	377	377	377	377
2,525	2,550	381	381	381	381
2,550	2,575	384	384	384	384
2,575	2,600	388	388	388	388
2,600	2,625	392	392	392	392
2,625	2,650	396	396	396	396
2,650	2,675	399	399	399	399
2,675	2,700	403	403	403	403

If line 5 (Form 1040EZ), line 22 (Form 1040A), or line 37 (Form 1040) is–		And you are–			
At least	But less than	Single	Married filing jointly	Married filing separately	Head of a house-hold
		Your tax is–			
2,700	2,725	407	407	407	407
2,725	2,750	411	411	411	411
2,750	2,775	414	414	414	414
2,775	2,800	418	418	418	418
2,800	2,825	422	422	422	422
2,825	2,850	426	426	426	426
2,850	2,875	429	429	429	429
2,875	2,900	433	433	433	433
2,900	2,925	437	437	437	437
2,925	2,950	441	441	441	441
2,950	2,975	444	444	444	444
2,975	3,000	448	448	448	448

3,000

At least	But less than	Single	Married filing jointly	Married filing separately	Head of a house-hold
3,000	3,050	454	454	454	454
3,050	3,100	461	461	461	461
3,100	3,150	469	469	469	469
3,150	3,200	476	476	476	476
3,200	3,250	484	484	484	484
3,250	3,300	491	491	491	491
3,300	3,350	499	499	499	499
3,350	3,400	506	506	506	506
3,400	3,450	514	514	514	514
3,450	3,500	521	521	521	521
3,500	3,550	529	529	529	529
3,550	3,600	536	536	536	536
3,600	3,650	544	544	544	544
3,650	3,700	551	551	551	551
3,700	3,750	559	559	559	559
3,750	3,800	566	566	566	566
3,800	3,850	574	574	574	574
3,850	3,900	581	581	581	581
3,900	3,950	589	589	589	589
3,950	4,000	596	596	596	596

4,000

At least	But less than	Single	Married filing jointly	Married filing separately	Head of a house-hold
4,000	4,050	604	604	604	604
4,050	4,100	611	611	611	611
4,100	4,150	619	619	619	619
4,150	4,200	626	626	626	626
4,200	4,250	634	634	634	634
4,250	4,300	641	641	641	641
4,300	4,350	649	649	649	649
4,350	4,400	656	656	656	656
4,400	4,450	664	664	664	664
4,450	4,500	671	671	671	671
4,500	4,550	679	679	679	679
4,550	4,600	686	686	686	686
4,600	4,650	694	694	694	694
4,650	4,700	701	701	701	701
4,700	4,750	709	709	709	709
4,750	4,800	716	716	716	716
4,800	4,850	724	724	724	724
4,850	4,900	731	731	731	731
4,900	4,950	739	739	739	739
4,950	5,000	746	746	746	746

Continued on next page

* This column must also be used by a qualifying widow(er).

19-- Tax Table—Continued

If line 5 (Form 1040EZ), line 22 (Form 1040A), or line 37 (Form 1040) is—		And you are—			
At least	But less than	Single	Married filing jointly *	Married filing separately	Head of a household
		Your tax is—			
5,000					
5,000	5,050	754	754	754	754
5,050	5,100	761	761	761	761
5,100	5,150	769	769	769	769
5,150	5,200	776	776	776	776
5,200	5,250	784	784	784	784
5,250	5,300	791	791	791	791
5,300	5,350	799	799	799	799
5,350	5,400	806	806	806	806
5,400	5,450	814	814	814	814
5,450	5,500	821	821	821	821
5,500	5,550	829	829	829	829
5,550	5,600	836	836	836	836
5,600	5,650	844	844	844	844
5,650	5,700	851	851	851	851
5,700	5,750	859	859	859	859
5,750	5,800	866	866	866	866
5,800	5,850	874	874	874	874
5,850	5,900	881	881	881	881
5,900	5,950	889	889	889	889
5,950	6,000	896	896	896	896
6,000					
6,000	6,050	904	904	904	904
6,050	6,100	911	911	911	911
6,100	6,150	919	919	919	919
6,150	6,200	926	926	926	926
6,200	6,250	934	934	934	934
6,250	6,300	941	941	941	941
6,300	6,350	949	949	949	949
6,350	6,400	956	956	956	956
6,400	6,450	964	964	964	964
6,450	6,500	971	971	971	971
6,500	6,550	979	979	979	979
6,550	6,600	986	986	986	986
6,600	6,650	994	994	994	994
6,650	6,700	1,001	1,001	1,001	1,001
6,700	6,750	1,009	1,009	1,009	1,009
6,750	6,800	1,016	1,016	1,016	1,016
6,800	6,850	1,024	1,024	1,024	1,024
6,850	6,900	1,031	1,031	1,031	1,031
6,900	6,950	1,039	1,039	1,039	1,039
6,950	7,000	1,046	1,046	1,046	1,046
7,000					
7,000	7,050	1,054	1,054	1,054	1,054
7,050	7,100	1,061	1,061	1,061	1,061
7,100	7,150	1,069	1,069	1,069	1,069
7,150	7,200	1,076	1,076	1,076	1,076
7,200	7,250	1,084	1,084	1,084	1,084
7,250	7,300	1,091	1,091	1,091	1,091
7,300	7,350	1,099	1,099	1,099	1,099
7,350	7,400	1,106	1,106	1,106	1,106
7,400	7,450	1,114	1,114	1,114	1,114
7,450	7,500	1,121	1,121	1,121	1,121
7,500	7,550	1,129	1,129	1,129	1,129
7,550	7,600	1,136	1,136	1,136	1,136
7,600	7,650	1,144	1,144	1,144	1,144
7,650	7,700	1,151	1,151	1,151	1,151
7,700	7,750	1,159	1,159	1,159	1,159
7,750	7,800	1,166	1,166	1,166	1,166
7,800	7,850	1,174	1,174	1,174	1,174
7,850	7,900	1,181	1,181	1,181	1,181
7,900	7,950	1,189	1,189	1,189	1,189
7,950	8,000	1,196	1,196	1,196	1,196

If line 5 (Form 1040EZ), line 22 (Form 1040A), or line 37 (Form 1040) is—		And you are—			
At least	But less than	Single	Married filing jointly *	Married filing separately	Head of a household
		Your tax is—			
8,000					
8,000	8,050	1,204	1,204	1,204	1,204
8,050	8,100	1,211	1,211	1,211	1,211
8,100	8,150	1,219	1,219	1,219	1,219
8,150	8,200	1,226	1,226	1,226	1,226
8,200	8,250	1,234	1,234	1,234	1,234
8,250	8,300	1,241	1,241	1,241	1,241
8,300	8,350	1,249	1,249	1,249	1,249
8,350	8,400	1,256	1,256	1,256	1,256
8,400	8,450	1,264	1,264	1,264	1,264
8,450	8,500	1,271	1,271	1,271	1,271
8,500	8,550	1,279	1,279	1,279	1,279
8,550	8,600	1,286	1,286	1,286	1,286
8,600	8,650	1,294	1,294	1,294	1,294
8,650	8,700	1,301	1,301	1,301	1,301
8,700	8,750	1,309	1,309	1,309	1,309
8,750	8,800	1,316	1,316	1,316	1,316
8,800	8,850	1,324	1,324	1,324	1,324
8,850	8,900	1,331	1,331	1,331	1,331
8,900	8,950	1,339	1,339	1,339	1,339
8,950	9,000	1,346	1,346	1,346	1,346
9,000					
9,000	9,050	1,354	1,354	1,354	1,354
9,050	9,100	1,361	1,361	1,361	1,361
9,100	9,150	1,369	1,369	1,369	1,369
9,150	9,200	1,376	1,376	1,376	1,376
9,200	9,250	1,384	1,384	1,384	1,384
9,250	9,300	1,391	1,391	1,391	1,391
9,300	9,350	1,399	1,399	1,399	1,399
9,350	9,400	1,406	1,406	1,406	1,406
9,400	9,450	1,414	1,414	1,414	1,414
9,450	9,500	1,421	1,421	1,421	1,421
9,500	9,550	1,429	1,429	1,429	1,429
9,550	9,600	1,436	1,436	1,436	1,436
9,600	9,650	1,444	1,444	1,444	1,444
9,650	9,700	1,451	1,451	1,451	1,451
9,700	9,750	1,459	1,459	1,459	1,459
9,750	9,800	1,466	1,466	1,466	1,466
9,800	9,850	1,474	1,474	1,474	1,474
9,850	9,900	1,481	1,481	1,481	1,481
9,900	9,950	1,489	1,489	1,489	1,489
9,950	10,000	1,496	1,496	1,496	1,496
10,000					
10,000	10,050	1,504	1,504	1,504	1,504
10,050	10,100	1,511	1,511	1,511	1,511
10,100	10,150	1,519	1,519	1,519	1,519
10,150	10,200	1,526	1,526	1,526	1,526
10,200	10,250	1,534	1,534	1,534	1,534
10,250	10,300	1,541	1,541	1,541	1,541
10,300	10,350	1,549	1,549	1,549	1,549
10,350	10,400	1,556	1,556	1,556	1,556
10,400	10,450	1,564	1,564	1,564	1,564
10,450	10,500	1,571	1,571	1,571	1,571
10,500	10,550	1,579	1,579	1,579	1,579
10,550	10,600	1,586	1,586	1,586	1,586
10,600	10,650	1,594	1,594	1,594	1,594
10,650	10,700	1,601	1,601	1,601	1,601
10,700	10,750	1,609	1,609	1,609	1,609
10,750	10,800	1,616	1,616	1,616	1,616
10,800	10,850	1,624	1,624	1,624	1,624
10,850	10,900	1,631	1,631	1,631	1,631
10,900	10,950	1,639	1,639	1,639	1,639
10,950	11,000	1,646	1,646	1,646	1,646

If line 5 (Form 1040EZ), line 22 (Form 1040A), or line 37 (Form 1040) is—		And you are—			
At least	But less than	Single	Married filing jointly *	Married filing separately	Head of a household
		Your tax is—			
11,000					
11,000	11,050	1,654	1,654	1,654	1,654
11,050	11,100	1,661	1,661	1,661	1,661
11,100	11,150	1,669	1,669	1,669	1,669
11,150	11,200	1,676	1,676	1,676	1,676
11,200	11,250	1,684	1,684	1,684	1,684
11,250	11,300	1,691	1,691	1,691	1,691
11,300	11,350	1,699	1,699	1,699	1,699
11,350	11,400	1,706	1,706	1,706	1,706
11,400	11,450	1,714	1,714	1,714	1,714
11,450	11,500	1,721	1,721	1,721	1,721
11,500	11,550	1,729	1,729	1,729	1,729
11,550	11,600	1,736	1,736	1,736	1,736
11,600	11,650	1,744	1,744	1,744	1,744
11,650	11,700	1,751	1,751	1,751	1,751
11,700	11,750	1,759	1,759	1,759	1,759
11,750	11,800	1,766	1,766	1,766	1,766
11,800	11,850	1,774	1,774	1,774	1,774
11,850	11,900	1,781	1,781	1,781	1,781
11,900	11,950	1,789	1,789	1,789	1,789
11,950	12,000	1,796	1,796	1,796	1,796
12,000					
12,000	12,050	1,804	1,804	1,804	1,804
12,050	12,100	1,811	1,811	1,811	1,811
12,100	12,150	1,819	1,819	1,819	1,819
12,150	12,200	1,826	1,826	1,826	1,826
12,200	12,250	1,834	1,834	1,834	1,834
12,250	12,300	1,841	1,841	1,841	1,841
12,300	12,350	1,849	1,849	1,849	1,849
12,350	12,400	1,856	1,856	1,856	1,856
12,400	12,450	1,864	1,864	1,864	1,864
12,450	12,500	1,871	1,871	1,871	1,871
12,500	12,550	1,879	1,879	1,879	1,879
12,550	12,600	1,886	1,886	1,886	1,886
12,600	12,650	1,894	1,894	1,894	1,894
12,650	12,700	1,901	1,901	1,901	1,901
12,700	12,750	1,909	1,909	1,909	1,909
12,750	12,800	1,916	1,916	1,916	1,916
12,800	12,850	1,924	1,924	1,924	1,924
12,850	12,900	1,931	1,931	1,931	1,931
12,900	12,950	1,939	1,939	1,939	1,939
12,950	13,000	1,946	1,946	1,946	1,946
13,000					
13,000	13,050	1,954	1,954	1,954	1,954
13,050	13,100	1,961	1,961	1,961	1,961
13,100	13,150	1,969	1,969	1,969	1,969
13,150	13,200	1,976	1,976	1,976	1,976
13,200	13,250	1,984	1,984	1,984	1,984
13,250	13,300	1,991	1,991	1,991	1,991
13,300	13,350	1,999	1,999	1,999	1,999
13,350	13,400	2,006	2,006	2,006	2,006
13,400	13,450	2,014	2,014	2,014	2,014
13,450	13,500	2,021	2,021	2,021	2,021
13,500	13,550	2,029	2,029	2,029	2,029
13,550	13,600	2,036	2,036	2,036	2,036
13,600	13,650	2,044	2,044	2,044	2,044
13,650	13,700	2,051	2,051	2,051	2,051
13,700	13,750	2,059	2,059	2,059	2,059
13,750	13,800	2,066	2,066	2,066	2,066
13,800	13,850	2,074	2,074	2,074	2,074
13,850	13,900	2,081	2,081	2,081	2,081
13,900	13,950	2,089	2,089	2,089	2,089
13,950	14,000	2,096	2,096	2,096	2,096

* This column must also be used by a qualifying widow(er).

Continued on next page

19-- Tax Table–*Continued*

14,000 / 15,000 / 16,000

At least	But less than	Single	Married filing jointly*	Married filing separately	Head of a household
14,000					
14,000	14,050	2,104	2,104	2,104	2,104
14,050	14,100	2,111	2,111	2,111	2,111
14,100	14,150	2,119	2,119	2,119	2,119
14,150	14,200	2,126	2,126	2,126	2,126
14,200	14,250	2,134	2,134	2,134	2,134
14,250	14,300	2,141	2,141	2,141	2,141
14,300	14,350	2,149	2,149	2,149	2,149
14,350	14,400	2,156	2,156	2,156	2,156
14,400	14,450	2,164	2,164	2,164	2,164
14,450	14,500	2,171	2,171	2,171	2,171
14,500	14,550	2,179	2,179	2,179	2,179
14,550	14,600	2,186	2,186	2,186	2,186
14,600	14,650	2,194	2,194	2,194	2,194
14,650	14,700	2,201	2,201	2,201	2,201
14,700	14,750	2,209	2,209	2,209	2,209
14,750	14,800	2,216	2,216	2,216	2,216
14,800	14,850	2,224	2,224	2,224	2,224
14,850	14,900	2,231	2,231	2,231	2,231
14,900	14,950	2,239	2,239	2,239	2,239
14,950	15,000	2,246	2,246	2,246	2,246
15,000					
15,000	15,050	2,254	2,254	2,254	2,254
15,050	15,100	2,261	2,261	2,261	2,261
15,100	15,150	2,269	2,269	2,269	2,269
15,150	15,200	2,276	2,276	2,276	2,276
15,200	15,250	2,284	2,284	2,284	2,284
15,250	15,300	2,291	2,291	2,291	2,291
15,300	15,350	2,299	2,299	2,299	2,299
15,350	15,400	2,306	2,306	2,306	2,306
15,400	15,450	2,314	2,314	2,314	2,314
15,450	15,500	2,321	2,321	2,321	2,321
15,500	15,550	2,329	2,329	2,329	2,329
15,550	15,600	2,336	2,336	2,336	2,336
15,600	15,650	2,344	2,344	2,344	2,344
15,650	15,700	2,351	2,351	2,351	2,351
15,700	15,750	2,359	2,359	2,359	2,359
15,750	15,800	2,366	2,366	2,366	2,366
15,800	15,850	2,374	2,374	2,374	2,374
15,850	15,900	2,381	2,381	2,381	2,381
15,900	15,950	2,389	2,389	2,389	2,389
15,950	16,000	2,396	2,396	2,396	2,396
16,000					
16,000	16,050	2,404	2,404	2,404	2,404
16,050	16,100	2,411	2,411	2,411	2,411
16,100	16,150	2,419	2,419	2,419	2,419
16,150	16,200	2,426	2,426	2,426	2,426
16,200	16,250	2,434	2,434	2,434	2,434
16,250	16,300	2,441	2,441	2,441	2,441
16,300	16,350	2,449	2,449	2,449	2,449
16,350	16,400	2,456	2,456	2,456	2,456
16,400	16,450	2,464	2,464	2,464	2,464
16,450	16,500	2,471	2,471	2,471	2,471
16,500	16,550	2,479	2,479	2,479	2,479
16,550	16,600	2,486	2,486	2,486	2,486
16,600	16,650	2,494	2,494	2,494	2,494
16,650	16,700	2,501	2,501	2,501	2,501
16,700	16,750	2,509	2,509	2,509	2,509
16,750	16,800	2,516	2,516	2,516	2,516
16,800	16,850	2,524	2,524	2,524	2,524
16,850	16,900	2,531	2,531	2,531	2,531
16,900	16,950	2,539	2,539	2,539	2,539
16,950	17,000	2,546	2,546	2,546	2,546

17,000 / 18,000 / 19,000

At least	But less than	Single	Married filing jointly*	Married filing separately	Head of a household
17,000					
17,000	17,050	2,554	2,554	2,554	2,554
17,050	17,100	2,561	2,561	2,561	2,561
17,100	17,150	2,569	2,569	2,569	2,569
17,150	17,200	2,576	2,576	2,576	2,576
17,200	17,250	2,584	2,584	2,584	2,584
17,250	17,300	2,591	2,591	2,591	2,591
17,300	17,350	2,599	2,599	2,599	2,599
17,350	17,400	2,606	2,606	2,606	2,606
17,400	17,450	2,614	2,614	2,614	2,614
17,450	17,500	2,621	2,621	2,621	2,621
17,500	17,550	2,629	2,629	2,629	2,629
17,550	17,600	2,636	2,636	2,636	2,636
17,600	17,650	2,644	2,644	2,644	2,644
17,650	17,700	2,651	2,651	2,651	2,651
17,700	17,750	2,659	2,659	2,659	2,659
17,750	17,800	2,666	2,666	2,666	2,666
17,800	17,850	2,674	2,674	2,674	2,674
17,850	17,900	2,681	2,681	2,681	2,681
17,900	17,950	2,689	2,689	2,692	2,689
17,950	18,000	2,696	2,696	2,706	2,696
18,000					
18,000	18,050	2,704	2,704	2,720	2,704
18,050	18,100	2,711	2,711	2,734	2,711
18,100	18,150	2,719	2,719	2,748	2,719
18,150	18,200	2,726	2,726	2,762	2,726
18,200	18,250	2,734	2,734	2,776	2,734
18,250	18,300	2,741	2,741	2,790	2,741
18,300	18,350	2,749	2,749	2,804	2,749
18,350	18,400	2,756	2,756	2,818	2,756
18,400	18,450	2,764	2,764	2,832	2,764
18,450	18,500	2,771	2,771	2,846	2,771
18,500	18,550	2,779	2,779	2,860	2,779
18,550	18,600	2,786	2,786	2,874	2,786
18,600	18,650	2,794	2,794	2,888	2,794
18,650	18,700	2,801	2,801	2,902	2,801
18,700	18,750	2,809	2,809	2,916	2,809
18,750	18,800	2,816	2,816	2,930	2,816
18,800	18,850	2,824	2,824	2,944	2,824
18,850	18,900	2,831	2,831	2,958	2,831
18,900	18,950	2,839	2,839	2,972	2,839
18,950	19,000	2,846	2,846	2,986	2,846
19,000					
19,000	19,050	2,854	2,854	3,000	2,854
19,050	19,100	2,861	2,861	3,014	2,861
19,100	19,150	2,869	2,869	3,028	2,869
19,150	19,200	2,876	2,876	3,042	2,876
19,200	19,250	2,884	2,884	3,056	2,884
19,250	19,300	2,891	2,891	3,070	2,891
19,300	19,350	2,899	2,899	3,084	2,899
19,350	19,400	2,906	2,906	3,098	2,906
19,400	19,450	2,914	2,914	3,112	2,914
19,450	19,500	2,921	2,921	3,126	2,921
19,500	19,550	2,929	2,929	3,140	2,929
19,550	19,600	2,936	2,936	3,154	2,936
19,600	19,650	2,944	2,944	3,168	2,944
19,650	19,700	2,951	2,951	3,182	2,951
19,700	19,750	2,959	2,959	3,196	2,959
19,750	19,800	2,966	2,966	3,210	2,966
19,800	19,850	2,974	2,974	3,224	2,974
19,850	19,900	2,981	2,981	3,238	2,981
19,900	19,950	2,989	2,989	3,252	2,989
19,950	20,000	2,996	2,996	3,266	2,996

20,000 / 21,000 / 22,000

At least	But less than	Single	Married filing jointly*	Married filing separately	Head of a household
20,000					
20,000	20,050	3,004	3,004	3,280	3,004
20,050	20,100	3,011	3,011	3,294	3,011
20,100	20,150	3,019	3,019	3,308	3,019
20,150	20,200	3,026	3,026	3,322	3,026
20,200	20,250	3,034	3,034	3,336	3,034
20,250	20,300	3,041	3,041	3,350	3,041
20,300	20,350	3,049	3,049	3,364	3,049
20,350	20,400	3,056	3,056	3,378	3,056
20,400	20,450	3,064	3,064	3,392	3,064
20,450	20,500	3,071	3,071	3,406	3,071
20,500	20,550	3,079	3,079	3,420	3,079
20,550	20,600	3,086	3,086	3,434	3,086
20,600	20,650	3,094	3,094	3,448	3,094
20,650	20,700	3,101	3,101	3,462	3,101
20,700	20,750	3,109	3,109	3,476	3,109
20,750	20,800	3,116	3,116	3,490	3,116
20,800	20,850	3,124	3,124	3,504	3,124
20,850	20,900	3,131	3,131	3,518	3,131
20,900	20,950	3,139	3,139	3,532	3,139
20,950	21,000	3,146	3,146	3,546	3,146
21,000					
21,000	21,050	3,154	3,154	3,560	3,154
21,050	21,100	3,161	3,161	3,574	3,161
21,100	21,150	3,169	3,169	3,588	3,169
21,150	21,200	3,176	3,176	3,602	3,176
21,200	21,250	3,184	3,184	3,616	3,184
21,250	21,300	3,191	3,191	3,630	3,191
21,300	21,350	3,199	3,199	3,644	3,199
21,350	21,400	3,206	3,206	3,658	3,206
21,400	21,450	3,214	3,214	3,672	3,214
21,450	21,500	3,225	3,221	3,686	3,221
21,500	21,550	3,239	3,229	3,700	3,229
21,550	21,600	3,253	3,236	3,714	3,236
21,600	21,650	3,267	3,244	3,728	3,244
21,650	21,700	3,281	3,251	3,742	3,251
21,700	21,750	3,295	3,259	3,756	3,259
21,750	21,800	3,309	3,266	3,770	3,266
21,800	21,850	3,323	3,274	3,784	3,274
21,850	21,900	3,337	3,281	3,798	3,281
21,900	21,950	3,351	3,289	3,812	3,289
21,950	22,000	3,365	3,296	3,826	3,296
22,000					
22,000	22,050	3,379	3,304	3,840	3,304
22,050	22,100	3,393	3,311	3,854	3,311
22,100	22,150	3,407	3,319	3,868	3,319
22,150	22,200	3,421	3,326	3,882	3,326
22,200	22,250	3,435	3,334	3,896	3,334
22,250	22,300	3,449	3,341	3,910	3,341
22,300	22,350	3,463	3,349	3,924	3,349
22,350	22,400	3,477	3,356	3,938	3,356
22,400	22,450	3,491	3,364	3,952	3,364
22,450	22,500	3,505	3,371	3,966	3,371
22,500	22,550	3,519	3,379	3,980	3,379
22,550	22,600	3,533	3,386	3,994	3,386
22,600	22,650	3,547	3,394	4,008	3,394
22,650	22,700	3,561	3,401	4,022	3,401
22,700	22,750	3,575	3,409	4,036	3,409
22,750	22,800	3,589	3,416	4,050	3,416
22,800	22,850	3,603	3,424	4,064	3,424
22,850	22,900	3,617	3,431	4,078	3,431
22,900	22,950	3,631	3,439	4,092	3,439
22,950	23,000	3,645	3,446	4,106	3,446

* This column must also be used by a qualifying widow(er).

Continued on next page

19-- Tax Table—Continued

At least	But less than	Single	Married filing jointly*	Married filing separately	Head of a household

Your tax is—

23,000

At least	But less than	Single	Married filing jointly*	Married filing separately	Head of a household
23,000	23,050	3,659	3,454	4,120	3,454
23,050	23,100	3,673	3,461	4,134	3,461
23,100	23,150	3,687	3,469	4,148	3,469
23,150	23,200	3,701	3,476	4,162	3,476
23,200	23,250	3,715	3,484	4,176	3,484
23,250	23,300	3,729	3,491	4,190	3,491
23,300	23,350	3,743	3,499	4,204	3,499
23,350	23,400	3,757	3,506	4,218	3,506
23,400	23,450	3,771	3,514	4,232	3,514
23,450	23,500	3,785	3,521	4,246	3,521
23,500	23,550	3,799	3,529	4,260	3,529
23,550	23,600	3,813	3,536	4,274	3,536
23,600	23,650	3,827	3,544	4,288	3,544
23,650	23,700	3,841	3,551	4,302	3,551
23,700	23,750	3,855	3,559	4,316	3,559
23,750	23,800	3,869	3,566	4,330	3,566
23,800	23,850	3,883	3,574	4,344	3,574
23,850	23,900	3,897	3,581	4,358	3,581
23,900	23,950	3,911	3,589	4,372	3,589
23,950	24,000	3,925	3,596	4,386	3,596

24,000

At least	But less than	Single	Married filing jointly*	Married filing separately	Head of a household
24,000	24,050	3,939	3,604	4,400	3,604
24,050	24,100	3,953	3,611	4,414	3,611
24,100	24,150	3,967	3,619	4,428	3,619
24,150	24,200	3,981	3,626	4,442	3,626
24,200	24,250	3,995	3,634	4,456	3,634
24,250	24,300	4,009	3,641	4,470	3,641
24,300	24,350	4,023	3,649	4,484	3,649
24,350	24,400	4,037	3,656	4,498	3,656
24,400	24,450	4,051	3,664	4,512	3,664
24,450	24,500	4,065	3,671	4,526	3,671
24,500	24,550	4,079	3,679	4,540	3,679
24,550	24,600	4,093	3,686	4,554	3,686
24,600	24,650	4,107	3,694	4,568	3,694
24,650	24,700	4,121	3,701	4,582	3,701
24,700	24,750	4,135	3,709	4,596	3,709
24,750	24,800	4,149	3,716	4,610	3,716
24,800	24,850	4,163	3,724	4,624	3,724
24,850	24,900	4,177	3,731	4,638	3,731
24,900	24,950	4,191	3,739	4,652	3,739
24,950	25,000	4,205	3,746	4,666	3,746

25,000

At least	But less than	Single	Married filing jointly*	Married filing separately	Head of a household
25,000	25,050	4,219	3,754	4,680	3,754
25,050	25,100	4,233	3,761	4,694	3,761
25,100	25,150	4,247	3,769	4,708	3,769
25,150	25,200	4,261	3,776	4,722	3,776
25,200	25,250	4,275	3,784	4,736	3,784
25,250	25,300	4,289	3,791	4,750	3,791
25,300	25,350	4,303	3,799	4,764	3,799
25,350	25,400	4,317	3,806	4,778	3,806
25,400	25,450	4,331	3,814	4,792	3,814
25,450	25,500	4,345	3,821	4,806	3,821
25,500	25,550	4,359	3,829	4,820	3,829
25,550	25,600	4,373	3,836	4,834	3,836
25,600	25,650	4,387	3,844	4,848	3,844
25,650	25,700	4,401	3,851	4,862	3,851
25,700	25,750	4,415	3,859	4,876	3,859
25,750	25,800	4,429	3,866	4,890	3,866
25,800	25,850	4,443	3,874	4,904	3,874
25,850	25,900	4,457	3,881	4,918	3,881
25,900	25,950	4,471	3,889	4,932	3,889
25,950	26,000	4,485	3,896	4,946	3,896

26,000

At least	But less than	Single	Married filing jointly*	Married filing separately	Head of a household
26,000	26,050	4,499	3,904	4,960	3,904
26,050	26,100	4,513	3,911	4,974	3,911
26,100	26,150	4,527	3,919	4,988	3,919
26,150	26,200	4,541	3,926	5,002	3,926
26,200	26,250	4,555	3,934	5,016	3,934
26,250	26,300	4,569	3,941	5,030	3,941
26,300	26,350	4,583	3,949	5,044	3,949
26,350	26,400	4,597	3,956	5,058	3,956
26,400	26,450	4,611	3,964	5,072	3,964
26,450	26,500	4,625	3,971	5,086	3,971
26,500	26,550	4,639	3,979	5,100	3,979
26,550	26,600	4,653	3,986	5,114	3,986
26,600	26,650	4,667	3,994	5,128	3,994
26,650	26,700	4,681	4,001	5,142	4,001
26,700	26,750	4,695	4,009	5,156	4,009
26,750	26,800	4,709	4,016	5,170	4,016
26,800	26,850	4,723	4,024	5,184	4,024
26,850	26,900	4,737	4,031	5,198	4,031
26,900	26,950	4,751	4,039	5,212	4,039
26,950	27,000	4,765	4,046	5,226	4,046

27,000

At least	But less than	Single	Married filing jointly*	Married filing separately	Head of a household
27,000	27,050	4,779	4,054	5,240	4,054
27,050	27,100	4,793	4,061	5,254	4,061
27,100	27,150	4,807	4,069	5,268	4,069
27,150	27,200	4,821	4,076	5,282	4,076
27,200	27,250	4,835	4,084	5,296	4,084
27,250	27,300	4,849	4,091	5,310	4,091
27,300	27,350	4,863	4,099	5,324	4,099
27,350	27,400	4,877	4,106	5,338	4,106
27,400	27,450	4,891	4,114	5,352	4,114
27,450	27,500	4,905	4,121	5,366	4,121
27,500	27,550	4,919	4,129	5,380	4,129
27,550	27,600	4,933	4,136	5,394	4,136
27,600	27,650	4,947	4,144	5,408	4,144
27,650	27,700	4,961	4,151	5,422	4,151
27,700	27,750	4,975	4,159	5,436	4,159
27,750	27,800	4,989	4,166	5,450	4,166
27,800	27,850	5,003	4,174	5,464	4,174
27,850	27,900	5,017	4,181	5,478	4,181
27,900	27,950	5,031	4,189	5,492	4,189
27,950	28,000	5,045	4,196	5,506	4,196

28,000

At least	But less than	Single	Married filing jointly*	Married filing separately	Head of a household
28,000	28,050	5,059	4,204	5,520	4,204
28,050	28,100	5,073	4,211	5,534	4,211
28,100	28,150	5,087	4,219	5,548	4,219
28,150	28,200	5,101	4,226	5,562	4,226
28,200	28,250	5,115	4,234	5,576	4,234
28,250	28,300	5,129	4,241	5,590	4,241
28,300	28,350	5,143	4,249	5,604	4,249
28,350	28,400	5,157	4,256	5,618	4,256
28,400	28,450	5,171	4,264	5,632	4,264
28,450	28,500	5,185	4,271	5,646	4,271
28,500	28,550	5,199	4,279	5,660	4,279
28,550	28,600	5,213	4,286	5,674	4,286
28,600	28,650	5,227	4,294	5,688	4,294
28,650	28,700	5,241	4,301	5,702	4,301
28,700	28,750	5,255	4,309	5,716	4,309
28,750	28,800	5,269	4,316	5,730	4,320
28,800	28,850	5,283	4,324	5,744	4,334
28,850	28,900	5,297	4,331	5,758	4,348
28,900	28,950	5,311	4,339	5,772	4,362
28,950	29,000	5,325	4,346	5,786	4,376

29,000

At least	But less than	Single	Married filing jointly*	Married filing separately	Head of a household
29,000	29,050	5,339	4,354	5,800	4,390
29,050	29,100	5,353	4,361	5,814	4,404
29,100	29,150	5,367	4,369	5,828	4,418
29,150	29,200	5,381	4,376	5,842	4,432
29,200	29,250	5,395	4,384	5,856	4,446
29,250	29,300	5,409	4,391	5,870	4,460
29,300	29,350	5,423	4,399	5,884	4,474
29,350	29,400	5,437	4,406	5,898	4,488
29,400	29,450	5,451	4,414	5,912	4,502
29,450	29,500	5,465	4,421	5,926	4,516
29,500	29,550	5,479	4,429	5,940	4,530
29,550	29,600	5,493	4,436	5,954	4,544
29,600	29,650	5,507	4,444	5,968	4,558
29,650	29,700	5,521	4,451	5,982	4,572
29,700	29,750	5,535	4,459	5,996	4,586
29,750	29,800	5,549	4,466	6,010	4,600
29,800	29,850	5,563	4,474	6,024	4,614
29,850	29,900	5,577	4,481	6,038	4,628
29,900	29,950	5,591	4,489	6,052	4,642
29,950	30,000	5,605	4,496	6,066	4,656

30,000

At least	But less than	Single	Married filing jointly*	Married filing separately	Head of a household
30,000	30,050	5,619	4,504	6,080	4,670
30,050	30,100	5,633	4,511	6,094	4,684
30,100	30,150	5,647	4,519	6,108	4,698
30,150	30,200	5,661	4,526	6,122	4,712
30,200	30,250	5,675	4,534	6,136	4,726
30,250	30,300	5,689	4,541	6,150	4,740
30,300	30,350	5,703	4,549	6,164	4,754
30,350	30,400	5,717	4,556	6,178	4,768
30,400	30,450	5,731	4,564	6,192	4,782
30,450	30,500	5,745	4,571	6,206	4,796
30,500	30,550	5,759	4,579	6,220	4,810
30,550	30,600	5,773	4,586	6,234	4,824
30,600	30,650	5,787	4,594	6,248	4,838
30,650	30,700	5,801	4,601	6,262	4,852
30,700	30,750	5,815	4,609	6,276	4,866
30,750	30,800	5,829	4,616	6,290	4,880
30,800	30,850	5,843	4,624	6,304	4,894
30,850	30,900	5,857	4,631	6,318	4,908
30,900	30,950	5,871	4,639	6,332	4,922
30,950	31,000	5,885	4,646	6,346	4,936

31,000

At least	But less than	Single	Married filing jointly*	Married filing separately	Head of a household
31,000	31,050	5,899	4,654	6,360	4,950
31,050	31,100	5,913	4,661	6,374	4,964
31,100	31,150	5,927	4,669	6,388	4,978
31,150	31,200	5,941	4,676	6,402	4,992
31,200	31,250	5,955	4,684	6,416	5,006
31,250	31,300	5,969	4,691	6,430	5,020
31,300	31,350	5,983	4,699	6,444	5,034
31,350	31,400	5,997	4,706	6,458	5,048
31,400	31,450	6,011	4,714	6,472	5,062
31,450	31,500	6,025	4,721	6,486	5,076
31,500	31,550	6,039	4,729	6,500	5,090
31,550	31,600	6,053	4,736	6,514	5,104
31,600	31,650	6,067	4,744	6,528	5,118
31,650	31,700	6,081	4,751	6,542	5,132
31,700	31,750	6,095	4,759	6,556	5,146
31,750	31,800	6,109	4,766	6,570	5,160
31,800	31,850	6,123	4,774	6,584	5,174
31,850	31,900	6,137	4,781	6,598	5,188
31,900	31,950	6,151	4,789	6,612	5,202
31,950	32,000	6,165	4,796	6,626	5,216

* This column must also be used by a qualifying widow(er).

Continued on next page

19-- Tax Table—*Continued*

If line 5 (Form 1040EZ), line 22 (Form 1040A), or line 37 (Form 1040) is—		And you are—			
At least	But less than	Single	Married filing jointly *	Married filing separately	Head of a household
		Your tax is—			

32,000

At least	But less than	Single	Married filing jointly	Married filing separately	Head of a household
32,000	32,050	6,179	4,804	6,640	5,230
32,050	32,100	6,193	4,811	6,654	5,244
32,100	32,150	6,207	4,819	6,668	5,258
32,150	32,200	6,221	4,826	6,682	5,272
32,200	32,250	6,235	4,834	6,696	5,286
32,250	32,300	6,249	4,841	6,710	5,300
32,300	32,350	6,263	4,849	6,724	5,314
32,350	32,400	6,277	4,856	6,738	5,328
32,400	32,450	6,291	4,864	6,752	5,342
32,450	32,500	6,305	4,871	6,766	5,356
32,500	32,550	6,319	4,879	6,780	5,370
32,550	32,600	6,333	4,886	6,794	5,384
32,600	32,650	6,347	4,894	6,808	5,398
32,650	32,700	6,361	4,901	6,822	5,412
32,700	32,750	6,375	4,909	6,836	5,426
32,750	32,800	6,389	4,916	6,850	5,440
32,800	32,850	6,403	4,924	6,864	5,454
32,850	32,900	6,417	4,931	6,878	5,468
32,900	32,950	6,431	4,939	6,892	5,482
32,950	33,000	6,445	4,946	6,906	5,496

33,000

At least	But less than	Single	Married filing jointly	Married filing separately	Head of a household
33,000	33,050	6,459	4,954	6,920	5,510
33,050	33,100	6,473	4,961	6,934	5,524
33,100	33,150	6,487	4,969	6,948	5,538
33,150	33,200	6,501	4,976	6,962	5,552
33,200	33,250	6,515	4,984	6,976	5,566
33,250	33,300	6,529	4,991	6,990	5,580
33,300	33,350	6,543	4,999	7,004	5,594
33,350	33,400	6,557	5,006	7,018	5,608
33,400	33,450	6,571	5,014	7,032	5,622
33,450	33,500	6,585	5,021	7,046	5,636
33,500	33,550	6,599	5,029	7,060	5,650
33,550	33,600	6,613	5,036	7,074	5,664
33,600	33,650	6,627	5,044	7,088	5,678
33,650	33,700	6,641	5,051	7,102	5,692
33,700	33,750	6,655	5,059	7,116	5,706
33,750	33,800	6,669	5,066	7,130	5,720
33,800	33,850	6,683	5,074	7,144	5,734
33,850	33,900	6,697	5,081	7,158	5,748
33,900	33,950	6,711	5,089	7,172	5,762
33,950	34,000	6,725	5,096	7,186	5,776

34,000

At least	But less than	Single	Married filing jointly	Married filing separately	Head of a household
34,000	34,050	6,739	5,104	7,200	5,790
34,050	34,100	6,753	5,111	7,214	5,804
34,100	34,150	6,767	5,119	7,228	5,818
34,150	34,200	6,781	5,126	7,242	5,832
34,200	34,250	6,795	5,134	7,256	5,846
34,250	34,300	6,809	5,141	7,270	5,860
34,300	34,350	6,823	5,149	7,284	5,874
34,350	34,400	6,837	5,156	7,298	5,888
34,400	34,450	6,851	5,164	7,312	5,902
34,450	34,500	6,865	5,171	7,326	5,916
34,500	34,550	6,879	5,179	7,340	5,930
34,550	34,600	6,893	5,186	7,354	5,944
34,600	34,650	6,907	5,194	7,368	5,958
34,650	34,700	6,921	5,201	7,382	5,972
34,700	34,750	6,935	5,209	7,396	5,986
34,750	34,800	6,949	5,216	7,410	6,000
34,800	34,850	6,963	5,224	7,424	6,014
34,850	34,900	6,977	5,231	7,438	6,028
34,900	34,950	6,991	5,239	7,452	6,042
34,950	35,000	7,005	5,246	7,466	6,056

35,000

At least	But less than	Single	Married filing jointly	Married filing separately	Head of a household
35,000	35,050	7,019	5,254	7,480	6,070
35,050	35,100	7,033	5,261	7,494	6,084
35,100	35,150	7,047	5,269	7,508	6,098
35,150	35,200	7,061	5,276	7,522	6,112
35,200	35,250	7,075	5,284	7,536	6,126
35,250	35,300	7,089	5,291	7,550	6,140
35,300	35,350	7,103	5,299	7,564	6,154
35,350	35,400	7,117	5,306	7,578	6,168
35,400	35,450	7,131	5,314	7,592	6,182
35,450	35,500	7,145	5,321	7,606	6,196
35,500	35,550	7,159	5,329	7,620	6,210
35,550	35,600	7,173	5,336	7,634	6,224
35,600	35,650	7,187	5,344	7,648	6,238
35,650	35,700	7,201	5,351	7,662	6,252
35,700	35,750	7,215	5,359	7,676	6,266
35,750	35,800	7,229	5,366	7,690	6,280
35,800	35,850	7,243	5,377	7,704	6,294
35,850	35,900	7,257	5,391	7,718	6,308
35,900	35,950	7,271	5,405	7,732	6,322
35,950	36,000	7,285	5,419	7,746	6,336

36,000

At least	But less than	Single	Married filing jointly	Married filing separately	Head of a household
36,000	36,050	7,299	5,433	7,760	6,350
36,050	36,100	7,313	5,447	7,774	6,364
36,100	36,150	7,327	5,461	7,788	6,378
36,150	36,200	7,341	5,475	7,802	6,392
36,200	36,250	7,355	5,489	7,816	6,406
36,250	36,300	7,369	5,503	7,830	6,420
36,300	36,350	7,383	5,517	7,844	6,434
36,350	36,400	7,397	5,531	7,858	6,448
36,400	36,450	7,411	5,545	7,872	6,462
36,450	36,500	7,425	5,559	7,886	6,476
36,500	36,550	7,439	5,573	7,900	6,490
36,550	36,600	7,453	5,587	7,914	6,504
36,600	36,650	7,467	5,601	7,928	6,518
36,650	36,700	7,481	5,615	7,942	6,532
36,700	36,750	7,495	5,629	7,956	6,546
36,750	36,800	7,509	5,643	7,970	6,560
36,800	36,850	7,523	5,657	7,984	6,574
36,850	36,900	7,537	5,671	7,998	6,588
36,900	36,950	7,551	5,685	8,012	6,602
36,950	37,000	7,565	5,699	8,026	6,616

37,000

At least	But less than	Single	Married filing jointly	Married filing separately	Head of a household
37,000	37,050	7,579	5,713	8,040	6,630
37,050	37,100	7,593	5,727	8,054	6,644
37,100	37,150	7,607	5,741	8,068	6,658
37,150	37,200	7,621	5,755	8,082	6,672
37,200	37,250	7,635	5,769	8,096	6,686
37,250	37,300	7,649	5,783	8,110	6,700
37,300	37,350	7,663	5,797	8,124	6,714
37,350	37,400	7,677	5,811	8,138	6,728
37,400	37,450	7,691	5,825	8,152	6,742
37,450	37,500	7,705	5,839	8,166	6,756
37,500	37,550	7,719	5,853	8,180	6,770
37,550	37,600	7,733	5,867	8,194	6,784
37,600	37,650	7,747	5,881	8,208	6,798
37,650	37,700	7,761	5,895	8,222	6,812
37,700	37,750	7,775	5,909	8,236	6,826
37,750	37,800	7,789	5,923	8,250	6,840
37,800	37,850	7,803	5,937	8,264	6,854
37,850	37,900	7,817	5,951	8,278	6,868
37,900	37,950	7,831	5,965	8,292	6,882
37,950	38,000	7,845	5,979	8,306	6,896

38,000

At least	But less than	Single	Married filing jointly	Married filing separately	Head of a household
38,000	38,050	7,859	5,993	8,320	6,910
38,050	38,100	7,873	6,007	8,334	6,924
38,100	38,150	7,887	6,021	8,348	6,938
38,150	38,200	7,901	6,035	8,362	6,952
38,200	38,250	7,915	6,049	8,376	6,966
38,250	38,300	7,929	6,063	8,390	6,980
38,300	38,350	7,943	6,077	8,404	6,994
38,350	38,400	7,957	6,091	8,418	7,008
38,400	38,450	7,971	6,105	8,432	7,022
38,450	38,500	7,985	6,119	8,446	7,036
38,500	38,550	7,999	6,133	8,460	7,050
38,550	38,600	8,013	6,147	8,474	7,064
38,600	38,650	8,027	6,161	8,488	7,078
38,650	38,700	8,041	6,175	8,502	7,092
38,700	38,750	8,055	6,189	8,516	7,106
38,750	38,800	8,069	6,203	8,530	7,120
38,800	38,850	8,083	6,217	8,544	7,134
38,850	38,900	8,097	6,231	8,558	7,148
38,900	38,950	8,111	6,245	8,572	7,162
38,950	39,000	8,125	6,259	8,586	7,176

39,000

At least	But less than	Single	Married filing jointly	Married filing separately	Head of a household
39,000	39,050	8,139	6,273	8,600	7,190
39,050	39,100	8,153	6,287	8,614	7,204
39,100	39,150	8,167	6,301	8,628	7,218
39,150	39,200	8,181	6,315	8,642	7,232
39,200	39,250	8,195	6,329	8,656	7,246
39,250	39,300	8,209	6,343	8,670	7,260
39,300	39,350	8,223	6,357	8,684	7,274
39,350	39,400	8,237	6,371	8,698	7,288
39,400	39,450	8,251	6,385	8,712	7,302
39,450	39,500	8,265	6,399	8,726	7,316
39,500	39,550	8,279	6,413	8,740	7,330
39,550	39,600	8,293	6,427	8,754	7,344
39,600	39,650	8,307	6,441	8,768	7,358
39,650	39,700	8,321	6,455	8,782	7,372
39,700	39,750	8,335	6,469	8,796	7,386
39,750	39,800	8,349	6,483	8,810	7,400
39,800	39,850	8,363	6,497	8,824	7,414
39,850	39,900	8,377	6,511	8,838	7,428
39,900	39,950	8,391	6,525	8,852	7,442
39,950	40,000	8,405	6,539	8,866	7,456

40,000

At least	But less than	Single	Married filing jointly	Married filing separately	Head of a household
40,000	40,050	8,419	6,553	8,880	7,470
40,050	40,100	8,433	6,567	8,894	7,484
40,100	40,150	8,447	6,581	8,908	7,498
40,150	40,200	8,461	6,595	8,922	7,512
40,200	40,250	8,475	6,609	8,936	7,526
40,250	40,300	8,489	6,623	8,950	7,540
40,300	40,350	8,503	6,637	8,964	7,554
40,350	40,400	8,517	6,651	8,978	7,568
40,400	40,450	8,531	6,665	8,992	7,582
40,450	40,500	8,545	6,679	9,006	7,596
40,500	40,550	8,559	6,693	9,020	7,610
40,550	40,600	8,573	6,707	9,034	7,624
40,600	40,650	8,587	6,721	9,048	7,638
40,650	40,700	8,601	6,735	9,062	7,652
40,700	40,750	8,615	6,749	9,076	7,666
40,750	40,800	8,629	6,763	9,090	7,680
40,800	40,850	8,643	6,777	9,104	7,694
40,850	40,900	8,657	6,791	9,118	7,708
40,900	40,950	8,671	6,805	9,132	7,722
40,950	41,000	8,685	6,819	9,146	7,736

* This column must also be used by a qualifying widow(er).

Continued on next page

19-- Tax Table—*Continued*

If line 5 (Form 1040EZ), line 22 (Form 1040A), or line 37 (Form 1040) is—		And you are—			
At least	But less than	Single	Married filing jointly *	Married filing separately	Head of a household
		Your tax is—			

41,000

At least	But less than	Single	Married filing jointly	Married filing separately	Head of a household
41,000	41,050	8,699	6,833	9,160	7,750
41,050	41,100	8,713	6,847	9,174	7,764
41,100	41,150	8,727	6,861	9,188	7,778
41,150	41,200	8,741	6,875	9,202	7,792
41,200	41,250	8,755	6,889	9,216	7,806
41,250	41,300	8,769	6,903	9,230	7,820
41,300	41,350	8,783	6,917	9,244	7,834
41,350	41,400	8,797	6,931	9,258	7,848
41,400	41,450	8,811	6,945	9,272	7,862
41,450	41,500	8,825	6,959	9,286	7,876
41,500	41,550	8,839	6,973	9,300	7,890
41,550	41,600	8,853	6,987	9,314	7,904
41,600	41,650	8,867	7,001	9,328	7,918
41,650	41,700	8,881	7,015	9,342	7,932
41,700	41,750	8,895	7,029	9,356	7,946
41,750	41,800	8,909	7,043	9,370	7,960
41,800	41,850	8,923	7,057	9,384	7,974
41,850	41,900	8,937	7,071	9,398	7,988
41,900	41,950	8,951	7,085	9,412	8,002
41,950	42,000	8,965	7,099	9,426	8,016

42,000

At least	But less than	Single	Married filing jointly	Married filing separately	Head of a household
42,000	42,050	8,979	7,113	9,440	8,030
42,050	42,100	8,993	7,127	9,454	8,044
42,100	42,150	9,007	7,141	9,468	8,058
42,150	42,200	9,021	7,155	9,482	8,072
42,200	42,250	9,035	7,169	9,496	8,086
42,250	42,300	9,049	7,183	9,510	8,100
42,300	42,350	9,063	7,197	9,524	8,114
42,350	42,400	9,077	7,211	9,538	8,128
42,400	42,450	9,091	7,225	9,552	8,142
42,450	42,500	9,105	7,239	9,566	8,156
42,500	42,550	9,119	7,253	9,580	8,170
42,550	42,600	9,133	7,267	9,594	8,184
42,600	42,650	9,147	7,281	9,608	8,198
42,650	42,700	9,161	7,295	9,622	8,212
42,700	42,750	9,175	7,309	9,636	8,226
42,750	42,800	9,189	7,323	9,650	8,240
42,800	42,850	9,203	7,337	9,664	8,254
42,850	42,900	9,217	7,351	9,678	8,268
42,900	42,950	9,231	7,365	9,692	8,282
42,950	43,000	9,245	7,379	9,706	8,296

43,000

At least	But less than	Single	Married filing jointly	Married filing separately	Head of a household
43,000	43,050	9,259	7,393	9,720	8,310
43,050	43,100	9,273	7,407	9,734	8,324
43,100	43,150	9,287	7,421	9,748	8,338
43,150	43,200	9,301	7,435	9,762	8,352
43,200	43,250	9,315	7,449	9,776	8,366
43,250	43,300	9,329	7,463	9,791	8,380
43,300	43,350	9,343	7,477	9,806	8,394
43,350	43,400	9,357	7,491	9,822	8,408
43,400	43,450	9,371	7,505	9,837	8,422
43,450	43,500	9,385	7,519	9,853	8,436
43,500	43,550	9,399	7,533	9,868	8,450
43,550	43,600	9,413	7,547	9,884	8,464
43,600	43,650	9,427	7,561	9,899	8,478
43,650	43,700	9,441	7,575	9,915	8,492
43,700	43,750	9,455	7,589	9,930	8,506
43,750	43,800	9,469	7,603	9,946	8,520
43,800	43,850	9,483	7,617	9,961	8,534
43,850	43,900	9,497	7,631	9,977	8,548
43,900	43,950	9,511	7,645	9,992	8,562
43,950	44,000	9,525	7,659	10,008	8,576

44,000

At least	But less than	Single	Married filing jointly	Married filing separately	Head of a household
44,000	44,050	9,539	7,673	10,023	8,590
44,050	44,100	9,553	7,687	10,039	8,604
44,100	44,150	9,567	7,701	10,054	8,618
44,150	44,200	9,581	7,715	10,070	8,632
44,200	44,250	9,595	7,729	10,085	8,646
44,250	44,300	9,609	7,743	10,101	8,660
44,300	44,350	9,623	7,757	10,116	8,674
44,350	44,400	9,637	7,771	10,132	8,688
44,400	44,450	9,651	7,785	10,147	8,702
44,450	44,500	9,665	7,799	10,163	8,716
44,500	44,550	9,679	7,813	10,178	8,730
44,550	44,600	9,693	7,827	10,194	8,744
44,600	44,650	9,707	7,841	10,209	8,758
44,650	44,700	9,721	7,855	10,225	8,772
44,700	44,750	9,735	7,869	10,240	8,786
44,750	44,800	9,749	7,883	10,256	8,800
44,800	44,850	9,763	7,897	10,271	8,814
44,850	44,900	9,777	7,911	10,287	8,828
44,900	44,950	9,791	7,925	10,302	8,842
44,950	45,000	9,805	7,939	10,318	8,856

45,000

At least	But less than	Single	Married filing jointly	Married filing separately	Head of a household
45,000	45,050	9,819	7,953	10,333	8,870
45,050	45,100	9,833	7,967	10,349	8,884
45,100	45,150	9,847	7,981	10,364	8,898
45,150	45,200	9,861	7,995	10,380	8,912
45,200	45,250	9,875	8,009	10,395	8,926
45,250	45,300	9,889	8,023	10,411	8,940
45,300	45,350	9,903	8,037	10,426	8,954
45,350	45,400	9,917	8,051	10,442	8,968
45,400	45,450	9,931	8,065	10,457	8,982
45,450	45,500	9,945	8,079	10,473	8,996
45,500	45,550	9,959	8,093	10,488	9,010
45,550	45,600	9,973	8,107	10,504	9,024
45,600	45,650	9,987	8,121	10,519	9,038
45,650	45,700	10,001	8,135	10,535	9,052
45,700	45,750	10,015	8,149	10,550	9,066
45,750	45,800	10,029	8,163	10,566	9,080
45,800	45,850	10,043	8,177	10,581	9,094
45,850	45,900	10,057	8,191	10,597	9,108
45,900	45,950	10,071	8,205	10,612	9,122
45,950	46,000	10,085	8,219	10,628	9,136

46,000

At least	But less than	Single	Married filing jointly	Married filing separately	Head of a household
46,000	46,050	10,099	8,233	10,643	9,150
46,050	46,100	10,113	8,247	10,659	9,164
46,100	46,150	10,127	8,261	10,674	9,178
46,150	46,200	10,141	8,275	10,690	9,192
46,200	46,250	10,155	8,289	10,705	9,206
46,250	46,300	10,169	8,303	10,721	9,220
46,300	46,350	10,183	8,317	10,736	9,234
46,350	46,400	10,197	8,331	10,752	9,248
46,400	46,450	10,211	8,345	10,767	9,262
46,450	46,500	10,225	8,359	10,783	9,276
46,500	46,550	10,239	8,373	10,798	9,290
46,550	46,600	10,253	8,387	10,814	9,304
46,600	46,650	10,267	8,401	10,829	9,318
46,650	46,700	10,281	8,415	10,845	9,332
46,700	46,750	10,295	8,429	10,860	9,346
46,750	46,800	10,309	8,443	10,876	9,360
46,800	46,850	10,323	8,457	10,891	9,374
46,850	46,900	10,337	8,471	10,907	9,388
46,900	46,950	10,351	8,485	10,922	9,402
46,950	47,000	10,365	8,499	10,938	9,416

47,000

At least	But less than	Single	Married filing jointly	Married filing separately	Head of a household
47,000	47,050	10,379	8,513	10,953	9,430
47,050	47,100	10,393	8,527	10,969	9,444
47,100	47,150	10,407	8,541	10,984	9,458
47,150	47,200	10,421	8,555	11,000	9,472
47,200	47,250	10,435	8,569	11,015	9,486
47,250	47,300	10,449	8,583	11,031	9,500
47,300	47,350	10,463	8,597	11,046	9,514
47,350	47,400	10,477	8,611	11,062	9,528
47,400	47,450	10,491	8,625	11,077	9,542
47,450	47,500	10,505	8,639	11,093	9,556
47,500	47,550	10,519	8,653	11,108	9,570
47,550	47,600	10,533	8,667	11,124	9,584
47,600	47,650	10,547	8,681	11,139	9,598
47,650	47,700	10,561	8,695	11,155	9,612
47,700	47,750	10,575	8,709	11,170	9,626
47,750	47,800	10,589	8,723	11,186	9,640
47,800	47,850	10,603	8,737	11,201	9,654
47,850	47,900	10,617	8,751	11,217	9,668
47,900	47,950	10,631	8,765	11,232	9,682
47,950	48,000	10,645	8,779	11,248	9,696

48,000

At least	But less than	Single	Married filing jointly	Married filing separately	Head of a household
48,000	48,050	10,659	8,793	11,263	9,710
48,050	48,100	10,673	8,807	11,279	9,724
48,100	48,150	10,687	8,821	11,294	9,738
48,150	48,200	10,701	8,835	11,310	9,752
48,200	48,250	10,715	8,849	11,325	9,766
48,250	48,300	10,729	8,863	11,341	9,780
48,300	48,350	10,743	8,877	11,356	9,794
48,350	48,400	10,757	8,891	11,372	9,808
48,400	48,450	10,771	8,905	11,387	9,822
48,450	48,500	10,785	8,919	11,403	9,836
48,500	48,550	10,799	8,933	11,418	9,850
48,550	48,600	10,813	8,947	11,434	9,864
48,600	48,650	10,827	8,961	11,449	9,878
48,650	48,700	10,841	8,975	11,465	9,892
48,700	48,750	10,855	8,989	11,480	9,906
48,750	48,800	10,869	9,003	11,496	9,920
48,800	48,850	10,883	9,017	11,511	9,934
48,850	48,900	10,897	9,031	11,527	9,948
48,900	48,950	10,911	9,045	11,542	9,962
48,950	49,000	10,925	9,059	11,558	9,976

49,000

At least	But less than	Single	Married filing jointly	Married filing separately	Head of a household
49,000	49,050	10,939	9,073	11,573	9,990
49,050	49,100	10,953	9,087	11,589	10,004
49,100	49,150	10,967	9,101	11,604	10,018
49,150	49,200	10,981	9,115	11,620	10,032
49,200	49,250	10,995	9,129	11,635	10,046
49,250	49,300	11,009	9,143	11,651	10,060
49,300	49,350	11,023	9,157	11,666	10,074
49,350	49,400	11,037	9,171	11,682	10,088
49,400	49,450	11,051	9,185	11,697	10,102
49,450	49,500	11,065	9,199	11,713	10,116
49,500	49,550	11,079	9,213	11,728	10,130
49,550	49,600	11,093	9,227	11,744	10,144
49,600	49,650	11,107	9,241	11,759	10,158
49,650	49,700	11,121	9,255	11,775	10,172
49,700	49,750	11,135	9,269	11,790	10,186
49,750	49,800	11,149	9,283	11,806	10,200
49,800	49,850	11,163	9,297	11,821	10,214
49,850	49,900	11,177	9,311	11,837	10,228
49,900	49,950	11,191	9,325	11,852	10,242
49,950	50,000	11,205	9,339	11,868	10,256

* This column must also be used by a qualifying widow(er).

Continued on next page

19-- Tax Table—*Continued*

Column headings for all tables below:

If line 37 (Form 1040) is— At least	But less than	Single	Married filing jointly*	Married filing separately	Head of a household
				Your tax is—	

50,000

At least	But less than	Single	Married filing jointly*	Married filing separately	Head of a household
50,000	50,050	11,219	9,353	11,883	10,270
50,050	50,100	11,233	9,367	11,899	10,284
50,100	50,150	11,247	9,381	11,914	10,298
50,150	50,200	11,261	9,395	11,930	10,312
50,200	50,250	11,275	9,409	11,945	10,326
50,250	50,300	11,289	9,423	11,961	10,340
50,300	50,350	11,303	9,437	11,976	10,354
50,350	50,400	11,317	9,451	11,992	10,368
50,400	50,450	11,331	9,465	12,007	10,382
50,450	50,500	11,345	9,479	12,023	10,396
50,500	50,550	11,359	9,493	12,038	10,410
50,550	50,600	11,373	9,507	12,054	10,424
50,600	50,650	11,387	9,521	12,069	10,438
50,650	50,700	11,401	9,535	12,085	10,452
50,700	50,750	11,415	9,549	12,100	10,466
50,750	50,800	11,429	9,563	12,116	10,480
50,800	50,850	11,443	9,577	12,131	10,494
50,850	50,900	11,457	9,591	12,147	10,508
50,900	50,950	11,471	9,605	12,162	10,522
50,950	51,000	11,485	9,619	12,178	10,536

51,000

At least	But less than	Single	Married filing jointly*	Married filing separately	Head of a household
51,000	51,050	11,499	9,633	12,193	10,550
51,050	51,100	11,513	9,647	12,209	10,564
51,100	51,150	11,527	9,661	12,224	10,578
51,150	51,200	11,541	9,675	12,240	10,592
51,200	51,250	11,555	9,689	12,255	10,606
51,250	51,300	11,569	9,703	12,271	10,620
51,300	51,350	11,583	9,717	12,286	10,634
51,350	51,400	11,597	9,731	12,302	10,648
51,400	51,450	11,611	9,745	12,317	10,662
51,450	51,500	11,625	9,759	12,333	10,676
51,500	51,550	11,639	9,773	12,348	10,690
51,550	51,600	11,653	9,787	12,364	10,704
51,600	51,650	11,667	9,801	12,379	10,718
51,650	51,700	11,681	9,815	12,395	10,732
51,700	51,750	11,695	9,829	12,410	10,746
51,750	51,800	11,709	9,843	12,426	10,760
51,800	51,850	11,723	9,857	12,441	10,774
51,850	51,900	11,737	9,871	12,457	10,788
51,900	51,950	11,751	9,885	12,472	10,802
51,950	52,000	11,767	9,899	12,488	10,816

52,000

At least	But less than	Single	Married filing jointly*	Married filing separately	Head of a household
52,000	52,050	11,782	9,913	12,503	10,830
52,050	52,100	11,798	9,927	12,519	10,844
52,100	52,150	11,813	9,941	12,534	10,858
52,150	52,200	11,829	9,955	12,550	10,872
52,200	52,250	11,844	9,969	12,565	10,886
52,250	52,300	11,860	9,983	12,581	10,900
52,300	52,350	11,875	9,997	12,596	10,914
52,350	52,400	11,891	10,011	12,612	10,928
52,400	52,450	11,906	10,025	12,627	10,942
52,450	52,500	11,922	10,039	12,643	10,956
52,500	52,550	11,937	10,053	12,658	10,970
52,550	52,600	11,953	10,067	12,674	10,984
52,600	52,650	11,968	10,081	12,689	10,998
52,650	52,700	11,984	10,095	12,705	11,012
52,700	52,750	11,999	10,109	12,720	11,026
52,750	52,800	12,015	10,123	12,736	11,040
52,800	52,850	12,030	10,137	12,751	11,054
52,850	52,900	12,046	10,151	12,767	11,068
52,900	52,950	12,061	10,165	12,782	11,082
52,950	53,000	12,077	10,179	12,798	11,096

53,000

At least	But less than	Single	Married filing jointly*	Married filing separately	Head of a household
53,000	53,050	12,092	10,193	12,813	11,110
53,050	53,100	12,108	10,207	12,829	11,124
53,100	53,150	12,123	10,221	12,844	11,138
53,150	53,200	12,139	10,235	12,860	11,152
53,200	53,250	12,154	10,249	12,875	11,166
53,250	53,300	12,170	10,263	12,891	11,180
53,300	53,350	12,185	10,277	12,906	11,194
53,350	53,400	12,201	10,291	12,922	11,208
53,400	53,450	12,216	10,305	12,937	11,222
53,450	53,500	12,232	10,319	12,953	11,236
53,500	53,550	12,247	10,333	12,968	11,250
53,550	53,600	12,263	10,347	12,984	11,264
53,600	53,650	12,278	10,361	12,999	11,278
53,650	53,700	12,294	10,375	13,015	11,292
53,700	53,750	12,309	10,389	13,030	11,306
53,750	53,800	12,325	10,403	13,046	11,320
53,800	53,850	12,340	10,417	13,061	11,334
53,850	53,900	12,356	10,431	13,077	11,348
53,900	53,950	12,371	10,445	13,092	11,362
53,950	54,000	12,387	10,459	13,108	11,376

54,000

At least	But less than	Single	Married filing jointly*	Married filing separately	Head of a household
54,000	54,050	12,402	10,473	13,123	11,390
54,050	54,100	12,418	10,487	13,139	11,404
54,100	54,150	12,433	10,501	13,154	11,418
54,150	54,200	12,449	10,515	13,170	11,432
54,200	54,250	12,464	10,529	13,185	11,446
54,250	54,300	12,480	10,543	13,201	11,460
54,300	54,350	12,495	10,557	13,216	11,474
54,350	54,400	12,511	10,571	13,232	11,488
54,400	54,450	12,526	10,585	13,247	11,502
54,450	54,500	12,542	10,599	13,263	11,516
54,500	54,550	12,557	10,613	13,278	11,530
54,550	54,600	12,573	10,627	13,294	11,544
54,600	54,650	12,588	10,641	13,309	11,558
54,650	54,700	12,604	10,655	13,325	11,572
54,700	54,750	12,619	10,669	13,340	11,586
54,750	54,800	12,635	10,683	13,356	11,600
54,800	54,850	12,650	10,697	13,371	11,614
54,850	54,900	12,666	10,711	13,387	11,628
54,900	54,950	12,681	10,725	13,402	11,642
54,950	55,000	12,697	10,739	13,418	11,656

55,000

At least	But less than	Single	Married filing jointly*	Married filing separately	Head of a household
55,000	55,050	12,712	10,753	13,433	11,670
55,050	55,100	12,728	10,767	13,449	11,684
55,100	55,150	12,743	10,781	13,464	11,698
55,150	55,200	12,759	10,795	13,480	11,712
55,200	55,250	12,774	10,809	13,495	11,726
55,250	55,300	12,790	10,823	13,511	11,740
55,300	55,350	12,805	10,837	13,526	11,754
55,350	55,400	12,821	10,851	13,542	11,768
55,400	55,450	12,836	10,865	13,557	11,782
55,450	55,500	12,852	10,879	13,573	11,796
55,500	55,550	12,867	10,893	13,588	11,810
55,550	55,600	12,883	10,907	13,604	11,824
55,600	55,650	12,898	10,921	13,619	11,838
55,650	55,700	12,914	10,935	13,635	11,852
55,700	55,750	12,929	10,949	13,650	11,866
55,750	55,800	12,945	10,963	13,666	11,880
55,800	55,850	12,960	10,977	13,681	11,894
55,850	55,900	12,976	10,991	13,697	11,908
55,900	55,950	12,991	11,005	13,712	11,922
55,950	56,000	13,007	11,019	13,728	11,936

56,000

At least	But less than	Single	Married filing jointly*	Married filing separately	Head of a household
56,000	56,050	13,022	11,033	13,743	11,950
56,050	56,100	13,038	11,047	13,759	11,964
56,100	56,150	13,053	11,061	13,774	11,978
56,150	56,200	13,069	11,075	13,790	11,992
56,200	56,250	13,084	11,089	13,805	12,006
56,250	56,300	13,100	11,103	13,821	12,020
56,300	56,350	13,115	11,117	13,836	12,034
56,350	56,400	13,131	11,131	13,852	12,048
56,400	56,450	13,146	11,145	13,867	12,062
56,450	56,500	13,162	11,159	13,883	12,076
56,500	56,550	13,177	11,173	13,898	12,090
56,550	56,600	13,193	11,187	13,914	12,104
56,600	56,650	13,208	11,201	13,929	12,118
56,650	56,700	13,224	11,215	13,945	12,132
56,700	56,750	13,239	11,229	13,960	12,146
56,750	56,800	13,255	11,243	13,976	12,160
56,800	56,850	13,270	11,257	13,991	12,174
56,850	56,900	13,286	11,271	14,007	12,188
56,900	56,950	13,301	11,285	14,022	12,202
56,950	57,000	13,317	11,299	14,038	12,216

57,000

At least	But less than	Single	Married filing jointly*	Married filing separately	Head of a household
57,000	57,050	13,332	11,313	14,053	12,230
57,050	57,100	13,348	11,327	14,069	12,244
57,100	57,150	13,363	11,341	14,084	12,258
57,150	57,200	13,379	11,355	14,100	12,272
57,200	57,250	13,394	11,369	14,115	12,286
57,250	57,300	13,410	11,383	14,131	12,300
57,300	57,350	13,425	11,397	14,146	12,314
57,350	57,400	13,441	11,411	14,162	12,328
57,400	57,450	13,456	11,425	14,177	12,342
57,450	57,500	13,472	11,439	14,193	12,356
57,500	57,550	13,487	11,453	14,208	12,370
57,550	57,600	13,503	11,467	14,224	12,384
57,600	57,650	13,518	11,481	14,239	12,398
57,650	57,700	13,534	11,495	14,255	12,412
57,700	57,750	13,549	11,509	14,270	12,426
57,750	57,800	13,565	11,523	14,286	12,440
57,800	57,850	13,580	11,537	14,301	12,454
57,850	57,900	13,596	11,551	14,317	12,468
57,900	57,950	13,611	11,565	14,332	12,482
57,950	58,000	13,627	11,579	14,348	12,496

58,000

At least	But less than	Single	Married filing jointly*	Married filing separately	Head of a household
58,000	58,050	13,642	11,593	14,363	12,510
58,050	58,100	13,658	11,607	14,379	12,524
58,100	58,150	13,673	11,621	14,394	12,538
58,150	58,200	13,689	11,635	14,410	12,552
58,200	58,250	13,704	11,649	14,425	12,566
58,250	58,300	13,720	11,663	14,441	12,580
58,300	58,350	13,735	11,677	14,456	12,594
58,350	58,400	13,751	11,691	14,472	12,608
58,400	58,450	13,766	11,705	14,487	12,622
58,450	58,500	13,782	11,719	14,503	12,636
58,500	58,550	13,797	11,733	14,518	12,650
58,550	58,600	13,813	11,747	14,534	12,664
58,600	58,650	13,828	11,761	14,549	12,678
58,650	58,700	13,844	11,775	14,565	12,692
58,700	58,750	13,859	11,789	14,580	12,706
58,750	58,800	13,875	11,803	14,596	12,720
58,800	58,850	13,890	11,817	14,611	12,734
58,850	58,900	13,906	11,831	14,627	12,748
58,900	58,950	13,921	11,845	14,642	12,762
58,950	59,000	13,937	11,859	14,658	12,776

* This column must also be used by a qualifying widow(er).

Continued on next page

19-- Tax Table– Continued

If line 37 (Form 1040) is–		And you are —			
At least	But less than	Single	Married filing jointly *	Married filing separately	Head of a household
		Your tax is—			

59,000

At least	But less than	Single	Married filing jointly	Married filing separately	Head of a household
59,000	59,050	13,952	11,873	14,673	12,790
59,050	59,100	13,968	11,887	14,689	12,804
59,100	59,150	13,983	11,901	14,704	12,818
59,150	59,200	13,999	11,915	14,720	12,832
59,200	59,250	14,014	11,929	14,735	12,846
59,250	59,300	14,030	11,943	14,751	12,860
59,300	59,350	14,045	11,957	14,766	12,874
59,350	59,400	14,061	11,971	14,782	12,888
59,400	59,450	14,076	11,985	14,797	12,902
59,450	59,500	14,092	11,999	14,813	12,916
59,500	59,550	14,107	12,013	14,828	12,930
59,550	59,600	14,123	12,027	14,844	12,944
59,600	59,650	14,138	12,041	14,859	12,958
59,650	59,700	14,154	12,055	14,875	12,972
59,700	59,750	14,169	12,069	14,890	12,986
59,750	59,800	14,185	12,083	14,906	13,000
59,800	59,850	14,200	12,097	14,921	13,014
59,850	59,900	14,216	12,111	14,937	13,028
59,900	59,950	14,231	12,125	14,952	13,042
59,950	60,000	14,247	12,139	14,968	13,056

60,000

At least	But less than	Single	Married filing jointly	Married filing separately	Head of a household
60,000	60,050	14,262	12,153	14,983	13,070
60,050	60,100	14,278	12,167	14,999	13,084
60,100	60,150	14,293	12,181	15,014	13,098
60,150	60,200	14,309	12,195	15,030	13,112
60,200	60,250	14,324	12,209	15,045	13,126
60,250	60,300	14,340	12,223	15,061	13,140
60,300	60,350	14,355	12,237	15,076	13,154
60,350	60,400	14,371	12,251	15,092	13,168
60,400	60,450	14,386	12,265	15,107	13,182
60,450	60,500	14,402	12,279	15,123	13,196
60,500	60,550	14,417	12,293	15,138	13,210
60,550	60,600	14,433	12,307	15,154	13,224
60,600	60,650	14,448	12,321	15,169	13,238
60,650	60,700	14,464	12,335	15,185	13,252
60,700	60,750	14,479	12,349	15,200	13,266
60,750	60,800	14,495	12,363	15,216	13,280
60,800	60,850	14,510	12,377	15,231	13,294
60,850	60,900	14,526	12,391	15,247	13,308
60,900	60,950	14,541	12,405	15,262	13,322
60,950	61,000	14,557	12,419	15,278	13,336

61,000

At least	But less than	Single	Married filing jointly	Married filing separately	Head of a household
61,000	61,050	14,572	12,433	15,293	13,350
61,050	61,100	14,588	12,447	15,309	13,364
61,100	61,150	14,603	12,461	15,324	13,378
61,150	61,200	14,619	12,475	15,340	13,392
61,200	61,250	14,634	12,489	15,355	13,406
61,250	61,300	14,650	12,503	15,371	13,420
61,300	61,350	14,665	12,517	15,386	13,434
61,350	61,400	14,681	12,531	15,402	13,448
61,400	61,450	14,696	12,545	15,417	13,462
61,450	61,500	14,712	12,559	15,433	13,476
61,500	61,550	14,727	12,573	15,448	13,490
61,550	61,600	14,743	12,587	15,464	13,504
61,600	61,650	14,758	12,601	15,479	13,518
61,650	61,700	14,774	12,615	15,495	13,532
61,700	61,750	14,789	12,629	15,510	13,546
61,750	61,800	14,805	12,643	15,526	13,560
61,800	61,850	14,820	12,657	15,541	13,574
61,850	61,900	14,836	12,671	15,557	13,588
61,900	61,950	14,851	12,685	15,572	13,602
61,950	62,000	14,867	12,699	15,588	13,616

62,000

At least	But less than	Single	Married filing jointly	Married filing separately	Head of a household
62,000	62,050	14,882	12,713	15,603	13,630
62,050	62,100	14,898	12,727	15,619	13,644
62,100	62,150	14,913	12,741	15,634	13,658
62,150	62,200	14,929	12,755	15,650	13,672
62,200	62,250	14,944	12,769	15,665	13,686
62,250	62,300	14,960	12,783	15,681	13,700
62,300	62,350	14,975	12,797	15,696	13,714
62,350	62,400	14,991	12,811	15,712	13,728
62,400	62,450	15,006	12,825	15,727	13,742
62,450	62,500	15,022	12,839	15,743	13,756
62,500	62,550	15,037	12,853	15,758	13,770
62,550	62,600	15,053	12,867	15,774	13,784
62,600	62,650	15,068	12,881	15,789	13,798
62,650	62,700	15,084	12,895	15,805	13,812
62,700	62,750	15,099	12,909	15,820	13,826
62,750	62,800	15,115	12,923	15,836	13,840
62,800	62,850	15,130	12,937	15,851	13,854
62,850	62,900	15,146	12,951	15,867	13,868
62,900	62,950	15,161	12,965	15,882	13,882
62,950	63,000	15,177	12,979	15,898	13,896

63,000

At least	But less than	Single	Married filing jointly	Married filing separately	Head of a household
63,000	63,050	15,192	12,993	15,913	13,910
63,050	63,100	15,208	13,007	15,929	13,924
63,100	63,150	15,223	13,021	15,944	13,938
63,150	63,200	15,239	13,035	15,960	13,952
63,200	63,250	15,254	13,049	15,975	13,966
63,250	63,300	15,270	13,063	15,991	13,980
63,300	63,350	15,285	13,077	16,006	13,994
63,350	63,400	15,301	13,091	16,022	14,008
63,400	63,450	15,316	13,105	16,037	14,022
63,450	63,500	15,332	13,119	16,053	14,036
63,500	63,550	15,347	13,133	16,068	14,050
63,550	63,600	15,363	13,147	16,084	14,064
63,600	63,650	15,378	13,161	16,099	14,078
63,650	63,700	15,394	13,175	16,115	14,092
63,700	63,750	15,409	13,189	16,130	14,106
63,750	63,800	15,425	13,203	16,146	14,120
63,800	63,850	15,440	13,217	16,161	14,134
63,850	63,900	15,456	13,231	16,177	14,148
63,900	63,950	15,471	13,245	16,192	14,162
63,950	64,000	15,487	13,259	16,208	14,176

64,000

At least	But less than	Single	Married filing jointly	Married filing separately	Head of a household
64,000	64,050	15,502	13,273	16,223	14,190
64,050	64,100	15,518	13,287	16,239	14,204
64,100	64,150	15,533	13,301	16,254	14,218
64,150	64,200	15,549	13,315	16,270	14,232
64,200	64,250	15,564	13,329	16,285	14,246
64,250	64,300	15,580	13,343	16,301	14,260
64,300	64,350	15,595	13,357	16,316	14,274
64,350	64,400	15,611	13,371	16,332	14,288
64,400	64,450	15,626	13,385	16,347	14,302
64,450	64,500	15,642	13,399	16,363	14,316
64,500	64,550	15,657	13,413	16,378	14,330
64,550	64,600	15,673	13,427	16,394	14,344
64,600	64,650	15,688	13,441	16,409	14,358
64,650	64,700	15,704	13,455	16,425	14,372
64,700	64,750	15,719	13,469	16,440	14,386
64,750	64,800	15,735	13,483	16,456	14,400
64,800	64,850	15,750	13,497	16,471	14,414
64,850	64,900	15,766	13,511	16,487	14,428
64,900	64,950	15,781	13,525	16,502	14,442
64,950	65,000	15,797	13,539	16,518	14,456

65,000

At least	But less than	Single	Married filing jointly	Married filing separately	Head of a household
65,000	65,050	15,812	13,553	16,533	14,470
65,050	65,100	15,828	13,567	16,549	14,484
65,100	65,150	15,843	13,581	16,564	14,498
65,150	65,200	15,859	13,595	16,580	14,512
65,200	65,250	15,874	13,609	16,595	14,526
65,250	65,300	15,890	13,623	16,611	14,540
65,300	65,350	15,905	13,637	16,626	14,554
65,350	65,400	15,921	13,651	16,642	14,568
65,400	65,450	15,936	13,665	16,657	14,582
65,450	65,500	15,952	13,679	16,673	14,596
65,500	65,550	15,967	13,693	16,688	14,610
65,550	65,600	15,983	13,707	16,704	14,624
65,600	65,650	15,998	13,721	16,719	14,638
65,650	65,700	16,014	13,735	16,735	14,652
65,700	65,750	16,029	13,749	16,750	14,666
65,750	65,800	16,045	13,763	16,766	14,680
65,800	65,850	16,060	13,777	16,781	14,694
65,850	65,900	16,076	13,791	16,797	14,708
65,900	65,950	16,091	13,805	16,812	14,722
65,950	66,000	16,107	13,819	16,828	14,736

66,000

At least	But less than	Single	Married filing jointly	Married filing separately	Head of a household
66,000	66,050	16,122	13,833	16,843	14,750
66,050	66,100	16,138	13,847	16,859	14,764
66,100	66,150	16,153	13,861	16,874	14,778
66,150	66,200	16,169	13,875	16,890	14,792
66,200	66,250	16,184	13,889	16,905	14,806
66,250	66,300	16,200	13,903	16,921	14,820
66,300	66,350	16,215	13,917	16,936	14,834
66,350	66,400	16,231	13,931	16,952	14,848
66,400	66,450	16,246	13,945	16,967	14,862
66,450	66,500	16,262	13,959	16,983	14,876
66,500	66,550	16,277	13,973	16,998	14,890
66,550	66,600	16,293	13,987	17,014	14,904
66,600	66,650	16,308	14,001	17,029	14,918
66,650	66,700	16,324	14,015	17,045	14,932
66,700	66,750	16,339	14,029	17,060	14,946
66,750	66,800	16,355	14,043	17,076	14,960
66,800	66,850	16,370	14,057	17,091	14,974
66,850	66,900	16,386	14,071	17,107	14,988
66,900	66,950	16,401	14,085	17,122	15,002
66,950	67,000	16,417	14,099	17,138	15,016

67,000

At least	But less than	Single	Married filing jointly	Married filing separately	Head of a household
67,000	67,050	16,432	14,113	17,153	15,030
67,050	67,100	16,448	14,127	17,169	15,044
67,100	67,150	16,463	14,141	17,184	15,058
67,150	67,200	16,479	14,155	17,200	15,072
67,200	67,250	16,494	14,169	17,215	15,086
67,250	67,300	16,510	14,183	17,231	15,100
67,300	67,350	16,525	14,197	17,246	15,114
67,350	67,400	16,541	14,211	17,262	15,128
67,400	67,450	16,556	14,225	17,277	15,142
67,450	67,500	16,572	14,239	17,293	15,156
67,500	67,550	16,587	14,253	17,308	15,170
67,550	67,600	16,603	14,267	17,324	15,184
67,600	67,650	16,618	14,281	17,339	15,198
67,650	67,700	16,634	14,295	17,355	15,212
67,700	67,750	16,649	14,309	17,370	15,226
67,750	67,800	16,665	14,323	17,386	15,240
67,800	67,850	16,680	14,337	17,401	15,254
67,850	67,900	16,696	14,351	17,417	15,268
67,900	67,950	16,711	14,365	17,432	15,282
67,950	68,000	16,727	14,379	17,448	15,296

* This column must also be used by a qualifying widow(er).

Continued on next page

19-- Tax Table—*Continued*

If line 37 (Form 1040) is—		And you are —			
At least	But less than	Single	Married filing jointly *	Married filing separately	Head of a household
		Your tax is —			

68,000

At least	But less than	Single	MFJ	MFS	HoH
68,000	68,050	16,742	14,393	17,463	15,310
68,050	68,100	16,758	14,407	17,479	15,324
68,100	68,150	16,773	14,421	17,494	15,338
68,150	68,200	16,789	14,435	17,510	15,352
68,200	68,250	16,804	14,449	17,525	15,366
68,250	68,300	16,820	14,463	17,541	15,380
68,300	68,350	16,835	14,477	17,556	15,394
68,350	68,400	16,851	14,491	17,572	15,408
68,400	68,450	16,866	14,505	17,587	15,422
68,450	68,500	16,882	14,519	17,603	15,436
68,500	68,550	16,897	14,533	17,618	15,450
68,550	68,600	16,913	14,547	17,634	15,464
68,600	68,650	16,928	14,561	17,649	15,478
68,650	68,700	16,944	14,575	17,665	15,492
68,700	68,750	16,959	14,589	17,680	15,506
68,750	68,800	16,975	14,603	17,696	15,520
68,800	68,850	16,990	14,617	17,711	15,534
68,850	68,900	17,006	14,631	17,727	15,548
68,900	68,950	17,021	14,645	17,742	15,562
68,950	69,000	17,037	14,659	17,758	15,576

69,000

At least	But less than	Single	MFJ	MFS	HoH
69,000	69,050	17,052	14,673	17,773	15,590
69,050	69,100	17,068	14,687	17,789	15,604
69,100	69,150	17,083	14,701	17,804	15,618
69,150	69,200	17,099	14,715	17,820	15,632
69,200	69,250	17,114	14,729	17,835	15,646
69,250	69,300	17,130	14,743	17,851	15,660
69,300	69,350	17,145	14,757	17,866	15,674
69,350	69,400	17,161	14,771	17,882	15,688
69,400	69,450	17,176	14,785	17,897	15,702
69,450	69,500	17,192	14,799	17,913	15,716
69,500	69,550	17,207	14,813	17,928	15,730
69,550	69,600	17,223	14,827	17,944	15,744
69,600	69,650	17,238	14,841	17,959	15,758
69,650	69,700	17,254	14,855	17,975	15,772
69,700	69,750	17,269	14,869	17,990	15,786
69,750	69,800	17,285	14,883	18,006	15,800
69,800	69,850	17,300	14,897	18,021	15,814
69,850	69,900	17,316	14,911	18,037	15,828
69,900	69,950	17,331	14,925	18,052	15,842
69,950	70,000	17,347	14,939	18,068	15,856

70,000

At least	But less than	Single	MFJ	MFS	HoH
70,000	70,050	17,362	14,953	18,083	15,870
70,050	70,100	17,378	14,967	18,099	15,884
70,100	70,150	17,393	14,981	18,114	15,898
70,150	70,200	17,409	14,995	18,130	15,912
70,200	70,250	17,424	15,009	18,145	15,926
70,250	70,300	17,440	15,023	18,161	15,940
70,300	70,350	17,455	15,037	18,176	15,954
70,350	70,400	17,471	15,051	18,192	15,968
70,400	70,450	17,486	15,065	18,207	15,982
70,450	70,500	17,502	15,079	18,223	15,996
70,500	70,550	17,517	15,093	18,238	16,010
70,550	70,600	17,533	15,107	18,254	16,024
70,600	70,650	17,548	15,121	18,269	16,038
70,650	70,700	17,564	15,135	18,285	16,052
70,700	70,750	17,579	15,149	18,300	16,066
70,750	70,800	17,595	15,163	18,316	16,080
70,800	70,850	17,610	15,177	18,331	16,094
70,850	70,900	17,626	15,191	18,347	16,108
70,900	70,950	17,641	15,205	18,362	16,122
70,950	71,000	17,657	15,219	18,378	16,136

71,000

At least	But less than	Single	MFJ	MFS	HoH
71,000	71,050	17,672	15,233	18,393	16,150
71,050	71,100	17,688	15,247	18,409	16,164
71,100	71,150	17,703	15,261	18,424	16,178
71,150	71,200	17,719	15,275	18,440	16,192
71,200	71,250	17,734	15,289	18,455	16,206
71,250	71,300	17,750	15,303	18,471	16,220
71,300	71,350	17,765	15,317	18,486	16,234
71,350	71,400	17,781	15,331	18,502	16,248
71,400	71,450	17,796	15,345	18,517	16,262
71,450	71,500	17,812	15,359	18,533	16,276
71,500	71,550	17,827	15,373	18,548	16,290
71,550	71,600	17,843	15,387	18,564	16,304
71,600	71,650	17,858	15,401	18,579	16,318
71,650	71,700	17,874	15,415	18,595	16,332
71,700	71,750	17,889	15,429	18,610	16,346
71,750	71,800	17,905	15,443	18,626	16,360
71,800	71,850	17,920	15,457	18,641	16,374
71,850	71,900	17,936	15,471	18,657	16,388
71,900	71,950	17,951	15,485	18,672	16,402
71,950	72,000	17,967	15,499	18,688	16,416

72,000

At least	But less than	Single	MFJ	MFS	HoH
72,000	72,050	17,982	15,513	18,703	16,430
72,050	72,100	17,998	15,527	18,719	16,444
72,100	72,150	18,013	15,541	18,734	16,458
72,150	72,200	18,029	15,555	18,750	16,472
72,200	72,250	18,044	15,569	18,765	16,486
72,250	72,300	18,060	15,583	18,781	16,500
72,300	72,350	18,075	15,597	18,796	16,514
72,350	72,400	18,091	15,611	18,812	16,528
72,400	72,450	18,106	15,625	18,827	16,542
72,450	72,500	18,122	15,639	18,843	16,556
72,500	72,550	18,137	15,653	18,858	16,570
72,550	72,600	18,153	15,667	18,874	16,584
72,600	72,650	18,168	15,681	18,889	16,598
72,650	72,700	18,184	15,695	18,905	16,612
72,700	72,750	18,199	15,709	18,920	16,626
72,750	72,800	18,215	15,723	18,936	16,640
72,800	72,850	18,230	15,737	18,951	16,654
72,850	72,900	18,246	15,751	18,967	16,668
72,900	72,950	18,261	15,765	18,982	16,682
72,950	73,000	18,277	15,779	18,998	16,696

73,000

At least	But less than	Single	MFJ	MFS	HoH
73,000	73,050	18,292	15,793	19,013	16,710
73,050	73,100	18,308	15,807	19,029	16,724
73,100	73,150	18,323	15,821	19,044	16,738
73,150	73,200	18,339	15,835	19,060	16,752
73,200	73,250	18,354	15,849	19,075	16,766
73,250	73,300	18,370	15,863	19,091	16,780
73,300	73,350	18,385	15,877	19,106	16,794
73,350	73,400	18,401	15,891	19,122	16,808
73,400	73,450	18,416	15,905	19,137	16,822
73,450	73,500	18,432	15,919	19,153	16,836
73,500	73,550	18,447	15,933	19,168	16,850
73,550	73,600	18,463	15,947	19,184	16,864
73,600	73,650	18,478	15,961	19,199	16,878
73,650	73,700	18,494	15,975	19,215	16,892
73,700	73,750	18,509	15,989	19,230	16,906
73,750	73,800	18,525	16,003	19,246	16,920
73,800	73,850	18,540	16,017	19,261	16,934
73,850	73,900	18,556	16,031	19,277	16,948
73,900	73,950	18,571	16,045	19,292	16,962
73,950	74,000	18,587	16,059	19,308	16,976

74,000

At least	But less than	Single	MFJ	MFS	HoH
74,000	74,050	18,602	16,073	19,323	16,990
74,050	74,100	18,618	16,087	19,339	17,004
74,100	74,150	18,633	16,101	19,354	17,018
74,150	74,200	18,649	16,115	19,370	17,032
74,200	74,250	18,664	16,129	19,385	17,048
74,250	74,300	18,680	16,143	19,401	17,063
74,300	74,350	18,695	16,157	19,416	17,079
74,350	74,400	18,711	16,171	19,432	17,094
74,400	74,450	18,726	16,185	19,447	17,110
74,450	74,500	18,742	16,199	19,463	17,125
74,500	74,550	18,757	16,213	19,478	17,141
74,550	74,600	18,773	16,227	19,494	17,156
74,600	74,650	18,788	16,241	19,509	17,172
74,650	74,700	18,804	16,255	19,525	17,187
74,700	74,750	18,819	16,269	19,540	17,203
74,750	74,800	18,835	16,283	19,556	17,218
74,800	74,850	18,850	16,297	19,571	17,234
74,850	74,900	18,866	16,311	19,587	17,249
74,900	74,950	18,881	16,325	19,602	17,265
74,950	75,000	18,897	16,339	19,618	17,280

75,000

At least	But less than	Single	MFJ	MFS	HoH
75,000	75,050	18,912	16,353	19,633	17,296
75,050	75,100	18,928	16,367	19,649	17,311
75,100	75,150	18,943	16,381	19,664	17,327
75,150	75,200	18,959	16,395	19,680	17,342
75,200	75,250	18,974	16,409	19,695	17,358
75,250	75,300	18,990	16,423	19,711	17,373
75,300	75,350	19,005	16,437	19,726	17,389
75,350	75,400	19,021	16,451	19,742	17,404
75,400	75,450	19,036	16,465	19,757	17,420
75,450	75,500	19,052	16,479	19,773	17,435
75,500	75,550	19,067	16,493	19,788	17,451
75,550	75,600	19,083	16,507	19,804	17,466
75,600	75,650	19,098	16,521	19,819	17,482
75,650	75,700	19,114	16,535	19,835	17,497
75,700	75,750	19,129	16,549	19,850	17,513
75,750	75,800	19,145	16,563	19,866	17,528
75,800	75,850	19,160	16,577	19,881	17,544
75,850	75,900	19,176	16,591	19,897	17,559
75,900	75,950	19,191	16,605	19,912	17,575
75,950	76,000	19,207	16,619	19,928	17,590

76,000

At least	But less than	Single	MFJ	MFS	HoH
76,000	76,050	19,222	16,633	19,943	17,606
76,050	76,100	19,238	16,647	19,959	17,621
76,100	76,150	19,253	16,661	19,974	17,637
76,150	76,200	19,269	16,675	19,990	17,652
76,200	76,250	19,284	16,689	20,005	17,668
76,250	76,300	19,300	16,703	20,021	17,683
76,300	76,350	19,315	16,717	20,036	17,699
76,350	76,400	19,331	16,731	20,052	17,714
76,400	76,450	19,346	16,745	20,067	17,730
76,450	76,500	19,362	16,759	20,083	17,745
76,500	76,550	19,377	16,773	20,098	17,761
76,550	76,600	19,393	16,787	20,114	17,776
76,600	76,650	19,408	16,801	20,129	17,792
76,650	76,700	19,424	16,815	20,145	17,807
76,700	76,750	19,439	16,829	20,160	17,823
76,750	76,800	19,455	16,843	20,176	17,838
76,800	76,850	19,470	16,857	20,191	17,854
76,850	76,900	19,486	16,871	20,207	17,869
76,900	76,950	19,501	16,885	20,222	17,885
76,950	77,000	19,517	16,899	20,238	17,900

* This column must also be used by a qualifying widow(er).

Continued on next page

19-- Tax Table–Continued

If line 37 (Form 1040) is– At least	But less than	Single	Married filing jointly *	Married filing separately	Head of a household
77,000					
77,000	77,050	19,532	16,913	20,253	17,916
77,050	77,100	19,548	16,927	20,269	17,931
77,100	77,150	19,563	16,941	20,284	17,947
77,150	77,200	19,579	16,955	20,300	17,962
77,200	77,250	19,594	16,969	20,315	17,978
77,250	77,300	19,610	16,983	20,331	17,993
77,300	77,350	19,625	16,997	20,346	18,009
77,350	77,400	19,641	17,011	20,362	18,024
77,400	77,450	19,656	17,025	20,377	18,040
77,450	77,500	19,672	17,039	20,393	18,055
77,500	77,550	19,687	17,053	20,408	18,071
77,550	77,600	19,703	17,067	20,424	18,086
77,600	77,650	19,718	17,081	20,439	18,102
77,650	77,700	19,734	17,095	20,455	18,117
77,700	77,750	19,749	17,109	20,470	18,133
77,750	77,800	19,765	17,123	20,486	18,148
77,800	77,850	19,780	17,137	20,501	18,164
77,850	77,900	19,796	17,151	20,517	18,179
77,900	77,950	19,811	17,165	20,532	18,195
77,950	78,000	19,827	17,179	20,548	18,210
78,000					
78,000	78,050	19,842	17,193	20,563	18,226
78,050	78,100	19,858	17,207	20,579	18,241
78,100	78,150	19,873	17,221	20,594	18,257
78,150	78,200	19,889	17,235	20,610	18,272
78,200	78,250	19,904	17,249	20,625	18,288
78,250	78,300	19,920	17,263	20,641	18,303
78,300	78,350	19,935	17,277	20,656	18,319
78,350	78,400	19,951	17,291	20,672	18,334
78,400	78,450	19,966	17,305	20,687	18,350
78,450	78,500	19,982	17,319	20,703	18,365
78,500	78,550	19,997	17,333	20,718	18,381
78,550	78,600	20,013	17,347	20,734	18,396
78,600	78,650	20,028	17,361	20,749	18,412
78,650	78,700	20,044	17,375	20,765	18,427
78,700	78,750	20,059	17,389	20,780	18,443
78,750	78,800	20,075	17,403	20,796	18,458
78,800	78,850	20,090	17,417	20,811	18,474
78,850	78,900	20,106	17,431	20,827	18,489
78,900	78,950	20,121	17,445	20,842	18,505
78,950	79,000	20,137	17,459	20,858	18,520
79,000					
79,000	79,050	20,152	17,473	20,873	18,536
79,050	79,100	20,168	17,487	20,889	18,551
79,100	79,150	20,183	17,501	20,904	18,567
79,150	79,200	20,199	17,515	20,920	18,582
79,200	79,250	20,214	17,529	20,935	18,598
79,250	79,300	20,230	17,543	20,951	18,613
79,300	79,350	20,245	17,557	20,966	18,629
79,350	79,400	20,261	17,571	20,982	18,644
79,400	79,450	20,276	17,585	20,997	18,660
79,450	79,500	20,292	17,599	21,013	18,675
79,500	79,550	20,307	17,613	21,028	18,691
79,550	79,600	20,323	17,627	21,044	18,706
79,600	79,650	20,338	17,641	21,059	18,722
79,650	79,700	20,354	17,655	21,075	18,737
79,700	79,750	20,369	17,669	21,090	18,753
79,750	79,800	20,385	17,683	21,106	18,768
79,800	79,850	20,400	17,697	21,121	18,784
79,850	79,900	20,416	17,711	21,137	18,799
79,900	79,950	20,431	17,725	21,152	18,815
79,950	80,000	20,447	17,739	21,168	18,830

If line 37 (Form 1040) is– At least	But less than	Single	Married filing jointly *	Married filing separately	Head of a household
80,000					
80,000	80,050	20,462	17,753	21,183	18,846
80,050	80,100	20,478	17,767	21,199	18,861
80,100	80,150	20,493	17,781	21,214	18,877
80,150	80,200	20,509	17,795	21,230	18,892
80,200	80,250	20,524	17,809	21,245	18,908
80,250	80,300	20,540	17,823	21,261	18,923
80,300	80,350	20,555	17,837	21,276	18,939
80,350	80,400	20,571	17,851	21,292	18,954
80,400	80,450	20,586	17,865	21,307	18,970
80,450	80,500	20,602	17,879	21,323	18,985
80,500	80,550	20,617	17,893	21,338	19,001
80,550	80,600	20,633	17,907	21,354	19,016
80,600	80,650	20,648	17,921	21,369	19,032
80,650	80,700	20,664	17,935	21,385	19,047
80,700	80,750	20,679	17,949	21,400	19,063
80,750	80,800	20,695	17,963	21,416	19,078
80,800	80,850	20,710	17,977	21,431	19,094
80,850	80,900	20,726	17,991	21,447	19,109
80,900	80,950	20,741	18,005	21,462	19,125
80,950	81,000	20,757	18,019	21,478	19,140
81,000					
81,000	81,050	20,772	18,033	21,493	19,156
81,050	81,100	20,788	18,047	21,509	19,171
81,100	81,150	20,803	18,061	21,524	19,187
81,150	81,200	20,819	18,075	21,540	19,202
81,200	81,250	20,834	18,089	21,555	19,218
81,250	81,300	20,850	18,103	21,571	19,233
81,300	81,350	20,865	18,117	21,586	19,249
81,350	81,400	20,881	18,131	21,602	19,264
81,400	81,450	20,896	18,145	21,617	19,280
81,450	81,500	20,912	18,159	21,633	19,295
81,500	81,550	20,927	18,173	21,648	19,311
81,550	81,600	20,943	18,187	21,664	19,326
81,600	81,650	20,958	18,201	21,679	19,342
81,650	81,700	20,974	18,215	21,695	19,357
81,700	81,750	20,989	18,229	21,710	19,373
81,750	81,800	21,005	18,243	21,726	19,388
81,800	81,850	21,020	18,257	21,741	19,404
81,850	81,900	21,036	18,271	21,757	19,419
81,900	81,950	21,051	18,285	21,772	19,435
81,950	82,000	21,067	18,299	21,788	19,450
82,000					
82,000	82,050	21,082	18,313	21,803	19,466
82,050	82,100	21,098	18,327	21,819	19,481
82,100	82,150	21,113	18,341	21,834	19,497
82,150	82,200	21,129	18,355	21,850	19,512
82,200	82,250	21,144	18,369	21,865	19,528
82,250	82,300	21,160	18,383	21,881	19,543
82,300	82,350	21,175	18,397	21,896	19,559
82,350	82,400	21,191	18,411	21,912	19,574
82,400	82,450	21,206	18,425	21,927	19,590
82,450	82,500	21,222	18,439	21,943	19,605
82,500	82,550	21,237	18,453	21,958	19,621
82,550	82,600	21,253	18,467	21,974	19,636
82,600	82,650	21,268	18,481	21,989	19,652
82,650	82,700	21,284	18,495	22,005	19,667
82,700	82,750	21,299	18,509	22,020	19,683
82,750	82,800	21,315	18,523	22,036	19,698
82,800	82,850	21,330	18,537	22,051	19,714
82,850	82,900	21,346	18,551	22,067	19,729
82,900	82,950	21,361	18,565	22,082	19,745
82,950	83,000	21,377	18,579	22,098	19,760

If line 37 (Form 1040) is– At least	But less than	Single	Married filing jointly *	Married filing separately	Head of a household
83,000					
83,000	83,050	21,392	18,593	22,113	19,776
83,050	83,100	21,408	18,607	22,129	19,791
83,100	83,150	21,423	18,621	22,144	19,807
83,150	83,200	21,439	18,635	22,160	19,822
83,200	83,250	21,454	18,649	22,175	19,838
83,250	83,300	21,470	18,663	22,191	19,853
83,300	83,350	21,485	18,677	22,206	19,869
83,350	83,400	21,501	18,691	22,222	19,884
83,400	83,450	21,516	18,705	22,237	19,900
83,450	83,500	21,532	18,719	22,253	19,915
83,500	83,550	21,547	18,733	22,268	19,931
83,550	83,600	21,563	18,747	22,284	19,946
83,600	83,650	21,578	18,761	22,299	19,962
83,650	83,700	21,594	18,775	22,315	19,977
83,700	83,750	21,609	18,789	22,330	19,993
83,750	83,800	21,625	18,803	22,346	20,008
83,800	83,850	21,640	18,817	22,361	20,024
83,850	83,900	21,656	18,831	22,377	20,039
83,900	83,950	21,671	18,845	22,392	20,055
83,950	84,000	21,687	18,859	22,408	20,070
84,000					
84,000	84,050	21,702	18,873	22,423	20,086
84,050	84,100	21,718	18,887	22,439	20,101
84,100	84,150	21,733	18,901	22,454	20,117
84,150	84,200	21,749	18,915	22,470	20,132
84,200	84,250	21,764	18,929	22,485	20,148
84,250	84,300	21,780	18,943	22,501	20,163
84,300	84,350	21,795	18,957	22,516	20,179
84,350	84,400	21,811	18,971	22,532	20,194
84,400	84,450	21,826	18,985	22,547	20,210
84,450	84,500	21,842	18,999	22,563	20,225
84,500	84,550	21,857	19,013	22,578	20,241
84,550	84,600	21,873	19,027	22,594	20,256
84,600	84,650	21,888	19,041	22,609	20,272
84,650	84,700	21,904	19,055	22,625	20,287
84,700	84,750	21,919	19,069	22,640	20,303
84,750	84,800	21,935	19,083	22,656	20,318
84,800	84,850	21,950	19,097	22,671	20,334
84,850	84,900	21,966	19,111	22,687	20,349
84,900	84,950	21,981	19,125	22,702	20,365
84,950	85,000	21,997	19,139	22,718	20,380
85,000					
85,000	85,050	22,012	19,153	22,733	20,396
85,050	85,100	22,028	19,167	22,749	20,411
85,100	85,150	22,043	19,181	22,764	20,427
85,150	85,200	22,059	19,195	22,780	20,442
85,200	85,250	22,074	19,209	22,795	20,458
85,250	85,300	22,090	19,223	22,811	20,473
85,300	85,350	22,105	19,237	22,826	20,489
85,350	85,400	22,121	19,251	22,842	20,504
85,400	85,450	22,136	19,265	22,857	20,520
85,450	85,500	22,152	19,279	22,873	20,535
85,500	85,550	22,167	19,293	22,888	20,551
85,550	85,600	22,183	19,307	22,904	20,566
85,600	85,650	22,198	19,321	22,919	20,582
85,650	85,700	22,214	19,335	22,935	20,597
85,700	85,750	22,229	19,349	22,950	20,613
85,750	85,800	22,245	19,363	22,966	20,628
85,800	85,850	22,260	19,377	22,981	20,644
85,850	85,900	22,276	19,391	22,997	20,659
85,900	85,950	22,291	19,405	23,012	20,675
85,950	86,000	22,307	19,419	23,028	20,690

* This column must also be used by a qualifying widow(er).

Continued on next page

19-- Tax Table—*Continued*

If line 37 (Form 1040) is–		And you are—			
At least	But less than	Single	Married filing jointly *	Married filing separately	Head of a household
		Your tax is—			

86,000

At least	But less than	Single	Married filing jointly	Married filing separately	Head of a household
86,000	86,050	22,322	19,433	23,043	20,706
86,050	86,100	22,338	19,447	23,059	20,721
86,100	86,150	22,353	19,461	23,074	20,737
86,150	86,200	22,369	19,475	23,090	20,752
86,200	86,250	22,384	19,489	23,105	20,768
86,250	86,300	22,400	19,503	23,121	20,783
86,300	86,350	22,415	19,517	23,136	20,799
86,350	86,400	22,431	19,531	23,152	20,814
86,400	86,450	22,446	19,545	23,167	20,830
86,450	86,500	22,462	19,559	23,183	20,845
86,500	86,550	22,477	19,574	23,198	20,861
86,550	86,600	22,493	19,589	23,214	20,876
86,600	86,650	22,508	19,605	23,229	20,892
86,650	86,700	22,524	19,620	23,245	20,907
86,700	86,750	22,539	19,636	23,260	20,923
86,750	86,800	22,555	19,651	23,276	20,938
86,800	86,850	22,570	19,667	23,291	20,954
86,850	86,900	22,586	19,682	23,307	20,969
86,900	86,950	22,601	19,698	23,322	20,985
86,950	87,000	22,617	19,713	23,338	21,000

87,000

At least	But less than	Single	Married filing jointly	Married filing separately	Head of a household
87,000	87,050	22,632	19,729	23,353	21,016
87,050	87,100	22,648	19,744	23,369	21,031
87,100	87,150	22,663	19,760	23,384	21,047
87,150	87,200	22,679	19,775	23,400	21,062
87,200	87,250	22,694	19,791	23,415	21,078
87,250	87,300	22,710	19,806	23,431	21,093
87,300	87,350	22,725	19,822	23,446	21,109
87,350	87,400	22,741	19,837	23,462	21,124
87,400	87,450	22,756	19,853	23,477	21,140
87,450	87,500	22,772	19,868	23,493	21,155
87,500	87,550	22,787	19,884	23,508	21,171
87,550	87,600	22,803	19,899	23,524	21,186
87,600	87,650	22,818	19,915	23,539	21,202
87,650	87,700	22,834	19,930	23,555	21,217
87,700	87,750	22,849	19,946	23,570	21,233
87,750	87,800	22,865	19,961	23,586	21,248
87,800	87,850	22,880	19,977	23,601	21,264
87,850	87,900	22,896	19,992	23,617	21,279
87,900	87,950	22,911	20,008	23,632	21,295
87,950	88,000	22,927	20,023	23,648	21,310

88,000

At least	But less than	Single	Married filing jointly	Married filing separately	Head of a household
88,000	88,050	22,942	20,039	23,663	21,326
88,050	88,100	22,958	20,054	23,679	21,341
88,100	88,150	22,973	20,070	23,694	21,357
88,150	88,200	22,989	20,085	23,710	21,372
88,200	88,250	23,004	20,101	23,725	21,388
88,250	88,300	23,020	20,116	23,741	21,403
88,300	88,350	23,035	20,132	23,756	21,419
88,350	88,400	23,051	20,147	23,772	21,434
88,400	88,450	23,066	20,163	23,787	21,450
88,450	88,500	23,082	20,178	23,803	21,465
88,500	88,550	23,097	20,194	23,818	21,481
88,550	88,600	23,113	20,209	23,834	21,496
88,600	88,650	23,128	20,225	23,849	21,512
88,650	88,700	23,144	20,240	23,865	21,527
88,700	88,750	23,159	20,256	23,880	21,543
88,750	88,800	23,175	20,271	23,896	21,558
88,800	88,850	23,190	20,287	23,911	21,574
88,850	88,900	23,206	20,302	23,927	21,589
88,900	88,950	23,221	20,318	23,942	21,605
88,950	89,000	23,237	20,333	23,958	21,620

89,000

At least	But less than	Single	Married filing jointly	Married filing separately	Head of a household
89,000	89,050	23,252	20,349	23,973	21,636
89,050	89,100	23,268	20,364	23,989	21,651
89,100	89,150	23,283	20,380	24,004	21,667
89,150	89,200	23,299	20,395	24,020	21,682
89,200	89,250	23,314	20,411	24,035	21,698
89,250	89,300	23,330	20,426	24,051	21,713
89,300	89,350	23,345	20,442	24,066	21,729
89,350	89,400	23,361	20,457	24,082	21,744
89,400	89,450	23,376	20,473	24,097	21,760
89,450	89,500	23,392	20,488	24,113	21,775
89,500	89,550	23,407	20,504	24,128	21,791
89,550	89,600	23,423	20,519	24,144	21,806
89,600	89,650	23,438	20,535	24,159	21,822
89,650	89,700	23,454	20,550	24,175	21,837
89,700	89,750	23,469	20,566	24,190	21,853
89,750	89,800	23,485	20,581	24,206	21,868
89,800	89,850	23,500	20,597	24,221	21,884
89,850	89,900	23,516	20,612	24,237	21,899
89,900	89,950	23,531	20,628	24,252	21,915
89,950	90,000	23,547	20,643	24,268	21,930

90,000

At least	But less than	Single	Married filing jointly	Married filing separately	Head of a household
90,000	90,050	23,562	20,659	24,283	21,946
90,050	90,100	23,578	20,674	24,299	21,961
90,100	90,150	23,593	20,690	24,314	21,977
90,150	90,200	23,609	20,705	24,330	21,992
90,200	90,250	23,624	20,721	24,345	22,008
90,250	90,300	23,640	20,736	24,361	22,023
90,300	90,350	23,655	20,752	24,376	22,039
90,350	90,400	23,671	20,767	24,392	22,054
90,400	90,450	23,686	20,783	24,407	22,070
90,450	90,500	23,702	20,798	24,423	22,085
90,500	90,550	23,717	20,814	24,438	22,101
90,550	90,600	23,733	20,829	24,454	22,116
90,600	90,650	23,748	20,845	24,469	22,132
90,650	90,700	23,764	20,860	24,485	22,147
90,700	90,750	23,779	20,876	24,500	22,163
90,750	90,800	23,795	20,891	24,516	22,178
90,800	90,850	23,810	20,907	24,531	22,194
90,850	90,900	23,826	20,922	24,547	22,209
90,900	90,950	23,841	20,938	24,562	22,225
90,950	91,000	23,857	20,953	24,578	22,240

91,000

At least	But less than	Single	Married filing jointly	Married filing separately	Head of a household
91,000	91,050	23,872	20,969	24,593	22,256
91,050	91,100	23,888	20,984	24,609	22,271
91,100	91,150	23,903	21,000	24,624	22,287
91,150	91,200	23,919	21,015	24,640	22,302
91,200	91,250	23,934	21,031	24,655	22,318
91,250	91,300	23,950	21,046	24,671	22,333
91,300	91,350	23,965	21,062	24,686	22,349
91,350	91,400	23,981	21,077	24,702	22,364
91,400	91,450	23,996	21,093	24,717	22,380
91,450	91,500	24,012	21,108	24,733	22,395
91,500	91,550	24,027	21,124	24,748	22,411
91,550	91,600	24,043	21,139	24,764	22,426
91,600	91,650	24,058	21,155	24,779	22,442
91,650	91,700	24,074	21,170	24,795	22,457
91,700	91,750	24,089	21,186	24,810	22,473
91,750	91,800	24,105	21,201	24,826	22,488
91,800	91,850	24,120	21,217	24,841	22,504
91,850	91,900	24,136	21,232	24,857	22,519
91,900	91,950	24,151	21,248	24,872	22,535
91,950	92,000	24,167	21,263	24,888	22,550

92,000

At least	But less than	Single	Married filing jointly	Married filing separately	Head of a household
92,000	92,050	24,182	21,279	24,903	22,566
92,050	92,100	24,198	21,294	24,919	22,581
92,100	92,150	24,213	21,310	24,934	22,597
92,150	92,200	24,229	21,325	24,950	22,612
92,200	92,250	24,244	21,341	24,965	22,628
92,250	92,300	24,260	21,356	24,981	22,643
92,300	92,350	24,275	21,372	24,996	22,659
92,350	92,400	24,291	21,387	25,012	22,674
92,400	92,450	24,306	21,403	25,027	22,690
92,450	92,500	24,322	21,418	25,043	22,705
92,500	92,550	24,337	21,434	25,058	22,721
92,550	92,600	24,353	21,449	25,074	22,736
92,600	92,650	24,368	21,465	25,089	22,752
92,650	92,700	24,384	21,480	25,105	22,767
92,700	92,750	24,399	21,496	25,120	22,783
92,750	92,800	24,415	21,511	25,136	22,798
92,800	92,850	24,430	21,527	25,151	22,814
92,850	92,900	24,446	21,542	25,167	22,829
92,900	92,950	24,461	21,558	25,182	22,845
92,950	93,000	24,477	21,573	25,198	22,860

93,000

At least	But less than	Single	Married filing jointly	Married filing separately	Head of a household
93,000	93,050	24,492	21,589	25,213	22,876
93,050	93,100	24,508	21,604	25,229	22,891
93,100	93,150	24,523	21,620	25,244	22,907
93,150	93,200	24,539	21,635	25,260	22,922
93,200	93,250	24,554	21,651	25,275	22,938
93,250	93,300	24,570	21,666	25,291	22,953
93,300	93,350	24,585	21,682	25,306	22,969
93,350	93,400	24,601	21,697	25,322	22,984
93,400	93,450	24,616	21,713	25,337	23,000
93,450	93,500	24,632	21,728	25,353	23,015
93,500	93,550	24,647	21,744	25,368	23,031
93,550	93,600	24,663	21,759	25,384	23,046
93,600	93,650	24,678	21,775	25,399	23,062
93,650	93,700	24,694	21,790	25,415	23,077
93,700	93,750	24,709	21,806	25,430	23,093
93,750	93,800	24,725	21,821	25,446	23,108
93,800	93,850	24,740	21,837	25,461	23,124
93,850	93,900	24,756	21,852	25,477	23,139
93,900	93,950	24,771	21,868	25,492	23,155
93,950	94,000	24,787	21,883	25,508	23,170

94,000

At least	But less than	Single	Married filing jointly	Married filing separately	Head of a household
94,000	94,050	24,802	21,899	25,523	23,186
94,050	94,100	24,818	21,914	25,539	23,201
94,100	94,150	24,833	21,930	25,554	23,217
94,150	94,200	24,849	21,945	25,570	23,232
94,200	94,250	24,864	21,961	25,585	23,248
94,250	94,300	24,880	21,976	25,601	23,263
94,300	94,350	24,895	21,992	25,616	23,279
94,350	94,400	24,911	22,007	25,632	23,294
94,400	94,450	24,926	22,023	25,647	23,310
94,450	94,500	24,942	22,038	25,663	23,325
94,500	94,550	24,957	22,054	25,678	23,341
94,550	94,600	24,973	22,069	25,694	23,356
94,600	94,650	24,988	22,085	25,709	23,372
94,650	94,700	25,004	22,100	25,725	23,387
94,700	94,750	25,019	22,116	25,740	23,403
94,750	94,800	25,035	22,131	25,756	23,418
94,800	94,850	25,050	22,147	25,771	23,434
94,850	94,900	25,066	22,162	25,787	23,449
94,900	94,950	25,081	22,178	25,802	23,465
94,950	95,000	25,097	22,193	25,818	23,480

* This column must also be used by a qualifying widow(er).

Continued on next page

19-- Tax Table—*Continued*

95,000

At least	But less than	Single	Married filing jointly *	Married filing separately	Head of a household
95,000	95,050	25,112	22,209	25,833	23,496
95,050	95,100	25,128	22,224	25,849	23,511
95,100	95,150	25,143	22,240	25,864	23,527
95,150	95,200	25,159	22,255	25,880	23,542
95,200	95,250	25,174	22,271	25,895	23,558
95,250	95,300	25,190	22,286	25,911	23,573
95,300	95,350	25,205	22,302	25,926	23,589
95,350	95,400	25,221	22,317	25,942	23,604
95,400	95,450	25,236	22,333	25,957	23,620
95,450	95,500	25,252	22,348	25,973	23,635
95,500	95,550	25,267	22,364	25,988	23,651
95,550	95,600	25,283	22,379	26,004	23,666
95,600	95,650	25,298	22,395	26,019	23,682
95,650	95,700	25,314	22,410	26,035	23,697
95,700	95,750	25,329	22,426	26,050	23,713
95,750	95,800	25,345	22,441	26,066	23,728
95,800	95,850	25,360	22,457	26,081	23,744
95,850	95,900	25,376	22,472	26,097	23,759
95,900	95,950	25,391	22,488	26,112	23,775
95,950	96,000	25,407	22,503	26,128	23,790

96,000

At least	But less than	Single	Married filing jointly *	Married filing separately	Head of a household
96,000	96,050	25,422	22,519	26,143	23,806
96,050	96,100	25,438	22,534	26,159	23,821
96,100	96,150	25,453	22,550	26,174	23,837
96,150	96,200	25,469	22,565	26,190	23,852
96,200	96,250	25,484	22,581	26,205	23,868
96,250	96,300	25,500	22,596	26,221	23,883
96,300	96,350	25,515	22,612	26,236	23,899
96,350	96,400	25,531	22,627	26,252	23,914
96,400	96,450	25,546	22,643	26,267	23,930
96,450	96,500	25,562	22,658	26,283	23,945
96,500	96,550	25,577	22,674	26,298	23,961
96,550	96,600	25,593	22,689	26,314	23,976
96,600	96,650	25,608	22,705	26,329	23,992
96,650	96,700	25,624	22,720	26,345	24,007
96,700	96,750	25,639	22,736	26,360	24,023
96,750	96,800	25,655	22,751	26,376	24,038
96,800	96,850	25,670	22,767	26,391	24,054
96,850	96,900	25,686	22,782	26,407	24,069
96,900	96,950	25,701	22,798	26,422	24,085
96,950	97,000	25,717	22,813	26,438	24,100

97,000

At least	But less than	Single	Married filing jointly *	Married filing separately	Head of a household
97,000	97,050	25,732	22,829	26,453	24,116
97,050	97,100	25,748	22,844	26,469	24,131
97,100	97,150	25,763	22,860	26,484	24,147
97,150	97,200	25,779	22,875	26,500	24,162
97,200	97,250	25,794	22,891	26,515	24,178
97,250	97,300	25,810	22,906	26,531	24,193
97,300	97,350	25,825	22,922	26,546	24,209
97,350	97,400	25,841	22,937	26,562	24,224
97,400	97,450	25,856	22,953	26,577	24,240
97,450	97,500	25,872	22,968	26,593	24,255
97,500	97,550	25,887	22,984	26,608	24,271
97,550	97,600	25,903	22,999	26,624	24,286
97,600	97,650	25,918	23,015	26,639	24,302
97,650	97,700	25,934	23,030	26,655	24,317
97,700	97,750	25,949	23,046	26,670	24,333
97,750	97,800	25,965	23,061	26,686	24,348
97,800	97,850	25,980	23,077	26,701	24,364
97,850	97,900	25,996	23,092	26,717	24,379
97,900	97,950	26,011	23,108	26,732	24,395
97,950	98,000	26,027	23,123	26,748	24,410

98,000

At least	But less than	Single	Married filing jointly *	Married filing separately	Head of a household
98,000	98,050	26,042	23,139	26,763	24,426
98,050	98,100	26,058	23,154	26,779	24,441
98,100	98,150	26,073	23,170	26,794	24,457
98,150	98,200	26,089	23,185	26,810	24,472
98,200	98,250	26,104	23,201	26,825	24,488
98,250	98,300	26,120	23,216	26,841	24,503
98,300	98,350	26,135	23,232	26,856	24,519
98,350	98,400	26,151	23,247	26,872	24,534
98,400	98,450	26,166	23,263	26,887	24,550
98,450	98,500	26,182	23,278	26,903	24,565
98,500	98,550	26,197	23,294	26,918	24,581
98,550	98,600	26,213	23,309	26,934	24,596
98,600	98,650	26,228	23,325	26,949	24,612
98,650	98,700	26,244	23,340	26,965	24,627
98,700	98,750	26,259	23,356	26,980	24,643
98,750	98,800	26,275	23,371	26,996	24,658
98,800	98,850	26,290	23,387	27,011	24,674
98,850	98,900	26,306	23,402	27,027	24,689
98,900	98,950	26,321	23,418	27,042	24,705
98,950	99,000	26,337	23,433	27,058	24,720

99,000

At least	But less than	Single	Married filing jointly *	Married filing separately	Head of a household
99,000	99,050	26,352	23,449	27,073	24,736
99,050	99,100	26,368	23,464	27,089	24,751
99,100	99,150	26,383	23,480	27,104	24,767
99,150	99,200	26,399	23,495	27,120	24,782
99,200	99,250	26,414	23,511	27,135	24,798
99,250	99,300	26,430	23,526	27,151	24,813
99,300	99,350	26,445	23,542	27,166	24,829
99,350	99,400	26,461	23,557	27,182	24,844
99,400	99,450	26,476	23,573	27,197	24,860
99,450	99,500	26,492	23,588	27,213	24,875
99,500	99,550	26,507	23,604	27,228	24,891
99,550	99,600	26,523	23,619	27,244	24,906
99,600	99,650	26,538	23,635	27,259	24,922
99,650	99,700	26,554	23,650	27,275	24,937
99,700	99,750	26,569	23,666	27,290	24,953
99,750	99,800	26,585	23,681	27,306	24,968
99,800	99,850	26,600	23,697	27,321	24,984
99,850	99,900	26,616	23,712	27,337	24,999
99,900	99,950	26,631	23,728	27,352	25,015
99,950	100,000	26,647	23,743	27,368	25,030

100,000 or over—use tax rate schedules

* This column must also be used by a qualifying widow(er).

19--
Tax Rate
Schedules

Caution: *Use **only** if your taxable income (Form 1040, line 37) is $100,000 or more. If less, use the **Tax Table.** Even though you cannot use the tax rate schedules below if your taxable income is less than $100,000, all levels of taxable income are shown so taxpayers can see the tax rate that applies to each level.*

Schedule X—Use if your filing status is **Single**

If the amount on Form 1040, line 37, is: Over—	But not over—	Enter on Form 1040, line 38	of the amount over—
$0	$21,450	.... 15%	$0
21,450	51,900	**$3,217.50 + 28%**	21,450
51,900	....	**11,743.50 + 31%**	51,900

Schedule Y-1—Use if your filing status is **Married filing jointly** or **Qualifying widow(er)**

If the amount on Form 1040, line 37, is: Over—	But not over—	Enter on Form 1040, line 38	of the amount over—
$0	$35,800	.... 15%	$0
35,800	86,500	**$5,370.00 + 28%**	35,800
86,500	....	**19,566.00 + 31%**	86,500

Schedule Y-2—Use if your filing status is **Married filing separately**

If the amount on Form 1040, line 37, is: Over—	But not over—	Enter on Form 1040, line 38	of the amount over—
$0	$17,900	.... 15%	$0
17,900	43,250	**$2,685.00 + 28%**	17,900
43,250	....	**9,783.00 + 31%**	43,250

Schedule Z—Use if your filing status is **Head of household**

If the amount on Form 1040, line 37, is: Over—	But not over—	Enter on Form 1040, line 38	of the amount over—
$0	$28,750	.... 15%	$0
28,750	74,150	**$4,312.50 + 28%**	28,750
74,150	....	**17,024.50 + 31%**	74,150

19-- Standard Deduction Tables

Caution: If you are married filing a separate return and your spouse itemizes deductions, or if you are a dual-status alien, you cannot take the standard deduction even if you were 65 or older or blind.

Table 20-1.   **Standard Deduction Chart for Most People***

If Your Filing Status is:	Your Standard Deduction Is:
Single	$3,600
Married filing joint return or Qualifying widow(er) with dependent child	6,000
Married filing separate return	3,000
Head of household	5,250

* DO NOT use this chart if you were 65 or older or blind, OR if someone can claim you as a dependent.

Table 20-2.   **Standard Deduction Chart for People Age 65 or Older or Blind***

Check the correct number of boxes below. Then go to the chart.

You   65 or older ☐   Blind ☐

Your spouse, if claiming spouse's exemption   65 or older ☐   Blind ☐

Total number of boxes you checked ☐

If Your Filing Status is:	And the Number in the Box Above is:	Your Standard Deduction is:
Single	1	$4,500
	2	5,400
Married filing joint return or Qualifying widow(er) with dependent child	1	6,700
	2	7,400
	3	8,100
	4	8,800
Married filing separate return	1	3,700
	2	4,400
	3	5,100
	4	5,800
Head of household	1	6,150
	2	7,050

* If someone can claim you as a dependent, use the worksheet in Table 20-3, instead.

Table 20-3.   **Standard Deduction Worksheet for Dependents***

If you were 65 or older or blind, check the correct number of boxes below. Then go to the worksheet.

You   65 or older ☐   Blind ☐

Your spouse, if claiming spouse's exemption   65 or older ☐   Blind ☐

Total number of boxes you checked ☐

1. Enter your **earned income** (defined below). If none, go on to line 3	1._____
2. Minimum amount	2. $600
3. Compare the amounts on lines 1 and 2. Enter the **larger** of the two amounts here	3._____
4. Enter on line 4 the amount shown below for your filing status. • Single, enter $3,600 • Married filing separate return, enter $3,000 • Married filing jointly or Qualifying widow(er) with dependent child, enter $6,000 • Head of household, enter $5,250	4._____
5. **Standard deduction.** a. Compare the amounts on lines 3 and 4. Enter the **smaller** of the two amounts here. If under 65 and not blind, stop here. This is your standard deduction. Otherwise, go on to line 5b	5a._____
b. If 65 or older or blind, multiply $900 ($700 if married or qualifying widow(er) with dependent child) by the number in the box above. Enter the result	5b._____
c. Add lines 5a and 5b. This is your standard deduction for 1992.	5c._____

Earned income *includes wages, salaries, tips, professional fees, and other compensation received for personal services you performed. It also includes any amount received as a scholarship that you must include in your income.*

* Use this worksheet ONLY if someone can claim you as a dependent.

ORGANIZATION OF THE FEDERAL RESERVE SYSTEM

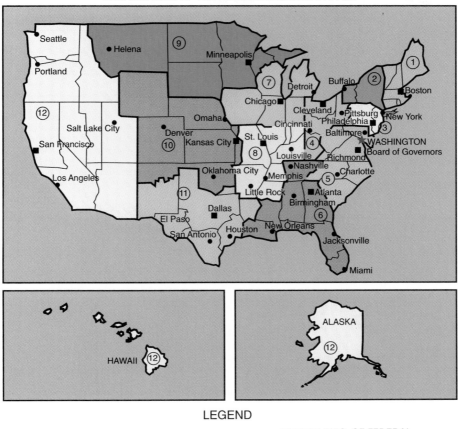

LEGEND

— BOUNDARIES OF FEDERAL
RESERVE DISTRICTS

★ BOARD OF GOVERNORS OF
THE FEDERAL RESERVE
SYSTEM

● CITY WHERE A BRANCH OF
FEDERAL RESERVE BANK
IS LOCATED

— BOUNDARIES OF FEDERAL
RESERVE BRANCH
TERRITORIES

■ CITY WHERE FEDERAL
RESERVE BANK IS LOCATED

③ FEDERAL RESERVE DISTRICT
NUMBER. THIS NUMBER APPEARS
ON THE CURRENCY ISSUED BY
THE FEDERAL RESERVE BANK IN
THE DISTRICT

Hypothetical Credit-Scoring Table

Fill out your credit profile by answering the nine questions below in Table 1. Circle the one response that applies to you, and then find your total score by adding up the points you got for each response. The points are found in the lower right-hand corner of each box. (For example: if you are 25 years old, you get 5 points.) Once you've totaled your score, look at Table 2 to find out how good a credit "bet" you may be.

1.

1.	age?	under 25 — 12	25–29 — 5	30–34 — 0	35–39 — 1	40–44 — 18	45–49 — 22	50 or over — 31
2.	time at address?	less than 1 yr. — 9	1yr. — 0	2–3 yrs. — 5	4–5 yrs. — 0	6–9 yrs. — 5	10 yrs. or more — 21	
3.	age of auto?	none — 0	0–1 yrs. — 12	2 yrs. — 16	3–4 yrs. — 13	5–7 yrs. — 3	8 yrs. or more — 0	
4.	monthly auto payment?	none — 18	less than $125 — 6	$126–$150 — 1	$151–$199 — 4	$200 or more — 0		
5.	housing cost?	less than $274 — 0	$275–$399 — 10	$400 or more — 12	owns clear — 12	lives with relatives — 24		
6.	checking and savings accounts	both — 15	checking only — 2	savings only — 2	neither — 0			
7.	finance company reference	yes — 0	no — 15					
8.	major credit cards?	none — 0	1 — 5	2 or more — 15				
9.	ratio of debt to income?	no debts — 41	1%–5% — 16	6%–15% — 20	16% or over — 0			

2.

A lender using this scoring table selects a cutoff point from a table like this, which gauges how likely applicants are to repay loans.

Total Score	Probability of Repayment
90	89 in 100
95	91 in 100
100	92 in 100
105	93 in 100
110	94 in 100
115	95 in 100
120	95.5 in 100
125	96 in 100
130	96.25 in 100

Source: Federal Reserve Board. Developed by Fair, Isaac, and Co., Inc. Modified to update.

ACCUMULATED CASH VALUE OF $100,000 WHOLE LIFE POLICY
AGE OF ISSUE: 25

Year	Person's Age	Cash Value	Year	Person's Age	Cash Value
1	25	$ 0	11	35	$10,187
2	26	700	12	36	11,501
3	27	1500	13	37	12,860
4	28	2300	14	38	14,246
5	29	3100	15	39	15,667
6	30	4020	16	40	17,094
7	31	5158	17	41	18,555
8	32	6349	18	42	20,014
9	33	7538	19	43	21,563
10	34	8898	20	44	23,197

ACCUMULATED CASH VALUE OF $100,000 WHOLE LIFE POLICY

Age at Issue	After Paying Premiums for:					At Age 65
	5 years	10 years	15 years	17 years	20 years	
20	$ 2,212	$ 7,105	$12,840	$15,371	$19,438	$57,716
25	3,100	8,898	15,667	18,555	23,197	55,890
30	3,999	11,027	18,837	22,098	27,339	53,533
35	5,151	13,369	22,203	25,868	31,759	50,453
40	6,367	15,921	26,016	30,001	36,407	46,426
45	7,737	18,715	29,992	34,586	41,181	41,181
50	9,210	21,707	34,080	38,556	45,751	34,080
55	10,842	24,827	37,980	42,562	49,928	24,827

MULTIPLES-OF-SALARY CHART

	Current Age							
	25 Years		35 Years		45 Years		55 Years	
Current Gross Earnings	75%	60%	75%	60%	75%	60%	75%	60%
$ 7,500	4.0	3.0	5.5	4.0	7.5	5.5	6.5	4.5
9,000	4.0	3.0	5.5	4.0	7.5	5.5	6.5	4.5
15,000	4.5	3.0	6.5	4.5	8.0	6.0	7.0	5.5
23,500	6.5	4.5	8.0	5.5	8.5	6.5	7.5	5.5
30,000	7.5	5.0	8.0	6.0	8.5	6.5	7.0	5.5
40,000	7.5	5.0	8.0	6.0	8.0	6.0	7.0	5.5
65,000	7.5	5.5	7.5	6.0	7.5	6.0	6.5	5.0

COMPARISON TABLE FOR TERM AND WHOLE LIFE PREMIUMS
Policy Face Value is $100,000

Age	Five-Year Renewable Term	Whole Life	First-Year Difference
20	$205	$ 775	$ 570
25	207	918	711
30	218	1112	894
35	254	1374	1120
40	363	1729	1366
45	562	2127	1565
50	878	2689	1811

EXPECTED DEATHS
PER 100,000 ALIVE AT SPECIFIED AGE

Age	Expected Deaths Within 1 Year	Expected to be Alive in 1 Year
15	63	99,937
16	79	99,921
17	91	99,909
18	99	99,901
19	103	99,897
20	106	99,894
21	110	99,890
22	113	99,887
23	115	99,885
24	117	99,883
25	118	99,882
26	120	99,880
27	123	99,877
28	127	99,873
29	132	99,868
45	315	99,685
46	341	99,659
47	371	99,629
48	405	99,595
49	443	99,557

Moonbeam	Dealer Cost	Sticker Price
Model A, 4-door wagon, 8 cylinder	$12,725	$14,722
Model B, 4-door sedan, 6 cylinder	14,062	16,295
Model C, 4-door sedan, 4 cylinder	11,053	12,755
Options		
Conventional spare tire	$ 62	$ 73
Air conditioning	695	817
Electronic climate control	850	1000
Antilock brake system	838	985
Rear window defroster	136	160
AM/FM radio and cassette player	132	155
Rear-facing third seat	132	155
Cruise control	178	210
Stripe, painted	51	61

AUTOMOBILE INSURANCE, SIX-MONTH BASIC RATE SCHEDULE

Car Class Rating	Collision Deductible			Comprehensive Deductible		
	$100	**$250**	**$500**	**$50**	**$250**	**$500**
1–10	$ 500	$ 430	$ 370	$140	$130	$110
11–20	820	750	690	350	310	280
21–30	1140	1040	940	450	410	370
31–40	1280	1190	1100	670	610	560

DRIVER RATING FACTOR (Part I)
Category: Married Youths

Age	Sex	Owner or Usual Driver	Driver Training	Drive to Work	Rating Factor
16–17	F	Yes	Yes	Yes	2.40
				No	2.25
			No	Yes	2.60
				No	2.45
		No	Yes	Yes	1.90
				No	1.75
			No	Yes	2.05
				No	1.90
	M	Yes	Yes	Yes	2.45
				No	2.30
			No	Yes	2.70
				No	2.55
		No	Yes	Yes	2.45
				No	2.30
			No	Yes	2.70
				No	2.55
18–21	F	Yes	Yes	Yes	1.95
				No	1.80
			No	Yes	2.10
				No	1.95
		No	Yes	Yes	1.55
				No	1.40
			No	Yes	1.65
				No	1.50
	M	Yes	Yes	Yes	2.00
				No	1.85
			No	Yes	2.15
				No	2.00
		No	Yes	Yes	2.00
				No	1.85
			No	Yes	2.15
				No	2.00

DRIVER RATING FACTOR (Part II)
Category: Unmarried Youths

Age	Sex	Owner or Usual Driver	Driver Training	Drive to Work	Rating Factor
16–17	F	Yes	Yes	Yes	2.40
				No	2.25
			No	Yes	2.60
				No	2.45
		No	Yes	Yes	1.90
				No	1.75
			No	Yes	2.05
				No	1.90
	M	Yes	Yes	Yes	3.75
				No	3.60
			No	Yes	4.20
				No	4.05
		No	Yes	Yes	2.60
				No	2.45
			No	Yes	2.85
				No	2.70
18–21	F	Yes	Yes	Yes	1.95
				No	1.80
			No	Yes	2.10
				No	1.95
		No	Yes	Yes	1.55
				No	1.40
			No	Yes	1.65
				No	1.50
	M	Yes	Yes	Yes	3.00
				No	2.85
			No	Yes	3.35
				No	3.20
		No	Yes	Yes	2.15
				No	2.00
			No	Yes	2.30
				No	2.15

DRIVER RATING FACTOR (Part III)
Category: All Adults

Age	Sex	Owner or Usual Driver	Drive to Work	Rating Factor
21–24	F	Yes	Yes	1.50
			No	1.35
		No	Yes	1.40
			No	1.25
	M	Yes	Yes	2.25
			No	2.10
		No	Yes	1.55
			No	1.40
25–29	Both	Yes	Yes	1.65
			No	1.50
30–49	Both	Yes	Yes	1.25
			No	1.00
50–64	Both	Yes	Yes	1.15
			No	0.90
65 +	Both	Yes	Yes	1.05
			No	0.80

United States Mileage Chart

	Atlanta, GA	Boston, MA	Cheyenne, WY	Chicago, IL	Cincinnati, OH	Cleveland, OH	Dallas, TX	Denver, CO	Des Moines, IA	Detroit, MI	Indianapolis, IN	Kansas City, MO	Louisville, KY	Memphis, TN	Milwaukee, WI	Minneapolis, MN	New Orleans, LA	Omaha, NE	Philadelphia, PA	Pittsburgh, PA	Portland, OR	St. Louis, MO	Salt Lake City, UT	San Francisco, CA	Seattle, WA	Toledo, OH	Tulsa, OK	Washington, DC	Wichita, KS
Albuquerque, NM	1381	2172	517	1281	1372	1560	638	417	977	1525	1266	782	1301	1010	1319	1190	1134	858	1899	1619	1371	1038	604	1115	1440	1469	645	1824	593
Amarillo, TX	1097	1897	511	1043	1096	1285	358	423	742	1315	991	547	1019	726	1084	975	1068	643	1624	687	1655	756	888	1399	1724	1210	361	1549	350
Atlanta, GA		1037	1442	674	461	714	795	1398	870	699	493	798	382	371	761	1084	479	980	750	687	2601	555	1878	2496	2618	717	772	608	903
Austin, TX	919	1911	994	1083	1110	1327	193	709	645	1286	877	603	1174	628	1161	1184	517	837	1615	1367	2069	741	823	1302	1748	1256	665	1537	548
Birmingham, AL	150	1165	1347	657	481	736	642	1286	787	724	475	695	391	246	364	941	342	898	869	741	2505	465	1781	2366	2535	665	647	736	778
Boston, MA	1037		1907	963	735	628	1748	1949	1280	695	906	1391	941	1296	1050	1368	1507	1412	296	561	3046	1141	2158	3095	2976	739	1537	429	1587
Charleston, SC	289	929	1722	877	603	730	1072	1678	1150	1078	627	1211	650	696	842	1050	696	1282	720	666	2881	821	2158	2785	2890	739	1061	500	1192
Cheyenne, WY	1442	1907		954	1279	1174	879	100	869	1257	1068	650	1161	1101	530	987	1390	495	1678	1390	1159	901	436	1302	1188	1228	650	1611	583
Chicago, IL	674	963	954		335	335	917	996	327	266	181	499	292	530	87	405	912	459	738	495	2083	289	1390	2142	2013	232	683	671	696
Cleveland, OH	672	628	1174	335	244		1174	1321	652	170	294	779	345	779	335	917	1159	917	405	129	1159	1028	552	1610	2142	111	925	346	975
Columbus, OH	533	735	1235	308	107	143	1028	1229	618	205	176	591	211	652	294	692	786	786	462	182	2391	406	1610	2423	2321	133	802	387	852
Dallas, TX	795	1748	879	917	920	1174		781	669	1143	865	489	819	452	976	936	496	644	1452	1411	2009	630	1242	1753	2078	1084	257	1319	365
Denver, CO	1398	1949	100	996	1164	1321	781		669	1253	1058	600	1120	1029	566	841	1273	537	1691	1411	1238	857	504	1235	1307	1218	549	1616	509
Des Moines, IA	870	1280	680	327	571	652	669	669		584	465	195	566	361	252	361	978	132	1051	763	1786	333	1063	1815	1749	549	443	984	392
Detroit, MI	699	695	1257	266	266	170	1143	1253	584		195	743	360	713	360	671	1045	716	573	287	2349	513	1647	2399	2279	59	909	506	940
Flagstaff, AZ	1704	2495	627	1748	1695	1949	961	781	1300	1848	1589	1105	1624	1333	1642	1481	1457	1171	2222	1942	1241	1361	511	792	1347	1792	968	2147	916
Harrisburg, PA	700	373	1579	639	468	314	1383	1592	952	428	474	666	569	726	569	931	1044	726	102	189	2722	769	2015	2767	2652	415	1165	107	1215
Indianapolis, IN	493	906	1068	181	106	294	865	1058	465	278		485	111	435	268	586	796	587	633	353	2227	235	1504	2256	2194	219	631	558	681
Jackson, MS	391	1406	1257	742	655	914	404	1169	809	1029	809	646	554	212	824	1036	178	845	1110	939	2401	495	1646	2157	2470	855	549	855	708
Kansas City, MO	798	1391	650	499	591	779	489	600	195	743	485		520	451	537	447	806	201	1118	838	1809	257	1086	1835	1839	687	248	1043	197
Knoxville, TN	193	911	1372	527	253	489	837	1328	800	530	360	346	253	385	614	932	596	916	615	511	2531	471	1808	2510	2540	453	786	482	871
Louisville, KY	382	941	1161	292	101	345	819	1120	566	360	111	520		367	379	697	685	687	668	388	2320	263	1597	2349	2305	301	659	582	710
Mackinaw City, MI	935	916	1291	387	495	439	1261	1341	673	284	460	864	562	880	368	508	1247	805	842	556	2128	651	1691	2443	2058	328	1047	775	1061
Miami, FL	655	1504	2097	1329	1095	1264	1300	2037	1525	1392	1352	1148	1148	1037	997	1416	860	1641	1208	1200	3256	856	2532	3053	3273	1293	1398	1078	1352
Minneapolis, MN	1068	1368	788	405	692	740	936	841	244	671	586	447	697	554	332		1382	357	1143	857	1678	552	1186	1940	1608	637	695	1076	644
New Orleans, LA	479	1507	1390	912	786	1030	496	1273	978	1045	796	806	700	390	826	1251		806	1211	1070	2505	673	1738	2249	2574	986	647	1078	816
Norfolk, VA	540	558	1764	831	617	496	1329	1758	1141	666	700	1162	642	877	1043	1236	918	1019	263	384	2914	905	2200	2952	2844	607	1278	188	1352
Pierre, SD	1361	1726	434	763	1390	1204	1098	518	287	943	1029	592	1055	1043	752	388	1394	391	1501	1215	1353	824	823	1575	1283	995	760	1434	578
Pittsburgh, PA	687	561	1390	452	287	129	1204	1411	763	287	360	492	388	838	539	857	1070	895	288		2535	588	1826	2578	2465	228	995	221	1034
Portland, ME	1139	106	1986	1042	707	707	1850	2028	1359	775	1148	1486	984	1398	1129	1447	1609	1491	398	663	3125	1236	2422	3174	3055	172	1632	531	1682
Portland, OR	2601	3046	1159	2083	2333	2418	2009	1238	1786	2349	2227	1809	2320	2259	2010	1678	2505	1654	2821	2535		2086	767	636	172	2315	1913	2754	1739
San Antonio, TX	983	1988	1027	1187	1160	1404	270	939	1098	1392	954	759	1114	759	1059	1261	550	914	1692	1444	2086	900	636	1319	1737	1333	527	1559	625
San Francisco, CA	2496	3095	1188	2013	2362	2467	1753	1235	1815	2399	2256	1835	2349	2290	2175	1940	2574	1683	2866	2578	636	2089	752		808	2364	1760	2799	1695
Seattle, WA	2618	2976	1228	2013	2348	2348	2078	1307	1749	2279	2194	1839	2305	2290	1940	1608	2574	1638	2751	2465	172	2081	836	808		2245	1982	2684	1808
Tulsa, OK	772	1537	765	683	683	925	257	681	443	909	631	248	659	401	757	695	647	387	1264	984	1913	396	1172	1760	2245	850		1189	182
Washington, DC	608	429	1611	671	481	346	1319	1616	984	506	607	1043	582	867	758	867	1075	1078	133	221	2754	793	2047	2799	2684	447	1189		1239
Wichita, KS	903	1587	583	696	787	975	365	509	392	940	681	197	710	532	734	644	815	298	1314	1034	1739	447	1003	1695	1808	884	182	1239	

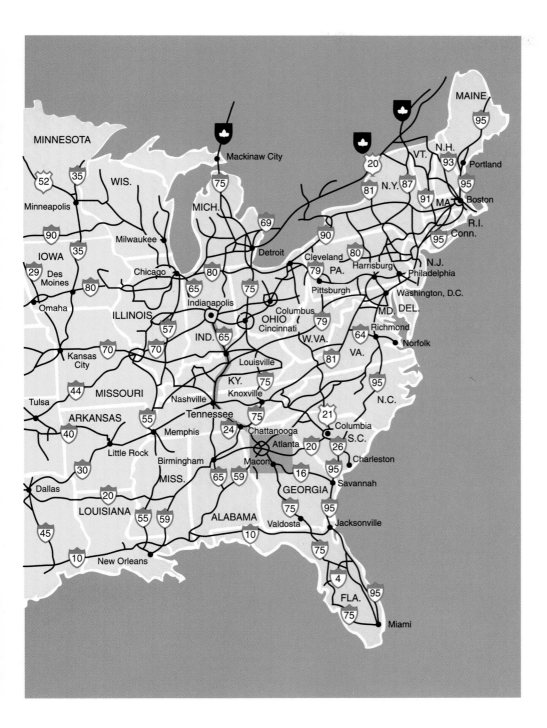

Time Zone World Map

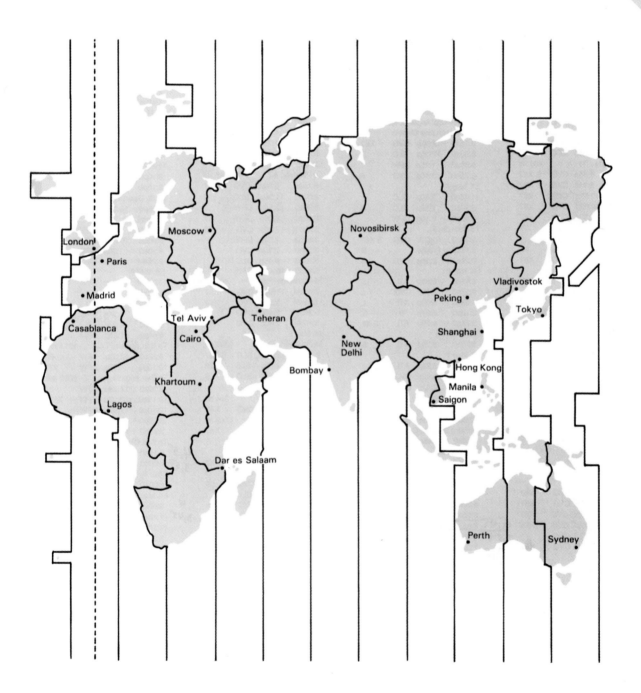

WINDWARD AIRLINES SCHEDULE

Leave	Arrive	Stops/Via	Rmks

From Chicago, IL

To Muscle Shoals, AL 491 mi
| 1:00p0 | 5:10p | ATL | L |
| 4:55p0 | 9:00p | ATL | D X6 |

To Myrtle Beach, SC 743 mi
| 8:10a0 | 1:18p | ATL | B |
| 9:54a0 | 3:15p | ATL | S |

Eff. May 15
| 1:00p0 | 6:59p | ATL | L |
| 6:44p0 | 11:45p | ATL | D X6 |

To Naples, FL 1136 mi
| 8:15a0 | 1:48p | MCO | B |
| 2:10p0 | 7:25p | MCO | S |

To Nashville, TN 401 mi
5:30a0	8:45a	CVG	S X67
6:15a0	9:24a	CVG	S X67
7:40aM	10:55a	CVG	X67
9:57a0	12:59p	CVG	X67
11:35aM	2:50p	CVG	X67
3:20pM	6:55p	CVG	X6
3:50p0	6:55p	CVG	
5:35p0	8:43p	CVG	X6
5:40pM	8:43p	CVG	X67

Eff. May 1
To Nassau, Bah 1301 mi
| 8:10a0 | 1:55p | ATL | |

To New Orleans, LA 831 mi
6:15a0	10:14a	CVG	B/S X67
8:10a0	12:11p	ATL	B
9:25aM	1:35p	CVG	L6

Eff. May 6
9:57a0	1:35p	CVG	L
10:18a0	2:45p	DFW	X67
11:35a0	3:55p	ATL	L
1:00p0	5:20p	ATL	L
3:14p0	7:20p	ATL	S6
3:20pM	7:45p	CVG	DX6
3:50p0	7:45p	CVG	D
6:44p0	10:35p	ATL	D
8:45p0	12:59a	DFW	S

**To New York, NY/
Newark, NJ** 734 mi
L-LaGuardia; J-Kennedy;
E-Newark
7:40aM	12:25pL	CVG	X67
11:35aM	4:40pL	CVG	X67
11:35aM	4:55pE	CVG	X67
3:20pM	7:45pE	CVG	D X6
5:35p0	10:20pL	CVG	
5:40pM	10:20pL	CVG	X67

Eff. May 1
**To Norfolk/Virginia Beach/
Williamsburg, VA** 707 mi
6:30a0	12:29p	ATL	B X7
9:54a0	3:10p	ATL	S
11:35aM	3:55p	CVG	X67
1:00p0	6:25p	ATL	L
5:35p0	9:40p	CVG	
5:40pM	9:40p	CVG	X67

To Oakland, CA 1843 mi
Also see San Francisco, San Jose
8:00a0	12:05p	SLC	B
11:45a0	3:45p	SLC	
3:10p0	7:10p	SLC	D
6:15p0	9:55p	SLC	

To Oklahoma City, OK 695 mi
8:30a0	12:35p	DFW	B
10:18a0	2:15p	DFW	X67
12:10p0	4:15p	DFW	X7
3:50p0	9:25p	DFW	S
8:45p0	12:35a	DFW	S

To Ontario, CA 1707 mi
Also see Los Angeles, Burbank,
Long Beach and Orange County
8:00a0	11:55a	SLC	B
8:30a0	1:10p	DFW	
11:45a0	3:25p	SLC	
12:10p0	4:20p	DFW	X7
3:50p0	8:10p	DFW	
6:15p0	9:45p		1

To Orange County, CA 1732 mi
Also see Los Angeles, Burbank,
Long Beach and Ontario
6:15a0	12:05p	CVG	X67
8:00a0	12:05p	SLC	B
8:30a0	1:15p	DFW	
9:25aM	3:40p	CVG	6

Eff. May 6
11:45a0	3:40p	SLC	
3:10p0	7:35p	SLC	S
3:50p0	8:10p	DFW	

To Orlando, FL 995 mi
5:30a0	10:00a	CVG	B X67
8:15a0	11:42a		B
9:25aM	2:35p	CVG	L6

Eff. May 6
9:45a0	2:54p	ATL	S
9:57a0	2:35p	CVG	L
11:35a0	5:00p	ATL	L
2:10p0	5:37p	0	S
3:14p0	8:29p	ATL	S X6

Disc. Apr. 30
| 3:14p0 | 8:29p | ATL | S |

Eff. May 1
3:14p0	8:29p	ATL	S6
3:20pM	8:52p	CVG	DX6
3:50p0	8:52p	CVG	D
4:55p0	9:44p	ATL	D
6:44p0	11:35p	ATL	D
7:59p0	12:55a	ATL	S

To Palm Springs, CA 1658 mi
| 8:00a0 | 11:45a | SLC | B |

Disc. Apr. 30
11:45a0	3:45p	SLC	
3:10p0	8:25p	SLC	S
6:15p0	10:25p	SLC	X5

To Panama City, FL 812 mi
6:30a0	10:15a	ATL	B X7
8:10a0	12:05p	ATL	B
9:54a0	1:35p	ATL	S
11:35a0	3:29p	ATL	L
1:00p0	5:07p	ATL	L
3:14p0	6:59p	ATL	S X6

Disc. Apr. 30
| 3:14p0 | 6:59p | ATL | S |

Eff. May 1
3:14p0	6:59p	ATL	S6
4:55p0	8:45p	ATL	D
6:44p0	10:30p	ATL	D X6

To Paris, Fra 4154 mi
Orly Airport
| 11:35aM | 7:25a | CVG | X67 |
Eff. May 8; aircraft change enroute
| 11:35aM | 7:25a | CVG | X67 |
Disc. May 5; aircraft change
enroute
| 1:00p0 | 7:25a | ATL | |

**To Pasco/Richland/
Kennewick, WA** 1579 mi
8:00a0	11:40a	SLC	B
3:10p0	7:33p	SLC	S
6:15p0	9:40p	SLC	

To Pensacola, FL 787 mi
8:10a0	11:53a	ATL	B
11:35a0	3:12p	ATL	L
3:14p0	6:50p	ATL	S X6

Disc. Apr. 30
| 3:14p0 | 6:50p | ATL | S |

Eff. May 1
| 3:14p0 | 6:50p | ATL | S6 |
| 6:44p0 | 10:20p | ATL | D |

To Philadelphia, PA 669 mi
6:15a0	11:50a	CVG	B/SX67
11:35a0	4:23p	CVG	X67
5:35p0	10:05p	CVG	
5:40pM	10:05p	CVG	X67

Eff. May 1
To Phoenix, AZ 1445 mi
6:15a0	9:50a	CVG	X67
8:00a0	11:30a	SLC	B
8:30a0	12:15p	DFW	
12:10p0	3:35p	DFW	X7
3:20pM	7:39p	CVG	X6
3:50p0	7:22p	DFW	
3:50p0	7:39p	CVG	
6:15p0	9:45p	SLC	
6:44p0	10:40p	DFW	

To Pittsburgh, PA 403 mi
7:40aM	11:45a	CVG	X67
11:35aM	3:50p	CVG	X67
5:35p0	9:40p	CVG	
5:40pM	9:40p	CVG	X67

Eff. May 1
To Pocatello, ID 1268 mi
8:00a0	12:55p	SLC	B X67
11:45a0	4:59p	SLC	X6
3:10p0	6:45p	SLC	S
6:15p0	10:25p	SLC	

To Portland, ME 894 mi
| 11:35aM | 6:30p | CVG | X67 |

To Portland, OR 1749 mi
6:15a0	11:00a	CVG	X67
8:00a0	12:00n	SLC	B
11:45a0	3:35p	SLC	
12:10p0	6:19p	DFW	X7
3:10p0	7:05p	SLC	D
3:20pM	10:20p	CVG	X6
6:15p0	10:20p	SLC	

To Raleigh/Durham, NC 636 mi
5:30a0	9:20a	CVG	B/SX67
6:15a0	12:43p	CVG	B/SX67
6:30a0	11:30a	ATL	B X7
7:40aM	12:43p	CVG	X67
9:54a0	2:55p	ATL	S
1:00p0	5:57p	ATL	L
3:20pM	7:50p	CVG	S X6
3:50p0	7:50p	CVG	S
4:55p0	9:50p	ATL	D

To Reno, NV/Tahoe 1680 mi
8:00a0	11:30a	SLC	B
11:45a0	4:27p	SLC	
3:10p0	6:20p	1	S
6:15p0	9:21p	SLC	

**To Richmond/
Williamsburg, VA** 632 mi
6:15a0	11:35a	CVG	B/SX67
6:30a0	11:25a	ATL	B X7
7:40aM	11:35a	CVG	X67
9:54a0	3:05p	ATL	S
11:35aM	3:55p	CVG	X67
1:00p0	6:40p	ATL	L
5:35p0	9:50p	CVG	
5:40pM	9:50p	CVG	X67

Eff. May 1
To Roanoke, VA 520 mi
5:30a0	9:50a	CVG	S X67
6:15a0	11:45a	CVG	S X67
7:40aM	11:45a	CVG	X67
11:35aM	3:50p	CVG	X367
11:35aM	4:25p	CVG	3
5:35p0	9:55p	CVG	X6
5:40pM	9:55p	CVG	X67

Eff. May 1
To Rock Springs, WY 1099 mi
| 11:45a0 | 4:30p | SLC | |

To Sacramento, CA 1790 mi
8:00a0	11:50a	SLC	B
11:45a0	3:40p	SLC	
3:10p0	6:40p	SLC	S
6:15p0	9:50p	SLC	

To St. George, UT 1414 mi
| 6:15p0 | 10:37p | SLC | X67 |

To Salt Lake City, UT 1257 mi
7:40aM	1:55p	CVG	X67
8:00a0	10:09a	0	B
8:30a0	1:36p	DFW	
9:25aM	2:14p	CVG	

Eff. May 6
11:35aM	5:28p	CVG	S X67
11:45a0	1:57p	0	
3:10p0	5:21p	0	S
3:20pM	8:34p	CVG	X6
6:15p0	8:25p	0	
6:44p0	11:41p	DFW	

To San Angelo, TX 1011 mi
| 8:30a0 | 12:58p | DFW | B |
| 12:10p0 | 4:30p | DFW | X7 |

To San Antonio, TX 1040 mi
8:30a0	12:38p	DFW	B
10:18a0	2:25p	DFW	X67
12:10p0	4:05p	DFW	X7
3:50p0	7:50p	DFW	S
8:45p0	12:40a	DFW	S

To San Diego, CA 1729 mi
| 6:15a0 | 10:20a | CVG | X67 |
| 8:00a0 | 11:50a | SLC | B |

Disc. Apr. 30
| 8:00a0 | 11:50a | SLC | B |

Eff. May 1
8:30a0	12:55p	DFW	
11:45a0	3:30p	SLC	
12:10p0	5:25p	DFW	W X7
3:20pM	9:11p	CVG	X6
3:50p0	7:40p	DFW	
6:15p0	10:00p	SLC	
6:44p0	11:05p	DFW	

To San Francisco, CA 1854 mi
Also see Oakland and San Jose
6:15a0	11:15a	CVG	X67
8:00a0	12:00n	SLC	B
8:30a0	1:20p	DFW	
11:45a0	3:45p	SLC	C
12:10p0	4:45p	DFW	X7
3:10p0	7:10p	SLC	D
3:20pM	8:35p	CVG	X6
3:50p0	8:35p	CVG	
3:50p0	8:50p	DFW	
6:15p0	10:00p	SLC	
6:44p0	11:45p	DFW	

To San Jose, CA 1837 mi
Also see San Francisco and
Oakland
| 7:40aM | 3:45p | CVG | X67 |
| 8:00a0 | 12:10p | SLC | B |

Measurement Equivalents and Conversion Table

U.S. Customary System

Length
1 inch
12 inches = 1 foot
36 inches = 3 feet = 1 yard

Volume or Capacity
1 pint = 1/2 quart = 1/8 gallon
2 pints = 1 quart   = 1/4 gallon
4 pints = 2 quarts = 1/2 gallon
8 pints = 4 quarts = 1 gallon

Dry Measure
1 pint
2 pints = 1 quart
16 pints = 8 quarts = 1 peck
64 pints = 32 quarts = 4 pecks = 1 bushel

Weight
1 ounce (oz)
16 ounces = 1 pound (lb)
2000 pounds = 1 ton

Conversion Factors

Metric System   **U.S. Customary System**

Meters	Yards	Inches
1	1.094	39.37
0.914	1	36

Centimeters	Inches	Feet
1	0.394	0.0328
2.54	1	0.0833
30.48	12	1

Kilometers	Miles	
1	0.621	
1.609	1	

Grams	Ounces	Pounds
1	0.035	0.0022
28.35	1	0.0625
453.59	16	1
1000	35.274	2.205

Kilograms	Ounces	Pounds
1	35.274	2.205
0.028	1	0.0625
0.454	16	1

Liters	Pints	Quarts	Gallons
1	2.114	1.057	0.264
0.473	1	0.5	0.125
0.946	2	1	0.25
3.785	8	4	1

Metric System

Length
1 meter (m) = 100 cm = 1000 mm

1 millimeter (mm)	= 0.001	m
1 centimeter (cm)	= 0.01	m
1 decimeter (dm)	= 0.1	m
1 dekameter (dkm)	= 10	m
1 hectometer (hm)	= 100	m
1 kilometer (km)	= 1000	m

Volume or Capacity
1 liter (L) = 100 cL = 1000 mL

1 milliliter (mL)	= 0.001	L
1 centiliter (cL)	= 0.01	L
1 deciliter (dL)	= 0.1	L
1 dekaliter (dkL)	= 10	L
1 hectoliter (hL)	= 100	L
1 kiloliter (kL)	= 1000	L

Weight
1 gram (g) = 100 cg = 1000 mg

1 milligram (mg)	= 0.001	g
1 centigram (cg)	= 0.01	g
1 decigram (dg)	= 0.1	g
1 dekagram (dkg)	= 10	g
1 hectogram (hg)	= 100	g
1 kilogram (kg)	= 1000	g

Common Conversion Factors
1 centimeter = 0.39 inches
1 meter = 39.4 inches
1 kilometer = 0.62 miles
1 gram = 0.035 ounces
1 kilogram = 2.20 pounds
1 liter = 1.06 quarts

Each formula is identified by lesson number.

1–1 Earnings Formula

$$E = rh + t$$

where E = earnings, r = hourly rate, h = number of hours, and t = tips

1–3 Commission Formula

$$C = prn$$

where C = commission, p = price of one item, r = commission rate, and n = number of items sold

1–3 Earnings with Commission Formula

$$E = s + rt$$

where E = earnings, s = monthly salary, r = commission rate, and t = total sales for the month

1–3 Piece Rate Formula

$$E = rn$$

where E = earnings, r = the piece or item rate, and n = the number of items

1–4 FICA taxes—Social Security and Medicare

If annual income is less than $57,600, then the deduction for FICA taxes is 7.65% of gross pay.

1–4 Take-home Pay

$$T = g - (w + p)$$

where T = take-home pay, g = gross pay, w = amount withheld for income taxes, and f = deduction for FICA taxes

2–1 Amount Earned or Charged for Checks

$$A = br - cn - s$$

where A = amount earned or charged, b = monthly average balance, r = monthly interest rate, n = number of checks, c = charge for each check, and s = service charge

3–1 Simple Interest Formula

$$i = prt$$

where i = interest, p = principal, r = interest rate, and t = time

3–2 Compound Interest Formula

$$B = p(1 + r)^n$$

where B = balance, p = original principal, r = interest rate for the time period, and n = total number of time periods

3–2 Rule of 72

$$\frac{72}{\text{Annual interest rate} \cdot 100} = \text{years to double}$$

3–3 Multiplier Effect

$$S = \frac{a}{1 - r}$$

where S = maximum amount of money a demand deposit can create, a = initial loan amount, and $1 - r$ = reserve requirement

$$M = \frac{D + S}{D}$$

where M = multiplier, D = original demand deposit, and S = maximum amount of money a demand deposit can create

4–1 FICA taxes

Social Security: 6.2% of all income at or under $57,600

Medicare: 1.45% of all income at or under $135,000

4–3   Cost Function

$$c = un + f$$

where c = total cost, u = unit cost, n = number of units, and f = fixed cost

4–3   Revenue Function

$$r = sn$$

where r = revenue, s = selling price per unit, or unit price, and n = number of units sold

4–3   Profit Function

$$p = r - c$$

where p = profit, r = revenue, and c = total cost

5–1   Monthly Payment Formula

$$M = \frac{Pr(1 + r)^n}{(1 + r)^n - 1}$$

where M = monthly payment, P = amount of loan, r = monthly interest rate, and n = number of payment periods

5–1   Amount Formula

$$P = \frac{M[(1 + r)^n - 1]}{r(1 + r)^n}$$

where P = amount of loan, r = monthly interest rate, n = number of payment periods, and M = monthly payment

5–3   Loans With Down Payment Formulas

$$D = rC$$

where D = down payment, r = percent for down payment, and C = cost

$$P = C - D$$

where P = loan amount, C = cost, and D = down payment

$$T = nM + D$$

where T = total amount, n = number of payments, M = monthly payment, and D = down payment

5–4   Amortization Schedule Formulas

$$
\begin{array}{ll}
I_1 = rL & I_2 = rB_1 \\
R_1 = P - I_1 & R_2 = P - I_2 \\
B_1 = L - R_1 & B_2 = B_1 - R_2
\end{array}
$$

where L = loan amount, r = monthly interest rate, P = payment amount, I_1 = interest due at end of month 1, R_1 = loan reduction at end of month 1, and B_1 = balance at end of month 1

5–4   Prepayment Formula

$$A = \frac{M[1 - (1 + r)^{-q}]}{r}$$

where M = monthly payment, r = monthly interest rate, q = number of remaining payment periods and A = prepayment amount

5–4   Rule of 78

The buyer pays a portion of the yearly interest equal to

$\frac{12}{78}$ in month 1

$\frac{11}{78}$ in month 2

$\frac{10}{78}$ in month 3

$\vdots$

$\frac{1}{78}$ in month 12

6–1   Effective Interest Rate Formula

$$^i\text{eff} = \left(1 + \frac{i}{12}\right)^{12} - 1$$

where ^ieff = effective interest rate, and i = APR (annual percentage rate)

6–2   Time-to-Pay-Off Formula

$$n = \frac{\log\left(\frac{M}{M - Pr}\right)}{\log\left(1 + r\right)}$$

where P = amount of loan, r = monthly interest rate, M = monthly payment, and n = number of payment periods

6–2   Monthly Payment Formula for Paying Off Balance

$$M = \frac{Pr(1 + r)^n}{(1 + r)^n - 1}$$

where P = current balance, r = monthly interest rate, n = number of payment periods, and M = monthly payment

6–5   Average Daily Balance Formula

$$b = \frac{s}{d}$$

where b = average daily balance, s = sum of the daily balances, and d = total number of days in the billing cycle

7–1   Replacement Life Insurance Formula

$$R = mS$$

where R = required replacement insurance, m = multiple from the table, and S = original gross salary or wage

7–2   Probability of an Event

$$P(E) = \frac{m}{n}$$

where $P(E)$ = the probability of an event E, m = the number of times the event occurs, and n = the number of all possible outcomes

7–2   Expected Value Formula

$$E = P_1 v_1 + P_2 v_2$$

where v_1 and v_2 are values and P_1 and P_2 are the corresponding probabilities

7–2   Profit on Insurance

$$P = R - B - C$$

where P = profit, R = revenue received as premiums, B = benefits paid out, and C = costs or expenses

7–3   Future Value of a Periodic Investment Formula

$$A = \frac{p[(1 + r)^n - 1]}{r}$$

where A = the future value of the investment, p = the investment made at the end of each period, r = the interest rate for the period, and n = the number of periods

8–2   Commission Cost Formula

$$c = npr$$

where c = cost of commission, n = number of shares, p = price of one share, and r = commission rate

8–2   Rate-of-Change Formula

$$r = \frac{|P_n - P_o|}{P_o}$$

where r = rate of increase or decrease, P_o = original price, and P_n = new price

9–1   Formulas for Tax Rate Schedule X (for single taxpayers)

i. $t = 0.15I$
    if $I \leq 21{,}450$
ii. $t = 3{,}217.50 + 0.28(I - 21{,}450)$
    if $21{,}450 < I \leq 51{,}900$
iii. $t = 11{,}743.50 + 0.31(I - 51{,}900)$
    if $51{,}900 < I$

where I = taxable income and t = tax on the income

9–1   Formulas for Tax Rate Schedule Y-1 (for married filing jointly or qualifying widow or widower)

i. $t = 0.15I$
    if $I \leq 35{,}800$
ii. $t = 5{,}370.00 + 0.28(I - 35{,}800)$
    if $35{,}800 < I \leq 86{,}500$
iii. $t = 19{,}566.00 + 0.31(I - 86{,}500)$
    if $86{,}500 < I$

where I = taxable income and t = tax on the income

10–3   Average Miles per Gallon

$$a = m \div g$$

where a = average miles per gallon, m = total miles driven, and g = number of gallons purchased

10–3   Average Cost per Mile

$$b = c \div m$$

where b = average cost per mile, c = total cost, and m = total miles driven

10–4   Annual Insurance Premium Formula

$$p = 2rb$$

where p = the annual premium paid to the company, r = the driver rating factor, and b = the six month basic rate for the car

11–1   Number of Gallons Needed

$$g = \frac{d}{a}$$

where g = number of gallons needed, d = total distance in miles, and a = average miles per gallon

11–1   Total Cost for Gas

$$t = gc$$

where t = total cost for gas, g = number of gallons, and c = average cost per gallon

12–4   Periodic Investment Formula

$$p = \frac{Ar}{(1 + r)^n - 1}$$

where p = periodic investment, A = future value of periodic investment, r = periodic interest rate, and n = number of periods

14–2   Common Conversion Factors

1 centimeter (cm) = 0.39 inches (in.)
1 meter (m) = 39.4 inches
1 kilometer (km) = 0.62 miles (mi)
1 gram (g) = 0.035 ounces (oz)
1 kilogram (kg) = 2.2 pounds (lb)
1 liter (l) = 1.06 quarts (qt)

SELECTED ANSWERS

CHAPTER 1

Algebra Refresher

1. 10.5
3. 11.8
5. 379.5
7. 5
9. 28.8
11. 35,175
13. $x = 11$
15. $t = 1377$
17. $x = 75$

Lesson 1–1

Algebra Review

1. 29
3. 137.5
5. 167.8
7. 120.5

Try Your Skills

1. $342.75
3. $397.38
5. Weekly: $714
   Monthly: $3,094
   Yearly: $37,128
7. Hourly: $17.50
   Monthly: $3,033.33
   Yearly: $36,400

Exercise Your Skills

1. Hourly employees are paid for exact hours worked including overtime. Salaried employees are paid a flat rate no matter how many hours they work.
3. Answers may vary. (Ex.: to complete assigned projects)
5. Answers may vary. (Ex.: 10–20 hours)
7. $339.75
9. $291.50
11. $393.75
13. $394.38
15. $327.25
17. $465.63
19. Weekly: $428
    Monthly: $1,854.67
    Yearly: $22,256
21. Hourly: $7.50
    Monthly: $1,300
    Yearly: $15,600
23. Hourly: $11.25
    Monthly: $1,950
    Yearly: $23,400

25. Hourly: $19.23
    Weekly: $769.23
    Monthly: $3333.33

Lesson 1–2

Algebra Review

1. 4.8
3. 12
5. 200
7. 175

Try Your Skills

1. a. $1698.00 b. $1200.00
3. a. $5954.50 b. $3807.69
5. a. $42,372.50; $41,680; $40,990 b. A
7. $3933.88

Exercise Your Skills

1. Answers may vary.
3. Answers may vary. (Ex.: What do you like most about your job? What skills do you use? What training did you need?
5. a. $1518 b. $600
7. a. $4542.00 b. $3033.33
9. a. $9273 b. $3675
11. a. $10,398.00 b. $4,766.67
13. a. $29,280; $33,000; $30,675 b. B

Mixed Review

1. $406.00
3. $464
5. Weekly: $250
   Monthly: $1,083.33
   Yearly: $13,000
7. Hourly: $21.92
   Weekly: $876.92
   Yearly: $45,600

Lesson 1–3

Algebra Review

1. $x = 2$
3. $a = 3$
5. $x = 2028$
7. $12.5x$
9. $10b$
11. $14y + 5$

Try Your Skills

1. **a.** $15.08 **b.** $y = 1.5075x$
   c. $15.08; $27.14; $36.18; $48.24
3. **a.** $8.23 **b.** $y = 1.64625x$
   c. $8.23; $19.76; $24.69; $29.63
5. $718.75; $1193.75
7. $690
9. 20 memberships

Exercise Your Skills

1. Since commission is a percent of sales, you will earn more.
3. No. Next month's sales will probably be back to average or below.
5. **a.** $6.75
   b. $y = 0.375x$
   c. $6.75; $10.50; $14.25; $18.00
7. **a.** $11.39
   b. $y = 0.75905x$
   c. $11.39; $18.98; $26.57; $34.16
9. $877.61
11. $3497
13. $E = 0.07t + 2400; E = 0.10t, t = $80,000$
15. $620

Mixed Review

1. $450
3. **a.** $5460.00 **b.** $4000.00
5. **a.** $8047.50 **b.** $3807.69

Lesson 1–4
Algebra Review

1. 0.05
3. 0.075
5. 0.0765
7. 30%
9. 150%
11. 6
13. 969.6
15. 376.38

Try Your Skills

1. $1930.40
3. $2850.90
5. $1182.08
7. $1710.00
9. $2168.92

	Income Tax Withholding	FICA Withholding	Take-Home Pay
11.	$286	$183.60	$1930.40
13.	289	260.10	2850.90
15.	0	97.92	1182.08
17.	137	153.00	1710.00
19.	186	195.08	2168.92

Exercise Your Skills

1. Answers may vary. (Ex.: no one likes tax increases, tax laws affect many people)
3. The workers in the future
5. Retirement income will not cover living expenses.
7. $2040.25
9. $1384.25
11. $1355.25
13. $3305.92
15. $2873.25

	Income Tax Withholding	FICA Withholding	Take-Home Pay
17.	268	191.25	2040.75
19.	1	114.75	1384.25
21.	30	114.75	1355.25
23.	542	318.75	3305.92
25.	359	267.75	2873.25

Mixed Review

1. $568.75
3. $3885
5. 75%; $1.50n \cdot 5 = 150 + 5n; n = 60$; 60 is 75% of 80.
7. $398

Chapter 1 Review

1. Hourly. You would be paid for overtime.
3. Deductions are subtracted.
5. $354.50
7. $1116.00
9. $144.38
11. $51,806
13. $10,215; $837.63
15. $2,925; $204.75
17. $544
19. $400
21. $193; $114.75; $1192.25
23. $83; $132.44; $1515.76
25. **a.** $3237.75 **b.** $1988.46

Cumulative Review

1. $520
3. $250
5. No. Some benefits, such as health insurance, are the same for all employees.
7. $12,500
9. $4851.00
11. $494; $228.74; $2267.26
13. $213; $168.30; $1818.70

CHAPTER 2

Algebra Refresher

1. $(5, -3)$
3. $(2.5, 1.5)$
5. $(4, 3)$
7. $(2, 1)$
9. $(3, 0)$

Lesson 2–1
Algebra Review

1. 3, 7.5, 21
3. $-3, -7.5, -21$; they are opposites.
5. $-4, -8.5, -22$; they are one less than the opposites.
7. $-2, 0, 2$
9. 67.7, 2.7, -1.3

Try Your Skills

1. $0.02 earned
3. $3.68 charged
5. $1.28 earned
7. $2.00; $1.80; $0; +$0.20
   $2.16; $0.40; $0; +$1.76
   $1.40; $1.20; $0; +$0.20

Exercise Your Skills

1. $A = br - cn - s$
3. Information is concise and easy to compare.
5. Writing large numbers of checks
9. C3, F3, G3
11. cn
13. $b = A3, r = B3, c = D3, A = G3$
15. $3.25; $3.00; $0.41; $0; $649.84
    $3.13; $3.00; $0.59; $0; $624.54
    $2.40; $3.00; $0.45; $0; $478.95
    $0.95; $3.00; $0.70; $0; $187.25
17. $A = br$
19. 0.4%

21. Bank S: $y = 4.4 - 0.07x$
    Bank T: $y = 2 - 0.025x$
    $x = 53.3$
    The account at Bank S will be more expensive if fewer than 53 checks are written and less expensive if more than 53 checks are written.

Mixed Review

1. $275.50
3. $3600
5. 20%
7. B5 = 0.0765*B4
9. B8 = B4 − B7

Lesson 2–2
Algebra Review

1. $y = 2x$
3. $y = 2x + 1$
5. $y = -2x + 1$
7. $y = 2x$
9. $y = 3x - 0.5$

Try Your Skills

1. Balance: $400.00
3. $196.50
5. $195.87

Exercise Your Skills

1. So you will not forget to enter it later
3. How much money remains in the account
9. 478.50
11. 428.50
13. 409.24
15. 279.20
17. 518.58
19. 367.15
21. 468.65
23. 419.65
25. 306.65
27. 301.65
29. 221.35
31. 311.85
33. 288.57
35. 276.52
37. 252.41
39. 145.70
41. Answers may vary. (Ex.: For deposit only. Latoya S. Marshall.)

43. Answers may vary. (Ex.: Pay to the order of Judy Bruns. L. Marshall.

Mixed Review

1. $14.10
3. $446.50
5. $258
7. $5625
9. $0.39 charged

Lesson 2–3
Algebra Review

1. $x = 2$
3. $x = 4$
5. $x = 1$
7. $x = 3$
9. $x = 4$

Try Your Skills

1. deposits not recorded
3. check fees, service charges, ATM withdrawal
5. $23.88; $40.00
7. $88.72

Exercise Your Skills

1. Answers may vary. (Ex.: convenient, open 24 hours)
3. additional deposits made after statement prepared, out-standing checks
5. 11
7. $140.22
9. $2.30
11. $326.68
13. $570.56

Mixed Review

1. $225.25
3. $1158
5. $0.20 charged
7. $4.35
9. $250.70
11. $39.76

Chapter 2 Review

1. $0.42
3. $423.57
5. Ridgewood Savings Bank: $y = 6.5 - 0.10x$
Second National Bank: $y = 3.5 - 0.04x$
$x = 50$
The account at Ridgewood Savings Bank will be more expensive if fewer than 50 checks are written and less expensive if more than 50 checks are written.

7. $455.78
369.83
331.30
323.85
309.09
255.19
360.07
343.79

9.

Beginning Balance	Number of Checks	Cost of Checks	Extra Charges	Service Charges	Interest Earned	New Balance
$155	11	$0.25	0	$5.00	$0.70	$150.45
278	17	0.38	0	5.00	1.25	273.87
505	21	0.47	$0.15	0	2.27	506.65
789	27	0.61	1.05	0	3.55	790.89

Cumulative Review

1. $64.00
3. $112.70
5. $9500
7. $154.38
9. $1.00
11. Answers may vary.
13.

Amount	Balance
	95.16
24.65	−24.65
	70.51

CHAPTER 3
Algebra Refresher

1. $11a + 12$
3. $x + 7y$
5. $4x + 5.97$
7. $31a - 20b$
9. $-x^2 + x + 1.06$

Lesson 3–1
Algebra Review

1. $A = 18.2$
3. $s = 20.6$
5. $i = 3375$
7. $B = 780.63$

Try Your Skills

1. $520; $18.20; $538.20
3. $1000; $35.00; $1035.00
5. $1500; $52.50; $1552.50
7. 17; $340
9. 19; $342
11. 10; $350
13. 6 wk
15. 20 wk

Exercise Your Skills

1. The bank has use of the funds for a guaranteed amount of time.
3. to avoid penalties for withdrawing the funds
5. $2400; $96.00; $2496.00
7. 10; $150
9. 6; $150
11. 14; $210
13. 8; $200
15. 10; $150
17. 6; $150
19. 8 wk
21. 4 wk
23. 5 wk; 5 wk; 3 wk.

Mixed Review

1. $13,000
3. $3.20; $3.50; $0.42; $799.28
5. $0.66; $3.50; $0.24; $162.92
7. $152.48
9. $183.75

Lesson 3–2
Algebra Review

1. x^4
3. x^8
5. $15y^8$
7. a^6
9. 15,609.06

Try Your Skills

1. First Period: $60.00; $1060.00
   Second Period: $63.60; $1123.60
3. $328.10
5. $55,093.29
7. 48 years

Exercise Your Skills

1. 12
3. 3
5. 1.5%; 3
7. $15.23; $1030.23
9. $15.69; $1061.37
11. $16.16; $1093.45
13. $16.65; $1126.50
15. $169.02
17. $2770.17; slightly higher interest rates yield much higher interest over time
19. 36 y
21. 21.6 y
23. 14 y

25. 16 y
27. Robert:
    $$2000(1.06)^{41} = \$21,805.72$$
    Amos:
    $$2000(1.06)^{30} = 11,486.98$$
    $$2000(1.06)^{20} = 6,414.27$$
    $$2000(1.06)^{10} = \underline{3,581.70}$$
    $$\$21,482.95$$

Mixed Review

1. $627.75; $1527.75
3. $80 deposit
5. $189.63
7. Bank A; Bank B is a better deal only if the number of checks reaches 188 checks a month.

Lesson 3–3
Algebra Review

1. yes; 0.5
3. no
5. yes; 0.04
7. yes; 0.07

Try Your Skills

1.

Level	Demand Deposit	Required Reserve	Loans and Investments
1	$500.00	$100.00	$400.00
2	400.00	80.00	320.00
3	320.00	64.00	256.00
4	256.00	51.20	204.80
5	204.80	40.96	163.84
6	163.84	32.77	131.07
7	131.07	26.21	104.86
8	104.86	20.97	83.89
9	83.89	16.78	67.11
10	67.11	13.42	53.69
		Total after 10 levels	$1785.26

3.

Level	Demand Deposit	Required Reserve	Loans and Investments
1	$500.00	$125.00	$375.00
2	375.00	93.75	281.25
3	281.25	70.31	210.94
4	210.94	52.74	158.20
5	158.20	39.55	118.65
6	118.65	29.66	88.99
7	88.99	22.25	66.74
8	66.74	16.69	50.05
9	50.05	12.51	37.54
10	37.54	9.39	28.15
		Total after 10 levels	$1415.51

5. $2,000

7. $1,500

9. 5

11. 4

13. The greater the reserve requirement, the less extra money generated.

Exercise Your Skills

1. No, the information is transmitted electronically.

3. A financial institution creates extra money or credit from excess reserves.

5.

Level	Demand Deposit	Required Reserve	Loans and Investments
1	$500.00	$75.00	$425.00
2	425.00	63.75	361.25
3	361.25	54.19	307.06
4	307.06	46.06	261.00
5	261.00	39.15	221.85
6	221.85	33.28	188.57
7	188.57	28.29	160.28
8	160.28	24.04	136.24
9	136.24	20.44	115.80
10	115.80	17.37	98.43
		Total after 10 levels	$2275.48

7.

Level	Demand Deposit	Required Reserve	Loans and Investments
1	$12,500.00	$1875.00	$10,625.00
2	10,625.00	1593.75	9,031.25
3	9,031.25	1354.69	7,676.56
4	7,676.56	1151.48	6,525.08
5	6,525.08	978.76	5,546.32
6	5,546.32	831.95	4,714.37
7	4,714.37	707.16	4,007.21
8	4,007.21	601.08	3,406.13
9	3,406.13	510.92	2,895.21
10	2,895.21	434.28	2,460.93
		Total after 10 levels	$56,888.06

9.

Level	Demand Deposit	Required Reserve	Loans and Investments
1	$12,500.00	$2500.00	$10,000.00
2	10,000.00	2000.00	8,000.00
3	8,000.00	1600.00	6,400.00
4	6,400.00	1280.00	5,120.00
5	5,120.00	1024.00	4,096.00
6	4,096.00	819.20	3,276.80
7	3,276.80	655.36	2,621.44
8	2,621.44	524.29	2,097.15
9	2,097.15	419.43	1,677.72
10	1,677.72	335.54	1,342.18
		Total after 10 levels	$44,631.29

11. $2833.33

13. $70,833.33

15. $50,000.00

17. 6.7

19. 6.7

21. 5

23. the greater the reserve requirement, the less extra money generated

25. $5250

Mixed Review

1. $30,600

3. $200; 4 wk

5. $180; 9 wk

7. $19,087.31

9. $368.66

11. 12 y

13. 6 y

15. $1380

Chapter 3 Review

1. $180

3. $1170

5. 11; $165

7. 4; $200

9. $1300.00; $39.00; $1339.00

11. $3400.00; $102.00; $3502.00

13. $50.00; $2050.00

15. $52.53; $2153.78

17. $25.00; $2025.00

19. $25.63; $2075.94

21. $26.27; $2128.16

23. $26.93; $2181.69

25. $2207.63; $2208.97; They only differ by one cent.

27. 283.41 pounds

29. 24 y

31. 9 y

33.

Level	Demand Deposit	Required Reserve	Loans and Investments
1	$2000.00	$400.00	$1600.00
2	1600.00	320.00	1280.00
3	1280.00	256.00	1024.00
4	1024.00	204.80	819.20
5	819.20	163.84	655.36
		Total after 10 levels	$5378.56

35.

Level	Demand Deposit	Required Reserve	Loans and Investments
1	$11,500.00	$1955.00	$9545.00
2	9,545.00	1622.65	7922.35
3	7,922.35	1346.80	6575.55
4	6,575.55	1117.84	5457.71
5	5,457.71	927.81	4529.90
		Total after 10 levels	$34,030.51

37. $8000
39. $56,147.06
41. 5
43. 5.9
45. $3644.44

Cumulative Review

1. **a.** $6742.50 **b.** $4500.00
3. $4340
5. $2.38
7. $14,116.50
9. Check Number Date To: (name of recipient) For: (purpose of payment) Amount: $243.50 Balance: $739.11
11. 18 y
13. 6.9 y

CHAPTER 4

Lesson 4–1

Algebra Review

1. $y = 2x - 1; m = 2$
3. $y = -x + 1; m = -1$
5. $y = x + 3; m = 1$
7. $y = 6.50x$
9. $m = 7.25$

Try Your Skills

1. $137.50
3. $10.52
5. $72.00
7. $5.51

Exercise Your Skills

1. Answers may vary. (Ex.: good management, good market, quality product)
3. (1) efforts and skills; (2) workers
5.

Employee	Gross Pay	Income Tax Withholding	FICA Withholding	Total Deductions	Take-Home Pay
Kendra	105.00	9.00	8.03	17.03	87.97
Paulo	73.80	0.00	5.65	5.65	68.15
Michael	176.90	19.00	13.53	32.53	144.37
Total	335.70	28.00	27.21	55.12	300.49

9.

Employee	Exemptions	Hourly Rate	Hours Worked	Gross Pay	Income Tax Withholding	FICA Withholding	Take-Home Pay
Adikes	1	$7.85	10	$ 78.50	$ 0.00	$ 6.01	$ 72.49
Carney	1	5.30	18	95.40	1.00	7.30	87.10
Inez	0	6.78	27	183.06	20.00	14.00	149.06
Sun-Li	0	9.36	17	159.12	16.00	12.17	130.95
Total				516.08	37.00	39.48	439.60

7. Week ending 4/4

Employee	Gross Pay	Income Tax Withholding	FICA Withholding	Total Deductions	Take-Home Pay
Catlyn	60.00	2.00	4.59	6.59	53.41
Sara	90.00	0.00	6.89	6.89	83.11
Joleen	208.00	23.00	15.91	38.91	169.09
Hernando	87.36	0.00	6.68	6.68	80.68
Total	445.36	25.00	34.07	59.07	386.29

Week ending 4/11

Employee	Gross Pay	Income Tax Withholding	FICA Withholding	Total Deductions	Take-Home Pay
Catlyn	52.50	1.00	4.02	5.02	47.48
Sara	126.00	5.00	9.64	14.64	111.36
Joleen	128.00	12.00	9.79	21.79	106.21
Hernando	62.40	0.00	4.77	4.77	57.63
Total	368.90	18.00	28.22	46.22	322.68

Week ending 4/18

Employee	Gross Pay	Income Tax Withholding	FICA Withholding	Total Deductions	Take-Home Pay
Catlyn	78.75	4.00	6.02	10.02	68.73
Sara	112.50	3.00	8.61	11.61	100.89
Joleen	128.00	12.00	9.79	21.79	106.21
Hernando	187.20	14.00	14.32	28.32	158.88
Total	506.45	33.00	38.74	71.74	434.71

Week ending 4/25

Employee	Gross Pay	Income Tax Withholding	FICA Withholding	Total Deductions	Take-Home Pay
Catlyn	120.00	11.00	9.18	20.18	99.82
Sara	81.00	0.00	6.20	6.20	74.80
Joleen	192.00	22.00	14.69	36.69	155.31
Hernando	124.80	4.00	9.55	13.55	111.25
Total	517.80	37.00	39.62	76.62	441.18

11. When the total gross salary has reached $135,000, there are no FICA taxes for the remainder of the year.

13. $4824.00 **15.** 5528.70

17. $135,000

Mixed Review

1. $225.60 **3.** $75

5. 10 wk **7.** $1071.91

9. $760

Lesson 4–2
Algebra Review

1. $x = 10$ **3.** $x = 6$

5. $x = 20$ **7.** $x = -1.5$

9. $x = -\frac{1}{3}$ **11.** $x = 4$

13. $m = 5$ **15.** $m = 3$

Try Your Skills

1. $60.00 **3.** $7.00

5. $7.00 **7.** $12.00

Exercise Your Skills

1. labor, advertising, energy and transportation, packaging

3. Yes, probably a lot

5. Labor: $202.50; Materials: $208.00; Advertising, energy, and transportation costs: $32.50; Cost: $443.00

7. Labor: $240.00; Materials: $200.00, $20.00; Packaging: $10.00; Advertising, energy, and transportation costs: $32.50; Cost: $502.50

9. Labor: $180.00; Materials: $200.00, $40.00; Advertising, energy, and transportation costs: $32.50; Cost: $452.50

11. Labor: $250.00; Materials: $150.00, $20.00; Packaging: $7.50; Advertising, energy, and transportation costs: $32.50; Cost: $460.00

13. Labor: $80.00; Materials: $215.00; Advertising, energy, and transportation costs: $32.50; Cost: $327.50

15. Labor: $157.50; Materials: $175.00, $25.00; Advertising, energy, and transportation costs: $32.50; Cost: $390.00

17. lower price, enhance quality of product

19. decrease price, find ways to decrease costs

Mixed Review

1. $55.77

3. $415.07

5. $8349.79

7.

Level	Demand Deposit	Required Reserve	Loans and Investments
1	$1800	$270.00	$1530.00
2	1530	229.50	1300.50

9. $530.03

11. $5528.70

13. 48 y

15. 27 y

17. 5.9

Lesson 4–3
Algebra Review

1. $c = 12g$

3. $c = hn$

5. $c = e \div 15$

7. Answers may vary. r = sales revenue, p = unit price, n = number units sold; $r = pn$

9. Answers may vary. x = unit price of first item, y = unit price of second item, a = average unit price; $a = (x + y) \div 2$

Try Your Skills

1. $c = 6.77(23) + 282 = 437.71

3. $4.20 **5.** $209.40

7. $p = 3.8n - 75$ **9.** $+A4*B4+C4$

11. $+F4-D4$

Exercise Your Skills

1. competition, demand

3. The cost is greater than the revenue.

5. It can drive some companies out of business. It can create unfair high prices.

7. $502.50; $732.50; $962.50

9. $500.00; $1000.00; $1500.00

11. ($2.50); $267.50; $537.50

13.

Unit Cost	Number Produced	Fixed Cost	Total Cost	Unit Price	Revenue	Profit (Loss)
$1.20	200	$212.50	$452.50	$2.00	$400.00	($52.50)
1.20	300	212.50	572.50	2.00	600.00	27.50
1.20	400	212.50	692.50	2.00	800.00	107.50

15. $c = 3.55n + 282.50$; $r = 10n$; $p = 6.45n - 282.50$

19.

Unit Cost	Number Produced	Fixed Cost	Total Cost	Unit Price	Revenue	Profit (Loss)
$2.75	100	$112.50	$387.50	$4.00	$ 400.00	$ 12.50
2.75	200	112.50	662.50	4.00	800.00	137.50
2.75	300	112.50	937.50	4.00	1200.00	262.50

21. $c = 4n + 190$; $r = 7.5n$; $p = 3.5n - 190$

Mixed Review

1. $5.00 per hour
3. $204.58
5. To: Farley Smith; For: VCR repair; Balance: $363.50
7. To: Abe's Supermarket; For: groceries; Balance: $668.73
9. 9 memberships

Lesson 4-4

Algebra Review

1. $y = -x + 10$; $m = -1$
3. $y = \frac{2}{3}x + 2$; $m = \frac{2}{3}$
7. $(10, -2)$
9. $(7.18, 0.53)$

Try Your Skills

1. $c = 1.04n + 242.5$
3. 253 stickers

Exercise Your Skills

1. cost function and revenue function
3. $c = 1.20n + 212.50$
5. 266 pennants
7. $r = 4n$
9. 90 cans
11. $r = 9n$
15. 25 book bags
17. $c = 2.30n + 272.50$
21. $p = 2.7n - 272.50$
23. 173 towels

Mixed Review

1. 12%
3.

Level	Demand Deposit	Required Reserve	Loans and Investments
1	$5000	$1000	$4000
2	4000	800	3200

5. 5
7. $172.50

Lesson 4-5

Algebra Review

3. $(3, 2)$ **7.** $(0, 6), (4, 4), (4, 6)$

Try Your Skills

1. $(4, 8)$
3. $(4, 8), (12, 8), (12, 4)$
5. A $(7.5, 45)$ represents the least number of hours
  (7.5 hours) you can work in a week to earn $45.
  B $(10, 45)$ represents the lowest rate you would accept
  $(45 \div 10 = \$4.50)$.
  C $(10, 60)$ represents the most you can earn ($60).

Exercise Your Skills

1. $x \le 200$; $y \le 400$; $x + y \ge 300$
3. A$(0, 300)$; B$(200, 100)$; C$(200, 400)$; D $(0, 400)$
5. $r = 2x + 2y$
7. 200 pennants; 400 stickers
9. $x \ge 4000$; $y \ge 6000$; $x + y \le 15,000$
11. A$(4000, 6000)$; B$(9000, 6000)$; C$(4000, 11,000)$
13. $c = 3.55x + 6.40y$
17. 40 mugs. 10 organizers
19. $x \ge 40$; $y \ge 50$; $x \le 80$; $y \le 2x$; $x + y = 210$
21. A$(40, 50)$; B$(80, 50)$; C$(80, 130)$; D$(70, 140)$, E$(40, 80)$
23. $c = 2.5x + 9y$
25. $p = 1.5x + y$
29. 5 pairs of earrings; 10 necklaces
31. 30 pairs of earrings; 15 necklaces

Mixed Review

1. 25
3. $1146.74
5. $13,150.13
7. $9.41; $94.59
9. $15.38; $157.62
11. $50

Chapter 4 Review

1. $127.30
3. $930.70
5. $5.00; $9.74; $112.56
7. $183.00; $71.20; $676.50
9. $162.90
11. $253.70
13. profit: $71.70
15. $117.50; $2.27; $162.90; $7.00; $140.00; ($22.90)
17. $117.50; $2.27; $253.70; $7.00; $420.00; $166.30
19. $r = 7n$

23. 25 kittens
25. Answers may vary. (Ex.: higher prices, higher production, lower costs, and so on)
27. $c = 2.27x + 3y$
29. $p = 4.73x + 4.5y$
33. 40 kittens and 10 puppies
35. 40 kittens and 20 puppies

Cumulative Review

1. $192.88
3. $19,972.68
5. $79.81
7. Answers may vary.
9. $43.75
11. $22,000
13. Loss: $10

CHAPTER 5

Algebra Refresher

1. $\frac{1}{64}$

3. $\frac{1}{531,441}$

5. 4096

7. $\frac{x^2 z^{11}}{y^7}$

9. $\frac{x^{12}}{y^4 z^{14}}$

11. 0.136

13. 0.0078
15. 16
17. 22.01
19. 6702.88

Lesson 5–1
Algebra Review

1. 0.89
3. 500
5. 1.03
7. 4800
9. 4.34
11. $\frac{y}{(1-r)^2}$

13. $\frac{3rq^{24}}{z(q+2)^{24}}$

Try Your Skills

1. $198.01; $11,880.72
3. $96.80; $76.09; $63.74
5. $161.34; $126.81; $106.24
7. $225.87; $177.54; $148.73
9. $4648.69; $5914.22; $7059.81

Exercise Your Skills

1. People do not like the sound of being "in debt."
3. The loan enables someone to acquire a home or a car long before he or she could save the full amount needed to purchase it.
5. repossess the item(s) bought with the credit
7. $323.27; $15,517.17
9. $217.45; $7,828.15
11. $146.24; $8,774.20
13. $252.51; $12,120.48
15. $200.72; $7,225.98
17. $134.99; $8,099.26
19. $318.96; $15,310.08
21. $585.44; $21,075.78
23. $393.71; $23,622.84
25. $309.66; $14,863.54
27. $293.56; $10,568.00
29. $197.42; $11,845.16
31. $318.00; $248.85; $207.58
33. $381.60; $298.62; $249.10
35. $445.20; $348.39; $290.62
37. $9434.04 for 3 years; $12,055.43 for 4 years; $14,452.01 for 5 years

Mixed Review

1. $2.54
3. $8164.22
5. 9 wk
7. $948
9. 78 items

Lesson 5–2
Algebra Review

1. 174.46
3. $z = 6978.40$
5. $s = 278.40$
7. $x = 44.70$
9. $z = 5659.61$
11. 48(359.28) is smaller.

Try Your Skills

3. $3,187.06; $687.06
5. $126.81; $6,087.02; $1087.02
7. $242.00; $8,712.14; $1212.14
9. $159.35; $9,561.17; $2061.17
11. $253.63; $12,174.04; $2174.04
13. Total cost is $2061.17 versus $808.09. You might consider total cost as a factor in your choice.

Exercise Your Skills

1. You could find yourself with excessive debt.
3. The stress and worry can undermine one's health.
5. $8,292.96; $1792.96
7. $317.81; $11,441.14; $1941.14
9. $213.73; $12,823.82; $3323.82
11. $159.48; $7,655.04; $1655.04
13. $401.44; $14,451.97; $2451.97
15. $269.98; $16,198.52; $4198.52
17. $465.15; $22,327.20; $4827.20
19. $389.73; $14,030.45; $2380.45
21. $262.10; $15,726.06; $4076.06
23. $233.24; $11,195.50; $2420.50
25.

Loan Amount	Number of Years	Monthly Payment	Total Payment	Total Cost
$20,000	5	$400.76	$24,045.54	$4045.54
20,000	6	345.80	24,897.76	4897.76
20,000	7	306.77	25,768.30	5768.30
20,000	8	277.68	26,657.03	6657.03
20,000	9	255.22	27,563.79	7563.79
20,000	10	237.40	28,488.42	8488.42

Mixed Review

1. $31.06
3. $238.75
5. 9 y

Lesson 5–3
Algebra Review

1. $1110
3. $1290
5. 2160
7. $z = 3600(0.6)$

Try Your Skills

1. $114.40; $4118.24; $106.48; $3833.34; $284.90
3. $62.40; $2246.31; $58.08; $2090.91; $155.40
5. $2490; $9,960; $80.20; $26,550.19; $1762.52

Exercise Your Skills

1. the income left after paying for basic needs
3. The financial institution makes higher profits.
5. $83.04; $2989.29; $79.50; $2861.98; $127.31
7. $33.21; $1195.72; $31.80; $1144.79; $50.93
9. $83.04; $2989.29; $1750.00; $58.13; $2842.50; $146.79
11. $33.21; $1195.72; $700.00; $23.25; $1137.00; $58.72
13. $8400.00; $33,600; $45,198.38; $46,304.47
15. $8400.00; $33,600; $46,276.65; $47,773.16

17. $8400.00; $33,600; $47,374.93; $49,277.21
19. 5 y; 8%; 10% down

Mixed Review

1. $21,500
3. $446.77 − 57.49 = 389.28
5. $29,900
7. The average-balance method; the minimum-balance method will result in extra charges even if your balance drops below the limit for one day.
9. 13 items

Lesson 5–4
Algebra Review

1. 1.25%
3. 44
5. 0.06
7. 19
9. 1.0016
11. 450
13. 3688.87

Try Your Skills

1. $1963.03
3. $46.97; $9.63; $37.34; $1888.54
5. $1515.26
7. $20.77

Exercise Your Skills

1. When the prepayment penalty is less than the interest.
3. It is a schedule that lists the interest portion of monthly loan payments.
5. 2; $231.92; $62.21; $169.71; $9161.70
7. 4; $231.92; $59.94; $171.98; $8818.88
9. 6; $231.92; $57.64; $174.28; $8471.47
11. 8; $231.92; $55.31; $176.61; $8119.42
13. 10; $231.92; $52.94; $178.98; $7762.65
15. 12; $231.92; $50.55; $181.37; $7401.11
17. $2937.82
19. $26,028.08
21. $40.38

Mixed Review

1. a. $3346.67; b. $2964.53; loan a is greater
3. $9566.82
5. $895
7. $x + y \leq 160; x \geq 40; x \leq 100; y > 60$
9. (40, 60), (40, 120), (100, 60)

Lesson 5–5
Algebra Review
1. 7.578
3. 1.572
5. y
7. $48y$

Try Your Skills
1. $33,600; $34,600
   $812.41; $765.85
   $40,064.17; $38,995.73; $36,760.69
   Plan 3
3. $86.27

Exercise Your Skills
1. Answers may vary
3. Some lending institutions will give you a better deal than others.
5. $10,500; $11,200; $12,000
   $353.79; $354.86; $359.65
   $12,736.31; $12,774.79; $12,947.43
   Plan 1
7. $6800; $7550; $7800
   $172.47; $183.43; $183.18
   $8278.35; $8804.78; $8792.79
   Plan 1
9. 0; $1884.09; $1198.72
   0; $600.00; $1200.00
   0; $2484.09; $2398.72
   Plan 2
11. 0; $1548.10; $990.82
    0; $300.00; $1500.00
    0; $1848.10; $2490.82
    Plan 3
13. 0; $1699.73; $861.78
    0; $1000.00; $2000.00
    0; $2699.73; $2861.78
    Plan 3
15. $719.10
17. $599.75

Mixed Review
1. $484.50
3. $70.38
5. $31
7. $2156
9. $75.10
11. $r = 4n$

Chapter 5 Review
1. $326.19
3. $52.87
5. $1766.95; $530,084.40
7. $33,600.54
9. $8431.20
11. $1096.77; $39,483.82; $3096.61
13. $709.67; $42,580.43; 0
15. $7,036.14; $6,575.98; $480.16
17. $13,305.26; $12,244.60; $1060.66
19. $936; $8424; $11,386.69
21. $2808; $6552; $10,936.31; $450.38
23. $773.79
25. $16.15
27. 0; $2918.71; $1194.62
    0; $500.00; $1750.00
    0; $3418.71; $2944.62
    Plan 2
29. No; sometimes you cannot afford the higher monthly payments.
31. Since the unpaid balance decreases each month, the interest due also decreases.

Cumulative Review
1. $4.00
3. $17,605
5. 16 years
7. $1546.68 - 597.99 = 948.69$
9. $5594.88
11. $29,877.31
13. $21.15
15. The amount of the loan is lower and therefore the interest payments are less.

CHAPTER 6
Algebra Refresher
1. 15
3. 22,765.04
5. 766,584
7. 27,399.68
9. d: {7, 3, 2}; r: {3, 7, 4}
11. Yes; each member of the domain is paired with a different member of the range.
13. 272
15. 2
17. 4
19. January: 6.89%; February: 8.41%; March: 9.85%; April: 8.96%; May: 7.01%; June: 8.76%; July: 9.67%; August: 9.00%; September: 8.53%; October: 7.10%; November: 8.24%; December: 7.58%

Lesson 6–1
Algebra Review

1. 0.125
3. 1.19
5. 3330%
7. 76%
9. 0.13

Try Your Skills

1. $0.00; $0.00; $2300.00
3. $2048.00; $0.00; $20.48; $275.00; $1793.48
5. 12.68%

Exercise Your Skills

1. the amount of time that elapses before you must pay interest on the purchase
3. VISA is a credit card with a monthly payment whereas American Express is a charge card with the entire balance expected on receipt of the bill.
5. $3000.00; $0.00; $40.00; $325.00; $2715.00
7. $0.00; $3215.00
9. $2907.24; $0.00; $33.68; $345.00; $2595.92
11. $2280.99; $0.00; $26.42; $345.00; $1962.41
13. $1640.14; $0.00; $19.00; $345.00; $1314.14
15. $984.36; $0.00; $11.40; $345.00; $650.76
17. $313.30; $0.00; $3.63; $316.93; $0.00
19. 14.37%; 1.13%
21. 17.23%; 1.33%
23. 13.1%
25. 15.68%

Mixed Review

1. $650.67
3. $8121.43
5. 16 wk
7. $150
9. $195.89
11. $r = 10x + 20y$
13. $540

Lesson 6–2
Algebra Review

1. d: {9, 3}; r: {2, 8}
3. No; the domain value of 9 is paired with range values of 2 and 8.

Try Your Skills

1. 2.44
3. 0.16
5. 10 mo
7. 9 mo
9. $1137.18

Exercise Your Skills

1. After 7 months, there would still be a small balance.
3. 2.22
5. 0.50
7. 15 mo
9. 18 mo
11. 18 mo
13. 15 mo
15. 23 mo
17. 19 mo
19. $118.53
21. $54.66
23. $46.64

Mixed Review

1. $3039.77
3. $43.35 per wk
5. 24 y
7. 39 items

Lesson 6–3
Algebra Review

1. 0.139
3. 2.45
5. 45,600%
7. 9%
9. $25.60
11. $797.98
13. 15%
15. 25
17. $52,000

Try Your Skills

1. $12.44; $1007.44; $50.00
3. $921.41; $11.52; $932.93; $47.00
5. $47.00
7. $906.44; $11.33; $917.77; $92.00
9. $752.09; $9.40; $761.49; $76.00
11. $3.51
13. $902.46; $6.77; $909.23; $91.00
15. $742.37; $5.57; $747.94; $75.00
17. $17.55

Exercise Your Skills

1. Answers may vary.
3. the balance due decreases, thus the interest on the balance also decreases
5. $366.00; $6.41; $372.41; $37.00
7. $307.28; $5.38; $312.66; $31.00

9. $257.59; $4.51; $262.10; $26.00
11. $216.23; $3.78; $220.01; $22.00
13. $181.48; $3.18; $184.66; $20.00
15. $147.54; $2.58; $150.12; $20.00
17. $407.00; $20.00
19. $373.77; $6.54; $380.31; $20.00
21. $346.62; $6.07; $352.69; $20.00
23. $318.51; $5.57; $324.08; $20.00
25. $289.40; $5.06; $294.46; $20.00
27. $259.26; $4.54; $263.80; $20.00
29. $68.07
33. $367.64; $37.00
35. $300.95; $3.01; $303.96; $30.00
37. $248.70; $2.49; $251.19; $25.00
39. $205.45; $2.05; $207.50; $21.00
41. $168.37; $1.68; $170.05; $20.00
43. $131.55; $1.32; $132.87; $20.00
45. $24.25

Mixed Review

1. a. $4753.14 **b.** $3608.78
Loan **b** has the greater total cost.
3. $145.86
5. 10 mo

Lesson 6–4
Algebra Review

1. 14.00   **3.** 3051.76
5. 7,112,448.00   **7.** 231.00
9. 15,616.26   **11.** 494.64

Try Your Skills

1. $56.00
3. $533.17; $8.00; $541.17; $54.00
5. $521.47; $7.82; $529.29; $53.00
7. $455.58; $6.83; $462.41; $46.00
9. $545.00; $8.18; $553.18; $55.00
11. $549.12; $8.24; $557.36; $56.00
13. $95.12

	Month	Purchases	Balance	Interest	Amount Owed	Payment
15.	2	$86.35	593.42	8.90	602.32	60.00
17.	4	40.00	535.45	8.03	543.48	54.00
19.	6	22.80	539.75	8.10	547.85	55.00
21.	8	97.88	557.27	8.36	565.63	57.00
23.	10	31.00	564.38	8.47	572.85	57.00
25.	12	67.00	598.58	8.98	607.56	61.00

Exercise Your Skills

1. Answers may vary.
3. Creditors cannot cancel a divorced or widowed person's credit unless the income drops dramatically.
5. $799.79; $12.00; $811.79; $81.00
7. $667.75; $10.02; $677.77; $68.00
9. $715.72; $10.74; $726.46; $73.00
11. $682.26; $10.23; $692.49; $69.00
13. $653.22; $9.80; $663.02; $66.00
15. $667.98; $10.02; $678.00; $68.00
17. $147.00
19. $1210.27; $18.15; $1228.42; $123.00
21. $1223.51; $18.35; $1241.86; $124.00
23. $1021.63; $15.32; $1036.95; $104.00
25. $851.94; $12.78; $864.72; $86.00
27. $1043.11; $15.65; $1058.76; $106.00
29. $204.44

	Month	Purchases	Balance	Interest	Amount Owed	Payment
31.	2	0.00	799.79	12.00	811.79	81.00
33.	4	0.00	686.05	10.29	696.34	70.00
35.	6	158.80	730.54	10.96	741.50	74.00
37.	8	125.00	734.51	11.02	745.53	75.00
39.	10	0.00	715.27	10.73	726.00	73.00
41.	12	123.00	719.80	10.80	730.60	73.00

	Month	Purchases	Balance	Interest	Amount Owed	Payment
43.	1	0.00	$1450.63	$ 21.76	$1472.39	$147.00
45.	3	0.00	1224.50	18.37	1242.87	124.00
47.	5	0.00	1249.39	18.74	1268.13	127.00
49.	7	0.00	1042.25	15.63	1057.88	106.00
51.	9	15.00	884.16	13.26	897.42	90.00
53.	11	45.00	1096.69	16.45	1113.14	111.00

55. Total Interest: $209.56

Mixed Review

1. $1981.35
3. $79.30
5. $182,438.25; Total cost for 15 years = $358,753.78
Total cost for 30 years = $541,192.03
7. $p = 0.2x + 0.25y$
9. $1775 when $x = 2000$ and $y = 5500$
11. $864
13. The option of Exercise 12 is better than the cost of the installment option, $794.76.

Lesson 6–5

Algebra Review

1. 51.4

3. 12.47

5. 290.73

Try Your Skills

1. 1; $567.00
15; $8505.00
30; $18,060.00
a. $602.00; **b.** $7.53; **c.** $574.53

3. 1; $580.00
$580.00; 5; $2900.00
$640.00; 1; $640.00
$640.00; 12; $7680.00
a. $641.94; **b.** $11.23; **c.** $651.23

Exercise Your Skills

1. to verify the customer's right to use the card

3. 1; $1850.50
23; $42,561.50
a. $1872.63; **b.** $28.09; **c.** $1878.59

5. 15; $27,193.05
1; $1724.87
15; $25,873.05
a. $1767.45; **b.** $26.51; **c.** $1751.38

7. 20; $33,878.00
1; $1614.90
10; $16,149.00
a. $1665.87; **b.** $24.99; **c.** $1639.89

9. 18; $28,599.30
1; $1517.85
12; $18,214.20
a. $1559.08; **b.** $23.39; **c.** $1541.24

11. 13; $19,443.19
1; $1431.63
16; $22,908.08
a. $1459.36; **b.** $21.89; **c.** $1453.52

13. $0.33

15. $0.64

17. $0.77

19. $0.62

21. $0.42

23. 18; $15,750.00
1; $785.00
$785; 6; $4710.00
$920.00; 1; $920.00
$920.00; 5; $4600.00
a. $863.39; **b.** $10.79; **c.** $930.79

25. 8; $10,200.00
1; $1500.00
$1500.00; 8; $12,000.00
$1325.00; 1; $1325.00
$1325.00; 12; $15,900.00
a. $1364.17; **b.** $17.05; **c.** $1342.05

Mixed Review

1. $1052.01

3. $104.04

5. $45.00

7. $200.92

Lesson 6–6

Algebra Review

1. 98

3. 83

5. Tom: B

7. Let S represent the test score, x represent the number of questions 1–10 answered correctly, y represent the number of questions 11–14 answered correctly, and z represent the number of questions 15–16 answered correctly. $S = 5x + 7y + 11z + 10$, if the bonus question is answered correctly and $S = 5x + 7y + 11z$, if the bonus question is not answered correctly.

Try Your Skills

1. 5

3. 13

5. 24

7. 15

9. 41

11. 96.25%

Exercise Your Skills

1. to secure credit for a large loan

3. Answers may vary.

5. 22; 21; 3; 18; 10; 15; 15; 15; 0; Score: 119; 95%

7. 18; 21; 12; 18; 12; 15; 15; 15; 20;
Score: 146; greater than 96.25%

9. 31; 21; 13; 18; 10; 15; 15; 15; 16;
Score: 154; greater than 96.25%

11. 12; 21; 12; 0; 0; 15; 15; 5; 16; Score: 96; 91%

Mixed Review

1. $1706.43

3. $778.22 - 105.89 = 672.33$

5. $20,832

Lesson 6–7

Algebra Review

1. $54.45

3. $1370.88

5. 0.78%

7. 37,500

9. 435.29

Try Your Skills

1. 21.6%; Yes, they are over the 20% limit.

Exercise Your Skills

1. Answers may vary.

3. Answers may vary.

5. 55.8%; Yes, they are in credit overload.

7. 17%; They are not in credit overload.

9. 11.4%; no

11. 18.3%; no

Mixed Review

1. 10 months

3. 18.62%

5. $8349.79

7. $129.40

9. $4.85

13. $3.90 added

Chapter 6 Review

1.

Month	Previous Balance	New Charges	Finance Charges	Payment Received	New Balance
1	0.00	$2315.00	0.00	0.00	$2315.00
2	$2315.00	0.00	$24.89	$250.00	2089.89
3	2089.89	0.00	22.47	250.00	1862.36
4	1862.36	0.00	20.02	250.00	1632.38
5	1632.38	0.00	17.55	250.00	1399.93
6	1399.93	0.00	15.05	250.00	1164.98
7	1164.98	0.00	12.52	250.00	927.50
8	927.50	0.00	9.97	250.00	687.47
9	687.47	0.00	7.39	250.00	444.86
10	444.86	0.00	4.78	250.00	199.64
11	199.64	0.00	2.15	201.79	0.00
12	0.00	0.00	0.00	0.00	0.00

3. 11 mo

5.

Month	Balance	Interest	Amount Owed	Payment
1	$875.00	$8.75	$883.75	$44
2	839.75	8.40	848.15	42
3	806.15	8.06	814.21	41
4	773.21	7.73	780.94	39

7.

Month	Balance	Interest	Amount Owed	Payment
1	$875.00	$8.75	$883.75	$88
2	795.75	7.96	803.71	80
3	723.71	7.24	730.95	73
4	657.95	6.58	664.53	66

9. $2.41

11.

Month	New Charges	Balance	Interest	Amount Owed	Payment
1	0.00	$980.12	$14.70	$994.82	$ 99.00
2	$ 87.35	983.17	14.75	997.92	100.00
3	0.00	897.92	13.47	911.39	91.00
4	0.00	820.39	12.31	832.70	83.00
5	25.50	775.20	11.63	786.83	79.00
6	100.00	807.83	12.12	819.95	82.00
7	53.33	791.28	11.87	803.15	80.00
8	0.00	723.15	10.85	734.00	73.00

13. $381.39; $6.67

15. $778.06; $9.73

17. 19.1%; No, they are not in credit overload but are approaching the 20% limit.

Cumulative Review

1. $355.25

3. $67,500

5. One hundred ten and $\frac{23}{100}$

7. $608.00

9. $3996

11. $1800

13. Plan 3; Plan 1: $18,260.93; Plan 2: $18,085.16; Plan 3: $17,776.15

CHAPTER 7

Algebra Refresher

1. {1, 3, 5}

3. {1, 2, 3, 4}

5. 0.5

7. $0.\overline{6}$

9. 0

11. $0 < P < 1$

Lesson 7–1

Algebra Review

1. 0.5

3. 0.125

5. 0.5

Try Your Skills

1. $487,500

3. $280,000

5. $918

7. $2748

9. $6381

Exercise Your Skills

1. to replace lost income
3. No, she probably does not support the family with income.
5. Answers may vary.
7. $200,000
9. $195,000
11. $105,750
13. $90,000
15. $556
17. $2593.50
19. $6722.50
21. Term; $447; $2049; $4527.50

Mixed Review

1. $453.75
3. $1380
5. $797.60; $5.98; $803.58; $80.00
7. $104.20

Lesson 7–2
Algebra Review

1. 0.458
3. 0.167
5. 0.583
7. 0.583
9. 0

Try Your Skills

1. $2.50
3. 0.00094
5. $E = 0.99906x + 0.00094\ (x - 60,000)$
7. $72,000; $1000x = 200(1000) - 1.03(100,000) - 25(1000)$
9. $767,750; $5000x = 350(5000) - 6.35(135,000) - 25(5000)$

Exercise Your Skills

1. The older a person becomes the more likely death may be.
3. anyone who is alive past his/her life expectancy
5. life expectancy changes over the years
7. $4.00
9. $0.40
11. $88.50
13. $380.00
15. $717.50
17. $195.50

19. $295
21. $146.25
23. no; $13 reduction
25. $133,333.33

Mixed Review

1. Five hundred and $\frac{00}{100}$
3. 30% down payment; the less owed, the less interest owed
5. $7.50; $1007.50; $50.00
7. $14.68
9. $355.25
11. 18 mo
13. $6000
15. $84

Lesson 7–3
Algebra Review

1. 3.03
3. 4183.63
5. 102,320.24
7. 204,887.17

Try Your Skills

1. $91,523.93
3. term: $8280; whole: $36,720
5. $4020
7. $17,094
9. $218
11. $894
13. $3999

Exercise Your Skills

1. no taxes paid on income invested
3. early withdrawal
5. $214,438.05
7. term: $7630; whole: $38,920
9. A person cannot borrow from an annuity.
11. $6537.92
13. IRA
15. when it is unlikely he/she will need to borrow the money in the future

Mixed Review

1. $9000
3. compound interest
5. 16.9%
7. $2160

Chapter 7 Review

1. to protect dependent family members
3. Premiums are lower than renewable term.
5. $240,000
7. $459
9. $726
11. 0.00094
13. $121.60
15. $34
17. $45,426.37
19. $61,830
21. $1366
23. $6367

Cumulative Review

1. $7351.50
3. $8500
5. $92.39
7.

Month	Balance	Interest	Amount Owed	Payment
1	$690.00	$12.08	$702.08	$70.00
2	632.08	11.06	643.14	64.00
3	579.14	10.13	589.27	59.00

CHAPTER 8

Algebra Refresher

1. good
3. excellent
5. $y = -3.07x + 1.19$
7. $y = 1.27x + 4.25$
9. $y = 0.98x + 0.66$

Lesson 8–1

Algebra Review

1. $x = 62.35$
3. $x = 345.38$
5. $x = 6.125$ or $6\frac{1}{8}$

Try Your Skills

1. $49,998.00
3. 5479; $49,995.88
5. 51
7. $651.70

Exercise Your Skills

1. If the company is not successful, part or all of the investment could be lost.
3. schools, hospitals, streets, and so on
5. 5,633; $49,992.88
7. 555; $49,950.00
9. 671; $49,989.50

11. The bond would be a safer, long-term investment if he is a beginning investor.
13. LukInd
15. MamaMi
17. $16\frac{1}{2}$
19. $64

Mixed Review

1. $300; $15,600
3. $360; $18,720
5. $9.37; $633.87; $63
7. $605.81; $9.09; $614.90; 61
9. $4800
11. $1223.19
13. $3105.06

Lesson 8–2

Algebra Review

1. 7
3. 2
5. $15
7. $30x$

Try Your Skills

1. $6337.50
3. $1462.50
5. $95.06; $6432.56
7. $21.94; $1484.44
9. $105 loss
11. 15.79% loss
13.

r	0	0.005	0.01	0.02	0.03	0.04	0.05	0.06	0.07	0.08
t	2000	2010	2020	2040	2060	2080	2100	2120	2140	2160

Exercise Your Skills

1. without patience, money may foolishly be lost
3. Answers may vary.
5. $2,203.75; $3,547.50; $1343.75 gain; 61%
7. $120,000.00; $109,500.00; $10,500 loss; 8.8%
9. $3,061.25; $5,347.50; $2286.25 gain; 74.7%
11. $256.28; $13,070.03
13. $255; $17,255
15. 0%; −7%
    $1080; 10%; 2%
    $1080; −5%; −12%
    $1080; 30%; 20%
    $1080; 75%; 62%
    $1080; 130%; 113%
    $1080; 220%; 196%
    $1080; 305%; 275%

Mixed Review

1. $6227
3. $1469.33
5. $199.08
7. $402.83
9. $317.04

Lesson 8–3

Algebra Review

1. c   3. a

Try Your Skills

3. (0, 10)   5. c
7. $20.70

Exercise Your Skills

1. to ensure sound trading practices
3. if the broker will sell only one stock or fund, makes promises for a quick, sure profit, has inside information, and urges you to buy before "prices go up"
7. $y = 2.98x + 19.97$; good, $r = 0.9797$
9. $48.28
11. $52.75
13. $34.84
15. The line of best fit is accurate in predicting stock prices.

Mixed Review

1. $6445.25
3. $9,960; $39,840; $320.80; $106,200.74; $7050.09
5. Helps to determine the credit risk of a borrower in an objective way that does not discriminate.
7. Chuck Lewis (Charles Lewis)
9. $4513.70
11. $700.00

Lesson 8–4

Algebra Review

1. 12   3. 5
5. 3

Try Your Skills

1. 5.4%
3. $\frac{10,000}{113.6} = \frac{x}{100}$ 5. 4%

   $x = \$8,802.82$
7. 0.85

Exercise Your Skills

1. to measure inflation or deflation
3. Food; explanations may vary.
5. $691.50
7. $35.18
9. 5.4%
11. 7.5%
13. 18.87 compared to 22.3
17. 80%

Mixed Review

1. $443.95; $375; ($68.95)
3. $775
5. $1777.50
9. 0.01671

Chapter 8 Review

1. through careful investments
3. It is a way to have a stake in many different corporations, rather than risking all your money on the future of just one or two corporations.
5. The auction method
7. Stocks are not instant profit and take time to develop.
9. 2622; $19,992.75
11. 133; $19,950
13. $46,250; $75,625; $29,375
15. $108,750; $222,500; $113,750
19. $23.14
21. close; model is $27.916 and actual is $30.00

Cumulative Review

1. $239.15
3. $1.50
5. 12 wk
7. $427.50; $9427.50
   $447.81; $9875.31
9. $2467.50; $25,370; $22,902.50
11. 14.93%

CHAPTER 9

Algebra Refresher

1. The number -1.5 is half way between -1 and -2.
3. The number $-\frac{8}{3}$ is a little less than half way between -3 and -2.
5. The number $\sqrt{2}$ is a little less than half way between 1 and 2.

Lesson 9–1

Algebra Review
1. 840
3. 3214.50
5. 11,740.70
7. 19,137.00

Try Your Skills
1. $309
3. $339
5. $1,654
7. $8,030
9. $t = 0.15I$; $336.75
11. $t = 11,743.50 + 0.31(I - 51,900)$; 27,894.50
13. $t = 11,743.50 + 0.31(I - 51,900)$; 36,574.50
15. $t = 11,743.50 + 0.31(I - 51,900)$; 57,654.50

Exercise Your Skills
1. Income tax is the largest revenue-producing tax the government has.
3. No; instead of the government telling us what we owe, we inform them
5. $11,217
7. $1,526
9. $35,686.00
11. $9,154.62
13. Taxpayer #11
15. all three
17. $18,660; $12,760; $1,916; $2,040; +$124
19. $28,536; $24,936; $4,191; $4,572; +$381
21. $42,000; $3,960; −$64
23. $63,600; $53,000; $10,193; $10,260; +$67
25. $21,360; $9,210; $1,384; $1,764; +380
27. $38,508; $24,058; $3,611; $3,444; −$167
29. $30,940.50

Mixed Review
1. $528
3. $29; $21.96; $236.04
5.

Level	Demand Deposit	Required Reserve	Loans and Investments
1	$1,200.00	$240.00	$ 960.00
2	960.00	192.00	768.00
3	768.00	153.60	614.40
4	614.40	122.88	491.52
5	491.52	98.30	393.22
		Total after 5 levels:	$3,227.14

7. $4,800
9. $1245.80
11. $196.00

Lesson 9–2

Try Your Skills
1. Refund: $225
3. Refund: $388
5. Tax owed: $114

Exercise Your Skills
1. Not everyone has the same income, number of dependents, investments, and so on.
3. It is easier than documenting your itemized deductions, and it saves the average person more money.
5. Refund: $501
7. Refund: $79
9. Tax owed: $169
11. Refund: $367
13. Refund: $597
15. Refund: $151

Mixed Review
1. $303.28; $18,196.52
3. 5-year term; $1565
5. 120; 95.5%
7. 6%

Lesson 9–3

Algebra Review
1. 330
3. 7750
5. 950

Try Your Skills
1. +$4000
3. None
5. +$4300
7. no effect

Exercise Your Skills
1. The IRS and tax preparers can make mistakes; it is good to be informed just in case.
3. When the taxpayers' deductions are larger than the standard deduction.
5. $11,054
7. $72,275
9. Refund: $945.00
11. Line 27
13. $222
15. No. The standard deduction would be higher.

Mixed Review
1. 5 mo
3. $2120
5. 2666; $49,987.50
7. $14,533
9. $820
11. $4781.95

Chapter 9 Review

1. how you submit your claim form—single, married filing jointly, married filing separately, head of household
3. It lowers the amount of taxes.
5. $19,729
7. $7,022
9. $2,929
11. $26,231.00
13. $28,742.04
15. $11,280; $0; $0; $60; +$60
17. $53,400; $38,950; $7176; $8928; +$1752
19. Refund: $823
21. Refund: $1134
23. Adjusted gross income: $64,035;
    Itemized deductions: $11,475; Tax owed: $79

Cumulative Review

1. A: $47,270; B: $53,340; C: $48,018
3. $314
5. $1413.73
7. $1387.71
9. $1,163,652.10 − $398,041.76 = $765,610.34
11. $22,721

CHAPTER 10

Algebra Refresher

1. mean: 77.7; median: 78; mode: 78
3. mean: 630; median: 320; mode: 220
5. mean: 80.2; median: 82; mode: 75

Lesson 10-1
Algebra Review

1. $1129.40   3. $105.06
5. 40%   7. $4905
9. $4876

Try Your Skills

1. $2432   3. $2081.25
5. 0.007   7. $14,623.11

Exercise Your Skills

1. 15.7%
3. 15.4%
5. 17.6%
7. 17.4%
9. $15,203; $17,637
11. $12,067; $13,948
13. $429.24

15. $15,732.99
17. $17,204
19. $228.10
21. 6% APR for 6 years with 20% down
23. In Exercise 19, although the high down payment decreased the amount financed, the increased term of the loan increased the total finance charge. The less expensive financing is in Exercise 20 because the term of the loan is decreased to 4 years. You would choose the financing in Exercise 19 if you could not afford a high monthly payment. If you could afford the higher monthly payments, it would be better to choose the financing in Exercise 20, and lower the total finance charge.
25. $145.86

Mixed Review

1. $3506.38
3. 72%
5. $17,141.34
7. $y = 0.998x + 8.942$
9. $18,250

Lesson 10-2
Algebra Review

5. 1154.39   7. 9.5
9. 5.4

Try Your Skills

1. $396.40   3. $319.60
5. $8000   7. $2364.34

Exercise Your Skills

1. For a new car, depreciation is greater and begins immediately after you purchase the car. For a used car, depreciation is more gradual.
3. $620.61
5. $518.61
11. 3 years
13. $60
17. $6000
19. Dena will need to choose 10% down at 6.5% annual interest over 5 years to make her monthly payments as low as possible.
21. $3200

Mixed Review

1. $62.40   3. $3.56
5. $14,666.75   7. $273

Lesson 10–3
Algebra Review
1. $6\frac{2}{3}$
3. 8.75 minutes per mile
5. 13 miles per gallon
7. (7, 15)

Try Your Skills
1. 22.7 mpg
3. $455.83
5. $376.95

Exercise Your Skills
1. the miles per gallon affect the annual operating cost
3. in order to put money aside to pay for the repairs needed
5. 417; 22.7; $0.060
7. 151; 15.0; 0.093
9. 211; 19.4; 0.068
11. $100.55
13. $0.499
17. 1428.6
19. Mark's car
21. lease

Mixed Review
1. $2000
3. $54,750
5. Refund: $396

Lesson 10–4
Algebra Review
1. 945
3. 5280
5. p decreases
7. 228,000
9. −7501.31

Try Your Skills
1. 2.10
3. $p = 2rb$; p represents the annual insurance premium; r represents the driver rating factor; b represents the six-month basic rate for the car

Exercise Your Skills
1. Any driver might have an accident.
3. The insured person pays the deductible amount of the bill resulting from the accident before the insurance company begins to pay. The premium is the amount one pays for the insurance coverage.
5. 1.65
7. 2.40

9. 1.25
11. $2340
13. $1152
15. $550
17. $610.60
19. expenses paid out (claims): number of people paying premiums; and overhead costs

Mixed Review
1. $4200
3. 10%
5. $1801.63
7. $4171.15
9. itemizing; $9770

Chapter 10 Review
1. factors that raise monthly payments: higher interest, longer term, higher amount financed
factors that lower monthly payments: lower interest, shorter term, smaller amount financed
3. loan payments followed by insurance costs
5. the driver rating factor, the car being insured, coverage and deductible
7. 20%
9. $2922.50; 19.8%
11. $408.87
13. Interest $67.20; Principal $341.67
15. $1680
17. $4704
19. $0.05 per mile
21. $1431; $119.25
27. 3.35
29. $2835
31. 1.65

Cumulative Review
1. $790
3. $43.33
5. $14.31
7. $28,499
9. $1564
11. $3600
13. **b** and **d**

CHAPTER 11
Algebra Refresher
1. $y = -2x - 3$; $y = -2x$; $y = -2x + 4$; $m = -2$

Lesson 11–1
Algebra Review
1. $g = 20$
3. $a = 10.71$
5. $c = 1.21$
7. $(a + b) \div 2$
9. d

Try Your Skills

1. North on 59, East on 10
3. 2010 miles
5. 18 gallons
7. $41.14
9. Answers may vary. (Ex.: $g = d \div a$; g = gallons of gas, d = total distance in miles, a = average miles per gallon)

Exercise Your Skills

1. Answers may vary.
3. Answers may vary.
5. east on 16; north on 95
7. northeast on 30; east on 40
9. 681 mi
11. 916 mi
13. 695 mi
15. 26.3; $33.93
17. 28.4; $35.22
19. 92.7; $125.15
21. Answers may vary.
23. Answers may vary.

Mixed Review

1.

Checks/Deposits	Amount	Balance
To: Grocery Mart		$778.22
For: Groceries	$105.89	− 105.89
		672.33

3. $16.55
5. $327.50
7. $7271

Lesson 11–2
Algebra Review

1. 11.2
5. $c + 1$
3. 10.0
7. (15, 75)

Try Your Skills

1. $50.06
3. $4.73, $5.44, $18.53

Exercise Your Skills

1. $181.44
3. $331.76
5. $85.95; $21.49
7. $124.70; $79.85; $93.17; $139.38; $437.10; $109.28
9. Answers may vary.
11. $y = 7500 + 180x$
13. 20 people
15. loss: $1800

17. 50
19. Gomezes want dinner prices to cover costs; patrons want an inexpensive dinner.
21. Lower prices may lead to more patrons; if so, revenues may increase.
23. $y = 445x$
25. Yes, if more than 41 people take the tour.

Mixed Review

1. $568.75
3. $502.25
5. Loan b; a. $16,753.14; b. $18,608.78
7. $11,000
9. 42 years
11. 24 years
13. $86,062.50
15. $900

Lesson 11–3
Algebra Review

1. 100 mi
3. 1400 mi

Try Your Skills

1. 8:15 A.M.
3. Yes
5. $625 is an expensive flight, so Ramón would likely save money by driving, given he has a car to use.

Exercise Your Skills

1. 9:54 A.M. or 11:35 A.M.
3. Midway
5. If she's traveling on the weekend, the second flight is not available.
7. Car travel costs include gas, lodging and meals; air travel costs include ticket and transportation to and from airport.
9. The ticket is probably not refundable so you can't change your plans.
11. 8:30 A.M.
13. 6:15 P.M.
15. 381 min
17. Central
19. Mountain
21. Mountain
23. 1749 mi
25. $415.40

Mixed Review

1. $20.95; $1696.95; $170
3. $1391.04; $17.39; $1408.43; $141

5. $16,430
7. 616; 31.6; $0.045
9. $630.75
11. $4560.83
13. $4016.25

Chapter 11 Review

1. faster
3. unexpected events
5. often less expensive, more flexible
7. 20 east, 55 south, 10 east
9. 20 east, 95 northeast
11. 30 northeast, 40 east, 55 north, 57 north
13. 1315 mi
15. 1448 mi
17. $81.14, $104.59, $89.35, $41.22
19. $341.28
21. $196.56
23. $85.28; $21.32
25. $130.30; $124.84; $91.13; $79.93; $426.20; $106.55
27. $y = 450x$
29. 15 people
31. $1600
33. pacific
35. central
37. 8:30 A.M.
39. 230 min
41. 1187 mi
43. $347.16

Cumulative Review

1. $8.50/h; $340/wk; $1473.33/mo
3. $220/wk; $953.33/mo; $11,440/y
5. $7045.45
7. 1026
9. $805
11. $319
13. 19
15. $5549
17. 27%

CHAPTER 12

Algebra Refresher

1. $x^2 + x - 2$
3. $x^2 + 6x + 9$
5. $x^2 - 4x - 5$
7. $(x - 1)(x + 2)$
9. $(x - 1)(x - 3)$
11. $(x + 5)(x - 4)$
13. $x = -4$ or $x = -1$
15. $x = 3$ or $x = -2$

Lesson 12–1

Algebra Review

1. 184.97
3. 10.78
5. −44.60
7. 268.20
9. 1449.78

Try Your Skills

1. $y = -4.74 + 0.0763x$
5. yes

Exercise Your Skills

1. Answers may vary.
3. Answers may vary.
5. yes; $r = -0.9873$
7. 19 years
9. $y = -1.8969 + 0.0151x$
13. no; $220,000 would be fair; $x = 223.11$
15. yes; $r = -0.9627$
17. about 1; $y = 1.059$

Mixed Review

1. $864.57
3. 17.8%
5. 12.1%
7. $218.40

Lesson 12–2

Algebra Review

1. 1988

Try Your Skills

1. $659,520
3. $48,296
5. $448,294; $24,548
7. $472,842; $0

Exercise Your Skills

1. down payment, closing costs, mortgage, additional costs
3. between 2 and 4 times
5. 2.5:1
7. $40,000; $160,000; $545,481
9. $40,000; $160,000; $588,538
11. $60,000; $140,000; $502,296
13. $60,000; $140,000; $539,971
15. $30,000; $90,000; $267,740
17. $30,000; $90,000; $290,698
19. $30,000; $90,000; $194,311
21. $30,000; $90,000; $184,816
23. lowest: 25% down, 8% interest, 15 years; highest: 20% down, 9% interest, 30 years
25. Plan C

Mixed Review

1. $1.19
3. $21.01; $1221.51; $122
$1099.51; $19.24; $1118.75; $112
$1006.75; $17.62; $1024.37; $102
5. 763
7. 16.2%

Lesson 12–3

Algebra Review

1. 4000
3. 1424
5. $540
7. 2222.22
9. $\dfrac{5ab^{36}}{z(c+2)^{36}}$

Try Your Skills

1. $115,149.42
3. yes; 20% is $28,787.36

Exercise Your Skills

1. the higher the interest rate, the higher the total cost of the loan
3. federal government
5. $700; $86,997.31; $108,746.64
7. $980; $121,796.23; $152,245.29
9. $742; $92,217.14; $115,271.43
11. $1468.60; $182,520.35; $228,150.44
13. $978.60; $121,622.23; $152,027.79
15. $1617.56; $201,033.37; $251,291.71
17. no; $160,000
19. yes
21. yes

Mixed Review

1. 33 wk
3. $456,238.19
5. $52.42
7. $469

Lesson 12–4

Algebra Review

1. 0.227
3. 0.0131
5. $7019.44
7. 95,244.28
9. 18.83

Try Your Skills

1. yes
3. $120.67
5. $2025; $17,901; $5012.28

Exercise Your Skills

1. Answers may vary.
3. document with photos and contact the insurance agent before beginning to clean
5. $1025
7. yes
9. $75
11. $11.91
13. $22.54
15. $5.47
17. $10.92
19. $23.18
21. $21.57
23. $3360; $13,985; $3915.80
25. $2562.50; $18,706.25; $5237.75
27. $867; $8,899.50; $2491.86

Mixed Review

1. 14 years
3. $880
5. 1615 mi
7. $116.60

Chapter 12 Review

1. Answers may vary.
3. 28%
7. 4
9. $534,332; $64,206
11. $576,890; $21,648
13. $896; $122,110.01; $152,637.51
15. $924; $125,925.95; $157,407.44
17. $1736; $236,588.15; $295,735.19
19. $665; $90,628.52; $113,285.65
21. $628.32; $85,629.65; $107,037.06
23. no; $125,000
25. $107.23
27. $2180.50; $19,168.50; $5367.18

Cumulative Review

1. $178.52
3. $21.70; $1261.70; $126
5. $1039.57; $18.19; $1057.76; $106
7. $240,000
9. 491.6 mi; 28.3 mi/gal; $0.04/mi

CHAPTER 13

Algebra Refresher

1. 25
3. 8
5. 5
7. 2
9. −2, 2
11. $\frac{9}{2}$

Lesson 13–1
Algebra Review
1. $2146.56
3. $7936.50
5. 375
7. F

Try Your Skills
1. $721.15
3. $437.50
5. $15,340
7. Yes. The lease is for $4800; food costs $4200; 9000 ÷ 2 = $4500. Jacob will save $1500 over a 12-month period by living in the apartment. Even if they leave the apartment after 9 months, their grocery expenses will only be $3150, lowering their apartment costs further.

Exercise Your Skills
1. sharing expenses, companionship
3. Answers may vary. (Ex.: rent—do not have to worry about maintenance)
5. $423.08
7. $569.23
9. $867.21
11. $346.15
13. $36,400
15. $40,300
17. $50,700
19. $33,540
21. $7979; $2929
23. Yes, in 2 months he will recover his moving cost and start saving $108/mo.
25. no; rent is $3480, food for 9 months is $1800; total: $5280

Mixed Review
1. $13,809.99
3. $16,637.50
5. 6:15 A.M.
7. $1281
9. 16.4%; no

Lesson 13–2
Algebra Review
1. 34.31
3. −138.84
5. $y = 2x + 3$
7. $y = 150x + 100$
   $y = $ cost; $x = $ number of rooms
9. $970.00

Try Your Skills
1. $y = 1.573550568 − 0.0026312761x$
3. 0.4 miles

Exercise Your Skills
1. A lease is usually a 1-year commitment.
3. Tenants will give you an inside view of the way the apartment building is run.
5. $y = −0.4737640598 + 0.0033466353x$
7. 1
9. $y = 141.1439994 + 1.076640466x$
11. 679 ft²
13. no; $r = 0.145972874$
15. yes; $r = 0.9593668769$

Mixed Review
1. $9667.50
3. $390,000
5. $256.34
7. $164,605.34
9. $828.85

Lesson 13–3
Algebra Review
1. 1575
3. 936.39
5. 1972.71
7. 1176.16
9. 236,208.80

Try Your Skills
1. $352,023.47
3. total financed price

Exercise Your Skills
1. It is responsible for management of the grounds, building, and common areas.
3. renting to get acquainted with the community; buying if you are already a long-term resident of the area
5. $283.23; purchase
7. $547.14; purchase
9. $901.18; purchase
11. rental payments
13. rental payments
15. rental payments
17. $142,709.49
19. $367,365.10
21. $804,088.76

Mixed Review
1. 16 years
3. $79.38
5. $2000; $10,215; $2860.20
7. $3971; $18,533; $5189.24

Chapter 13 Review

1. Advantages: shared expenses, companionship
   Disadvantages: lack of privacy
3. Advantages: interest on mortgage is tax deductible, mortgage payments provide equity
   Disadvantages: may be difficult to sell, requires large outlay of cash for purchase, extra fees to condo association, property taxes
5. $817.31
7. $1538.46
9. $740.38
11. $1841.35
13. $68,640
15. $43,680
17. $25,480
19. $20,280
21. yes; in $1\frac{1}{2}$ months moving costs will be covered
23. no; rent = $5520; food = 9 • $150 = $1350; total = $6870; dorm is less expensive
25. yes, $r = 0.9883815163$
27. yes
29. rental payments
31. $609.23

Cumulative Review

1. $72,750
3. Profit: $185.75
5. Plan 2
7. $19,762.50
9. $1067.31
11. $424.84, rental

CHAPTER 14

Algebra Refresher

1. 10 or −10
3. 7 or −5
5. 4 or −4

Lesson 14–1

Algebra Review

1. $6 = 0.05x$
3. $2.5 = 500x$
5. $x = 0.5\%$
7. $2.8 = 0.05x; x = 56$
9. $t = 0.035c$

Try Your Skills

1. 18.3 mills
3. 1.6 mills
5. $1521

7.

	Month 1	Month 2	Month 3	Month 4	Total	Average
Energy	$120	$108	$130	$126	$ 484	$121.00
Electricity	31	28	37	40	136	34.00
Water	12	13.50	15.50	21	62	15.50
Telephone	80	36	47	51	214	53.50
Home Maint.	43	72	18	10	143	35.75
Total	286	257.50	247.50	248	1039	259.75

Exercise Your Skills

1. It is determined by the local government and not by the owner.
3. It is satisfactory for expenses not affected by seasons such as telephone and water but not for heating/air conditioning.
5. $139.40
7. 8.3 mills
9. 15.9 mills
11. $220; $55.00
    $292; $73.00
    $98; $24.50
    $130; $32.50
    $222; $55.50
    $134; $230; $292; $306; $962; $240.50
13. $638.50

Mixed Review

1. $14,750
3. $53\frac{5}{8}$
5. $2272
7. 1537 mi

Lesson 14–2

Algebra Review

1. $\frac{ay}{b}$
3. $\frac{bx}{y}$
5. $\frac{14}{x}$
9. graphing calculator shows more points
11. 12
13. 24

Try Your Skills

1. $0.456 per ounce
3. 20 oz for $2.15
5. 485.714 grams

Exercise Your Skills

1. Answers may vary.
3. Answers may vary.
5. 18 oz; $1.79
7. 12 oz; $1.98
9. 14.08 lb
11. 3.77 liters
13. $3.63; $2.69; $2.38; $2.22

Mixed Review

1. $3669.23
3. $36,574.50
5. yes; $y = 200.62$
7. $28
9. 15
11. $12
13. 17th floor since predicted value is $456 while predicted value on 26th floor is $493

Lesson 14–3

Algebra Review

1. 120°
3. 80°
5. $200; Miscellaneous

Try Your Skills

1. $4950
3. $1125
5. $900
7. $450
9. $2475
11. yes

Exercise Your Skills

1. 20%
3. If one budget area takes too large a percentage of the income, it may mean trouble in other areas.
5. 100%
7. 33.2%
9. 25.8%
11. 5.3%
13. 0.4%
15. −$873
17. +$838
19. −$253
21. +$28
23. Answers may vary.

Mixed Review

1. It does not allow the check to be cashed.
3. $144.02

5. payment due is $400
7. It might show a pattern, for example a linear pattern, that would suggest the fund's future performance.
9. $181,572 + $45,393 = $226,965
11. $634.70; the rental option would be higher

Chapter 14 Review

1. $3349
3. 11.2 mills
5. 18.9 mills
7. $481; $120.25
$184; $46.00
$89; $22.25
$226; $56.50
$217; $54.25
$237; $346; $277; $337; $1197; $299.25
9. 16 oz; $1.29
11. 12 oz; $1.69
13. 5.66 liters
15. 5.45 kg
17. $y = 950 + 2.4x$
19. $15.07
21. $9.71
23. 24.8%
25. 7.3%
27. 4.2%
29. 1.8%
31. −$85
33. $0
35. −$38
37. +$16
39. −$25
41. $7466
43. $1600
45. $1866
47. $533

Cumulative Review

1. $28,053.32
3. 77
5. 2666
7. $32,409.80
9. $176,468.58
11. 9.1%; no
13. $2543.75; $2117.50; −$426.25; −16.8%

GLOSSARY

A

add-on costs Expenses that add to the price of a product as it goes through processing, marketing, and distribution. Examples include labor, packaging, advertising, energy, and transportation.

adjustable-rate mortgage (ARM) Mortgage loan with an interest rate that can be adjusted up or down during the life of the loan

adjusted gross income Total of income earned as well as income from interest and other sources minus IRA deductions

advance purchase Paying ahead of time; if you pay for an airline ticket ahead of time you may get a lower price.

advertising Materials and activities used to promote the sales of a product

all-purpose bank cards Credit cards widely accepted by businesses that may not have their own charge accounts; examples include MasterCard and VISA.

amortize To pay off; a mortgage is usually amortized over a period from 15 to 30 years

amortization schedule A list showing the amounts of principal and interest that are paid with each monthly payment of a loan

annual percentage rate (APR) The percent cost of credit on a yearly basis

annuity Investment plan that provides income upon retirement

appraised value The approximate selling price of a house as determined by an independent agent

assessed value Amount that a local government assigns to a house or other property for tax purposes

assumability clause Statement in a mortgage allowing a new buyer to take over the mortgage from the specified holder

automated teller machines (ATMs) Computerized machines that perform banking services quickly and automatically at convenient locations and during nonbanking hours

average-balance account Checking account in which a fee is charged if the average balance for the month falls below a stated minimum

average daily balance The average of daily balances over the month, found by figuring the unpaid balance for each day of the billing cycle, adding these balances together, and dividing the total by the number of days

B

bait-and-switch Unethical sales practice in which one product is used to attract customers and another, more expensive product is then urged on them

balloon mortgage A form of mortgage in which the monthly payments are low, but there is a large final payment

bankruptcy The condition in which an individual or business declares the inability to pay its debts; assets must be sold in return for a discharge from outstanding debts

bar graph A graph in which the lengths of bars represent amounts

base salary Wages earned regardless of commissions or piece-work rates

bear market Condition in which stock prices are falling

beneficiary The person designated to receive the death benefit in a life insurance policy

blank endorsement Your signature on the back of a check with no additional message

bodily injury liability Insurance that covers injuries to others in an automobile accident caused by the insured person

bond A loan to a government or corporation for which the holder receives interest during the life of the bond, and the full amount at the maturity date

book value Standard estimate of the resale value of a car

break-even point (*also* **break-even value**) The point at which cost and revenue are equal

broken line graph A graph in which line segments connecting points represent data

broker Salesperson who specializes in buying and selling stocks and bonds

budget A plan showing categories and projected amounts of expenses or income

bull market Condition in which stock prices are rising

canceled checks Checks, drawn on an account, that have cleared through the bank and been returned to the account holder for record keeping

cap A limit to the total amount of increase allowable in an adjustable-rate mortgage

capital gain Increase in the market value of a security or other asset; increase in the value of property during the time that it is owned

cash discount A lower price given to a customer who pays cash instead of using credit

cash shortage Insufficient funds to cover expenses

cash-value insurance Insurance plus savings plan; one of two main types of life insurance

cell In a spreadsheet, the intersection of a row and column

certificate of deposit (CD) Special form of savings in which a fixed amount of money is deposited for a specified length of time

charge card A card that allows you to make purchases for later billing

check register A record of checks written, fees charged, and deposits made in a checking account

checks Forms filled out and signed by the owner of a checking account authorizing the bank to pay the amount shown to the person designated

closing costs Charges and fees associated with the transfer of ownership of a home to a buyer

collateral Something of value offered as assurance that a loan will be repaid; (*See* **security**)

collection agency A firm hired by creditors to collect overdue debts or repossess items not paid for

collision Insurance that pays for damage to the insured person's car in an accident

commercial banks Banks offering a wide variety of services in addition to savings accounts: checking accounts, credit cards, loans, financial counseling, safe deposit boxes, traveler's checks, money orders and transfers, and trust and investment services

commission Money paid a salesperson for each item sold; a percentage of the cost of the item. Also, broker's fee for buying and selling securities

commodity An item that has not yet been enhanced, refined, or packaged to become a consumer good

common logarithm function Function of the form $y = \log_{10} x$, where y represents the exponent to which 10 must be raised to obtain x.

common ratio The ratio between two consecutive terms in a geometric series

common stock Stock that entitles the owner to receive dividends and vote for the company's board of directors

competition Rivalry among sellers for consumers' purchases, or rivalry among producers when seeking the rights for production

complement In probability, if E is an event then "E does not occur" or "not E" is the complement of E

compound interest Interest paid on the principal and also on previously earned interest, assuming that the interest is left in the account

compounded quarterly Interest calculated and paid four times a year, every three months

compounded semiannually Interest calculated and paid twice a year, every six months

comprehensive physical damage Insurance that pays for damage to a car that results from fire, falling objects, theft, storm, flood, earthquake, mischief, flying objects, and vandalism

condominium Living space in which individual units are privately owned and common spaces are jointly owned

connection Change of planes during a trip

constraints Conditions that limit business activities

Consumer Price Index (CPI) An economic tool for comparing prices from one era to another

consumers Purchasers of goods and users of services; all the members of a society who need to buy things

corporate bond Bond issued by a private company to raise money for plant expansion or other operations

corporation A business made up of a number of owners who have shares of stock in the company

cosign Parents or others can cosign a loan with a borrower; this makes the cosigner responsible for repaying the loan if the borrower does not

cost-per-check account Checking accounts that charge a small fee for every check that clears during the month

coverage The items and dollar amounts for which an insurance company will compensate a policy holder; homeowner's insurance coverage may include dwelling, personal property, loss of use, personal liability, and medical payments

credit A form of debt that occurs when cash, goods, or services are provided in exchange for a promise to pay at a future date

credit card A small plastic card that identifies the holder and gives the holder privileges of making purchases on credit; the holder is not required to pay the total outstanding bill every month, but to pay a minimum required amount

credit counseling Advice about how to solve credit problems

credit limit The maximum value a credit card holder can charge on an account; once the balance is partly paid back the holder can again make purchases up to the limit

credit rating An indication of a person's ability to secure goods, services, and money in return for the promise to pay

credit risk A customer who is unlikely to be able to pay for purchases made on credit is referred to as a poor credit risk

credit unions Not-for-profit savings and lending financial institutions in which all members belong to the same social or professional organization

credit-scoring systems Award points for various factors to determine credit worthiness

creditor Business, bank, or individual who extends credit

debit card A card that enables the card holder to pay through a direct deduction from a checking account

decreasing term insurance Kind of term insurance in which the death benefits decrease over time and the premium is lower than for renewable term insurance

deductible The portion of a loss that the insured person pays; the insurance company pays the costs above the deductible

deduction **1.** An amount of money subtracted from your gross pay for such items as FICA taxes, income taxes, insurance premiums, and union dues **2.** An amount that you may deduct from your gross income to find the amount on which you owe taxes; state and local taxes, interest paid on a mortgage, and charitable contributions are deductible

deferred payment price The amount including interest that you pay for an item you buy on installment

deflation A general decrease in prices

demand deposits Checks

dependent A person supported by another; a tax payer's dependents include children and spouse supported by the tax payer

deposit ticket A form used by banks for deposits to accounts

depreciation The difference between the original purchase price of a car or other asset, and its resale price at a given time

direct flight A plane flight that has one or more stops but you will not change planes

disability insurance Benefits paid through social security to those who can no longer work because of injury or illness

discount broker A broker in securities who charges much lower commissions than a full-service broker, but provides no investment advice

discretionary income Income remaining after you pay for basic needs such as food, clothing, and shelter

diversify Acquiring a balanced variety of securities rather than owning too much of any one; diversification cuts down the risk of investment

dividend The part of the profits of a corporation that each stockholder receives

down payment The amount of a purchase you make in cash before taking out a loan for the remainder

drawer The person who writes and signs a check

driver rating factor The multiple applied to a base premium to determine the insurance costs for an individual; in general, the factor for a young person is higher than for an older person, so the younger person pays higher premiums

easy-money policy An easy-money policy results when the reserve requirement is lowered and money is thus freed

effective interest rate Interest rate when the result of compounding is included

efficient Working well; a business is efficient when it is producing a product with a minimum of energy, expense, and waste.

endorsement Signature and directions to the bank on the back of a check

endowment policy Type of cash-value life insurance which builds up value quickly through high premiums

entrepreneurship Putting together the ingredients of a business, including making plans, hiring workers, obtaining raw materials, and obtaining equipment

Equal Credit Opportunity Act Law that prohibits discrimination in granting credit on the basis of sex, marital status, race, color, religion, age, or national origin

equity The difference between what a home, car, or other object is worth and what the buyer still owes on it in the form of a mortgage or loan

excess reserves Money remaining from a deposit after the required reserves have been subtracted; money available for investment or loans by a bank

exemption A tax deduction taken for dependents

expected value The amount of money to be won or lost in the long run

exponential regression equation The equation of the form $y = ab^x$ that best fits a set of points

face value of a bond The original loan amount that appears on the bond

face value of the policy The amount of the death benefit in a life insurance policy

Fair Credit Billing Act Law that allows people to preserve their good credit rating while settling a dispute with a store or credit card company

Fair Credit Reporting Act Law giving people the right to examine their own credit files and have incorrect information removed

Fair Debt Collection Practices Act Law prohibiting violence, the threat of violence, harassment, and certain other methods in the collection of debts

Federal Housing Administration loans Loans in which federal insurance is provided

Federal Reserve note Main type of U.S. currency

Federal Reserve System A system of banks under the direction of a board of governors that directs the purchases and sales by the reserve banks of federal government securities and other obligations.

FICA Federal Insurance Contributions Act (*See* **Social Security**)

FICA deduction An amount withheld from a paycheck and remitted to the federal government as part of the FICA tax

FICA tax A percentage of income paid to the federal government by employees and employers to pay for pensions and Medicare

finance charge 1. A charge levied on the unpaid balance of a credit account **2.** The amount over the loan amount that you pay to a lending institution for a loan; also called interest

financing Borrowing money needed for a purchase

financial responsibility laws Laws requiring drivers to pay, either with insurance or with personal funds, for damage to other people or to property

fixed costs A company's costs that remain constant over a period of time; may include labor, transportation, advertising, energy, and so on

fixed-rate mortgage (FRM) A mortgage paid off in equal monthly payments on a regular schedule over the years of the loan; (*Compare* **adjustable rate mortgage**)

Form 1040 Federal income tax form that is used when you are itemizing deductions and/or reporting self-employment income

Form 1040A Basic form for calculating and filing individual or joint federal income tax

Form 1040 EZ Simplified federal tax form for single individuals with no dependents

Form W-2 Form from an employer that provides a record of wages earned as well as federal, state, local, and other taxes withheld during the past year

Form W-4 Filled out at the beginning of employment, the form provides the IRS with information used to determine how much of the worker's wages to withhold for income tax

401(k) plan A pension plan approved by the IRS through which an employer can make tax-sheltered contributions

free checking account Has no minimum balance requirement or service charges

fringe benefits Benefits offered to workers in addition to wages or salaries; these include paid vacations, life and health insurance, and retirement plans

full endorsement Your signature on the back of a check made out to you, with directions to the bank to transfer the check to someone else

full service broker Broker offering reports on companies and advice about buying and selling securities

future value of a periodic investment The value, at a given time in the future, of money periodically invested in an interest-bearing account

garnishment To hold part of a debtor's earnings in order to pay a debt; a legal method of debt collection

geometric series A sum of terms in which the ratio of two consecutive terms is constant

grace period The time between when a bill is received and when interest is charged

graduated commission Commission plan in which a salesperson receives a higher rate of commission for sales above a certain amount

gross pay Total pay before deductions have been taken

group life insurance Insurance purchased through the place of employment, covering employees as a group

health insurance Hospital and medical insurance; for senior citizens, Medicare

homeowner's insurance Insurance covering a dwelling and personal property against damage or loss from fire and lightning, theft, storm, explosion, and so on. Mortgage lenders typically require the homeowner to take out this insurance

hourly rate Amount of money paid for each hour of work

income The money a person receives from work or investments

income security Guarantee that income will continue

individual income tax Tax on an individual's earnings from wages, salary, tips, interest, rents, dividends, and capital gains; in the U.S. the largest revenue producing tax for the federal government. The U.S. income tax is based on ability to pay, voluntary compliance, and paying-as-you-earn through withholding

Individual Retirement Arrangement (IRA) An account in which money is invested for retirement and taxes are deferred until retirement

inflation A general increase in prices

installment loan Borrowed money that is paid back in installments, often monthly; a common way to buy expensive items

interest 1. The amount over the loan amount that you pay to a lending institution for a loan; also called finance charge. **2.** Money paid to you by a financial institution for the privilege of using your money to make investments and loans to others.

interest due Interest to be paid, found by multiplying the previous month's unpaid balance by the monthly interest rate

interest rate The percent that is the basis for interest earned or paid

interstate highways Major, controlled access highways linking cities in the U.S.

inventory Items produced or purchased and held for resale; stock on hand

investing Purchasing securities, real estate, or some other item with the hope that it will increase in value

item rate *See* **piece rate**

itemized deductions Amounts listed on Schedule A of Form 1040 and deducted from the adjusted gross income

J

joint return Income tax form completed by a married couple showing combined incomes and deductions

K

Keogh plan A tax-sheltered pension plan, similar to an IRA, for self-employed persons

L

labor Work; the use of the efforts and skills of workers in manufacturing a product or extending a service

late payment penalties Possible extra charge imposed if payment on a credit account, loan, or mortage is late

lease Rental agreement between renter and landlord

leasing An arrangement under which a car is hired for a monthly fee

legal tender Money that by law must be accepted for paying debts and taxes

liability legal responsibility for the results of an accident

life insurance A contract to pay a specified amount of money to a designated person upon the death of the insured

linear programming A mathematical method for planning within given constraints

limited payment life insurance Cash-value life insurance which accumulates value more quickly than ordinary life because higher premium payments are made over a shorter period of time

liquidation bankruptcy Outright bankruptcy; an extreme measure in which a debtor who cannot pay must sell most assets and then is left free of debt; the person cannot reapply for credit for seven years.

linear regression equation The equation, $y = ax + b$, of the line that best fits a given set of points

liquidity The ease and speed with which you can withdraw money from deposit

loan sharks Illegal money lenders who charge extremely high interest for short-term cash loans and may use violence to collect the debt

loss region On a graph displaying revenue and cost, the region between the lines and below the break-even point

low-balance account Checking account in which a fee is charged if the balance falls below the minimum any time during the month

low-load insurance policy An insurance policy with a very low commission

M

maintainance fee *See* **service charge**

management Leadership in a company; those people who make decisions affecting labor, production, advertising, distribution, and so on

market basket List of goods and services that are tracked in the CPI

market economy An economic system in which profit is the incentive

market rate The interest rate currently charged by lending institutions such as banks and credit unions

markup The difference between the price at which a retailer buys an item and sells it

maximize To find the greatest number possible within given constraints

mean Average, found by adding scores and dividing by the total number of scores

median The middle score when scores are ordered from least to greatest

medical payments Insurance coverage for medical costs resulting from an automobile accident

Medicare A federally funded health insurance program for people over 65

metric system A standard system of measurement based on the decimal system. Some basic units are the meter, gram, and liter

mileage chart Chart showing distances between cities or towns

miles per gallon The average number of miles that a car drives on one gallon of gas

mill rate Amount of tax per thousand dollars of assessed value

minimize To find the least number possible within given constraints

minimum-balance account Checking account that requires you to keep a minimum amount on deposit

minimum monthly payment The least amount that a credit card holder may pay on a monthly bill

mobile home A portable living space designed to be used without a permanent foundation, intended as a year-round dwelling

mode In a set of scores, the score that appears most frequently

money market account An account with restricted access in exchange for a higher rate of interest

money supply The amount of money (coins, currency, and demand deposits) that is in circulation; this amount is regulated by the Federal Reserve System

monthly interest rate Percent used to calculate interest for one month, the annual percentage rate divided by 12

mortgage Loan for the purpose of buying property; usually paid in monthly payments of principal and interest, over a period of from 15 to 30 years

multiplier effect In the multiplier effect, money is loaned or invested over and over again, by different institutions, each institution keeping the required portion in reserve but passing on the larger portion for additional investment or loan making

multi-purpose travel and entertainment cards Charge cards often used for travel and entertainment for which an annual fee is charged and for which the total balance must be paid each month; examples include American Express, Diners Club, and Carte Blanche

municipal bond Issued by states, cities, counties, school districts, or other governmental bodies to raise money for schools, hospitals, streets, and so on

mutual fund A means of pooling funds with many other investors to acquire a wide variety of stocks, bonds, and other kinds of investments

nonrefundable A purchase is nonrefundable when you cannot return it for money

non-stop flight A plane flight that does not make any stops between two points

note reduction An amount found by subtracting the interest due from the monthly payment

NOW account Negotiable Order of Withdrawal account; checking account in which the balance earns interest but which may charge a service or per-check fee if the balance falls below a required minimum

objective function The equation used to find a maximum or minimum in linear programming

odd lot A stock trade of fewer than 100 shares

odometer The counter on the dashboard of a car that shows the number of miles the car has been driven

outstanding checks Checks that have been written but have not yet cleared through the bank

over-the-counter market A national network of dealers and brokers in securities

overtime hours Hours worked in addition to regular hours; pay for overtime is usually higher than for regular hours

owner's association The group of owners of individual units in a condominium; responsible for maintenance of building, grounds, and common areas

passbook or regular savings account An account that allows deposits and withdrawals at any time

pay-as-you-earn The payroll withholding feature of the income tax laws

payment number The number of months that have elapsed since an amount was borrowed

pawnbroker A person who will make a cash loan for about 40 percent of the value of an item left with the lender

payee The person or organization to whom a check is written

piece rate Pay based on the number of items produced

points One-time charge by the lending institution at the time of obtaining a mortgage; each point is equal to one percent of the mortgage value

preferred stock Stock which entitles the owner to receive dividends ahead of common stock owners, but which does not usually entitle the owner to vote for the board of directors

premium An amount of money paid on a regular basis for an insurance policy

prepayment clause Statement allowing a mortgage holder to make additional payments toward the principal

prepayment penalty Charge that may be imposed when a person pays back the full amount of a loan ahead of the payment schedule

prime interest rate The interest rate that large banks give their best customers

principal An amount of money invested or borrowed

probability of an event $P(\text{probability}) = \frac{m}{n}$ where m is the number of times the event occurs; n is the number of all possible outcomes.

producers Those who manufacture goods

profit The amount of money a business makes above the cost of the products or services it produces

profit region On a graph displaying revenue and cost, the region between the lines and above the break-even point

progressive tax schedule The means through which higher income is taxed at a higher rate

property damage liability Insurance that pays for damage done to other people's property

push money Money paid to a salesperson by a manufacturer to sell its goods ahead of others

quality of life The degree of satisfaction that people experience in their lives

rate of interest *See* **interest rate**

real-estate agents Individuals licensed by the state to help people find apartments

real wages The value of wages in terms of the goods and services that can be purchased. In times of inflation, real wages may fall while the dollar amount of wages rises

rebate An amount paid back to an individual purchasing a car or other item, either in cash or as a reduction of the purchase price.

reconciliation The process of finding the correct balance in a checking account by comparing the bank statement with the check register

refund An amount of money sent back to a taxpayer when the amount withheld is greater than the total payments due

regular hours The hours an employee is required to work each week

regular savings accounts An account that allows deposits and withdrawals at any time

renewable convertible term insurance Life insurance covering a person for a period of time such as one, five, or ten years and can be renewed without a medical examination

rent Money paid to a landlord, usually by the month, for the use of an apartment or house

rent-to-own plan A purchase plan in which you pay in monthly installments to rent an item; later you can choose to buy the item, applying the payments already made to the purchase price.

replacement cost Insurance coverage which compensates owners for lost or damaged items of personal property at the amount it would cost to replace them

required reserve Percentage of deposits a bank must hold in reserve to repay depositors; the bank may not use this amount for loans or investments

reservation A hotel's or restaurant's agreement to hold a room or a table for the person making the reservation; sometimes a deposit, or part of the price, is required to hold the reservation

restrictive endorsement Your signature on the back of a check with the words "for deposit only," protects the check against being cashed by someone else

revenue Money received for the sale of goods and services

round lot A stock trade of 100 shares

Rule of 72 Method for determining the time it will take for money invested to double at given interest rates: Divide 72 by the annual interest rate times 100, the quotient is the doubling time in years; for 9%, $72 \div 9 = 8$; an investment will double in eight years

Rule of 78 Rule used by lending institutions to determine payment for a buyer who purchases merchandize on credit and wishes to prepay

salary Weekly, monthly, or yearly rate of pay

savings banks Banks offering a variety of services, including checking and savings accounts, and real estate loans

savings and loan associations Banking institutions that specialize in lending money for home purchases and construction

scatter plot A graph in which a finite number of points represent data

secondary roads Well-marked federal and state highways that pass through towns

Securities and Exchange Commission (SEC) An independent federal agency whose purpose is to prevent unsound stock selling practices

security Something of value offered as assurance that a loan will be repaid; also collateral

securities stocks and bonds

security deposit Money, often equal to one month's rent, given to the landlord when a lease is signed; the deposit is refunded when the renter leaves the property in good condition

self-employment income Income earned by people working for themselves—such as physicians, farmers, writers, and consultants

service charge (*also* **maintainance fee**) A fee charged by a bank for a checking account

shares Units of an investment such as "100 shares of stock"

shareholders Those who own a company by owning shares of its stock

simple interest Interest paid one time a year at the end of the year on the total balance in a savings account

Simplified Employee Pension Plan (SEP) A plan that allows a person to contribute to an IRA without all of the usual IRA rules and limitations

single-purpose credit cards Credit cards issued by companies and businesses for use in purchasing their products and services; examples include oil companies and department stores

Social Security Compulsory federal insurance program which provides a level of income security for workers who have retired, surviving dependents of workers, and others

spreadsheet program A computer program that arranges data in rows and columns so that calculations can be made.

standard deduction One of the amounts you are permitted to deduct from your income to determine your taxable income; alternatively you may choose to itemize deductions for such things as charitable contributions, medical expenses, and interest paid on a mortgage loan

standard of living Economic level at which an individual or family lives; the affordable goods and services owned or used by an individual

statement Record of a month's activities in a checking account, credit card or other similar account sent monthly to the holder of the account

sticker price The price shown in a car window in a show room; usually higher than the dealer actually expects to receive

stocks Shares in the ownership of a corporation

sum of an infinite geometric series If a geometric series having an infinite number of terms has a sum, that is the sum of the series. For example, $1 + \frac{1}{2} + \frac{1}{4} + \frac{1}{8} + \ldots = 2$

survivors insurance Money for living expenses paid to survivors of workers who would have been eligible for social security

take-home pay Amount of money a worker receives after deductions have been made for social security, income tax, and other purposes

tax liability The amount of income tax you owe

tax preparation software A computer program used to compute taxes

Tax Rate Schedules Formulas from which taxes are calculated on the basis of fixed amounts and percentages for given incomes

tax shelter An investment or retirement plan that reduces or defers the payment of taxes

tax tables Charts that indicate the income tax owed based on your taxable income

taxable income Gross income minus personal exemptions (for self and dependents) and the standard deduction or itemized deductions; the amount you use to look up your taxes in the tax table

technology The development of new machines or labor-saving methods; technology requires better trained workers to fill new types of positions

term insurance Life insurance that pays death benefits for a stated number of years; one of two major kinds; (*Compare* **whole life**)

tight-money policy A tight-money policy results when the reserve requirement is higher so that less money is available for investments and loans

time zone A region within which all places have the same time

title Document stating the ownership of a car

total financed price Total purchase price of a house plus the financing costs

total income All income from wages, tips, interest, dividends, and other sources

total payment Amount of income tax to be paid

trade-in Selling a used car to the dealer from whom one purchases a new car

Truth in Lending Act Law that requires creditors to tell customers the exact cost of buying on credit

turnover Rate at which people move out and new people move into apartments or other rental units

uninsured and underinsured motorist protection Insurance that pays for damage to your car if you are in an accident caused by a driver who had little or no bodily injury insurance

unit cost The cost to produce one item

unit price The price per standard unit of a product; the cost used to compare different size packages of the product, such as cost per ounce, per pound, per square foot, and so on

U.S. Savings Bonds Bonds issued and backed by the U.S. government

universal life insurance Gives whole life protection and permits savings

unpaid balance The amount of borrowed money remaining to be paid

utilities Suppliers of heat, water, electricity, and natural gas to households; customers pay periodic bills, usually monthly, for these services. The products themselves are also called utilities

variable costs A company's costs that vary depending upon the number of items produced; may include materials, packaging, and the like

Veterans Administration loans Loans with special privileges and insurance given to veterans who served during wars or at other specified times

wage Hourly or daily rate of pay

Wage Earner Plan—Chapter XIII A plan under Chapter XIII of the Federal Bankruptcy Act in which people can restructure their debt without declaring bankruptcy

warranty The guarantee to keep a car in good repair; the period during which the guarantee is in force

whole life insurance Life insurance in which premiums are paid for your entire life

In this index, *See* cross-references indicate the word under which page numbers are listed. *See also* cross-references indicate the page numbers of related topics or more detailed breakdowns of a topic.

CHAPTER 1

pp. 2-3: Photo by Ian Crysler; **p. 4:** © Stephen Whalen/Picture Perfect USA, Inc.; **p. 6:** © Campbell & Boulanger/Picture Perfect USA, Inc.; **p. 12:** © SuperStock, Inc.; **p. 14:** © Charles Gupton/Stock, Boston; **p. 16:** © C. Orrico/SuperStock, Inc.; **p. 23:** © SuperStock, Inc.; **p. 24:** Location courtesy of Northland Volkswagen/Cincinnati, Ohio; **p. 30:** Jeff Greenberg, Photographer; **p. 35:** © C. J. Allen/Stock, Boston; **p. 36:** © Gala/SuperStock, Inc.; **p. 37:** © Charles Gupton/Stock, Boston.

CHAPTER 2

p. 51: top: © FPG International Corp./J. McNee, **bottom:** © FPG International Corp./Ron Chapple; **p. 52:** Photo by Jim Whitmer; **p. 53:** © Bob Daemmrich/Stock, Boston; **p. 56:** © Rick McClain/Picture Perfect USA, Inc.; **p. 63:** © Bob Daemmrich/Stock, Boston; **p. 67:** © Warren Morgan/Westlight; **p. 70:** © R. Chen/SuperStock, Inc.; **p. 73:** © David R. Frazier/Tony Stone Images; **p. 74:** © Poulides & Thatcher/Tony Stone Images; **p. 79:** © Bob Daemmrich/Stock, Boston.

CHAPTER 3

pp. 90-91: bottom: © Comstock, Inc./Michael Thompson; **p. 92:** © Billy E. Barnes/Picture Perfect USA, Inc.; **p. 93:** © Walter Hodges/Westlight; **p. 97:** © G. Brettnacher/SuperStock, Inc.; **p. 102:** © H. Richard Johnston/Tony Stone Images; **p. 103:** © Melanie Carr/Picture Perfect USA, Inc.; **p. 111:** © SuperStock, Inc.; **p. 114:** Photo by Jim Whitmer; **p. 123:** © P. Cantor/SuperStock, Inc.

CHAPTER 4

p. 133: Photos by Ian Crysler; **p. 134:** © W. Cody/Westlight; **p. 135:** © Richard Pasley/Stock, Boston; **p. 140:** © SuperStock, Inc.; **p. 143:** © Robert Frerck/Tony Stone Images; **p. 147:** © Gene Stein/Westlight; **p. 151:** © Charles Gupton/Stock, Boston; **p. 152:** © Nik Wheeler/Westlight; **p. 158:** © Wendt WorldWide; **p. 161:** © Digital Art/Westlight; **p. 163:** © Bob Daemmrich/Stock, Boston; **p. 169:** © SuperStock, Inc.; **p. 175:** © Lawrence Manning/Westlight.

CHAPTER 5

p. 185: Photo by David Michael Allen; **p. 186:** Jeff Greenberg, Photographer; **p. 187:** © J. Trotter/SuperStock, Inc.; **p. 190:** © Wendt WorldWide; **p. 196:** © Jim Richardson/Westlight; **p. 197:** © Mark E. Gibson; **p. 205:** © Warren Morgan/Westlight; **p. 206:** © Don Smetzer/Tony Stone Images; **p. 210:** © Guy Motil/Westlight; **p. 214:** © William Johnson/Stock, Boston; **p. 216:** © Graham Pym/Picture Perfect USA, Inc.; **p. 222:** © Robert E. Daemmrich/Tony Stone Images; **p. 223:** © Spencer Grant/Stock, Boston; **p. 225:** © Andy Sacks/Tony Stone Images; **p. 226:** Location courtesy of Chevrolet; **p. 233:** © SuperStock, Inc.

CHAPTER 6

pp. 242–243: Photo by David Michael Allen; **p. 244:** Location courtesy of Lazarus; **p. 245:** © William Johnson/Stock, Boston; **p. 247:** © Mark E. Gibson; **p. 253:** © Walter Hodges/Westlight; **p. 257:** © Digital Art/Westlight; **p. 259:** © Charles Thatcher/Tony Stone Images; **p. 260:** © S. William/SuperStock, Inc.; **p. 265:** Photo by Jim Whitmer; **p. 271:** © Graham Lawrence/Picture Perfect USA, Inc.; **p. 272:** © Tony Stone Images; **p. 280:** © Michael Neveux/Westlight; **p. 281:** © Mark E. Gibson; **p. 291:** © John Maher/Stock, Boston; **p. 292:** © Mark E. Gibson; **p. 295:** © Mark E. Gibson; **p. 299:** Photo by Jim Whitmer; **p. 301:** © Ellis Herwig/Stock, Boston.

CHAPTER 7

p. 319: Photo by David Michael Allen; **p. 320:** © K. Coppieters/SuperStock, Inc.; **p. 321:** © Steve Chenn/Westlight; **p. 327:** © Wendt WorldWide; **p. 328:** Photo by Jim Whitmer; **p. 330:** © John Fortunato/Tony Stone Images; **p. 333:** © Ong & Associates/SuperStock, Inc.; **p. 336:** © SuperStock, Inc.; **p. 340:** Photo by Jim Whitmer; **p. 341:** Photo by Jim Whitmer; **p. 344:** © Gary A. Bartholomew/Westlight.

CHAPTER 8

p. 357: © Comstock, Inc.; **p. 358:** © Walter Hodges/Westlight; **p. 359:** © SuperStock, Inc./P. Cantor; **p. 361:** © Wendt WorldWide; **p. 368:** © Greg Pease/Tony Stone Images; **p. 369:** © R. Llewellyn/SuperStock, Inc.; **p. 374:** © Frank Siteman/Stock, Boston; **p. 379:** © Tangent/Picture Perfect USA, Inc.; **p. 380:** © Tony Stone Images; **p. 382:** © SuperStock, Inc.; **p. 388 :** © FPG International Corp./Telegraph Colour Library; **p. 389:** © Tony Stone Images; **p. 392:** © SuperStock, Inc.

CHAPTER 9

p. 405: © FPG International Corp./Dick Luria; **p. 406:** © Robert E. Daemmrich/Tony Stone Images; **p. 413:** © Bill Horsman/Stock, Boston; **p. 419:** © Adamsmith Productions/Westlight; **p. 420:** © Mark E. Gibson/Picture Perfect USA, Inc.; **p. 430:** © Walter Hodges/Westlight; **p. 434:** © SuperStock, Inc.; **p. 435:** © Walter Hodges/Westlight; **p. 437:** © Steve Chenn/Westlight.

CHAPTER 10

p. 453: © Donovan Reese/Tony Stone Images; **p. 454:** Photo by Jim Whitmer; **p. 455:** © SuperStock, Inc.; **p. 457:** © Donald Johnston/Tony Stone Images; **p. 463:** © Rhoda Sidney/Stock, Boston; **p. 464:** © Bill Ross/Westlight; **p. 475:** © Robert Landau/Westlight; **p. 477:** © Joseph Pobereskin/Tony Stone Images; **p. 478:** © Dick Wade/Picture Perfect USA, Inc.; **p. 484:** © Mark E. Gibson; **p. 485:** © Eric Neurath/Stock, Boston; **p. 487:** © Tim Andrew/Picture Perfect USA, Inc.